Key Thinkers on Space and Place

THIRD EDITION

Edited by
Mary Gilmartin, Phil Hubbard, Rob Kitchin and Susan M. Roberts

Key Thinkers on Space and Place

Sage

Sage

1 Oliver's Yard
55 City Road
London EC1Y 1SP

2455 Teller Road
Thousand Oaks
California 91320

Unit No 323-333, Third Floor, F-Block
International Trade Tower, Nehru Place
New Delhi 110 019

8 Marina View Suite 43-053
Asia Square Tower 1
Singapore 018960

Editor: Natalie Aguilera
Editorial assistant: Sarah Moorhouse
Production editor: Victoria Nicholas
Marketing manager: Ben Sherwood
Cover design: Bhairvi Vyas
Typeset by: KnowledgeWorks Global Ltd
Printed and bound by CPI Group (UK) Ltd, Croydon, CR0 4YY

Library of Congress Control Number: 2023947363

British Library Cataloguing in Publication data

A catalogue record for this book is available from the British Library

ISBN 978-1-5297-3256-6
ISBN 978-1-5297-3255-9 (pbk)

Contents

Archived Entries

Chapters on the following key thinkers are available in the second edition of the book and are archived at https://keythinkersonspaceplace.wordpress.com. These entries are ***bold italicized*** in the text for information.

Marc Augé
Trevor Barnes
Jean Baudrillard
Zygmunt Bauman
Ulrich Beck
Brian Berry
Homi Bhabha
Anne Buttimer
Manuel Castells
Stuart E. Corbridge
Michel de Certeau
Mike Davis
Michael Dear
Peter Dicken
Anthony Giddens
Reginald Golledge
Derek Gregory
Torsten Hägerstrand
Peter Haggett
Brian Harley
Peter Jackson
David Ley
Kevin Lynch
Allan Pred
Andrew Sayer
David Sibley
Edward Soja
Michael Storper
Peter Taylor
Gerard Toal
Waldo Tobler
Paul Virilio
Immanuel Wallerstein
Michael Watts
Benno Werlen
Raymond Williams
Alan Wilson
Iris Marion Young

International Advisory Board

About the Contributors

Peter Adey, Department of Geography, Royal Holloway, University of London, United Kingdom.

Unusah Aziz, School of Geography, University of Leeds, United Kingdom.

Simon P.J. Batterbury, School of Geography, Earth and Atmospheric Sciences, University of Melbourne, Australia.

Christian Berndt, Department of Geography, University of Zurich, Switzerland.

Sage Brice, Department of Geography, Durham University, United Kingdom.

Gary Bridge, School of Geography and Planning, Cardiff University, United Kingdom.

Tim Bunnell, Department of Geography and Asia Research Institute, National University of Singapore, Singapore.

Jordan T. Camp, American Studies, Trinity College, Hartford, Connecticut, United States.

Lauren Cape-Davenhill, School of Geography, University of Leeds, United Kingdom.

Noel Castree, Department of Geography, University of Manchester, United Kingdom, and University of Technology Sydney, Australia.

David B. Clarke, Department of Geography, Swansea University, United Kingdom.

Daniel Clayton, School of Geography and Sustainable Development, University of St. Andrews, Scotland.

Nir Cohen, Department of Geography and Environment, Bar Ilan University, Israel.

Deirdre Conlon, School of Geography, University of Leeds, United Kingdom.

Ruth Craggs, Department of Geography, King's College London, United Kingdom.

Tim Cresswell, School of GeoSciences, University of Edinburgh, Scotland.

Andrew Crooks, Department of Geography, University at Buffalo, State University of New York, United States.

Patricia Daley, School of Geography and the Environment, University of Oxford, United Kingdom.

Sarah de Leeuw, Northern Medical Program, University of Northern British Columbia, Canada.

Marcus Doel, Department of Geography, Swansea University, United Kingdom.

Kelly Dombroski, School of People, Environment and Planning, Massey University, Aotearoa/New Zealand.

Andrew C. Dwyer, Information Security Group, Royal Holloway, University of London, United Kingdom.

LaToya E. Eaves, Department of Geography and Sustainability, University of Tennessee, Knoxville, United States.

Kelsey Emard, Geography and Geospatial Science, College of Earth, Ocean, and Atmospheric Science, Oregon State University, United States.

Cristina Faiver-Serna, Departments of Geography and Women's and Gender Studies, University of New Hampshire, Durham, United States.

Caroline Faria, Department of Geography and the Environment, University of Texas at Austin, United States.

James Faulconbridge, Lancaster Management School. Lancaster University, United Kingdom.

Jude L. Fernando, Department of Sustainability and Social Justice, Clark University, United States.

Mary Gilmartin, Department of Geography, Maynooth University, Ireland.

Martin Gren, Department of Cultural Sciences, Linnaeus University, Sweden.

Laam Hae, Department of Politics, York University, Canada.

Euan Hague, Department of Geography and GIS, DePaul University, United States.

Sarah Hall, Department of Geography, University of Cambridge, United Kingdom.

Suzanne Hall, Department of Sociology, London School of Economics and Political Science, United Kingdom.

Stephen Healy, Department of Geography, Planning and Urban Studies, Western Sydney University, Australia.

Phil Hubbard, Department of Geography, King's College London, United Kingdom.

Rik Huizinga, Human Geography and Spatial Planning, Utrecht University, The Netherlands.

Rhys Jones, Department of Geography and Earth Sciences, Aberystwyth University, Wales.

Mikko Joronen, Regional Studies/Tampere Institute for Advanced Study, Tampere University, Finland.

Hamish Kallin, School of GeoSciences, University of Edinburgh, Scotland.

Eden Kinkaid, Department of Geography, University of Arizona, United States.

Eben Kirksey, Department of Anthropology and Museum Ethnography, Oxford, United Kingdom.

Rob Kitchin, Department of Geography and Maynooth University Social Sciences Institute, Maynooth University, Ireland.

Sneha Krishnan, Department of Geography and the Environment, Oxford, United Kingdom.

Wendy Larner, Vice-Chancellor, Cardiff University, United Kingdom.

Eric Laurier, School of GeoSciences, University of Edinburgh, United Kingdom.

Taibat Lawanson, Department of Urban and Regional Planning, University of Lagos, Nigeria.

Mary Lawhon, School of GeoSciences, University of Edinburgh, United Kingdom.

Keith Lilley, Geography, School of Natural and Built Environment, Queens University of Belfast, United Kingdom.

Hayden Lorimer, School of GeoSciences, University of Edinburgh, United Kingdom.

Jamie Lorimer, Department of Geography and the Environment, Oxford, United Kingdom.

Sophia Maalsen, School of Architecture, Design and Planning, University of Sydney, Australia.

Katharine McKinnon, Centre for Sustainable Communities, University of Canberra, Australia.

Katherine McKittrick, Department of Gender Studies, Queen's University, Ontario, Canada.

Jessica McLean, Macquarie School of Social Sciences, Macquarie University, Australia.

Phil McManus, School of Geosciences, University of Sydney, Australia.

Mahboobe Safaei Mehr, Department of Geography and the Environment, University of Texas at Austin, United States.

Lucas Melgaço, Department of Criminology, Vrije Universiteit Brussel, Belgium.

Peter Merriman, Department of Geography and Earth Sciences, Aberystwyth University, United Kingdom.

Mara Miele, School of Geography and Planning, Cardiff University, United Kingdom.

Robina Mohammad, robina-therapies.com, Kuala Lumpar and Singapore.

Amber Murrey, School of Geography and the Environment, University of Oxford, United Kingdom.

Reecia Orzeck, Department of Geography, Geology, and the Environment, Illinois State University, United States.

Kimberley Peters, Helmholtz Institute for Functional Marine Biodiversity at the University of Oldenburg, Germany.

Chris Philo, Department of Geography, University of Glasgow, United Kingdom.

MM Ramírez, Department of Geography, Simon Fraser University, Burnaby, British Columbia, Canada.

Susan M. Roberts, Department of Geography, University of Kentucky, United States.

Rory Rowan, Department of Geography, Trinity College Dublin, Ireland.

Jonathan Rutherford, LATTS, Ecole des Ponts ParisTech, CNRS, Université Gustave Eiffel, France.

David Seitz, Department of Humanities, Social Sciences and the Arts, Harvey Mudd College, United States.

Rob Shields, Human Geography and Sociology, University of Alberta, Canada.

James D. Sidaway, Department of Geography, National University of Singapore, Singapore.

Kirsten Simonsen, Department of People and Technology, Roskilde University, Denmark.

Tracey Skelton, School of Geography, University of Otago, Aotearoa/New Zealand.

Jen Rose Smith, Department of American Indian Studies and Department of Geography, University of Washington, United States.

Sara Smith, Department of Geography and Environment, University of North Carolina at Chapel Hill, United States.

Rachael Squire, Department of Geography, Royal Holloway University of London, United Kingdom.

***tebrakunna* country and Emma Lee,** National Centre for Reconciliation, Truth, and Justice, Federation University, Australia.

Bettina van Hoven, University College Groningen and Department of Cultural Geography, University of Groningen, The Netherlands.

Margaret Walton-Roberts, Geography and Environmental Studies, Wilfrid Laurier University, Ontario, Canada.

Barney Warf, Department of Geography, University of Kansas, United States.

Marion Werner, Department of Geography, University at Buffalo – State University of New York, United States.

Dominica Whitesell, Department of Geography and the Environment, University of Texas at Austin, United States.

Matthew W. Wilson, Department of Geography, University of Kentucky, United States.

Haim Yacobi, The Bartlett Planning Development Unit, University College London, United Kingdom.

Acknowledgements

The editors wish to thank several individuals without whom this volume would not have been possible. Firstly, we are grateful for the patient support of Natalie Aguilera and Sarah Moorhouse at SAGE Publishers throughout the duration of this project. Secondly, we are thankful for the guidance and advice supplied by our international advisory board and to all those who provided us with suggestions via social media and email for thinkers to include. Thirdly, we would like to thank the large number of contributors who patiently dealt with our stream of editorial enticements, cajolements and suggestions with good humour and grace. The intellectual rigour, quality and incisiveness of the contributions is testament to their work. We hope that everyone who has aided us is as delighted with the book as we are.

Preface to Third Edition

Key Thinkers on Space and Place was conceived at the turn of the millennium, at a time when geography lacked a focus on the lives and work of its key practitioners. That volume had to tread a delicate balance between a future orientation and looking back to those who had made the discipline what it was. Ultimately, it focused on key contemporary thinkers, those whose ideas still circulated widely in (Anglo-American) human geography and who seemed to exercise influence on disciplinary debates. Rather than selecting people who have been honoured or celebrated for their contributions to teaching and learning, or their wider role in communicating geographical ideas, the focus was on those whose *ideas* have shaped human geography. The goal was to profile thinkers from both within and beyond Geography whose ideas about space and place had come to influence others studying a broad range of geographical themes and topics.

In this third edition of the book, entries that have appeared in earlier editions have been refined or omitted to reflect how ideas develop over time and how their reception changes. What was identified as a contemporary or current idea in an earlier edition may no longer be viewed in this way, while other ideas have now emerged to become more significant and more influential. We have hence 'retired' over half the entries from the second edition and replaced them with a new selection of key thinkers. As we detail in our introductory essay, recent years have witnessed Anglo-American geography grappling with a range of complex 'wicked' problems – climate change, environmental racism, housing crises – which demand new conceptual frameworks and ways of approaching space and place. Moreover, the academy has undergone demographic changes, shifting in terms of gender and race, and arguably becoming more international with influential contributions from around the world gaining prominence.

The entries in this book do not aim to offer a thorough or balanced overview of the entire career of each thinker. Instead, each entry is intended to provide a starting point for an engagement with a line of spatial thought, providing a summary of a thinker's key contributions. Each entry follows a common format. The opening section provides an overview of the thinker with some basic biographical information. This contextual material is followed by a summary of how each has conceived of space and place, aiming to identify why they are regarded as an important and influential thinker in debates on space and place. In a final section, each contributor offers a critical reflection on the work of the key thinker. Each entry concludes with two reference lists. The first is a guide to 'key works', while the second contains minor texts as well as a range of secondary sources. Throughout we highlight other thinkers with entries in the book in **bold** while those indicated in ***bold italics*** are key thinkers whose entry can be found in previous editions.

Planning for the third edition began in early 2019. Ten years had passed since the second edition was published, and Sage were anxious to capture the significant intellectual and contextual changes that had occurred in the period. The first key decision was to expand the editorial team to include Mary Gilmartin and Sue Roberts. Next, we established an international advisory board to help inform our decision about which key thinkers would be retained in the third edition, and which new thinkers would be added. The international advisory board included people based in a range of geographical contexts, with a diversity of disciplinary expertise. In addition, we also put out a call on social media for suggestions for new thinkers to be included. We shared that call widely. In total, over 100 people responded, with suggestions that varied from one to fifteen key thinkers each. We consolidated recommendations from social media and from the board, which resulted in a list of over 240 new thinkers for possible inclusion.

We also needed to decide which of the people profiled in the second edition would not be retained in the third edition. In making this difficult decision, as editors, we first discussed the existing list and identified thinkers who we believed needed to be retained given their ongoing important influence (25 in total). We then canvassed the views of the international advisory board, asking them to indicate ten of the remaining 41 entries that they felt could be removed from the new edition. Again, we consolidated these views, and made a preliminary selection of 38 key thinkers whose entries would not be updated for the third edition. This in turn provided the space for up to 38 new entries.

As editors, we met to discuss the long list of suggested new entries for the third edition. We wanted to ensure that the third edition was more diverse with respect to gender and race, to counter the overrepresentation of white male thinkers in previous editions. Moreover, we aimed to ensure better representation of thinkers from around the world and to broaden the disciplinary base of the scholars included. We also wanted to make sure that the selection of key thinkers represented the breadth of thinking on space and place and was not overly concentrated on particular philosophical approaches or areas of scholarship. This led to difficult, and at times subjective, decisions on our part. Faced with a number of key thinkers who focused on similar areas, we had to choose a smaller number as being the most 'key'. We shared our suggested list with the international advisory board, got their feedback, and decided on the final list of 66 key thinkers on space and place (one of which is a community – **Bawaka Country**).

This difficult process took a while and by the time we were ready to begin inviting contributions the COVID-19 pandemic struck. We were all struggling to do our jobs and to respond to the challenges of a global pandemic, so we felt it was inappropriate to continue with the book under these circumstances. With the agreement of Sage, we paused the project, and only took it up again over two years later, in June 2022. With our already agreed list of key thinkers, we contacted people who had contributed to earlier editions to see if they wanted to update their entries and in some cases we commissioned new versions of old entries. We also recruited new contributors for the newly added entries. In the end, the book is slightly shorter than the previous edition as four entries were not submitted by commissioned authors for various reasons.

The book is, we think, very different to its preceding editions: while it still has its omissions and exclusions, it is more diverse in terms of the range of contributors and profiles it includes. As such, we hope it informs, entertains and provokes as much as previous editions, while reaching different and new audiences.

Introduction

Mary Gilmartin, Phil Hubbard, Rob Kitchin and Susan M. Roberts

The Importance of Space and Place

The twenty-first century has already been characterised as one defined by a series of global 'crises'. Since the turn of the millennium, we have witnessed at least three major pandemics (SARS, COVID-19 and swine flu), economic shocks and recessions, migration crises (in Europe and elsewhere) and warfare and bloodshed in Palestine, Sudan, Syria, Ukraine, and Yemen, as well as countless other conflicts. In world cities there is a crisis of affordability as working people struggle to secure housing, whilst racialised inequality continues to be endemic. Precarious working is the norm for many, and though technology is advancing to create a more democratic public sphere in which individual voices and minoritised populations can be heard, there are concerns that the new media remains controlled by vested interests and the global capitalist class. The super-rich appear increasingly cut off from the rest of society, insulated in new elite enclaves that are distanced from the workers their prosperity relies upon. And all the time we are reminded of an ongoing climate crisis which ultimately will displace many from their homelands but which politicians and policymakers seem unable to tackle with any degree of effectiveness or conviction.

Human geography is about all of these things – and many more besides. More so than any other social science, it offers an integrative understanding of the world that uses methods derived from both science and the humanities, combining qualitative and quantitative methods to understand how the world shapes society, and how society shapes the world. As such, it is a heterodox discipline, one not afraid to borrow ideas and methods from other disciplines. Indeed, human geographers often work with concepts and techniques derived from other social sciences – most notably, Economics, Gender Studies, Politics, and Sociology – and contribute to ongoing debates in Anthropology, Cultural Studies, Ethnic and Migration Studies, International Relations, Psychology and Urban Studies, amongst others. Many who teach and research in Geography departments were often trained in these disciplines rather than Geography *per se*.

This given, Human Geography is a highly diverse, contested and complex discipline that is often accused of lacking a clear history or identity. Yet at the heart of the human geographical 'tradition' is a particular interest in the role of space and place in the making of society. This focus on space and place, and an insistence on understanding social, economic and political processes as always grounded in specific milieux, gives human geography its *raison d'être*. Other terms – region, environment, locale, landscape, mobility, scale – ripple out from these, but space and place are the 'building blocks' of geographical thinking. In essence, contemporary human geography develops theories about the social world in which space and place need to be taken seriously in explaining phenomena. As a discipline, it is not simply concerned with 'mapping' the distribution of human activities – though this is an important part of the discipline's methodological toolkit – rather, it takes place and space seriously as constitutive of the world, imbuing them with explanatory power.

Yet what space and place are remains a matter of debate. Not all geographers subscribe to the same definitions of these terms. For some, space and place are seen as interchangeable. Others are at pains to conceptually distinguish between space and place, though again the ways in which they are differentiated vary and have been the matter of sometimes fierce dispute. One version characterises space as objective and measurable, and place as subjective and intangible. Another describes space as nomothetic (concerning the general) and place as ideographic (concerning the specific). Yet another version sees space as progressive and open, with place as conservative and bounded. And others still, while recognising differences between space and place, emphasise their relationality and interconnectedness, viewing both as produced contingently through social relations and practices.

Moreover, both space and place are often understood through their relations to time. Geographer **Nigel Thrift**, for example, fused the concept of space with time to speak of *timespace*, recognising these as mutually constituted rather than separate but related entities. This emphasises that space is always *becoming*, having rhythmical qualities. Place, too, has complex relations with ideas of time (and memory). For example, place can be understood as an assemblage that makes it distinctive and meaningful at a specific time only for its essence to be destroyed subsequently (through forms of 'place annihilation' that are often experienced as traumatic). Here, relational understandings of place emphasise how different places are connected through unfolding processes such as capitalism and globalisation, as in the 'progressive politics of place' approach taken by **Doreen Massey** or **Cindi Katz**, noting the capacity of places to change and alter the processes that shape them. Alternatively, place can be understood through the personal and embodied experiences that unfold over people's lives, as examined by **Yi-Fu Tuan** and **Tim Cresswell**, noting that 'sense of place' is both individual and sometimes idiosyncratic, changing over people's lifetimes.

These debates about space and place are then about more than just semantics. Space and place can be conceptualised and thought about differently, and whilst for some geographers this is not of particular concern, most think carefully about how they use these terms, with geographical theories being based on particular understandings of them. This matters because it changes the way we think about the world, and the conclusions that we draw about the processes shaping it. Take the COVID-19 pandemic, for example. As the global spread and ubiquity of this pandemic became clear, the value of a geographical perspective became obvious. Nationally compiled data on mortality and morbidity was assembled, and mapped, allowing everyone to become an 'amateur geographer', poring over data to try to understand the aetiology (i.e., the natural history) of the virus as it spread across the globe. Geographers, however, generally offered more sophisticated analyses. Some, for example, drew on understandings of space as the surface across which the virus was transmitted, developing spatial diffusion models that could explain the pattern of spread, helping inform strategies of lockdown designed to halt its seemingly inevitable ubiquity. Others worked with a more striated view of the world, thinking about the global corridors of transmission and the networks that connect seemingly distanced places (e.g., through aeromobility), and noting the border closure strategies designed to rupture these networks, albeit only temporarily. Yet others focused more on the particular risks and exposures that occurred in specific places such as public transport hubs, supermarkets and sporting stadia, some bringing this down to the scale of the household to reveal the importance of density of living in making some more likely to become ill: see, for example, the special issues of *Dialogues in Human Geography* (Rose-Redwood et al., 2020), *Urban Studies* (Orford et al., 2023) and *Social & Cultural Geography* (Maddrell et al., 2023) for contrasting overviews. Different conceptualisations of space and place can then result in very different understandings of the world, leading us to emphasise different materialities, meanings, networks, differences, solidarities, (in)justices and (in)equalities.

Why Key Thinkers?

There are many ways of understanding how the 'geographical tradition' has unfolded (Livingstone, 1992). One is to focus on the *sites* where geographical thought has been developed and circulated, and where it has maintained a dialogue about its form and content: the field, the lab, the lecture room, the library and so on. This would be a story of geography's institutionalisation, and its incorporation within higher education as a legitimate sphere of study (see, for example, Berg et al., 2022). Another is to focus on geographical *technologies* and devices which allow us to see the world in different ways (Brunn et al., 2004). For example, a history can be written of the way in which the emergence of supercomputing in the 1960s led to geography's 'quantitative revolution', with statistical and mathematical conceptualisations of space leading to particular ways of modelling spatial processes. In the twenty-first century new forms of remote sensing, wearable technology, satellite technology, virtual mapping and various forms of spatial media (e.g., virtual reality, geolocated smartphone apps) allow us to see our world in new ways, again changing conceptions of space and place (Kitchin et al., 2017). Yet perhaps the dominant approach to writing the history of geography is to adopt a *paradigm* approach that considers its dominant intellectual framing and philosophy: this considers the emergence, and then decline, of successive forms of thinking that have come to be accepted as legitimate ways of understanding the world (see, for example, Johnston and Sidaway, 2016).

The paradigm model is attractive, and widely taught, as it gives the impression of order and progression: a paradigm emerges as a new way of thinking, replacing the preceding status quo in knowledge production, only for it to be critiqued and discounted. The discipline moves forward and develops, often at a fast pace. For example, many histories of geography trace a dizzying progression through the quantification of the 1950s and 1960s and the rise of a positivist paradigm (based on the scientific method and certain scientific principles such as a rejection of metaphysical questions) through a humanistic, behavioural and radical critique of positivism in the 1970s and

1980s. At this time, humanistic geographers like **Yi-Fu Tuan** rejected spatial science in favour of a more phenomenological approach to how people produce and experience space and place, and radical geographers promoted critical historical and geographical materialist approaches that challenged the predominant understanding of space as merely a backdrop for human activity. Instead, theorists of space and spatiality – in and beyond Geography – argued that space and social relations were intimately connected. **Henri Lefebvre,** for example, conceived of space as a (social) product, articulating an understanding of space as shaped by, and in turn shaping, broader social processes, such as capitalism and urbanisation. Related perspectives emerged in the work of Marxist geographer **David Harvey** and his one-time student **Neil Smith** on the uneven production of space, and also in **Milton Santos'** conception of space as an active moment. 'New' cultural geographers, such as **Denis Cosgrove**, invigorated discussions of place in the 1980s by drawing attention to relations of power, inspired by the work of cultural theorists like **Stuart Hall**. Later, a renewed focus on place as embodied brought attention to the power dynamics of embodied experiences in registers that included gender, race, class and sexuality, evident in the work of **Gill Valentine**, for example, on sexual geographies, or **Linda McDowell's** work on the gendered performativity of doing business in the City of London.

In the twenty-first century, human geography has been shaped by two major challenges. The first may be broadly understood as a move towards decolonisation that draws attention to and challenges the colonial nature of knowledge production within Geography, within the academy, and within the English-speaking world. These challenges are clearly articulated by thinkers based in former colonies in the so-called Global South, such as **Akin Mabogunje** and **Milton Santos**, and by others who emphasise the spatiality of the violence associated with colonial projects, such as **Arturo Escobar**, **Achille Mbembe** and **Eyal Weizman**. Thinkers such as **Sara Ahmed**, **Gloria Anzaldúa** and **Ruth Wilson Gilmore,** highlight the lasting legacies of colonialism in the experiences of racial and ethnic minorities in diverse contexts. Other thinkers, such as **bell hooks**, **Audrey Kobayashi**, **Katherine McKittrick** and **Eve Tuck**, draw upon feminist approaches born from the experiences of those who have been or are marginalised to articulate new and liberatory epistemologies.

The second major challenge may be framed as a response to growing environmental and societal crises and a challenge to the human-centred thinking that has amplified these crises. Political ecology, advocated for by **Paul Robbins** among others, attends to nature–society relations, developing critical analyses of power relations that intersect and shape access to natural resources, while **Laura Pulido** emphasises the connections between racialisation and environmental justice. Meanwhile, more-than-human theorisations of space and place can be seen in the multispecies ethnography of **Anna Tsing**, in the 'inhuman' geography of **Kathryn Yusoff**, and in the Indigenous epistemologies and ontologies articulated by **Bawaka Country.**

Described in this way, the development of thinking on space and place is then a story of so many -isms – positivism, humanism, behaviouralism, critical realism, Marxism, feminism, postmodernism, post-structuralism, postcolonialism, materialism, vitalism – all presented as if they are successive episodes in the discipline's development. There are of course clear limits to such an approach: it is hard to say that any given approach is dominant at a given point in time considering the sheer diversity of ways in which geography is taught and practised globally, and suggesting that only the 'latest' knowledge is legitimate or useful often does violence to forms of scholarship that have value in the context in which they are pursued. The paradigm idea, originally developed in relation to the natural sciences, appears much less applicable in geography because of this multiplicity.

This book is designed to engage with theoretical debates in human geography but does so through a focus on certain thinkers who exercise a degree of influence over the shape and form of contemporary thinking. We do this because we believe that the *biographical approach* offers important insights into how ideas are shaped by contexts, and how those ideas in turn effect change. Following the often convoluted career paths of key thinkers offers insight into how people's theoretical discussions of space and place are shaped by ongoing (inter)disciplinary debates and may be linked to important political moments. It shows how theoretical understandings evolve, shift and change. It also highlights the connections between different thinkers, whose ideas are developed in collaboration with or in reaction to others. The biographical approach situates thinkers in geographical, social and intellectual contexts. It highlights the material realities of knowledge production (such as the political economy of the University sector, and the particular pressures on academic time which can steer them in particular directions). And it demonstrates that spatial thought is never developed in a vacuum, but is always constructed by individuals and groups of people located in particular institutional and social structures, with their own sets of personal and political beliefs. A biographical approach reveals how individual thinkers draw on a rich legacy of ideas from past and contemporary generations. The courses they took as students, the discussions they had with peers and mentors and the texts they read have played roles in shaping their intellectual development. Similarly, key thinkers on space and place are shaped by broader political struggles that influence how they experience, think about and engage with the world.

The biographical approach is very useful for demonstrating the genealogy of intellectual ideas, revealing, for instance, the ways in which personal history affects intellectual development. For example, **Edward Said's** experiences of being born into an Arab Christian family in British-occupied Palestine, and his subsequent fight throughout his adult life for Palestinian causes, undoubtedly shaped his thinking about the relationship between culture and imperialism. Likewise, **bell hooks** has attributed her work to theorise the problems of Black patriarchy, sexism, and gender subordination, as well as her reconceptualisation of home and belonging, to her experiences of growing up as a young Black woman in Kentucky (US) during the 1950s and early 1960s. **Anssi Paasi's** thinking on regions and regional geography has been shaped by the nature of Finnish academia and his strong empirical focus on Finland, while **Jennifer Robinson's** contributions to the spatiality of theory and comparison draw from her experience of growing up in and researching South Africa, as well as her subsequent migration to the UK. A biographical approach then enables an appreciation of both the roots (origins) and routes (directions they have evolved) of thinking on space and place.

A biographical approach also helps us understand how individuals travel not just between places but also within and between academic disciplines. Many of the key thinkers on space and place included in this collection identify, or are identified, as academic geographers, yet began their career in different disciplines or have enjoyed sojourns where they worked outside of academia. Others have never formally studied Geography, but rather have worked in other disciplinary or interdisciplinary contexts, including Architecture, Anthropology, Cultural Studies, Development Studies, Education, Ethnic Studies, Literature, Planning, Political Science, Sociology, Women's Studies and Urban Studies. The fact that over half the thinkers profiled here are not conventionally defined as 'geographers' is an acknowledgement of the by now well-established centrality of spatial thinking in a wide range of disciplines.

For all the strengths of a biographical approach, it also has limitations. We are particularly mindful of the pitfalls of constructing a hagiography or perpetuating the 'great man' approach to telling the stories of academic disciplines and intellectual thought. Here, the concern is that certain individuals (most often, white, Anglophone male scholars) are singled out as having a disproportionate influence on shaping thought and approaches, while the work of others (for example, Black or female scholars, or scholars from the Global South) is marginalised and dismissed. For these reasons, it is important to acknowledge that around the world, not every thinker has access to formal education or to the institutions that constitute academia and that, even among those who do, there are vast inequalities in opportunity and access to resources (including language) that still shape who gets considered to be among the ranks of 'key thinkers'. Moreover, concepts are usually developed by several people working in concert or competition, engaging and critiquing each other's ideas, while also drawing on a rich intellectual heritage. Indeed, it should be clear from the cross-referencing between entries that no theorist develops their view of the world in an intellectual vacuum. Ideas develop across thinkers and generations and charting these connections reveals a genealogy of thinking. For example, **Gillian Rose's** ideas about the privileging of male ways of conceiving of space and place have been heavily influenced by psychoanalytic and post-structural writing. One major source of inspiration was the works of the feminist philosopher **Judith Butler**. Judith Butler, in turn, while again drawing from a diverse set of philosophical texts, has extensively utilised the writings of **Michel Foucault**. Likewise, when developing his critical philosophy, Foucault was influenced by (amongst others) the German philosophers Friedrich Nietzsche and Martin Heidegger. Of course, Gillian Rose's thinking is not the end point in this lineage but is rather a node in a complex web of interconnections, and it is no surprise that her ideas are cited as influential sources in other entries. Equally, it is important to challenge the exclusions and marginalisations of the past by acknowledging a wider range of key thinkers and their significance, such as **J.K. Gibson-Graham**, **Richa Nagar**, **Kathryn Yusoff** and **Bawaka Country** who actively seek to co-create knowledge in collaboration with others both within and beyond the academy.

Though not all would agree with the list of 'key thinkers' profiled in this book, the chapters capture, we feel, the exciting reorientation of the discipline of Geography towards new forms of knowledge production in the last decade, as well as the expanded register of spatial thinking that has emerged in relation to crucial societal challenges. We hope that each entry inspires you to explore the references and develop your own take on the varied geographical imaginations deployed by these key thinkers on space and place.

References

Berg, L., Best, U., Gilmartin, M. and Larsen, H.G. (eds) (2022) *Placing Critical Geographies: Historical Geographies of Critical Geography*. London: Routledge.

Brunn, S., Cutter, S. and Harrington, J.W. (2004) *Geography and Technology*. Dordrecht: Springer.

Johnston, R. and Sidaway, J.D. (2016) *Geography & Geographers: Anglo-American Human Geography Since 1945*. London: Routledge.

Kitchin, R., Lauriault, T.P. and Wilson, M. (eds) (2017) *Understanding Spatial Media*. London: Sage.

Livingstone, D.N. (1992) *The Geographical Tradition*. Oxford: Blackwell.

Maddrell, A., Ho, E.L-E. and Lobo, M. (eds) (2023) 'Special Issue: Social and cultural geographies of COVID-19', *Social & Cultural Geography*, 24 (3–4): 385–718.

Orford, S., Fan, Y. and Hubbard, P. (2023) 'Urban public health emergencies and the COVID-19 pandemic. Part 1: Social and spatial inequalities in the COVID-city', *Urban Studies*, 60 (8): 1329–45.

Rose-Redwood, R., Kitchin, R., Apostolopoulou, E., Rickards, L., Blackman, T., Crampton, J., Rossi, U. and Buckley, M. (eds) (2020) 'Special Issue: Geographies of the COVID-19 pandemic', *Dialogues in Human Geography*, 10 (2): 97–295.

1
Sara Ahmed

Kirsten Simonsen, Roskilde University

Biographical Details and Theoretical Context

Sara Ahmed is a British-Australian author and independent scholar, known for her contributions to feminism, queer theory, critical race theory and postcolonialism. She was born in Salford, England, in 1969, to a father from Pakistan and an English mother. In 1974, the whole family migrated to Adelaide, Australia.

Ahmed studied at the University of Adelaide and took her BA there in 1989 in English, Philosophy and History, returning to Britain in 1991 to finish her PhD at the Centre for Critical and Cultural Theory at University of Cardiff. Ahmed subsequently held positions at Lancaster University in Women's Studies, and at Goldsmiths, University of London, in Race and Cultural Studies. In 2016, she resigned from her position as Director of the Centre for Feminist Research at Goldsmiths to draw attention to neglected sexual harassment complaints lodged by students. From that point, she has worked as an independent scholar whose writings concern how power is experienced and challenged in everyday life and institutional cultures. Her subsequent work appeared in three key books (e.g., Ahmed, 2017; 2019a; 2021) and in her blog, *Feminist Killjoys* (feministkilljoys.com).

From an early age, Ahmed was aware of her gendered and racialised position. In an autobiographical chapter (1997), she wrote from her experience as a mixed race, middle-class and migrant woman. She did this in a recognition of the subject as an embodied and located entity immersed within a social and political context, needing to traverse the impossible distinction between the psychic and the social. As an illustration of the complex processes of identification and disidentification in play, she referred to an incident when she was 14 years old, walking around the streets of Adelaide without any shoes on. She was stopped and addressed by two policemen. They asked her what she was doing ('I am walking', she answered), then why she wasn't wearing shoes (she didn't remember her answer, only that the question was infringing on her rights). One of them then asked if she was Aboriginal (an indignant 'no'). The other policeman interrupted, gave her a wink and said, 'it's a suntan, isn't it?' She just smiled. Then they asked where she lived. She told them, and mentioned her school (a private girls' school). They said that was fine, but that she ought to wear shoes. They told her that they were checking into some recent break-ins in the area. Ahmed explains how the incident involved complex identifications and disidentifications. In the first instance, the policemen identified her as working class (from dress), and Aboriginal (from her skin colour), in this way reading her as a suspect – a potential robber. Throughout the interaction, a shift occurred that forced a disidentification from her identity as suspect. However, the whole incident left her with mixed feelings; on the one hand very angry, on the other ashamed that she had disavowed being Aboriginal, in this way getting herself implicated in a structure of racial identification that she sensed was wrong.

Ahmed's (1995) PhD thesis – *Structural Differences: Feminist Theory and Postmodernism* – was an exploration of feminism and included attention to other embodied identities, in particular race (see Ahmed, 1999). Subsequently, Ahmed focused on sexuality and disability, in addition to gender and race. In doing so, she turned towards phenomenology, most explicitly in *Queer Phenomenology* (Ahmed, 2006) and the article 'A Phenomenology of Whiteness' (Ahmed, 2007), where she drew on Husserl, Heidegger and Merleau-Ponty. The leading concept in this work was *orientation,* connected to the central thesis of phenomenology – that consciousness is always directed towards something (human or non-human) – and its emphasis on the lived experience of inhabiting a body. Orientations hence help us to 'find our way' and 'feel at home'.

Ahmed explored fissures in these arguments in order to trace 'failed orientations' and familiarities, demonstrating that orientations are not pre-given. With Merleau-Ponty (1962), she describes bodily horizons as 'sedimented histories' (Ahmed, 2006: 56) and discusses 'moments of disorientation' where orientation happens to fail. In this sense, she argued that the question of queer can be seen from the point of view of orientation – of how we align with and/or fail to align with straightness (i.e., the heteronormative). This understanding of the queering of orientation both takes inspiration from and, in turn, has relevance for phenomenological analyses of race, gender, and disability.

Ahmed's theoretical approach at this time has been designated 'critical' phenomenology, in that she mobilises phenomenological analysis in the service of reflexive inquiry into how power relations infuse experience (see Guenther, 2017; Simonsen and Koefoed, 2020; Weiss et al., 2020). The work of this period saw Ahmed recognised with The Phenomenology Roundtable Award (2010) for her 'outstanding contribution to the field of phenomenological research' and The Feminist and Women's Studies Network Book Prize for *The Promise of Happiness,* heralded for its 'ingenuity and scholarship in the fields of feminism, gender or women's studies' in 2012.

In 2016, Ahmed left Goldsmiths, turning her attention to power relations within institutions, especially universities, extending her work on the complexities of diversity, diversity work, and complaint.

Spatial Contributions

Ahmed's key contributions to the discussion of space primarily evolve from two concepts. The first one is the above-mentioned idea of *orientation.* In *Queer Phenomenology: Orientations, Objects, Others* (Ahmed, 2006) and the book chapter 'Orientations matter' (Ahmed, 2010a), she discusses how orientations influence our relation with objects and others. They shape how we inhabit the surrounding space, how we apprehend this shared inhabitancy, and 'who' and 'what' we direct our energy and attention toward.

This understanding of orientation refers to the spatiality of the lived. Heidegger (1962: 144), for example, sees orientation as a question of the familiarity of the world, stating 'I necessarily orient myself both in and from my being already alongside a world which is "familiar"'. Familiarity is connected to the given point from which the body gains the capacity to orientate itself in one way or another. Familiarity is neither delimited nor static. The work of inhabiting space involves a negotiation between what is familiar and unfamiliar, such that it is still possible to incorporate new impressions depending on which way we turn and what is within reach. These impressions shape the body as well as the 'inhabited' space, such that the body *produces* itself in space and simultaneously produces that space – an insight that Ahmed develops from **Henri Lefebvre** (1991). When bodies orientate towards objects in the world, the surroundings tend to take shape around the (material and immaterial) practices in play. Orientation starts from our body, and it is from 'here' that it reaches out towards 'something else' or 'somebody else'. It can give occasion for both an extension of our reach (of what we can have, and what we can do) and an opening towards the other. However, none of that is given. The formation of the surroundings can happen through the way 'spaces and tools take shape by being oriented around some bodies more than other bodies' (Ahmed, 2006: 132) – that is, some bodies' formation of space blocks others' possibilities and practices.

Adopting a social view on orientation, then, renders obvious that it can take many forms; we orient ourselves in time and space, but we also assume political, sexual, cultural and social orientations. We can follow the lines (material and immaterial) which exist as paths formed by past journeys, or we can arrive into situations opening new lines. Following a line is never neutral: it takes time, resources, energy and social investment. A lifeline also expresses our identity. However, life is not linear as it includes crises, turning points, setbacks or possibilities, opening up new worlds.

For Ahmed this can lead to 'moments of disorientation'. Moments of disorientation turn our world upside down. Disorientation is a bodily feeling that can shatter one's sense of confidence in the grounding of one's existence. It is a situation that can make bodies react defensively, as they reach out for support or search for a place to reground and re-orientate their relation to the world. The feeling of being shattered might persist and become a crisis, or it may pass if the ground returns. Moments of disorientation, then, are destabilising and undermining but they can also be productive, leading to new hopes and new directions. An example of a moment of disorientation is (forced) migration, which can be described as both processes of disorientation and reorientation as bodies both 'move away' and 'arrive'. Oppositely, arriving bodies can be perceived as 'out of place', disturbing the picture. These are situations of failed orientations where bodies come to inhabit spaces that do not extend their reach, where they live with social experiences of restriction, uncertainty and blockage. It is therefore clear that disorientation is unevenly distributed, and that some bodies, more than others, experience a crisis in their involvement in the world.

The second concept in Ahmed's work with spatial resonance is 'encounter', developed extensively in the book *Strange Encounters: Embodied Others in Post-Coloniality* (2000). As the title suggests, this discussion largely concerns cross-cultural meetings, and here Ahmed takes much inspiration from Frantz Fanon's *Black Skin, White Masks* (1967). The term encounter suggests a meeting, but a meeting that involves surprise and sometimes conflict because of its inevitable content of similarities and difference and processes of inclusion and exclusion that constitute the boundaries of bodies or communities (Ahmed, 2000). Encounters are unpredictable and premised with the unknown that does not allow us to control them or predict their outcome. Encounters are *temporal* and *spatial* because they involve at least two subjects approaching each other, and their repetition can shift the boundaries of the familiar. That means encounters are not about different pre-existent 'cultures' meeting each other: difference is produced *in* the encounters, meaning encounters can make and transform difference, destabilising, reworking or reproducing it in unpredictable ways (see also Darling and Wilson, 2016: 6).

At one level, encounters refer to face-to-face meetings as we experience them in everyday life – 'eye-to-eye' or 'skin-to-skin' (Ahmed, 2000: 7). They are, however, also rendered temporal and spatial through historical-geographical mediation. They presuppose encounters with other bodies, other spaces and other times. In this way, they reopen histories of encounter and geo-political imaginations of the Other and incorporate them in the encounters as traces of broader social relationships. Encounters between embodied subjects always dwell between the domain of the particular and that of the general, with the encounter framed by broader relationships of power and antagonism. As such, this understanding of encounters adds to Merleau-Ponty's ideas of the social body as a body opening up and intertwining with the world (i.e., other bodies and materialities) by accentuating that this is not a general but rather a *differentiated* world. In such a world, what is meant by the social body is more often than not 'precisely the effect of being with some others over other others' (Ahmed, 2000: 49).

Hence, the constitution of Others involves spatial negotiations over mobility and home, (imagined) communities, boundaries and bridges. 'Like bodies' and 'unlike bodies' do not precede encounters of inclusion or expulsion. Rather, likeness and unlikeness as characteristics of bodies are produced through these encounters. Part of that process is the experience of being exposed to oppressive visions or emotions. Here, Ahmed follows Fanon (1967) when he describes the phenomenology of incorporating otherness and the development of a 'double consciousness' due to the enculturation of the body. Men and women of colour, Fanon says, develop a third-person consciousness trying to reconcile their own experiences with the operation of a 'historical-racial schema' within which their corporeal schema is supposed to fit.

Encounters, then, are deeply charged with emotions. The 'strange encounter' is played out *on* the body and *with* emotions (Ahmed, 2000: 39). It is basically a sensuous process involving an affective opening out of bodies to other bodies. However, encounters with Other bodies also involve practices and techniques of differentiation. Various familial relations involve particular forms of emotion and ways of touch, while the recognition of somebody as a stranger – a body that is 'out of place' – might involve disgust or fear of touching. Bodily encounters are processes where bodies come together and co-mingle in different ways, slide away from each other again and become re-lived or reformed in their apartness. Within Geography, these ideas have been influential and widely discussed in works exploring 'geographies of encounter' (e.g., Valentine, 2008; Simonsen, 2010; Leitner, 2012; Merrifield, 2013; Darling and Wilson, 2016; Wilson, 2017; Koefoed et al., 2017).

Key Advances and Controversies

A major contribution of Ahmed's work concerns the ways in which identity has to be understood in social theory. She approaches difference from a standpoint in which phenomenology needs post-structuralism's deconstructive critique of power, while post-structuralism needs phenomenology's recognition of embodiment and lived experience. While Ahmed's thinking, then, is not phenomenological in the strictest sense, it provides us with a new embodied approach to understanding race, gender and sexuality – as well as any other way in which bodies are made to feel othered in certain spaces – to show 'how bodies are gendered, sexualized and raced by how they extend into space' (Ahmed, 2006: 5).

It is from this position Ahmed runs into discussions with contemporary authors approaching similar questions within schools of thought labelled 'post-humanist' or 'new materialist' (Ahmed, 2008; Davis, 2009). The disagreements revolve around the status of biology and matter within feminist theory. Ahmed's point is that since most feminism today does not ignore the body as a real, living physical and biological entity, new materialist's accusations of biophobia are misplaced. The question is whether a total 'entanglement' of biological and social theory is possible without returning to biological determinism.

Similar discussions revolve around the different facets and framings of the issue of emotion and affect. In a conversation on what has been called the 'affective turn', Ahmed (2014: 97) says: 'I am reluctant to use the word affect [...] And for me, the reason why I picked up emotion as my word was partly because I was interested in this idea of movement that is explicit in its etymology'. She has developed this standpoint in the books *The Cultural Politics of Emotion* (Ahmed, 2004) and *The Promise of Happiness* (Ahmed, 2010b). She argues that affect might be a form of stickiness and contrasts this with the influential (post-humanist) work of Brian Massumi (2002), which suggests that affects are autonomous from any persons and distinct from emotions. She thinks that the distinction between affect/emotion tends to under-describe the work of emotions. For her, emotions are not simply about subjective content, nor are they socio-linguistic 'after-thoughts' of feelings. They are orientated and directed towards somebody or something and are as such relational and a part of world-making. Here we again see in Ahmed's thought the traces of phenomenology.

However, the most outspoken controversies for Ahmed sprouted up when she turned her attention towards institutional life at the universities. In her book, *On Being Included: Racism and Diversity in Institutional Life* (2012), Ahmed examined how universities developed processes designed to manage diversity and racism. Ahmed's research blog *Feminist Killjoys* invited public engagement on the essays and lectures that later formed her book *Living a Feminist Life* (2017), which included the *Killjoy Manifesto*. This manifesto develops feminist concepts and principles for the critique of institutional everyday life, expanding ideas first outlined by Ahmed (2010b) in *The Promise of Happiness*. In 2017 Ahmed announced a related research project on complaint, in order to support students testifying in multiple inquiries into sexual harassment and sexual misconduct (Ahmed, 2019b). It was designed to be a qualitative empirical study of what Ahmed calls 'complaint biographies', following the paths and destinies of complainants and those of the individuals considering the doing or non-doing of formal complaint. Two influential books grew out of this project: *What's the Use? On the Uses of Use* (Ahmed, 2019a) and *Complaint!* (Ahmed, 2021).

Ahmed's Key Works

Ahmed, S. (2000) *Strange Encounters: Embodied Others in Post-coloniality*. London: Routledge.

Ahmed, S. (2004) *The Cultural Politics of Emotion*. Edinburgh: Edinburgh University Press.

Ahmed, S. (2006) *Queer Phenomenology: Orientations, Objects, Others*. Durham, NC: Duke University Press.

Ahmed, S. (2010b) *The Promise of Happiness*. Durham, NC: Duke University Press.

Ahmed, S. (2017) *Living a Feminist Life*. Durham, NC: Duke University Press.

Secondary Sources and References

Ahmed, S. (1997) '"It's a sun-tan, isn't it?" Auto-biography as an identificatory practice', in H.S. Mirza (ed), *Black British Feminism*. London: Routledge. pp. 153–67.

Ahmed, S. (1999) *Differences That Matter: Feminist Theory and Postmodernism*. Cambridge: Cambridge University Press.

Ahmed, S. (2007) 'A phenomenology of whiteness', *Feminist Theory*, 8 (2): 149–68.

Ahmed, S. (2008) 'Open forum imaginary prohibitions: Some preliminary remarks on the founding gestures of the "New Materialism"', *European Journal of Women's Studies*, 15 (1): 23–39.

Ahmed, S. (2010a) 'Orientations matter', in D. Cool and S. Frost (eds), *New Materialisms: Ontology, Agency, and Politics*. Durham, NC: Duke University Press. pp. 234–58.

Ahmed, S. (2012) *On Being Included: Racism and Diversity in Institutional Life*. Durham, NC: Duke University Press.

Ahmed, S. (2014) 'Affect/Emotion: Orientation Matters. A Conversation between Sigrid Schmitz and Sara Ahmed', *Freiburger Zeitschrift für GeschlechterStudien*, 20 (2): 97–108.

Ahmed, S. (2019a) *What's the Use? On The Uses of Use*. Durham, NC: Duke University Press.

Ahmed, S. (2019b) 'A complaint biography', *Biography*, 42 (3): 514–23.

Ahmed, S. (2021) *Complaint*! Durham, NC: Duke University Press.

Darling, J. and Wilson, H.F. (eds) (2016) *Encountering the City: Urban Encounters from Accra to New York*. London: Routledge.

Davis, N. (2009) 'New materialism and feminism's anti-biologism. A response to Sara Ahmed', *European Journal of Women's Studies*, 16 (1): 67–80.

Fanon, F. (1967) *Black Skin, White Mask*. New York, NY: Grove Press.

Guenther, L. (2017) 'A critical phenomenology of solidarity and resistance in the 2013 California Prison hunger strike', in L. Dolezal and D. Petherbridge (eds), *Body/Self/Other: The Phenomenology of Social Encounters*. Albany, NY: State University of New York Press. pp. 47–74.

Heidegger, M. (1962) *Being and Time*. Oxford: Blackwell.

Koefoed, L., Christensen, M.D. and Simonsen, K. (2017) 'Mobile encounters: Bus 5A as a cross-cultural meeting place', *Mobilities*, 12 (5): 726–39.

Lefebvre, H. (1991) *The Production of Space*. Oxford: Blackwell.

Leitner, H. (2012) 'Spaces of encounters: Immigration, race, class, and the politics of belonging in small-town America', *Annals of the Association of American Geographers*, 102 (4): 828–46.

Massumi, B. (2002) *Parables for the Virtual: Movement, Affect, Sensation*. Durham, NC: Duke University Press.

Merleau-Ponty, M. (1962) *The Phenomenology of Perception*. London: Routledge and Keagan Paul.

Merrifield, A. (2013) *The Politics of Encounter: Urban Theory and Protest under Planetary Urbanization*. Atlanta, GA: University of Georgia Press.

Simonsen, K. (2010) 'Encountering O/other bodies: Practice, emotion and ethics', in B. Anderson and P. Harrison (eds), *Taking-Place: Non-Representational Theories and Geography*. Farnham: Ashgate. pp. 221–40.

Simonsen, K. and Koefoed, L. (2020) *Geographies of Embodiment. Critical Phenomenology and the World of Strangers*. London: Sage.

Valentine, G. (2008) 'Living with difference: Reflections on the geographies of encounter', *Progress in Human Geography*, 32 (3): 323–37.

Weiss, G., Murphy, A.V. and Salamon, G. (2020) *50 Concepts for a Critical Phenomenology*. Evanston, IL: Northwestern University Press.

Wilson, H.F. (2017) 'On geography and encounter: Bodies, border and difference', *Progress in Human Geography*, 41 (4): 451–71.

2
Louise Amoore

Andrew C. Dwyer, Royal Holloway, University of London

Biographical Details and Theoretical Context

Louise Amoore completed an MA in European Studies (1995) and a PhD in International Relations (1998) at the University of Newcastle in the North-East of England. Her doctoral dissertation investigated the 'social roots of global change' in industrial relations between British and German firms in the 1990s. She was appointed as a Lecturer in Politics in Newcastle in 2000 after a brief time at Northumbria University, before moving to take up a position as Lecturer in Geography at nearby Durham University in 2005. She was promoted to Professor of Political Geography in Durham in 2011. She has held visiting positions at the Australian National University, Copenhagen Business School and the University of Wisconsin-Madison, and has been an editor of the prestigious journal *Progress in Human Geography*. In 2018, she was appointed to the Biometric and Forensic Ethics Group, a UK advisory non-departmental public body, to advise on the ethics of algorithms and technologies being deployed in policing, security, and border controls.

Although her early career was rooted in the political economy of work and the politics of globalisation, she is best known for her critical contributions to algorithmic governance and the 'War on Terror' from the early 2000s onwards, and in the process, has developed a sustained analysis of the ethical and political stakes of computational life and its inherent spatialities. Motivated by the response to Al-Qaeda's attacks against the US in 2001, otherwise referred to as '9/11', the long and persistent shadow of the 'War on Terror' has been central to Amoore's development as a thinker and researcher. Her empirical and theoretical work has included an examination of numerous forms of (statistical) measurement, how people and objects become modelled and ordered, the modes of governance that are utilised by states and corporations, and the deployment of big data analysis and algorithmic processing. Her initial work on the post 9/11 biometric processes to secure the state (Amoore, 2006), theorising probability in the algorithmic age (Amoore, 2011; 2013), and inductive processes of algorithmic detection (Amoore, 2017) have become an essential bedrock to analyses of security and society. This initial body of work – particularly on the performance of the (state) border – provides resources for understanding the dynamics of how computational processes and big data move, and are contorted and twisted, so that states and corporations have increasing power to articulate the future.

In later work, especially 'Cloud Ethics' (2021), Amoore's contributions explored the rippling aftereffects of the 'War on Terror' and have developed a deeper attention to and engagement with Black thought and the enduring colonialism and racism of statistics as they have become computationally mediated across *algorithmic societies* (the title of a 2020 European Research Council advanced investigator award). In such analysis, she has been careful not to become constrained by an overt focus on the technological, always ensuring her analysis considers the profound impacts of digital systems upon marginalised and othered peoples and communities. Moreover, she has been clear on how space and place are being constructed in new ways, where mobility and belonging are not just about the interrogation of the 'secrets' of the body and its attributes, but about a new cloud of attributes – an ontology of association – that form new political communities which the state, and other organisations, can use for inductive reasoning. Indeed, in the most radical of her *ethico-political* work, as she terms it, she has rallied against an aperture of the future. Here, she claims that the space for questioning, or doubting, the future becomes foreclosed as algorithms produce singular outcomes and predictions of a future predicated on past attributes (Amoore, 2019). In this sense, Amoore's work questions linear, geometric, and anthropocentric views of space and place through examining the role of computation in generating new forms of social, cultural, and political geographies, albeit in ways that do not betray an overbearing conceptual baggage. As such, her numerous theoretical and empirical contributions unsettle notions of space and place at their intersection with digital and computational phenomena – often at the boundaries of the discipline of Geography. Indeed, just as her career has traversed disciplines, so too has her analysis and her work been widely engaged with across the social sciences. Perhaps unsurprisingly, Amoore describes herself as both a 'social scientist and political theorist' (Dwyer et al., 2021: 48).

Spatial Contributions

Following a feminist praxis to open the political, Amoore has sought to encourage a questioning of the 'orthodox' of political and cultural geography. This has animated where and how *the political* becomes expressed, in what spaces, and the transformations of place that occur (Amoore and Hall, 2013). In broad terms, her spatial contributions relate to four key figures: the *border*, the *body*, the *cloud*, and the *deep*. These figures are reflective, albeit in coarse terms, of both her long-standing research commitments and of the scholarly environment and wider debates on space and place that have occurred from the 2000s onwards.

The *border* is an early matter of concern for Amoore, often explored alongside her friend and collaborator Marieke de Goede (Amoore and de Goede, 2005; 2008; 2021). The examination of the border has always been essential to understanding the formation and performance of state power and how certain places become sites of ordering, filtering, and often state violence. However, in a post-9/11 landscape where data was becoming an ever-more important feature of assessing perceived risk, and thus, processes at state borders, Amoore engaged with the changing dynamics of *where* the border lies, *how* it is enacted, and what this meant for the practices of states increasingly reliant on algorithmic processing. Rather than simply decisions being made by a border guard, Amoore traces how data and algorithms became mediators for such decisions. The admission at the border, in this reading, becomes increasingly distributed to everyday, lived patterns of life, in addition to frequently racialised and colonial practices upon which attributes are deemed to be suspicious (which countries are visited, the types of financial transactions, and so on). The border then becomes a place where data about us is shared, transferred, and manipulated to provide a data assessment of whether a body should be permitted through a state border or not. Amoore informs us that the border has not simply expanded beyond the boundary of the state, but is increasingly regulated through financial data brokers and passenger travel records. She suggested pre-emptive assessments of risk were being made on individuals in the early years after the start of the War on Terror, with the spatialisation of the state extending well beyond the drawn borderline into a sequence of different data-points scattered far beyond the sovereignty of the state.

Meanwhile, *the body* became an important site of investigation post-9/11 for Amoore, particularly its examination through Backscatter X-ray scanners installed by the UK and the US at their borders (Amoore and Hall, 2009). Here, the body is something which is examined, dissected, and exposed by the state in order to reveal the potential undesirable object. Rather than the body being simply something to be externally viewed, it becomes a site of algorithmic scrutiny. As Amoore and co-author Alexandra Hall (2009: 460) note, 'the history of knowing the body via taking apart has always been a history of security or, more exactly, of securability'. This practice of seeking to expose the body is twinned with the need to monitor the body in association with others, as Amoore (2006) details in a landmark paper on 'biometric borders'. In this piece, she makes two claims. First, that there was a significant change in the management of bodies in the post-9/11 environment, and second, drawing on **Michel Foucault**, that there has been a turn to biopower whereby the body becomes encoded and carries the border itself. Therefore, the body becomes an important site, where the border becomes manifest through a change in its governance and interactions with data. Today it is almost unimaginable to travel internationally without biometric information being shared and stored. The body is then both to be examined in movement and is to be made visible as a harbinger of the security state.

The discussion of *the cloud* is one of Amoore's (2020a) later signature contributions. In a critique of prior approaches to the study of cloud computing, she makes a claim that we should not simply aim to geometrically map the distribution and networks of collections of servers, such as the huge datacentres owned by Amazon Web Services (AWS) or Microsoft Azure, that she characterises as 'Cloud I', but rather we should focus on 'Cloud II'. This second formation of the cloud articulates how the connections processed through big data and algorithms are forming new bonds and ethico-political communities. This means that the human–computer interaction with a surgeon, how a face becomes identified, or the assessment of whether someone is moving suspiciously, are not as simple as one may initially expect. We may understand the cloud as part of a burgeoning interest in volume and computing capacity within the study of space and place, but that would be a mischaracterisation of Amoore's novel contribution. Rather, Amoore is seeking to build – albeit in her decidedly feminist tradition – a distinctive re-evaluation of how computation is knitting new collectives and associations, and often ones that we are unfamiliar with. For example, how does the collection of our bodily movement inform attributes within datasets that are then used to produce assessments of suspiciousness for the arrest of an individual? The cloud as a spatialisation proposes that we are intrinsically connected to one another through data collection practices, and one that holds strong ethico-political responsibilities for us as a society and for how we wish to govern ourselves.

The *deep* is one way in the which the cloud becomes operationalised – both in how society is being *deeply* reshaped by computational algorithms and, in the process, how these algorithms are creating new ethico-political communities and responsibilities. Algorithms take many different forms, with the 'neural network' machine learning architecture being prevalent in a wide range of applications – ranging from generative outputs to visual recognition, as much as predictive analytics. At its most simple and abstracted, such machine-learning algorithms produce 'feature spaces', where associations between data points are plotted and associations made. Although this is not new – given it has been a function of historical legacies of statistics – what Amoore demonstrates is how such abstract spaces are generative of new associations based on the collective societal 'weights' of past data collection. The deepness of these machine-learning algorithms derives from a recursive process whereby in their creation, inputs are examined across multiple layers (through backpropagation) to craft models that output features so that new data can be separated according to its similarities and dissimilarities to the model. Within 'self-learning' variants of these algorithms, the algorithms create their own features by generating attributes about the data they process.

It is in these new, abstract, spaces where the distinctiveness of Amoore's contribution becomes clear. Rather than reducing these algorithmic feature spaces down to mathematics or statistics, she articulates their ethico-political frame. For example, exploring machine-learning algorithms at various sites – whether in a hospital theatre room or in a police control room – she argues what may appear as neutral renditions of abstract mathematical representations are in fact weighted with the racist, colonial, misogynistic, and queerphobic practices of states, corporations, and others under capitalism. The big data that inform the machine-learning models hold within them the 'weights' of our society. That is, while machine-learning models – for example, profiling at the border – may not use socially identifiable data labels such as 'race' in their training data, through feature spaces, they craft similar outcomes through other attributes for, and associations of, race. This poses a significant dilemma for the liberal democratic approach – how does one say no to a set of associations and attributes that may not be sensible to representational approaches?

As Amoore had intricately analysed more than a decade earlier in the then-new data regimes of the 'War on Terror', her analysis of feature space explored a new evolving practice of the manipulation of different spaces for security – and increasingly within other areas of everyday life. Rather than the suspected body or pre-defined features of earlier algorithms for tracing the individual in the 2000s, these new forms of feature space not only determine who is purportedly suspicious, but articulate new communities of association. That is, the individual, as such, can be associated with others in ways unknown, through their similarities to other attributes, which could be drawn from protest videos, the police camera, as much as the supermarket self-check-out. This produces a new form of algorithmically mediated spatiality that raises a series of ethico-political concerns.

Key Advances and Controversies

Through the figures of the *border,* the *body,* the *cloud,* and the *deep,* Amoore has developed substantive critique of the role of the state, its security practices, and the move to increasing automation and algorithmic processing. Her more recent work is less explicit in its focus on security, but this should be read as a continuation of Amoore's interest in the methods of sorting and ordering that has pervaded various lines of enquiry (Amoore, 2022). That is, the 'War on Terror' can be understood to have introduced some of the early data practices that are today part of everyday life. So, one should not be surprised that Amoore's attention on the post-9/11 landscape has increasingly moved into more everyday spaces. From her work concerning the role of the *politics of possibility* through to the examination of data derivatives, to her studies of how computer science (and data science) come to understand and process varying concepts and terms, Amoore invokes a lens of ethico-political responsibility.

Unquestionably, one of Amoore's key insights has been on the role that prediction, probability, and possibility have had since the early 2000s in articulating the future – as computational power and processing have increased alongside a move to use data analytics for the 'efficiency' of the state and corporate operations. Here, the attention paid to the 'War on Terror' has advanced significant debates across many disciplines: no more so than on the data derivative, which has become a common referent point for contemporary societies' growing reliance on algorithmic outputs (Aradau and Blanke, 2016; Seaver, 2017). Being able to articulate novel spatial relations that work through computation, Amoore has demonstrated what has been so attractive to those who deploy algorithms – a rekindling of association itself, albeit often with a lack of critical attention to what types of futures are, and are not, being made possible. This is captured best in her articulation of the algorithmic aperture as 'an opening that is simultaneously a narrowing, a closure, and an opening onto a scene' (Amoore, 2020a: 16). Here, the spaces, bodies, and other materials upon which data have been generated are openings on the future, which are then foreclosed by the output of the algorithm. Here, then, is perhaps one of the most important implications of Amoore's career thus far: the future is increasingly narrowing. This work on possibility and risk has been enormously influential across diverse disciplines and fields – from borders and security (Hoijtink, 2022), to political theory (Panagia, 2021), questions of doubt (Taylor, 2020), to the role of prediction and data analytics in recursive practices of contemporary society (Beer, 2022).

In addition, Amoore's focus on marginalised and disadvantaged peoples and communities has been key to her enduring work on ordering, sorting, futures, probability, and prediction. Through a pursuit of feminist praxis and development, she has become more deeply involved in questions of race. As much as this has been present in her earlier work on the 'War on Terror' pointing out the racist associations made in early prediction and algorithmic practice, she has developed this analysis further in relation to deep algorithmic associations. Following Amoore's engagement with thinkers such as **Stuart Hall** (2021 [1997]), **Achille Mbembe** (2017) and Ramon Amaro (2022), she has examined how race becomes (re) constituted through deep-machine learning algorithms that are indifferent to the individual (Phan and Wark, 2021; Sharma and Nijjar, 2023).

In pursuing such crucial issues for our *algorithmic societies,* Amoore has not simply been a passive observer in this algorithmic aperture of the future. She has actively contributed to, and written on, debates about the implications of the use of algorithms, borders, and prediction. Whether on the use of facial recognition cameras by various UK police forces, or being the chair of the Facial Recognition Working Group of the Biometrics and Forensics Ethics Group (Hallowell et al., 2021), to intervening in the UK debate about predictive models for assigning grades to high school students during the COVID-19 pandemic, Amoore has been an important critical voice. Her call to 'ditch the algorithm' as a new form of political protest in her rendition of new practices of ordering, sorting, and calculation (Amoore, 2020b) has been the defining characteristic of her work, exemplifying both her feminist ethico-political project at its core as much as drawing on her earliest interests in political economy.

Amoore's Key Works

Amoore, L. (2006) 'Biometric borders: Governing mobilities in the war on terror', *Political Geography*, 25 (3): 336–51.

Amoore, L. (2009) 'Algorithmic war: Everyday geographies of the War on Terror', *Antipode*, 41 (1): 49–69.

Amoore, L. (2011) 'Data derivatives: On the emergence of a security risk calculus for our times', *Theory, Culture & Society*, 28 (6): 24–43.

Amoore, L. (2013) *The Politics of Possibility: Risk and Security Beyond Probability*. Durham, NC: Duke University Press.

Amoore, L. (2019) 'Doubt and the algorithm: On the partial accounts of machine learning', *Theory, Culture & Society*, 36 (6): 147–69.

Amoore, L. (2020a) *Cloud Ethics: Algorithms and the Attributes of Ourselves and Others*. Durham, NC: Duke University Press.

Amoore, L. (2022) 'Machine learning political orders', *Review of International Studies*, 49 (1): 1–17.

Amoore, L. and de Goede, M. (2005) 'Governance, risk and dataveillance in the war on terror', *Crime, Law and Social Change*, 43 (2): 149–73.

Amoore, L. and Hall, A. (2009) 'Taking people apart: Digitised dissection and the body at the border', *Environment and Planning D: Society and Space*, 27 (3): 444–64.

Amoore, L. and Hall, A. (2013) 'The clown at the gates of the camp: Sovereignty, resistance and the figure of the fool', *Security Dialogue*, 44 (2): 93–110.

Secondary Sources and References

Amaro, R. (2022) *The Black Technical Object: On Machine Learning and the Aspiration of Black Being*. London: Sternberg Press.

Amoore, L. (1998) 'The social roots of global change: States, firms and the restructuring of work', PhD Dissertation, University of Newcastle-upon-Tyne. Available from https://theses.ncl.ac.uk/jspui/handle/10443/422 (accessed October 25, 2023).

Amoore, L. (2017) 'What does it mean to govern with algorithms?', *Antipode Online, Intervention Symposium – Algorithmic Governance. 2017*. https://web.archive.org/web/20230906190110/https://antipodeonline.org/wp-content/uploads/2017/05/2-louise-amoore.pdf (accessed October 25, 2023).

Amoore, L. (2020b) 'Why "Ditch the algorithm" is the future of political protest", *The Guardian*, 19 Aug. https://www.theguardian.com/commentisfree/2020/aug/19/ditch-the-algorithm-generation-students-a-levels-politics (Accessed October 25, 2023).

Amoore, L. and de Goede, M. (2008) 'Transactions after 9/11: The banal face of the preemptive strike', *Transactions of the Institute of British Geographers*, 33 (2): 173–85.

Amoore, L. and de Goede, M. (2021) 'Datawars: Reflections twenty years after 9/11', *Critical Studies on Terrorism*, 14 (4): 425–9.

Aradau, C. and Blanke, T. (2016) 'Politics of prediction: Security and the time/space of governmentality in the age of big data', *European Journal of Social Theory*, 20 (3): 373–91.

Beer, D. (2022) 'The problem of researching a recursive society: Algorithms, data coils and the looping of the social', *Big Data & Society*, 9 (2). doi.org/10.1177/20539517221104997

Dwyer, A.C., O'Grady, N., Thornton, P., Straube, T., Gilbert, E. and Amoore, L. (2021) 'Cloud ethics: Algorithms and the attributes of ourselves and others', *The AAG Review of Books*, 9 (3): 36–49.

Hall, S. (2021 [1997]) 'Race, the floating signifier: What more is there to say about "race"?', in P. Gilroy and R. Wilson Gilmore (eds), *Selected Writings on Race and Difference*. Durham, NC: Duke University Press. pp. 359–73.

Hallowell, N., Amoore, L., Caney, S., Guest, R., Waggett, P. and Ni Loideain, N. (2021) *Briefing Note on the Ethical Issues Arising from Public–Private Collaboration in the Use of Live Facial Recognition Technology*. Biometrics and Forensic Ethics Group. https://assets.publishing.service.gov.uk/government/uploads/system/uploads/attachment_data/file/953359/LFR_briefing_note_18.1.21.final.pdf (accessed October 25, 2023).

Hoijtink, M. (2022) '"Prototype warfare": Innovation, optimisation, and the experimental way of warfare', *European Journal of International Security*, 7 (3): 322–36.

Mbembe, A. (2017) *Critique of Black Reason*. Durham, NC: Duke University Press.

Panagia, D. (2021) 'On the possibilities of a political theory of algorithms', *Political Theory*, 49 (1): 109–33.

Phan, T. and Wark, S. (2021) 'Racial formations as data formations', *Big Data & Society*, 8 (2). doi.org/10.1177/20539517211046377

Seaver, N. (2017) 'Algorithms as culture: Some tactics for the ethnography of algorithmic systems', *Big Data & Society*, 4 (2). doi.org/10.1177/2053951717738104

Sharma, S. and Nijjar, J.S. (2023) 'Post-racial politics, pre-emption and in/security', *European Journal of Cultural Studies*. doi.org/10.1177/13675494231168177

Taylor, L. (2020) 'The price of certainty: How the politics of pandemic data demand an ethics of care', *Big Data & Society*, 7 (2). doi.org/10.1177/2053951720942539

3

Benedict Anderson

Euan Hague, DePaul University

Biographical Details and Theoretical Context

Author of one of the most important concepts in political geography, that of nations being 'imagined communities', Benedict Richard O'Gorman Anderson was born in Kunming, China in 1936. He died in Indonesia in 2015, and his memoir, *A Life Beyond Boundaries* (Anderson, 2016), first available in Japanese in 2009, was translated, updated, and published posthumously. Brother of political theorist Perry Anderson, his father worked in the Chinese Maritime Customs Service, which led to a peripatetic childhood, including time in California, Ireland, and England. Anderson attended Eton College and, subsequently, Cambridge University, from which he graduated with a First Class degree in Classics in 1957 (Breuilly et al., 2016). In 1958, Anderson moved to Cornell University to pursue PhD research under the guidance of George Kahin, John Echols, and Claire Holt, completing fieldwork in Indonesia between 1961 and 1964 (Anderson, 1998; 2016).

Anderson's familiarity with the country meant that in 1965, when Indonesia's military leader, Suharto, foiled an alleged coup attempt by communist soldiers, purged the army, and massacred civilians, he and other graduate students analysed Suharto's version of events, questioning their veracity. This monograph, known as the 'Cornell Paper', reached the Indonesian military who in 1967 and 1968 invited Anderson to the country to persuade him of its errors. Failing to be convinced, Indonesian authorities denounced Anderson and, following formal publication of the allegations (Anderson et al., 1971), he was barred from Indonesia for what became the duration of Suharto's regime. With the exception of one clandestine visit in 1972, Anderson only returned to Indonesia following the dictator's 1998 resignation (Breuilly et al., 2016).

Anderson completed his PhD in 1967 on *The Pemuda Revolution: Indonesian Politics, 1945–46* (published as Anderson, 1972). He subsequently taught in the Department of Government at Cornell University until retirement in 2001 (Anderson, 2016). An expert on Southeast Asia, Anderson edited the interdisciplinary journal *Indonesia* between 1966 and 1984. He explored a range of topics pertaining to Indonesia, such as its government and politics (e.g., Anderson, 1964), and addressed committees of the United Nations and US Congress regarding Indonesia's human rights record and role in East Timor (e.g., Anderson, 1980). Exclusion from Indonesia led Anderson to examine other military conflicts and dictatorships of the 1960s and 1970s in

Southeast Asia, explorations that led him to analyse the importance of, and political attraction to, nationalist politics, resulting in *Imagined Communities – Reflections on the Origin and Spread of Nationalism* (Anderson, 1983; 1991; 2006; see also Breuilly et al., 2016; Calhoun, 2016; Goswami, 2020).

Imagined Communities, described by Chatterjee (1999: 128) as 'one of the most influential books of the late twentieth century', was translated into dozens of languages. In it, Anderson maintained that:

> Nation, nationality, nationalism – all have proved notoriously difficult to define, let alone analyse. In contrast to the immense influence that nationalism has exerted on the modern world, plausible theory about it is conspicuously meagre. (Anderson, 2006: 3)

For Anderson, Marxism was particularly culpable in this regard, and *Imagined Communities* was written as an intervention into debates animating *New Left Review* in the 1970s (see Anderson, 2006: 208–9; 2016: 119–27). In this climate, Anderson (2006: 3; original emphasis) argued that 'nationalism has proved an uncomfortable *anomaly* for Marxist theory and, precisely for that reason, has largely been elided, rather than confronted'. *Imagined Communities* sought to reconcile theories of Marxism and nationalism, and counter what Anderson envisaged as Eurocentric assessments of nationalism that overlooked colonial antecedents of modern nationalist politics. Central to Anderson's conceptualisation was the importance of interdisciplinarity and comparative case studies that drew on sources from anthropology to fiction to address how nationalisms developed around the world. Drawing on examples from the Americas and Indonesia, Anderson (2006: 5-6) proposed 'the following definition of the nation: it is an imagined political community – and imagined as both inherently limited and sovereign'.

After his retirement from teaching, Anderson advanced his theoretical work (e.g., 2005), anthropological field analyses (e.g., 2012), and studies of Southeast Asia (e.g., 2014). Reflecting on Anderson's impact, Breuilly et al. (2016: 626) note that *Imagined Communities* has the distinction of being 'by far the most cited text in the study of nationalism'.

Spatial Contributions

Anderson's concept of nations being 'imagined communities' is regularly commented on in books reviewing geographical thought (e.g. Crang, 1998; Anderson et al., 2003; Oakes and Price, 2008; Smith, 2020). The contention that a nation is 'imagined' does not mean that a nation is false, unreal, or to be distinguished from true (unimagined) communities. Rather, Anderson is proposing that a nation and nationality are constructed from shared processes and practices:

> It is *imagined* because the members of even the smallest nation will never know most of their fellow-members, meet them, or even hear of them, yet in the minds of each lives the image of their communion (Anderson, 2006: 6; original emphasis).

This understanding both shapes and is shaped by political and cultural institutions as people 'imagine' they share general beliefs, attitudes, and recognise a collective national populace as having similar opinions and sentiments to their own. Secondly:

> The nation is imagined as *limited* because even the largest of them, encompassing

> perhaps a billion living human beings, has finite, if elastic, boundaries, beyond which lie other nations. (Anderson, 2006: 7; original emphasis)

To have one nation means there must be another nation against which self-definition can be constructed. Anderson is thus arguing for the social construction of nations as political entities that have a limited spatial and demographic extent, rather than being organic, eternal entities. Further:

> It is imagined as *sovereign* because the concept was born in an age in which Enlightenment and Revolution were destroying the legitimacy of the divinely-ordained, hierarchical dynastic realm… nations dream of being free… The gage and emblem of this freedom is the sovereign state. (Anderson, 2006: 7; original emphasis)

Anderson argues that the concept of the nation emerged in the late-eighteenth century as a societal structure to replace previous monarchical or religious orders. In this manner, a nation was a new way of conceptualising state sovereignty and rule. This rule would be limited to a defined population and territory over which the state, in the name of nationality, could exercise power:

> Finally, it is imagined as a *community*, because, regardless of the actual inequality and exploitation that may prevail in each, the nation is always conceived as a deep, horizontal comradeship. Ultimately it is this fraternity that makes it possible, over the past two centuries, for so many millions of people, not so much to kill, as willingly to die for such limited imaginings. (Anderson, 2006: 7; original emphasis)

Nations hold such power over imaginations, claims Anderson, that patriotic calls to arms are understood as the duty of national residents. Further, in war, national citizens are equal and class boundaries are eroded in the communal struggle for national survival and greatness.

For Anderson (2006: 47), 'creole pioneers' are key to the development of nationalism. In both North and South America, those who fought for national independence in the eighteenth and nineteenth centuries had the same ancestries, languages, and traditions as the colonising powers they opposed. Anderson argues these 'creole pioneers' developed nationalist politics *before* Europe, because colonies were largely self-administering territorial units. Thus, residents conceived of their belonging to a common and potentially sovereign community, a sentiment enhanced by provincial newspapers raising debate about intercontinental political and administrative relationships. Anderson stakes much of his thesis on 'print-capitalism'; novels and newspapers, he claims, 'made it possible for rapidly growing numbers of people to think about themselves, and relate themselves to others, in profoundly new ways' (Anderson, 2006: 36). In addition, a standardised national calendar, language, and clock time generated a sense of simultaneous national experience and national distinction from elsewhere. Disparate occurrences became bound together as *national* experiences as people felt that all national residents were reading the same publications at the same time. Thus, 'the convergence of capitalism and print technology […] created the possibility of a new form of imagined community, which in its basic morphology set the stage for the modern nation' (Anderson, 2006: 46).

The worldwide impact of *Imagined Communities* across academic disciplines led to revised editions in 1991 and 2006. In the enlarged 1991 edition Anderson noted that he had '[*become*] uneasily aware that what I had believed to be a significantly new contribution to thinking about nationalism – changing apprehensions of time – patently lacked its necessary coordinate: changing apprehensions of space' (2006: xiii–xiv). Utilising Southeast Asian examples, Anderson corrected this omission by including chapters addressing the construction of national memories and the roles of national censuses, museums, biographies, and maps. Drawing on a 1988 PhD dissertation by Thongchai Winichakul about nineteenth-century Siam/Thailand (published as Winichakul, 1994), Anderson (2006: xiv) argued that maps contribute to the 'logoization of political space' and their myriad reproductions familiarise people with the limitations of national sovereignty and community.

Having examined mass communication with his thesis of 'print-capitalism', Anderson turned to the legacy of migration:

> The two most significant factors generating nationalism and ethnicity are both closely linked to the rise of capitalism. They can be described summarily as mass communication and mass migrations. (Anderson, 1992: 7)

Maintaining that nationalist movements were (and are) often initiated by expatriates, noting again the 'creole pioneers' of the Americas and financial contributions from overseas to the Irish Republican Army and ethno-nationalist factions in the Balkan Wars of the early 1990s, Anderson assessed:

> It may well be that we are faced here with a new type of nationalist: the 'long-distance nationalist' one might perhaps call him (fn. 'Him' because this type of politics seems to attract males more than females). For while technically a citizen of the state in which he comfortably lives, but to which he may feel little attachment, he finds it tempting to play identity politics by participating (via propaganda, money, weapons, any way but voting) in the conflicts of his imagined *Heimat* – now only fax-time away. (Anderson, 1992: 13)

Nationalism exists, therefore, in an 'undivorceable marriage to internationalism' (Anderson, 2006: 207). Comparing anti-colonial nationalisms in Southeast Asia, Anderson (1998; 2005) demonstrates that since the nineteenth century, political activists have engaged in multilingual global debates about the possibilities of nationalist revolution. Often expatriates, these individuals imagined their putative national communities by, amongst other things, writing novels and anthropological treatises that articulated the belief that indigenous peoples, often understood by colonial powers to be divided by tribal and ethnic difference, comprised a nation with common roots, traditions and aspirations.

Although not a geographer by training or career, Anderson's focus on issues of space, territory, and place, critical to nationalist politics, has led to his work being widely utilised within geographical scholarship.

Key Advances and Controversies

Imagined Communities received little attention from geographers upon its initial publication. Largely without review in geography journals, Anderson's concepts entered geographical debate through their impact on interdisciplinary studies of nationalism. Indeed, Spencer and Wollman (2002: 37) claim that such is the regularity with which articles about nationalism routinely cite *Imagined Communities* that Anderson's conceptualisation 'has become one of the commonest clichés of the literature', the result being that 'invocation has, in some cases, been a substitute for analysis'. Although the most sustained geographical engagement with Anderson's concepts came in the 1990s, geographers have not been immune to this (see, *inter alia*, Jackson and Penrose, 1993; Smith and Jackson, 1999), and prolonged geographical assessments of Anderson's contentions are few: Blaut (1987), for example, does not assess Anderson's work in his review of Marxist theories of nationalism. Some geographers, however, have built upon Anderson's contentions: Radcliffe and Westwood (1996: 118), exploring a national imagined community in Ecuador, maintain that Anderson's 'geographical imagination [...] permits him to link themes of space, mobility and the nation', although comment that Anderson fails to fully acknowledge or develop the implications of this within his work. Martin (1997: 90), in turn, suggests that although 'intellectuals have been given the power to "imagine" the nation or national community, [...] the material dimension, or political economy, of nationalism and the nation have been ignored'. Assessing late-nineteenth century Irish nationalism, Martin (1997: 90) argued for 'a corporeal approach to the nation' to interrogate how gender roles were constructed in the Irish nationalist imagination. '[I]f lived bodies are placed at the center in analyses of nationalism and the nation', Martin (1997: 91) suggests, 'the nation itself can be conceptualized as not only an imagined entity belonging to some, but as an identity differently or similarly embodied in disparate individuals'. Studying the construction of a Swiss heritage community in New Glarus, Wisconsin, Steven Hoelscher (1999: 538) invoked Anderson to explain that specific 'forms of imagining' are utilised by elites to produce place and community identities, and these local 'imagined communities' are often contested by non-elite groups, and Sidaway (2002) offers an exploration of 'imagined regional communities'.

Although Cheah (1999: 4) could maintain that 'Anderson's ideas are rarely thoroughly discussed', his subsequent collection offering a dialogue with Anderson about his thought (Cheah and Culler, 2003), postcolonial scholars strongly challenged Anderson's conceptualisations. Drawing a contrast to Anderson's spatially bounded explorations of nations, **Stuart Hall** (2008: 273) offers a fluid understanding of 'diaspora' as 'an alternative framework for thinking about "imagined communities"' which recognises that many individuals feel allegiances to numerous locations and connections with people who are dispersed globally, not solely nationally. **Edward Said** (1993) further contends that Anderson is too linear in claiming that political structures and institutions change from dynasties to sovereign nations through the standardising influence of print-capitalism. Partha Chatterjee (1993), in turn, argues that although processes of print-capitalism were important, Anderson's formulation of them as standardising language, time, and territorial extent, is too simplistic to impose on the diverse, multilingual, and asymmetrical power relations of the colonial situation. A similar contention animates Castro-Klarén and Chasteen's (2003) collection exploring nationalism in Latin America.

Arguably Anderson's most consistent critic, Chatterjee (1993) contends that European colonialism has necessarily limited the imagination of political communities. Anticolonial nationalisms typically opposed colonialism using the same nationalist arguments as the colonists, and postcolonial nations operate within institutional limits formulated in Europe. The postcolonial nation can, therefore, only be conceptualised within these European strictures; its sovereignty and territory already imagined by the colonisers. 'Even our imaginations', asserts Chatterjee (1993: 5) 'must remain forever colonized'. Chatterjee (1999: 130) also takes to task Anderson's (1998) subsequent work for advancing a conceptualisation that distinguishes 'the residual goodness of nationalism and the unrelieved nastiness of ethnic politics'.

A second major critique of *Imagined Communities* comes from a feminist perspective. With a focus on the 'fraternity' experienced by members of a nation (Anderson, 2006: 7), the protagonists in Anderson's conceptions of nationalism are typically assumed to be male. Mayer (2000: 6) argues that Anderson envisions 'a hetero-male project [...] imagined as a brotherhood', eliding gender, class, and racial structures within and between national communities and **Linda McDowell** (1999: 195) demonstrates that although seemingly neutral, 'the very term horizontal comradeship [...] brings with it connotations of masculine solidarity'. Subsequently, McClintock (1995: 353) laments that sustained 'explorations of the gendering of the national imagination have been conspicuously paltry'.

A further challenge comes from perspectives such as that offered by Mitchell, who argues that as well as *imagining* communities, there must be attention paid to:

> The *practices* and exercises of power through which these bonds are produced and reproduced. The questions this raises are ones about who defines the nation, how it is defined, how that definition is reproduced and contested, and, crucially, how the nation has developed and changed over time [...] The question is not what common imagination *exists*, but what common imagination is *forged*. (Mitchell, 2000: 269; original emphasis)

This echoes Winichakul (1994: 15) who suggests that 'Anderson seems too concerned with the imagination, the conceivability of a nation' rather than with 'the social institutions and practices which perpetuate the operation and reproduction of the imagined community in actual human relations'. The issue, therefore, must be to show the work needed to produce and maintain that imagination, how this impacts upon people's lives, and how power to enforce the national community that is imagined shapes behaviours across time and space.

In the forty years since Anderson (1983) first offered *Imagined Communities* as a theoretical basis to understand nationalism, the concept has become widely influential. Chatterjee (1999: 130) can thus legitimately claim that 'Anderson, more than anyone else in recent years, has inspired the study of the material instruments of literary and cultural production that made possible the imagining of modern political communities in virtually every region of the world'.

Anderson's Key Works

Anderson, B., McVey, R.T. and Bunnell, F.P. (1971) *A Preliminary Analysis of the October 1, 1965 Coup in Indonesia*. Ithaca: Cornell Modern Indonesia Project, Publication No. 52.

Anderson, B. (1972) *Java in a Time of Revolution: Occupation and Resistance, 1944–1946*. Ithaca: Cornell University Press.

Anderson, B. (1983) *Imagined Communities: Reflections on the Origin and Spread of Nationalism*. London: Verso.
Anderson, B. (1991) *Imagined Communities: Reflections on the Origin and Spread of Nationalism* (revised and enlarged edition). London: Verso.
Anderson, B. (1992) 'The new world disorder', *New Left Review*, 193: 3–13.
Anderson, B. (1998) *The Spectre of Comparisons: Nationalism, Southeast Asia and the World*. London: Verso.
Anderson, B. (2005) *Under Three Flags: Anarchism and the Anti-Colonial Imagination*. London: Verso.
Anderson, B. (2006) *Imagined Communities: Reflections on the Origin and Spread of Nationalism* (revised edition). London: Verso.
Anderson, B. (2016) *A Life Beyond Boundaries*. London and New York: Verso.

Secondary Sources and References

Anderson, B. (1964) 'Indonesia and Malaysia', *New Left Review*, 28: 4–32.
Anderson, B. (1967) *The Pemuda Revolution: Indonesian Politics, 1945–1946*. Unpublished PhD dissertation, Ithaca, Cornell University.
Anderson, B. (1980) 'Prepared testimony on human rights in Indonesia and East Timor', in *Human Rights in Asia: Noncommunist Countries*. Washington, DC: U.S. Government Printing Office. pp. 231–62 and 275–7.
Anderson, B. (2012) *The Fate of Rural Hell: Asceticism and Desire in Buddhist Thailand*. London: Seagull Books.
Anderson, B. (2014) *Exploration and Irony in Studies of Siam over Forty Years*. Ithaca, NY: Southeast Asia Program Publications, Cornell University.
Anderson, K., Domosh, M., Thrift, N. and Pile, S. (eds) (2003) *The Handbook of Cultural Geography*. London: Sage.
Blaut, J.M. (1987) *The National Question: Decolonising the Theory of Nationalism*. London: Zed Books.
Breuilly, J., Eriksen, T.H., Hearn, J., Leerssen, J., Green, E. and Sidel, J.T. (2016) 'Benedict Anderson's *Imagined Communities*: A symposium', *Nations and Nationalism*, 22 (4): 625–59.
Calhoun, C. (2016) 'The importance of *Imagined Communities* – and Benedict Anderson', *Debats – Journal on Culture, Power and Society*, 1: 11–16. Available from https://dialnet.unirioja.es/descarga/articulo/5715162.pdf (accessed October 25, 2023).
Castro-Klarén, S. and Chasteen, J.C. (eds) (2003) *Beyond Imagined Communities: Reading and Writing the Nation in Nineteenth-century Latin America*. Baltimore, MA: Johns Hopkins University Press.
Chatterjee, P. (1993) *The Nation and its Fragments: Colonial and Postcolonial Histories*. Princeton, NJ: Princeton University Press.
Chatterjee, P. (1999) 'Anderson's utopia', *Diacritics*, 29 (4): 128–34.
Cheah, P. (1999) 'Grounds of comparison: Around the work of Benedict Anderson', *Diacritics*, 29 (4): 2–18.
Cheah, P. and Culler, J. (eds) (2003) *Grounds of Comparison: Around the work of Benedict Anderson*. London: Routledge.
Crang, M. (1998) *Cultural Geography*. London: Routledge.
Goswami, M. (2020) 'Benedict Anderson, *Imagined Communities* (1983)', *Public Culture*, 32 (2): 441–8.
Hall, S. (2008 [1995]) 'New cultures for old', in T.S. Oakes and P.L. Price (eds), *The Cultural Geography Reader*. London: Routledge. pp. 264–74.

Hoelscher, S. (1999) 'From sedition to patriotism: Performance, place, and the reinterpretation of American ethnic identity', *Journal of Historical Geography*, 25 (4): 534–58.

Jackson, P. and Penrose, J. (1993) 'Introduction: Placing "race" and nation', in P. Jackson and J. Penrose (eds), *Constructions of Race, Place and Nation*. Minneapolis, MN: University of Minnesota Press. pp. 1–23.

Martin, A.K. (1997) 'The practice of identity and an Irish sense of place', *Gender, Place and Culture*, 4: 89–113.

Mayer, T. (2000) 'Gender ironies of nationalism: Setting the stage', in T. Mayer (ed), *Gender Ironies of Nationalism: Sexing the Nation*. London: Routledge. pp. 1–22.

McClintock, A. (1995) *Imperial Leather: Race, Gender and Sexuality in the Colonial Contest*. London: Routledge.

McDowell, L. (1999) *Gender, Identity and Place: Introducing Feminist Geographies*. Minneapolis, MN: University of Minnesota Press.

Mitchell, D. (2000) *Cultural Geography: A Critical Introduction*. Oxford: Blackwell.

Oakes, T.S. and Price, P.L. (eds) (2008) *The Cultural Geography Reader*. London: Routledge.

Radcliffe, S. and Westwood, S. (1996) *Remaking the Nation: Place, Identity and Politics in Latin America*. London: Routledge.

Said, E.W. (1993) *Culture and Imperialism*. New York, NY: Vintage.

Sidaway, J.D. (2002) *Imagined Regional Communities: Integration and Sovereignty in the Global South*. London: Routledge.

Smith, G. and Jackson, P. (1999) 'Narrating the nation – the "imagined community" of Ukrainians in Bradford', *Journal of Historical Geography*, 25 (3): 367–87.

Smith, S. (2020) *Political Geography: A Critical Introduction*. Hoboken, NJ: Wiley Blackwell.

Spencer, P. and Wollman, H. (2002) *Nationalism: A Critical Introduction*. London: Sage.

Winichakul, T. (1994) *Siam Mapped: A History of the Geo-Body of a Nation*. Honolulu: University of Hawaii Press.

4
Gloria Anzaldúa

M.M. Ramírez, Simon Fraser University

Biographical Details and Theoretical Context

Gloria Evangelina Anzaldúa was born in the Rio Grande Valley of south Texas in 1942 to sixth-generation Mexican American tenant farmers. A self-described 'chicana dyke-feminist, tejana patlache poet, writer, and cultural theorist' (Keating and Anzaldúa, 1993: 105), Anzaldúa has been a prolific presence in queer Chicana feminism, whose writings have reverberated far beyond disciplinary boundaries, influencing generations of scholars.

Anzaldúa's childhood experiences profoundly shaped her work. Her theorising is rooted in her retellings of her youth: how she was marked as 'other' from a young age as a smart Chicana kid in a segregated school system, as a rebellious queer woman who broke from cultural norms, and as a person who struggled with severe health problems from childhood onward. Anzaldúa's writings reveal an intimate awareness of how power asymmetries are experienced, especially by those marked as 'different'. Her insistence on embodiment roots her thinking in space.

Despite the alienation she felt within Texas' segregated school system, Anzaldúa loved reading and learning. After her father's passing when she was 14, she worked in the fields to help support her family as she attended high school. She was named valedictorian in 1962, and went on to obtain a teaching certificate, later earning an MA at the University of Texas, Austin in English and Education while working for the Texas school system. When she went on to work with the children of migrant farm workers in Indiana in 1973–74, she took a creative writing course at Indiana University, which inspired her to seek a doctorate at UT Austin in Comparative Literature in fall 1974. While taking courses, Anzaldúa participated in various activist movements including the Chicano movement, farmworkers' and women's movements, while devoting significant time to her creative writing. Anzaldúa struggled in graduate school due to economic precarity and an ongoing feeling of alienation. In the fall of 1977, she put her degree on hold; 'gave up the idea of becoming a university professor' (Anzaldúa and Keating, 2000: 101) and drove out to California to devote herself to writing.

As a queer, working-class Chicana, Anzaldúa was often made to feel inferior in feminist writing circles during this period. Spaces such as the Feminist Writers Guild were predominantly white and deeply classed. These experiences of exclusion made her determined to edit a collection exclusively featuring pieces by women of colour, and it was around this time that Anzaldúa met another

Chicana writer, Cherrie Moraga, in the Feminist Writers Guild. The two went on to live together in San Francisco, creating a poetry series called *El Mundo Zurdo* ('the left-handed world') for queer and women of colour poets, hosting readings in cafes and bookstores across the city in the late 1970s. This friendship and poetic dialogue resulted in publication of the foundational collection *This Bridge Called My Back: Writings by Radical Women of Color* (Moraga and Anzaldúa, 1981). Anzaldúa and Moraga were intentional in their framing of 'women of color' in the collection, presenting different racialised identities in relation to one another to build political alliance and solidarity. As Anzaldúa (1981: 209) wrote: 'Third World women, lesbians, feminists, and feminist-oriented men of all colors are banding and bonding together to right that balance. Only together can we be a force. I see us as a network of kindred spirits, a kind of family'. *This Bridge* was a ground-breaking poetic and political project, the first of its kind.

After living on the East Coast of the US for a few years in the early 1980s, Anzaldúa moved to Santa Cruz, California in 1986. Around this time, she submitted a 100-page poetry manuscript called *Borderlands* to Aunt Lute Books Press. After expanding the manuscript extensively, *Borderlands/La Frontera: The New Mestiza* was published in 1987. As her friend and biographer AnaLouise Keating described, 'With its code-switching, multigenre, feminist lens, *Borderlands* arrived in the world exactly on time. It was well-received by progressive academics and scholars, especially those interested in Chicana, multicultural, queer, and/or feminist thought' (2022: 28). The book's reception pushed Anzaldúa's career in a new direction, and she was appointed the Distinguished Visiting Professor in Women's Studies at the University of California, Santa Cruz (UCSC) in 1988. In her courses, she taught texts by third world and women of colour feminists. Frustrated by the lack of anthologies celebrating Third World genealogies, she began editing a second collection, *Making Face, Making Soul/Haciendo Caras: Creative and Critical Perspectives by Women of Color* (Anzaldúa, 1990) around this time, followed by *This Bridge We Call Home: Radical Visions for Transformation* (Anzaldúa and Keating, 2002).

Anzaldúa once wrote that 'the allure of learning and knowledge' led her back to graduate school. She joined the American Literature PhD programme at UCSC (after being rejected from the Comparative History of Ideas programme), taking classes with **Donna Haraway**, Teresa de Lauretis, and José David Saldívar. Anzaldúa continued to struggle with anxiety and alienation in the university environment: 'now I find myself *entremedias*, in between, occupying that strange space of having been first a lecturer and now a graduate student. From this position I can see clearly that the university for me is often a haven, a refuge, sometimes a prison and a hell. Es una área de conflicto y enfrentamiento en donde el racismo cobra vida (It is an area of conflict and confrontation where racism comes alive)' (Anzaldúa, cited in Keating, 2022: 30).

Anzaldúa struggled to complete her dissertation project, which focused on the cultural representations of La Llorona, a Mexican mythical figure that

> symbolized alienation: estrangement from self, estrangement from others (especially men) and from groups, estrangement from social environment, rebellion against cultural norms and in time estrangement from institutions of higher learning and from bodies of knowledge. (Anzaldúa, cited in Keating, 2022: 32)

Her analysis focused on the forms of embodiment of Chicana and Mestiza women in how they are situated in geographic space. Dissatisfied with western spatial analytics, Anzaldúa began to theorise her own spatial concept of *nepantla*.

During this time, Anzaldúa faced mental and physical health challenges. In 1992, she was diagnosed with diabetes, a diagnosis that profoundly altered her life. Managing diabetes not only demanded much of Anzaldúa's time, but the illness provoked a series of secondary ailments that made daily life challenging. Yet Anzaldúa needed to keep 'producing', as she was a self-employed writer who could not live on the royalties from her books; she needed to continually take on speaking engagements to get by, which took up a tremendous amount of time and energy. Despite this, Anzaldúa continued to write in every spare moment, taking on a multitude of projects, meticulously writing drafts and continuing to return to her dissertation, which she envisioned as her second major book project; she didn't want *Borderlands* to be her magnum opus, she wanted to achieve 'something greater' (Anzaldúa, cited in Keating, 2022: 40).

Gloria Anzaldúa died in May 2004 due to complications from diabetes. The essays she wrote in the final decade of her life primarily live in two posthumously published works: *The Gloria Anzaldúa Reader* (Anzaldúa and Keating, 2009) and *Light in the Dark/Luz en lo Oscuro* (Anzaldúa and Keating, 2015). Anzaldúa's essays in these collections fluctuate between theories of transformation, expansive methodological innovations, and profoundly intimate reflections on her anxieties around writing.

Spatial Contributions

Anzaldúa is widely recognised as having a tremendous influence on Chicana and women of colour feminisms. The spatiality of Anzaldúa's work is undeniable, both in how she theorised the multiple, embodied effects of power asymmetries, and how a subject's identity informs their experience of place. Anzaldúa's theories ground power relations at the scale of the body, articulating not only difference but also the physical, emotional, psychological and spiritual effects of racism, classism, patriarchy and ableism. Her focus on embodiment, through the exploration of the Chicana subject, insisted that we cannot divorce theory or intellectual pursuits from questions of embodiment; Anzaldúa's writings embraced the at times messy nature of inhabiting a body in a deeply unjust world.

Aligned with the thinking of Barbara Smith, Audre Lorde, **bell hooks**, Chela Sandoval, and other women of colour feminists from this era (many of whose writings were republished in Moraga and Anzaldúa, 1981) the 'theories in the flesh' that Anzaldúa wove with Moraga in *This Bridge Called My Back* have proven foundational in understanding how theory emerges from lived experience. As they write, 'a theory in the flesh means one where the physical realities of our lives – our skin colour, the land or concrete we grew up on, our sexual longings – all fuse to create a politic born of necessity' (Moraga and Anzaldúa, 1981: 23). They name the societal structures and colonial processes that have dispossessed Black, Indigenous, Chicanx and Asian peoples in North America, while also acknowledging the intimate struggles that occur within homes and familial relationships. The work demonstrates a multi-scalar feminist analysis that puts racism, colonialism, and heteropatriarchy in society in relation to the body. Anzaldúa and Moraga's writings reject Cartesian ontologies, refusing a dislocation of mind and body and recognising the ways that one's embodied experiences inform theory. They insist on a politics of relation, across subjectivities and geographies, to pursue 'a society that uses flesh and blood experiences to concretise a vision' of transformation (Moraga and Anzaldúa, 1981: 23).

'Theories in the flesh' provided the seed for conceptualisations that Anzaldúa introduced in *Borderlands* and in her later writings. Perhaps the best known is that of a 'new mestiza consciousness', an articulation that challenges Western philosophical traditions of dualism by introducing plural subjectivities (Lugones, 1992). The 'mestiza', a subject positioning that Anzaldúa theorises from her own embodied experience as a Chicana and Tejana, is a subject that straddles two distinct colonial cultures, the US and Mexico. As she articulates in *Borderlands/La Frontera*, the mestiza subject emerges from the violent geographies of the US–Mexico border, 'una herida abierta [an open wound] where the Third World grates against the first and bleeds' (Anzaldúa, 1987: 25). The mestiza does not fully inhabit either world, for she belongs to neither; she inhabits the real and the imagined space of the borderland, a liminal space between cultures and geographies. Living amid the contradictions of these two worlds, she develops 'a tolerance for ambiguity […] she operates in a pluralistic mode […] turns the ambivalence into something else' (Anzaldúa, 1987: 101). The mestiza subject thus develops a 'new mestiza consciousness': drawing energy 'from continual creative motion that keeps breaking down the unitary aspect of each new paradigm', her work is to 'break down the subject-object duality that keeps her prisoner and show in the flesh and through the images in her work how duality is transcended' (Anzaldúa, 1987: 102). This consciousness is actively produced and spatialised as a 'piece of ground to stand on' (Anzaldúa, 1987: 23). Spaces of transformative possibility that emerge through experiencing immense pain are a reoccurring theme in Anzaldúa's writing. *Borderlands* not only offered a feminist rendering of the embodied impacts of colonial heteropatriarchy and the multiple oppressions produced by US empire, but also opened up expansive conversations around gendered and racialised subjectivities in the colonial present.

Anzaldúa's theorisations of a new mestiza consciousness travelled widely and were taken up in conversations around third space, hybridity, and liminality across academic disciplines, including geography. With a few exceptions from feminist geographers (Wright, 1998; Nagar et al., 2002; Mountz and Hyndman, 2006), *Borderlands* became one of the token citations signifying postcolonial subjectivity and 'difference' in geography in the 1990s and early 2000s, often without substantial engagement with the geographic underpinnings of her work. As Chicana theorists have argued, detaching Anzaldúa's conceptualisations of borderlands from the political geographies of the US–Mexico border appropriates the theories and effectively disembodies them (Yarbro Bejarano, 1994).

After the publication of *Borderlands*, Anzaldúa continued to expand and nuance her theorising moving away from the use of 'borderlands' because she felt people were using it in 'a more limited sense' than she had intended. In her later writings, Anzaldúa delves into a theorisation of *nepantla*, a Nahuatl word that she describes as 'an in-between state, neither/nor, that uncertain terrain one crosses when moving from one place to another' (Keating, 2022: 94). Nepantla is a space of becoming; 'a site of transformation, the place where […] you question the basic ideas, tenets, and identities inherited from your family, your education, your different cultures' (Anzaldúa and Keating, 2015: 127). Anzaldúa thus frames nepantla as a geographic space of transformation that is created through a shift in consciousness. In *Light in the Dark*, Anzaldúa embedded nepantla into a broader theory she called conocimento, made up of seven stages of transformation that also included el arrebato, the Coatlicue state, and the Coyolxauhqui process (Anzaldúa and Keating, 2015). Each stage incorporates concepts Anzaldúa had theorised over the years, articulating a process of traversing intense realities, bursting paradigms, and a painful becoming into a social, spiritual and psychological space of something new, something otherwise.

Her articulation of autohistoria-teoría (autohistory-theory) – the methodology Anzaldúa employed throughout her life's work – was as a mode of 'self-reflective theorizing' in which a 'coherent story is pieced together from remembered fragments' and utilised to weave theories that are spatially and temporally expansive (Anzaldúa cited in Keating 2022: 83). An autohistoria-teoría delves into one's embodied life experiences, making sense of these experiences in relation to the spatial and cultural context from where one lives, weaving theories that reflect this relationship between subjectivity and place. This attention to the relationship between identity, embodiment and place makes Anzaldúa's autohistoria-teoría a spatial method, perhaps even a Latinx geographic method.

Key Advances and Controversies

Anzaldúa's individual and collective writings have been foundational for the development of Chicana feminisms, as well as for the understanding of feminist theory as encompassing a plurality of subjectivities. Yet Anzaldúa's major works have received their share of critique over the years, particularly around her uncritical privileging of the mestiza subject, the use of Nahua mythology in her writing, and the absence of Blackness in her spatial and temporal frames.

In *Borderlands*, the mestiza subject takes centre stage as the actor whose bifurcated cultural identity grounds Anzaldúa's theories. Critics argue that Anzaldúa romanticised a mestiza subjectivity without interrogating mestizaje as a racial ideology (Saldaña-Portillo, 2001; Guidotti-Hernandez, 2011). As a Mexican nationalist political project, mestizaje subsumes Indigeneity as central to its racial logics, fixing Indigenous peoples in the past and effectively erasing Indigenous subjectivities in contemporary Mexico, thus positioning mestizos as the modern Mexican subject (Saldaña-Portillo, 2001). By centring the mestiza subject, Anzaldúa perpetuates an erasure of Indigenous and Black subjectivities in Mexico that is engrained in the ideology of mestizaje. As Saldaña-Portillo notes,

> when she resuscitates this particular representation of indigenous subjectivity to be incorporated into contemporary mestiza consciousness, she too does so to the exclusion and, indeed, erasure of contemporary indigenous subjectivity and practices on both sides of the border. (Saldaña-Portillo, 2001: 416)

Furthermore, as Guidotti-Hernandez articulates, Anzaldúa's focus on indigenísmo to theorise a mestiza consciousness produces a troublesome 'chain of equivalence: if Chicano, then Mexican; if Mexican, then mestizo; if mestizo, then indigenous; if indigenous, then resistant' (Guidotti-Hernandez, 2011: 19).

These critiques extend to the engagement with Nahua myth which Anzaldúa employed throughout her work. This type of engagement, Saldaña-Portillo (2001: 420) notes, 'reproduces liberal models of choice that privilege her position as a US Chicana: she goes through her backpack and decides what to keep and what to throw out, and she chooses to keep signs of indigenous identity as ornamentation and spiritual revival'. Anzaldúa's focus on mestiza positionalities and 'indigenizing' practices thus reproduce logics of mestizaje beyond the Mexican context, romanticising Nahua myth and thereby perpetuating an imaginary of Indigenous presence as only existing in the distant past. This erasure of Indigenous subjectivities in Anzaldúa's work occurs not only in her centring of the mestiza subject, but also in how she depicts the geographic space of the US–Mexico borderlands. While she rightly frames the space of the borderlands as a geography fraught with violent colonial encounters, past and present, Guidotti-Hernández (2011: 17–18) argues that 'by privileging that "Indian essence", mestizaje fetishises a residual, abstract, de-historicised Indian identity that obscures Mexican, Mexican Indian, and American Indian participation in genocide and violence against other American Indians and Mexicans in the US-Mexico borderlands'. Anzaldúa's focus on the mestiza subject as existing in the in-between space between US (read Anglo/white) and Mexican (read Hispanic/mestizo) cultures thus oversimplifies the violent nature of racial formation in the borderlands and within each colonial nation-state, and erases the sovereign Indigenous nations whose territories the border cuts across.

Anzaldúa's centring of mestiza subjectivities also negates Black geographies and subjectivities in the borderlands. As Latina feminist geographer Madelaine Cahuas (2019: n.p.) argues, 'Anzaldúa does not engage with how transatlantic slavery, the plantation economy or fears of Black uprising played a role in the occupation of Mexico and Anglo oppression in the borderlands'. To describe the herida abierta of the border as a geography that begins with Cortés' conquest of Mexico overlooks the undeniable influence of slavery in the creation of the US and Mexican economies and nation-states. This oversight is an ahistorical rendering of the borderlands that reproduces the anti-Black, anti-Indigenous and eugenicist logics embedded in the ideologies of mestizaje, rather than building a relational understanding of how the violence of the border forged the racial formations of the region.

Notwithstanding these valid and necessary critiques of her work, Anzaldúa's influence is undeniable. Her contributions to feminist and queer theory reverberate widely, offering embodied conceptualisations of how people experience the power asymmetries embedded in place. The spatial dimensions of Anzaldúa's theories have much to offer, and though somewhat overlooked (Brady, 2002), have been increasingly explored by Latinx geographers (Muñoz 2016; Ramírez, 2020; Cahuas, 2021; Herrera, 2023). Such explorations acknowledge that Anzaldúa's work can inform geographic understandings of the production of both space and experience, as well as an appreciation of how to build new relations amid the oppressive structures of the present.

Anzaldúa's Key Works

Anzaldúa, G. (1981) 'La Prieta', in C. Moraga and G.E. Anzaldúa (eds), *This Bridge Called My Back: Writings by Radical Women of Color.* New York, NY: Kitchen Table, Women of Color Press. pp. 198–209.

Anzaldúa, G. (1987) *Borderlands/La Frontera: The New Mestiza.* San Francisco, CA: Aunt Lute.

Anzaldúa, G. (1990) *Making Face, Making Soul/ Haciendo Caras: Creative and Critical*

Perspectives by Feminists of Color. San Francisco, CA: Aunt Lute.
Anzaldúa, G. and Keating, A. (2002) *This Bridge We Call Home: Radical Visions for Transformation*. New York, NY: Routledge.
Anzaldúa, G. and Keating, A. (2009). *The Gloria Anzaldúa Reader.* Durham, NC: Duke University Press.
Anzaldúa, G. and Keating, A. (2015) *Light in the Dark/Luz en lo Oscuro: Rewriting Identity, Spirituality, Reality*. Durham, NC: Duke University Press.
Moraga, C. and Anzaldúa, G. (eds) (1981) *This Bridge Called My Back: Writings by Radical Women of Color*. New York, NY: Kitchen Table, Women of Color Press.

Secondary Sources and References

Anzaldúa, G. and Keating, A. (2000) *Interviews = Entrevistas*. New York, NY: Routledge.
Brady, M.P. (2002) *Extinct Lands, Temporal Geographies: Chicana Literature and the Urgency of Space*. Durham, NC: Duke University Press.
Cahuas, M.C. (2019) 'Interrogating absences in Latinx theory and placing Blackness in Latinx geographical thought: A critical reflection', *Society & Space Online*. https://www.societyandspace.org/articles/interrogating-absences-in-latinx-theory-and-placing-blackness-in-latinx-geographical-thought-a-critical-reflection (accessed October 26, 2023).
Cahuas, M.C. (2021) 'Reaching for El Mundo Zurdo: Imagining-creating-living Latinx decolonial feminist geographies in Toronto', *Gender, Place and Culture: A Journal of Feminist Geography,* 28 (9): 1213–33.
Herrera, J. (2023) 'Compassionate solidarities: Nos/Otras and a Nepantla praxis of care', in V. Lawson, S. Elwood and M. Daigle (eds), *Abolishing Poverty: Toward Pluriverse Futures and Politics*. Athens, GA: University of Georgia Press. pp. 117–40.
Guidotti-Hernández, N.M. (2011) *Unspeakable Violence Remapping U.S. and Mexican National Imaginaries*. Durham, NC: Duke University Press.
Keating, A. (2022) *The Anzaldúan Theory Handbook*. Durham, NC: Duke University Press.
Keating, A. and Anzaldúa, G. (1993) 'Writing, politics, and las lesberadas: Platicando con Gloria Anzaldúa', *Frontiers (Boulder),* 14 (1): 105–30.
Lugones, M. (1992) 'On Borderlands/La Frontera: An interpretive essay', *Hypatia,* 7 (4): 31–7.
Mountz, A., and Hyndman, J. (2006) 'Feminist approaches to the global intimate', *Women's Studies Quarterly,* 34 (1/2): 446–63.
Muñoz, L. (2016) 'BAR TASCO: Latina immigrant vendors' "Mestiza consciousness"', *Chicana/Latina Studies,* 16 (1): 110–32.
Nagar, R., Lawson, V., McDowell, L. and Hanson, S. (2002) 'Locating globalization: Feminist (re)readings of the subjects and spaces of globalization', *Economic Geography,* 78 (3): 257–84.
Ramírez, M.M. (2020) 'City as borderland: Gentrification and the policing of Black & Latinx geographies in Oakland', *Environment & Planning D: Society & Space,* 38 (1): 147–66.
Saldaña-Portillo, M.J. (2001) 'Who's the indian in Aztlán? Re-writing Mestizaje, Indianism, and Chicanismo from the Lacandón', in I. Rodríguez (ed), *The Latin American Subaltern Studies Reader*. Durham, NC: Duke University Press. pp. 402–23.
Wright, M. (1998) 'Maquiladora mestizas and a feminist border politics: Revisiting Anzaldúa', *Hypatia,* 13 (3): 114–31.
Yarbro-Bejarano, Y. (1994) 'Gloria Anzaldúa's Borderlands/La Frontera: Cultural Studies, "difference", and the non-unitary subject', *Cultural Critique,* 28 (28): 5–28.

5
Michael Batty

Andrew Crooks, University at Buffalo, New York

Biographical Details and Theoretical Context

(John) Michael Batty was born in Liverpool, England in 1945, where he grew up. He enrolled in the University of Manchester in 1962 to study Town and Country Planning, gaining a BA degree with First Class honours. While at Manchester, Batty was influenced by George Chadwick who introduced him to the 'science of cities' and the 'science of planning', including the works of William Alonso (1964), Britton Harris (1965) and Walter Isard (1956). He was also introduced to systems theory and approaches by Brian McLoughlin (1969) in which cities were viewed as systems of interacting parts whose mechanisms functioned as in the physical world. After completing his degree in 1966 he stayed at the University of Manchester for two years as an assistant lecturer in the Department of Town and Country Planning.

Batty then moved to the University of Reading to become a research assistant and then lecturer in Geography (1969–74), where he worked with Peter Hall in the Urban Systems Research Unit, developing large-scale urban models. After a short visit to the University of Waterloo, Canada, where he was an Assistant Professor of Civil Engineering (1974–75), Batty returned to the University of Reading and became Reader in Geography. While at Reading, Batty continued to develop his urban modelling research moving from static representations of cities to more dynamic models (i.e., adding temporal dynamics). In 1979, Batty moved to the University of Wales Institute of Science and Technology (UWIST, now part of Cardiff University) as a Professor of City and Regional Planning. During his time at UWIST he continued to work on urban models but also started to explore spatial analysis methods along with exploring how models could be used for design and decision-making processes. He also completed his PhD with a thesis entitled *'Pseudo Dynamic Urban Models'* (Batty, 1984). He became co-editor of the journal *Environment and Planning B* in 1981, then editor in 1985 – a role he was still performing in 2023.

In 1990, he moved to the State University of New York at Buffalo to become the director of the National Center for Geographic Information and Analysis (NCGIA). While at Buffalo, he further developed his interests in GIS and spatial analysis (Longley and Batty, 1997), as well as exploring how cities could be modelled using ideas from complexity theory and techniques from cellular automata and artificial life (Batty and Longley, 1994). He returned to the UK in 1995 to found the Centre for Advanced Spatial Analysis (CASA) at University College London. Under Batty's leadership between 1995 and 2010, CASA became world renowned for its work on the science of cities. In addition, Batty was the Bartlett Professor of Planning (2004–2010) and now holds the position of Emeritus Bartlett Professor of Planning. Batty has been elected a Fellow of the British Academy, the Academy of Social Sciences and the Royal Society, and in 2004 was made a Commander of the Most Excellent Order of the British Empire (CBE). He is a recipient of the Lauréat Prix International de Géographie Vautrin Lud, the Founders Medal from the Royal Geographical Society, and the Gold Medal of the Royal Town Planning Institute.

Spatial Contributions

Batty is one of the most distinguished urban geographers and planners of our time and has pioneered approaches to studying how cities emerge from the interactions of entities within them. His research focuses on making cities understandable from a scientific point of view by leveraging ideas from complexity thinking, with a clear line of thought that can be traced back to his time at Manchester. In the 1960s it was argued that cities and planning were domains that required systematic thinking. The approach taken, however, was a 'top-down view' in which cities were seen as being in balance and thus controllable through planning and models; typically using aggregate statistical models utilising ideas from social physics and location theory (Batty, 2008a), such as gravitational forces to capture the flows of movement (e.g., traffic, migration, employment) between places (Batty, 2007b). However, Batty questioned the utility of such models because they were relatively static and treated cities as if they were in equilibrium. In 1971 he outlined a design for a dynamic spatial model, illustrating its utility through a yearly model run over 15 years (from 1951–1966) (Batty, 1971). Separating population, employment and land uses, Batty used the Reading region, UK, to explore the dynamics of cities and the short-term consequences of planning decisions. Ideas explored included how, if the city expanded in one direction, would this impact population, employment, or the length of trips? His ideas on dynamic modelling were consolidated in the ground-breaking book *Urban Modelling: Algorithms, Calibrations, Predictions* (Batty, 1976).

Given his interest and approach, Batty has long been a strong proponent of computer modelling. With the advent of cheaper and more powerful computers in the early 1980s, Batty started to explore how computer graphics could be used to visualise models, with a focus on digital maps and user interactions. His early efforts are evident in *Microcomputer Graphics: Art Design and Creative Modelling* (Batty, 1987) in which he demonstrated how one could take simple shapes on the screen to make more complex patterns. The exploration of computer graphics also led him into the field of fractals, as they are often used as mathematical renderings in computer graphics (Batty, 2014). While fractals had been applied to many fields (e.g., geomorphology, biology), Batty along with Paul Longley was a pioneer in applying fractal geometry to the study of cities (Batty and Longley, 1986). In their book *Fractal Cities* they argue that 'cities are fractal in form, and that much of our pre-existing urban theory is a theory of the fractal city' (Batty and Longley, 1994: vi). They showed how they could develop 'realistic-looking patterns' at different levels of detail or hierarchies (i.e., self-similarity), such as the morphological structure of Greater London, through models based on diffusion-limited aggregation. This hints at Batty's interest in urban morphology: the form and function of cities. As he later observes, 'cities are thus classic examples of fractals in that their form reflects a statistical self-similarity or hierarchy of clusters' (Batty, 2008b: 770).

The realisation that cities are complex systems and that one could generate their form through models also led Batty to explore cellular automata (CA) models of cities. In this form of modelling, the morphological and aggregate structure of cities is based on simple rules from the bottom up. CA models are based on the notion that aggregate conditions seen within cities emerge from the interaction of many elements at a local scale through self-organisation, feedback loops and path dependence. Emergent phenomena are characterised by stable macroscopic patterns arising from local interaction of individual entities. This bottom-up approach became crucial in urban planning (Batty, 1995), in stark contrast with the top-down models of the past. However, in the 1990s the notion of top-down planning was also changing as societies became more decentralised. As Batty (2008b: 769) writes, 'cities are no longer regarded as being disordered systems. Beneath the apparent chaos and diversity of physical form, there is strong order and a pattern that emerges from the myriad of decisions and processes required for a city to develop and expand physically'. To this end, Batty, with Yichun Xie, proposed a general class of CA models for urban growth simulation (Batty and Xie, 1994). Their models, compared to past efforts, were inheritably dynamic. For example, in their dynamic urban evolutionary model (DUEM), hypothetical cities grow where their land uses (e.g., housing, industry and services) are structured through their life cycles (e.g., initiating, mature and declining), which was later applied to study urban sprawl in Ann Arbor, Michigan, in the US (Batty and Xie, 2005).

While CA models could capture the form of cities, they have difficulty in capturing individual level human decision making. This is because in CA models, all cells are treated the same way through various transition rules, which motivated Batty to explore agent-based modelling (ABM). The main difference today between agent-based and CA models is heterogeneity in behaviour within agent-based models, in the sense that all agents can have different behaviours. In such models, it is the interaction of many individuals that lead to the emergence of a specific phenomenon. For example, one of his first papers on agent-based modelling was how the small-scale movement of pedestrians can lead to crowds and also allow one to study the movement of people (e.g., at the Notting Hill carnival, London; Batty et al., 2003). By linking CA and ABM one can link form to function by linking the physical structure of the city to how it is used.

Batty's research in geography and planning has been enabled through developments in computing power and (big) data handling. Latterly, he has leveraged these to explore the creation of smart cities (Batty et al., 2012) and digital twins (Batty, 2018a). Here he has examined how streaming data can help with monitoring cities through city dashboards (Batty, 2015), how mobility data can aid with understanding travel patterns (Batty, 2013a; Reades et al., 2016) and how it is possible to simulate and visualise long-term structural change and dynamics in cities (Longley and Batty, 2003). Visualisations have ranged from developing a 3D city model of London (Batty and Hudson-Smith, 2005) to web-based models exploring land use and transportation (Batty and Milton, 2021).

Batty's book *The New Science of Cities* (Batty, 2013b) used complexity theory to examine networks and flows within cities, arguing that location is no longer the key to how cities function but rather the spatial interactions within them. Like his previous work, modelling, visualisation and simulation are a central focus, but not as tools for prediction. This sentiment was echoed in *Inventing Future Cities* (Batty, 2018b) which again emphasises that prediction is difficult due to complexity. Cities are unpredictable, while at the same time technological change keeps changing cities. Now, he argues, new data sources and the growth in urban science techniques are giving us a new lens to study the form and function of cities.

Key Advances and Controversies

There is a saying often attributed to George Box that '*All models are wrong, some are useful*' and Batty is overtly aware of this. He has been careful not to overpromise with respect to his models but rather notes their shortcomings. For example, Batty (1994: 10) notes that in earlier models the 'systems analysis being developed then was too simplistic dealing with well-behaved outcomes and simple equilibrium that bore little resemblance to the cities of the real world'. While this notion has changed with the realisation that cities are complex systems, challenges faced with earlier models with respect to 'particular, problems of estimation and calibration were broached and largely solved within the limits of the techniques available' (Batty, 1994: 11). However, with new styles of models which can explore cities from the bottom-up, like CA and ABM, challenges of validation and calibration remain (Batty and Torrens, 2005; Crooks et al., 2008). One reason for this is that, as Batty (2018a: 817) notes, 'models are, by definition, simplifications of the real thing and in that sense, do not aim to replicate the original system in the same detail as that system'. With this notion, Batty is hinting that no model is perfect. All models abstract the complexities of the real world to make the model tractable, and modellers have to focus on the key features that are of interest. With respect to agent-based modelling, Batty (2007a: 516) notes that we need to be cautious in using them as policy tools and emphasises that 'this approach to urban simulation is not designed to produce predictions that form the basis of policy'. He goes further in *Inventing Future Cities* (Batty, 2018b: 150) stating 'the future city can only be invented, not predicted'. Here, the notion is that one must accept that cities are unpredictable, but through the lens of complexity theory one can understand them. But it also raises questions with respect to prediction. Many researchers and practitioners see the role of modelling as a tool for prediction. But models are much more than this and there are many reasons to model (see Epstein, 2008). While some of his earlier works might be criticised as overly simplistic (while at the time they were cutting edge), one can look at these with hindsight as the product of limited computation.

Batty's research has been widely cited by urban planners and geographers, and it is difficult to find major criticisms of his oeuvre. As a quantitative urban geographer, someone more qualitative-minded might question his analytical approach to studying cities, but in general his writings have received very strong praise. His systematic study of cities has inspired a whole generation of researchers. His work on CA and ABM has inspired the work of many of his former students and colleagues. For example, Itzhak Benenson and Paul Torrens' (2004) work on the formalisation of geographic automata, David O'Sullivan and George Perry's work (2013) demonstrating how ideas from complexity theory and ABM can explain many spatial patterns, and Andrew Crooks' work on integrating GIS with various models to explore geographical systems (e.g., Crooks et al., 2019). His work on networks, technology, and virtual geographies (Batty, 1997) can be seen to have influenced the work of Martin Dodge and Rob Kitchin in mapping cyberspace and its spatial dimensions (e.g., Dodge and Kitchin, 2001). More generally his efforts pertaining to spatial analysis and making it accessible to a wider audience have influenced research in virtual geographic environments (e.g., Hudson-Smith et al., 2007; Lin and Batty, 2009), urban informatics (see Shi et al., 2021) and spatial analysis (see Maguire et al., 2005).

While Batty's conceptualisation of cities has changed from a top-down perspective to more of a bottom-up one, he has argued and demonstrated that cities are a product of millions of decisions and arise without coordination other than regulations and structures placed upon them by societies which themselves are emergent self-organising structures (Batty, 2013b). Batty's work has attempted to unravel the complexity of cities via the systematic use of science and technology. With the growth of digital data and computation, which now permeate our daily lives, we now have a body of knowledge, thanks to Batty, with which to study the form and function of cities.

Batty's Key Works

Batty, M. (1976) *Urban Modelling: Algorithms, Calibrations, Predictions*. Cambridge: Cambridge University Press.

Batty, M. (2007a) *Cities and Complexity: Understanding Cities with Cellular Automata, Agent-Based Models, and Fractals*. Cambridge, MA: MIT Press.

Batty, M. (2008b) 'The size, scale, and shape of cities', *Science*, 319 (5864): 769–71.

Batty, M. (2013b) *The New Science of Cities*. Cambridge, MA: MIT Press.

Batty, M. (2018b) *Inventing Future Cities*. Cambridge, MA: MIT Press.

Batty, M. and Longley, P.A. (1994) *Fractal Cities: A Geometry of Form and Functions*. London: Academic Press.

Secondary Sources and References

Alonso, W. (1964) *Location and Land Use: Toward a General Theory of Land Rent*. Cambridge, MA: Harvard University Press.

Batty, M. (1971) 'Modelling cities as dynamic systems', *Nature*, 231 (5303): 425–8.

Batty, M. (1984) *Pseudo-Dynamic Urban Models*. PhD thesis, The University of Wales, Cardiff, UK. http://www.casa.ucl.ac.uk/PhD.pdf (accessed October 26, 2023).

Batty, M. (1987) *Microcomputer Graphics: Art, Design and Creative Modelling*. London: Chapman and Hall.

Batty, M. (1994) 'A chronicle of scientific planning: The Anglo-American modelling experience', *Journal of the American Planning Association*, 60 (1): 7–16.

Batty, M. (1995) 'New ways of looking at cities', *Nature*, 377 (6550): 574.

Batty, M. (1997) 'Virtual geography', *Futures*, 29 (4/5): 337–52.

Batty, M. (2007b) 'Model cities', *Town Planning Review*, 78 (2): 125–78.

Batty, M. (2008a) 'Fifty years of urban modelling: Macro-statics to micro-dynamics', in S. Albeverio, D. Andrey, P. Giordano and A. Vancheri (eds), *The*

Dynamics of Complex Urban Systems: An Interdisciplinary Approach. New York, NY: Springer. pp. 1–20.

Batty, M. (2013a) 'Big data, smart cities and city planning', *Dialogues in Human Geography,* 3 (3): 274–9.

Batty, M. (2014) 'A science of cities in a science of planning'. Paper presented at The Evolution of Planning Thought Symposium. Vienna University of Technology, Austria, 19–23 May. http://www.complexcity.info/files/2014/09/BATTY-Biographical-Essay.pdf (accessed October 26, 2023).

Batty, M. (2015) 'A perspective on city dashboards', *Regional Studies, Regional Science,* 2 (1): 29–32.

Batty, M. (2018a) 'Digital twins', *Environment and Planning B,* 45 (5): 817–20.

Batty, M. and Hudson-Smith, A. (2005) 'Urban Simulacra', *Architectural Design,* 75 (6): 42–7.

Batty, M. and Longley, P.A. (1986) 'The fractal simulation of urban structure', *Environment and Planning A,* 11 (9): 1143–79.

Batty, M. and Milton, R. (2021) 'A new framework for very large-scale urban modelling', *Urban Studies,* 58 (15): 3071–94.

Batty, M. and Torrens, P.M. (2005) 'Modelling and prediction in a complex world', *Futures,* 37 (7): 745–66.

Batty, M. and Xie, Y. (1994) 'From cells to cities', *Environment and Planning B,* 21 (7): S31–S48.

Batty, M. and Xie, Y. (2005) 'Urban growth using cellular automata models', in D.J. Maguire, M. Batty and M.F. Goodchild (eds), *GIS, Spatial Analysis and Modelling*. Redlands, CA: ESRI Press. pp. 151–72.

Batty, M., Axhausen, K.W., Giannotti, F., Pozdnoukhov, A., Bazzani, A., Wachowicz, M., Ouzounis, G. and Portugali, Y. (2012) 'Smart cities of the future', *The European Physical Journal Special Topics,* 214 (1): 481–518.

Batty, M., Desyllas, J. and Duxbury, E. (2003) 'Safety in numbers? Modelling crowds and designing control for the Notting Hill Carnival', *Urban Studies,* 40 (8): 1573–90.

Benenson, I. and Torrens, P.M. (2004) *Geosimulation: Automata-Based Modelling of Urban Phenomena*. London: John Wiley.

Crooks, A.T., Castle, C.J.E. and Batty, M. (2008) 'Key challenges in agent-based modelling for geo-spatial simulation', *Computers, Environment and Urban Systems,* 32 (6): 417–30.

Crooks, A.T., Malleson, N., Manley, E. and Heppenstall, A.J. (2019) *Agent-based Modelling and Geographical Information Systems: A Practical Primer*. London: Sage.

Dodge, M. and Kitchin, R. (2001) *Mapping Cyberspace*. London: Routledge.

Epstein, J.M. (2008) 'Why model?', *Journal of Artificial Societies and Social Simulation,* 11 (4): 12.

Harris, B. (1965) 'New tools for planning', *Journal of the American Institute of Planners,* 31 (2): 90–5.

Hudson-Smith, A., Milton, R., Dearden, J. and Batty, M. (2007) *Virtual Cities: Digital Mirrors into a Recursive World*. London. Centre for Advanced Spatial Analysis (University College London), Working Paper 124. Available from https://discovery.ucl.ac.uk/id/eprint/15178/ (accessed October 26, 2023).

Isard, W. (1956) *Location and Space-Economy: A General Theory Relating to Industrial Location, Market Areas, Land Use, Trade and Urban Structure*. Cambridge, MA: MIT Press.

Lin, H. and Batty, M. (eds) (2009) *Virtual Geographic Environments*. Beijing, PRC: Science Press.

Longley, P., and Batty, M. (eds) (1997) *Spatial Analysis: Modelling in a GIS Environment*. Chichester: Wiley.

Longley, P. and Batty, M. (eds) (2003) *Advanced Spatial Analysis: The CASA Book of GIS*. Redlands, CA: ESRI Press.

Maguire, D.J., Batty, M. and Goodchild M.F. (eds) (2005) *GIS, Spatial Analysis and Modelling*. Redlands, CA: ESRI Press.

Mcloughlin, J.B. (1969) *Urban and Regional Planning: A Systems Approach*. London: Faber and Faber.

O'Sullivan, D. and Perry, G.L. (2013) *Spatial Simulation: Exploring Pattern and Process*. Chichester: John Wiley and Sons.

Reades, J., Zhong, C., Manley, E.D., Milton, R. and Batty, M. (2016) 'Finding pearls in London's oysters', *Built Environment*, 42 (3): 365–81.

Shi, W., Goodchild, M., Batty, M., Kwan, M.-P. and Zhang, A. (eds) (2021) *Urban Informatics*. New York, NY: Springer.

6
Bawaka Country

***tebrakunna country* and Emma Lee, Federation University**

Biographical Details and Theoretical Context

Bawaka Country is located in north-east Arnhem Land, Northern Territory, Australia, and is one of the many homelands – the unceded sovereign territories – of the Yolŋu (Yolngu) Peoples. While Yolŋu is the broader name for the collective of Indigenous Peoples across north-east Arnhem Land, Bawaka Country is lived and loved by the Gumatj people, and other clans related through kinship, connections and responsibilities. Bawaka Country is coastal and comes together with many neighbouring clan homelands under the Dhimurru Indigenous Protected Area (a region approximately 5,500 km^2 in size) to take care of the lands, seas and skies through ancestral Law.

The definition of Bawaka Country does not only reside within its physical geography; Bawaka Country is conscious, sensate, and relational. For Yolŋu, to exist in Bawaka Country – as human or non-human entity – is to *be* Bawaka Country. There can be no inanimate parts of Bawaka Country, nor atomisation of connections, knowledges and responsibilities, as everything is related through its story, spirit and song. Bawaka Country is not fully dependent upon humans to give it meaning, as it exists in and of itself, and values the 'co-emergence' of humans and non-humans working together to care as Country.

Bawaka Country is also an articulation of Rom, the ancestral Law of Yolŋu Peoples. Rom has always existed and continues to emerge in Bawaka Country, with Yolŋu, and thrives despite the coming of the colonisers. Rom is one of the oldest forms of Law in the world because Yolŋu Peoples are one of the oldest populations, and continuing cultures, in the world. Rom is the way in which the relationship between people and Bawaka Country is governed and managed. Rom is expressed in many ways, but, as outsiders, we begin to understand it through song and dance, through bodies, canvas, bark and rock walls painted in ochre-pigment, through kinship and reciprocity, through holding responsibility to keeping culture and people alive, and through the breath of every living thing.

Bawaka Country's contributions to the field of geography are known through the Bawaka Collective. Established in 2006, the Bawaka Collective was a coming together of women to write, exhibit, live and share research. They are four Yolŋu sisters from Bawaka, Laklak Burarrwaŋa, Ritjilili Ganambarr, Merrkiyawuy Ganambarr-Stubbs, Banbapuy Ganambarr, and their daughter Djawundil Maymuru; and three Western-heritage researchers from the east coast of Australia, Sarah Wright, Sandie Suchet-Pearson and Kate Lloyd. The Bawaka Collective is a living entity that continues to spiral across the generations to grow new kin and knowledge sharing in Bawaka Country. In recent years, Western-heritage researcher Lara Daley joined the Bawaka Collective; Laklak's eldest son, Djäwa Burarrwaŋa, and Yolŋu songman, Rrawun Maymuru, the son of Banbapuy, as well as daughters Rosealee Pearson and Siena Mayutu Stubbs, and other family members, also play important roles in supporting the Collective's work. There are two main themes to the Collective's work: intergenerational knowledge of the songspirals that nurture Yolŋu life, and Yolŋu mathematics, the 'complex matrix of patterns, relationships, shapes, motions and rhythms of time and space' (Bawaka Country including Lloyd et al., 2016: 1) that define how Yolŋu understand, know, relate to and experience Bawaka Country.

Dr Laklak Burarrwaŋa (a special and particular woman as the eldest daughter of the eldest daughter and holder of sacred knowledge, songs and behaviours) has led the work of the Collective to help educate non-Yolŋu in knowing Bawaka Country and knowing more-than-human ways of connecting with the world. The Bawaka Collective also relies on Dr Burarrwaŋa's vision for passing on knowledge to younger Yolŋu and caring for Bawaka Country by singing songspirals, also known as songlines. Songspirals are 'about a person, but you can only understand that person through Country, as Country' (Gay'Wu Group of Women, 2019: 45). The purpose of songspirals is to re-make and reaffirm the boundaries of the lands and waters through *milkarri* (keening), *gurruṯu* (kinship), *djalkiri* (foundations) and *märr* (love).

Here, songspirals are the *raki* – the string – that binds a people and culture together to be Bawaka Country. Songspirals are the language, Law and sacredness of Bawaka Country. They are the 'essence of people in this land, the essence of every clan [...] [*they*] spiral out and spiral in, they go up and down, round and round, forever' (Gay'Wu Group of Women, 2019: xvi). Dr Burarrwaŋa tells us that women have a particular role in the songspirals, when they 'cry the songspirals, we keen the songspirals – that is what we call milkarri [...] Milkarri is an ancient song, an ancient poem, a map, a ceremony and a guide, but it is more than all this too' (Gay'Wu Group of Women, 2019: xvi). Songspirals keen Bawaka Country; they are a way for Yolŋu to map Country and know who they are in relation to place, people and knowledge.

Spatial Contributions

Bawaka Country guides the Bawaka Collective to both decolonise geography research and promote Indigenous epistemologies and ontologies. Decolonising work here follows on from Linda Tuhiwai Smith (2012: x), a grandmother of decolonising studies, by seeking to 'disrupt relationships between researchers (mostly non-indigenous) and researched (indigenous), between a colonizing institution of knowledge and colonized peoples whose own knowledges was subjugated [...]'. The Bawaka Collective, writing under the authorship of the Gay'Wu Group of Women, are firstly assisting to decolonise geography research by speaking for themselves:

> Most of the books written about songlines have been written by white people, and mainly white men [...] Often when we read about Yolŋu, it is not from Yolŋu people. The words are not chosen and controlled by us. They often leave things out and get things wrong. Sometimes we are offended by what other people write about us. We also find that most of what's written talks

> about men's knowledge and doesn't pay attention to or value women's knowledge and milkarri. (Gay'Wu Group of Women, 2019: xxv)

Yolŋu are taking back the space to speak of geography as experts, as owners, as the true knowledge holders of Bawaka Country because they *are* Bawaka Country. Secondly, the Bawaka Collective and the Gay'Wu Group of Women are telling us that Country is held through gendered responsibilities. For many Indigenous Peoples across the world, knowledge and behaviours are often characterised as men's and women's business. The Bawaka Collective write that although 'milkarri is for women, and we share much women's knowledge, men's and women's knowledge balances each other. It is entwined, not separate' (Gay'Wu Group of Women, 2019: xxi). Gendered knowledge is not the same as universalizing Western feminist approaches, as Yolŋu women's knowledge resides within the songspirals and Yolŋu mathematics of Bawaka Country. As such, it constitutes a means of decolonising geography to illuminate the richness of relationality and connections to place. The women of the Bawaka Collective are calling geographers to move away from patriarchal, Western knowledge systems and are telling the White men to be quiet in Bawaka Country, to not speak for them or Rom or Country, and instead to listen.

In listening to, and hearing, the songspirals of Bawaka Country, the Bawaka Collective are inviting everyone to enact geography research as a 'co-becoming'. Co-becoming starts as a 'concept of humans as Country, rather than humans in Country' (Bawaka Country including Suchet-Pearson et al., 2013: 186) and is an Indigenous methodology that expands geography thinking. Co-becoming is a way of understanding that Bawaka Country has its own agency and 'author-ity' (Bawaka Country including Wright et al., 2015) and that the relationships of being as Country are an act of creation that emerges with *gurrutu*. Summarising the work of Bawaka Collective, then, is to co-become with Bawaka Country because Bawaka Country helps shape the arrangement and process of writing. Indeed, Bawaka Country becomes the leading author of Bawaka Collective writings through a:

> [...] more-than-human co-becoming of space/place that is not human-centric. In doing so, we, the human and more-than-human authors, elucidate a place/space that co-becomes – that is constituted through human–more-than-human relationships [...] a space/place that draws attention to the ways in which dynamic, ongoing, intense webs of connection actually author an academic paper. (Bawaka Country including Wright et al., 2016: 458)

Country as an author is a powerful revelation. What should seem intuitive within geographic research and practice – that place is alive and wonderfully present and can decide for itself – has largely been absent in the literature. Western systems of knowledge focus on the self, the human being, as the centre of scholarship and acknowledgement. The Bawaka Collective are extraordinary in translating a different way of seeing place; they have brokered the spaces where Country and the more-than-human are recognised entities that influence thinking, writing, acting and being.

By honouring Bawaka Country as a lead author, the Bawaka Collective:

> [...] decentres the privileging of human authors as the only beings able to control and create, as the sole deciders

> of content and structure, and opens up opportunities for reimaging and co-creating not only how we write but how we think and practise. So we are a mixed group as we talk with you. We are human and more-than-human, tangible and intangible and everything in between – especially those relationships which constantly bring us into being. (Bawaka Country including Wright et al., 2015: 271)

Bawaka Country and the Bawaka Collective do not discriminate against Western knowledge systems, but prioritise Country as central, where everything flows out from that complex and grounded web of relationships. In living Rom, Yolŋu have *djalkiri*, 'these are the footprints, ancestral imprints on our country that are our spiritual foundations. The djalkiri is our library which we learn from' (Wopurruwuy Gondarra, 2023). Knowing *djalkiri* is to learn how to share *wetj*: a way of thinking, feeling, doing, being and responding to the human and more-than-human as 'entangled becomings' (Bawaka Country including Suchet-Pearson et al., 2013: 187). *Wetj* is a deep sharing that links 'Yolŋu with each other and with the diverse beings of Country in a web of mutuality and responsiveness/responsibility' (Bawaka Country including Suchet-Pearson et al., 2013: 186–7). In sharing Yolŋu culture and knowledge as *wetj* with outsiders, Bawaka Collective highlight two-way learning, and show Yolŋu sovereign rights, knowledges and lifeworlds as of equal value to mainstream knowledges.

The concept of *gänma* helps to explain two-way learning and is especially important for understanding Yolŋu mathematics. *Gänma* encourages 'a mixing, a meeting and mingling, that brings difference together without erasing it. *Gänma* thus means new life and new ideas. It evokes knowledges coming together' (Bawaka Country including Lloyd et al., 2016: 1). While outsiders may understand *gänma* to mean that it is for Yolŋu to incorporate Western knowledge, *gänma* is best thought of as a way for non-Yolŋu to feel songspirals and co-becoming.

Yolŋu mathematics, as *gänma*, is a situated knowledge that is dependent upon relationships between everything:

> For Yolngu, relationships and relationality take precedence over ideas of discreteness or abstract quantity. Mathematical patterns bind people to each other and to Country in relationships of co-becoming and as manifestation of strong ethical frameworks [...] In Yolŋu mathematics, people are not separated from nature. The earth is not separated from the sky. Songs and stories are not separated from people and objects. All these things exist as part of one another and come into being together. It is a form of relativity. (Bawaka Country including Lloyd et al., 2016: 6)

Yolŋu mathematics expresses and nourishes *gurruṯu*. The patterns of Bawaka Country, as mathematics, are also the threads that weave the kinship rules and behaviours between humans, Country and more-than-humans. Co-becoming in Bawaka Country is governed not only by the emotional and relational connections to Country, but also by each thing (human, animal and other) having its own mathematics – a language, pattern and kinship – that alerts Yolŋu to their *wetj* responsibilities. Yolŋu cannot be Bawaka Country without the essential elements of Yolŋu mathematics giving form and structure to knowledge and behaviour.

In thinking about Bawaka Country and letting ourselves, as outsiders, begin to explore the notions of co-becoming and having Bawaka Country influence how we write, connect to and know its author-ity, we must also have love in our hearts. The *gurruṯu* of Yolŋu mathematics that connects everything, the *wetj* of generational responsibility to deeply share, the *milkarri* that Yolŋu women possess to re-make Country – all these things are about love. Colonisation in Australia has tried to destroy Indigenous Peoples by denying our capacity to love, but the Bawaka Collective are reasserting the centrality of *märr* (love) as Country (Bawaka Country including Wright et al., 2019). The Gay'Wu Group of Women (2019: 98-9) tell us that *märr* is about 'connection, the way our gurruṯu and our djalkiri, our foundations, overlap, how we work together and how we all overlap with the land [...] love emerged and emerges, where we are, whatever we do. It spirals out and round, in our connection with each other and the land'.

Geography, then, as a relational entity must encompass love and belonging. To know place, to sing boundaries and patterns of everything, to enact kinship and sharing, to respect and care for ourselves, and to cherish the more-than-human is, for Yolŋu, *märr*.

Key Advances and Controversies

When the Bawaka Collective started writing Bawaka Country in 2006, the comingling – the *gänma* – between Western and Yolŋu lifeworlds produced a unique view putting Country and Indigenous Peoples more central within Western geography. In 2012, Bawaka Country became, perhaps, the first sovereign Indigenous territory to be credited as a lead co-author in an international academic journal. In over ten years, while the pushback has lessened, the struggle to recognise Bawaka Country as the lead author and the Yolŋu co-authors as legitimate contributors has been significant. This has primarily occurred 'behind the scenes' with the peer reviewers, editors and publishers who require justification to name Bawaka Country as lead author and who struggle with the lack of Western literature cited over Yolŋu knowledges, while metadata cannot correctly cope with, or cite, Bawaka Country as author name (Bawaka Country including Burarrwaŋa et al., 2023). The controversy of authorship is not with Bawaka Country, it is with a system of Western knowledge that has actively denied Indigenous Peoples inclusion. Consequently, it is only in the past few years that Indigenous methodologies have been somewhat accepted into Geography. The slow uptake of Indigenous methodologies means there is little academic analysis, and the substance of Indigenous knowledges remains barely recognised or respected within mainstream learning.

The Bawaka Collective have themselves noted where particular controversy may lie for others in the ongoing struggle against colonisation for Yolŋu. They make clear that Yolŋu knowledge is not for everyone to know or share:

> There are many layers of knowledge and it is only appropriate we share some of the top layer [...] Yolŋu knowledge must always remain with Yolŋu people, it belongs with Country, and Country is the keeper of knowledge. This is some of what knowledge (and data) sovereignty means for us; it is important for researchers to learn that not all knowledge is for everyone. (Bawaka Country including Burarrwaŋa et al., 2022b: 3)

This idea of knowledge belonging to Country and not randomly held runs counter to prevalent Western knowledge systems that operate under broad access to everyone. In Bawaka

Country's call to own its intellectual property, others may misunderstand how Yolŋu knowledge is situated, earned and cannot be known through a solitary text: knowledge is relational and kin (see Tynan, 2020). However, developing a romanticised view would misinterpret Bawaka Country and culture as static, unblemished or unchanging (Bawaka Country including Suchet-Pearson et al., 2018) and reinforce a colonial attitude of superiority over Yolŋu knowledge. This is crucial because the colonisation of Bawaka Country is ongoing. The Collective tell a songspiral for sky country to show that 'outer space' is not empty or separate and that Yolŋu care for it (Bawaka Country including Mitchell et al., 2020). The commodification of outer space, premised on Western concepts of *terra nullius*, or no one's land, dispossesses Yolŋu of their rights to *gurruṯu* and *milkarri* for sky country. The same can be said for freshwaters of Yolŋu sovereign territories. As in outer space, terrestrial research and practice deny the agency of rivers, clouds, rain and waters and ignore the Wukun songspiral in favour of Western management and development principles (Bawaka Country including Burarrwaŋa et al., 2022b).

To combat and fend off these harms, the Bawaka Collective extend an invite to researchers and practitioners to hold positive aims rather than strive for methodological neutrality (Bawaka Country including Suchet-Pearson, 2020). Co-becoming with Bawaka Country cannot be at arms-length, but should advance research kinship and reciprocity as a driver to repair and heal damaging Western ideologies. The Bawaka Collective want to 'move beyond [...] cultural pluralism towards relationships [...] that refuse colonial hierarchies' (Bawaka Country including Burarrwaŋa et al., 2022a: 6–7). Refusing colonisation is not necessarily a protesting action, but instead a means to reflect on geography research and practice as 'response-abilities as, not responsibility to or for' Bawaka Country (Bawaka Country including Suchet-Pearson et al., 2018: 695).

Allied geographers, especially those in colonised country, are drawn to the Bawaka Collective's work to unsettle colonial supremacy (Castree, 2015; MacDonald, 2017; Howitt, 2020). They realise that while Yolŋu ways of being, like other Indigenous Peoples, are singular and cannot be replicated elsewhere (Sharp et al., 2022), the scholarship has opened intense opportunities for non-Indigenous researchers to see Country as a place of belonging rather than a site of geographic theoretical utility (Clement, 2019).

Above all, Bawaka Country and the Bawaka Collective have provided comfort and cultural safety for people like me to follow suit in holding Country as co-author too (Martuwarra RiverOfLife et al., 2022; *tebrakunna* country and Lee, 2022). Bawaka Country, and the Collective's outputs, gift us a safety to represent ourselves as we see ourselves – as Country. We are legitimised as Indigenous geographers because the cultural affiliations that give us life are affirmed through the rights of co-authorship with Country. The Bawaka Collective grows Indigenous methodologies and researchers, tells Yolŋu culture through songspirals and strengthens Bawaka Country's role in shifting geography thinking to a place of listening to more-than-human influences.

Bawaka Country's Key Works

Bawaka Country, Burarrwanga, L., Ganambarr, R., Ganambarr-Stubbs, M., Ganambarr, B., Maymuru, D., Lloyd, K., Daley, L., Suchet-Pearson, S., Wright, S.,

Tofa, M. and Hammersley, L. (2022a) 'Bala ga' lili: Communicating, relating and co-creating balance through relationships of reciprocity', *Social & Cultural Geography*, 24 (7): 1203–23.

Bawaka Country including Burarrwanga, L., Ganambarr, R., Ganambarr-Stubbs, M., Ganambarr, B., Maymuru, D., Wright, S., Suchet-Pearson, S., Lloyd, K. and Daley, L. (2022b) 'Gapu, water, creates knowledge and is a life force to be respected', *PLOS Water*, 1 (4): article #e0000020.

Bawaka Country including Burarrwanga, L., Ganambarr, R., Ganambarr-Stubbs, M., Ganambarr, B., Maymuru, D., Wright, S., Suchet-Pearson, S., Lloyd, K. and Daley, L. (2023) 'Author-ity of/as Bawaka Country', *Australian Archaeology* 89 (1): 75–7.

Bawaka Country including Lloyd, K., Wright, S., Suchet-Pearson, S., Burarrwanga, L., Ganambarr, R., Ganambarr-Stubbs, M., Whitehead, B. and Maymuru, D. (2016) '*Morrku mangawu – Knowledge on the Land:* mobilising Yolŋu mathematics from Bawaka, north east Arnhem Land, to reveal the situatedness of all knowledges', *Humanities*, 5 (3): article #61.

Bawaka Country including Mitchell, A., Wright, S., Suchet-Pearson, S., Lloyd, K., Burarrwanga. L., Ganambarr, R., Ganambarr-Stubbs, M., Ganambarr, B., Maymuru, D. and Maymuru, R. (2020) 'Dukarr lakarama: Listening to Guwak, talking back to space colonization', *Political Geography*, 81: article #102218.

Bawaka Country including Suchet-Pearson, S., Wright, S., Lloyd, K. and Burarrwanga, L. (2013) 'Caring as Country: Towards an ontology of co-becoming in natural resource management', *Asia Pacific Viewpoint*, 54: 185–97.

Bawaka Country including Suchet-Pearson, S., Wright, S., Lloyd, K., Tofa, M., Burarrwanga, L., Ganambarr, R., Ganambarr-Stubbs, M., Ganambarr, B. and Maymuru, D. (2020) 'Bunbum ga dhä-yu t agum: To make it right again, to remake', *Social & Cultural Geography*, 21 (7): 985–1001.

Bawaka Country including Suchet-Pearson, S., Wright, S., Lloyd, K., Tofa, M., Sweeney, J., Burarrwanga, L., Ganambarr, R., Ganambarr-Stubbs, M., Ganambarr, B. and Maymuru, D. (2018) 'Goŋ Gurtha: Enacting response-abilities as situated co-becoming', *Environment and Planning D, Society & Space*, 37 (4): 682–702.

Bawaka Country including Wright, S., Suchet-Pearson, S., Lloyd, K., Burarrwanga, L., Ganambarr, R., Ganambarr-Stubbs, M., Ganambarr, B. and Maymuru, D. (2015) 'Working with and learning from Country: Decentring human author-ity', *Cultural Geographies*, 22 (2): 269–83.

Bawaka Country including Wright, S., Suchet-Pearson, S., Lloyd, K., Burarrwanga, L., Ganambarr, R., Ganambarr-Stubbs, M., Ganambarr, B., Maymuru, D. and Graham, M. (2019) 'Everything is love: Mobilising knowledges, identities, and places as Bawaka', in N. Gombay and M. Palomino-Schalscha (eds), *Indigenous Places and Colonial Spaces: The Politics of Intertwined Relations*. London: Routledge. pp. 51–71.

Bawaka Country including Wright, S., Suchet-Pearson, S., Lloyd, K., Burarrwanga, L., Ganambarr, R., Ganambarr-Stubbs, M., Whitehead, B., Maymuru, D. and Sweeney, J. (2016) 'Co-becoming Bawaka: Towards an emergent understanding of place/space', *Progress in Human Geography*, 40 (4): 455–75.

Gay'Wu Group of Women including Burarrwanga, L., Ganambarr, R., Ganambarr-Stubbs, M., Ganambarr, B., Maymuru, D., Wright, S., Suchet-Pearson, S. and Lloyd, K. (2019) *Songspirals: Sharing Women's Wisdom of Country through Songlines*. Crows Nest: Allen & Unwin.

Secondary Sources and References

Castree, N. (2015) 'Geography and global change science: Relationships necessary, absent, and possible', *Geographical Research*, 53 (1): 1–15.

Clement, V. (2019) 'Beyond the sham of the emancipatory Enlightenment: Rethinking the relationship of Indigenous epistemologies, knowledges, and geography through decolonizing paths', *Progress in Human Geography* 43 (2): 276–94.

Howitt, R. (2020) 'Unsettling the taken (for granted)', *Progress in Human Geography*, 44 (2): 193–215.

MacDonald, K. (2017) 'My experiences with Indigenist methodologies', *Geographical Research*, 55: 369–78.

Martuwarra RiverOfLife, Unamen Shipu Romaine River, Poelina, A., Wooltorton, S., Guimond, L. and Sioui Durand, G. (2022) 'Hearing, voicing and healing: Rivers as culturally located and connected', *River Research and Applications*, 38 (3): 422–34.

Sharp, E.L., Brierley, G.J., Salmond, J. and Lewis, N. (2022) 'Geoethical futures: A call for more-than-human physical geography', *EPF, Philosophy, Theory, Models, Methods and Practice*, 1 (1): 66–81.

Smith, L.T. (2012) *Decolonizing Methodologies: Research and Indigenous Peoples*. London: Zed Books.

tebrakunna country and Lee, E. (2022) 'Reclaiming the first person voice', in *tebrakunna* country and E. Lee and J. Evans (eds), *Indigenous Women's Voices: 20 Years on from Linda Tuhiwai Smith's Decolonizing Methodologies*. London: Zed. pp. 137–50.

Tynan, L. (2020) 'Thesis as kin: Living relationality with research', *AlterNative: An International Journal of Indigenous Peoples*, 16 (3): 163–70.

Wopurruwuy Gondarra, R. (2023) 'The Bawaka weaving story', *Bawaka Aboriginal Corporation*. https://bawakaexperience.com.au/the-bawaka-weaving-story/ (accessed October 27, 2023).

7
Lauren Berlant

David K. Seitz, Harvey Mudd College

Biographical Details and Theoretical Context

Lauren Gail Berlant (they/them) was born 31 October 1957, in Philadelphia to Joanne Bauer Berlant, an interior decorator, and Nathan Berlant, a negligence lawyer. With Russian communists on one side of their family and Israeli bourgeois nationalists on the other, Berlant's upbringing in an upscale Philadelphia suburb provided a milieu in which talk of law, politics, aesthetics, and citizenship were available resources to process the social turbulence of the 1960s and 1970s (Berlant, 2019). Despite an affluent childhood, their parents' divorce and declaration of bankruptcy while they were a student at Oberlin College meant Berlant had to scramble to afford college tuition. They later remarked that their decision to attend graduate school (English at Cornell University) immediately upon graduating from Oberlin (1979) was partly informed by the imperative of avoiding student loan repayment (Berlant, n.d.).

Facing familial disappointment and financial precarity, and yearning for alternatives to the failures of the American Dream, Berlant became an active participant in feminist and other radical collectives in and beyond the classroom, including Oberlin's Experimental College. At Oberlin, Berlant began to elaborate a collaborative, hybridised approach to critique, pedagogy, political activity, and aesthetic production that synthesised elements of Marxism, anarchism, feminism, psychoanalysis, sociology, and film and literary theories without being reducible to any of them. Berlant (1997a) described being routinely dismissed as an 'unserious' graduate student for their perceived failure to perform sufficient disinterestedness, and intimated that participation in feminist classrooms and collectivities was crucial to sustaining their 'will to become possible' (1997a: 159). They began teaching English at the University of Chicago in 1984 and earned a doctorate in English literature from Cornell the following year.

Berlant's first three monographs – what they would come to call their 'national sentimentality trilogy' – offered an original approach to the study of US cultures. These works built on extant Marxist critiques of ideology, citizenship, and nationalism, but also offered unique insight into people's contradictory (often racialised and gendered) affective investments in these political forms, and the simultaneously resistant and complicit possibilities of popular memory, local political communities, and everyday life.

Berlant's first book, *The Anatomy of National Fantasy: Hawthorne, Utopia, and Everyday Life*

(1991), was a probing examination of 'the interarticulation, now four hundred years old, of the United States and America, the nation and utopia' (Berlant, 1991: 217). Berlant argued it was a felt, not entirely conscious and often incoherent promise of an egalitarian community ('America') that sustained many people's ambivalent attachments to political institutions ('The United States') that both advanced and betrayed their (sometimes-conflicting) material interests. The book identified a double-bind that would echo throughout Berlant's work. On the one hand, Berlant was a compassionate diagonistician of the lure of idealising 'America' as utopian, noting 'the incredible hope behind every [...] act of political faith, including the willingness to forget, in advance, its betrayal in the national public sphere' (1991: 217). Yet, on the other, they also worried that the politically-necessary but painful deidealisation of the US – 'the end of the seemingly eternal coupling of the United States and Utopia' – might end badly, to say the least (1991: 217).

The trilogy's second entry, *The Queen of America Goes to Washington City: Essays on Sex and Citizenship* (1997b), turned from nineteenth-century literature to late-twentieth-century popular culture, mapping the subjective and ideological effects of neoliberalism's hollowing-out of the US public sphere and its privileging of the white, affluent, heteronormative nuclear family as the locus of social solidarity. Berlant proposed a theory of 'infantile citizenship', arguing the 'naïve citizen' earnestly invested in the promise of 'America' – and its pairing with 'paralyzed cynical apathy' as 'the only positions a normal or moral American can assume' – were central to a 'strong and enduring [*structure of*] belief' (Berlant, 1997b: 28–9). Tracking the contracting public sphere, Berlant presciently critiqued the fetishisation of the foetus, which was figured by Reagan-era family value discourse as 'the utopian Other [...] of everything imaginable' and cast 'as a philanthropic object for its own mother' (Berlant, 1997b: 137; 101). The book's core chapter, on the testimony of legal scholar Anita Hill about her sexual harassment by then-US Supreme Court nominee Clarence Thomas, offered a theory of 'diva citizenship', tracing the constrained possibilities for critical pedagogy through individual performance on the national stage.

The trilogy's final component, *The Female Complaint: The Unfinished Business of Sentimentality in America* (Berlant, 2008a) returned to the late nineteenth and early twentieth centuries, examining the unprecedented emergence of US 'women's culture' – mass commercial literary entertainment, largely written for and by white, middle-class women. While critical of this culture's constitutive exclusions, Berlant was interested in the formal particularities of these 'intimate publics', which mediated community for subjects 'marked by the historical burden of being harshly treated in a generic way' by 'making an aesthetic and spiritual scene that generates *relief from the political*' (2008a: 10; original emphasis). Although 'what is personal has itself been threaded through mediating institutions and social hierarchy', Berlant observed, such publics demarcated 'the nonpolitical situation of most ordinary life as it is lived as a space of continuity and optimism and self-cultivation' (Berlant, 2008a: 10).

Between monographs, Berlant wrote numerous influential essays. Key among them was 'Sex in public', co-authored with Michael Warner, which argued that heteronormativity, or the sense of social rightness attached to heterosexuality, is not natural but the fetishised result of an elaborate material and ideological production of conventions governing intimacy (Berlant and Warner, 1998). '68, or something' (Berlant, 1994) defended the revolutionary 1960s from malignment in the neoliberal 1990s, grappled with the contradictions of national allegory in Black radical and feminist discourses, and raged against the competition between identity-based disciplines imposed by University administrations. 'The subject of true feeling' (Berlant, 1999) scrutinised the terms on which politically-inflicted collective suffering becomes admissible as personal (yet generic) pain in national public spheres. And 'Unfeeling Kerry' (Berlant, 2005) offered a post-mortem on the 2004 John Kerry presidential campaign, arguing that Kerry's own 'self-betrayal' – as an anti-war activist turned pro-war politician – undermined his campaign's affective resonance (Berlant, 2019).

Berlant's national sentimentality trilogy and essays established them as a central intellectual figure in the emergence of both queer theory and what came to be called the 'affective turn'. But it is *Cruel Optimism* (2011), which is considered Berlant's magnum opus, that has enjoyed the widest reach. Returning to the contemporary period to examine neoliberalism's corrosive effects on 'conventions of reciprocity' in the US and Europe (Berlant, 2011: 7), Berlant observed that 'a relation of cruel optimism exists when something you desire is actually an obstacle to your flourishing' (Berlant, 2011: 1). While objects of cruel-optimistic attachment are wide-ranging, Berlant pointed to the trappings of the normative good life – economic security and lasting intimacy – as at once increasingly elusive *and* objects of sustained pursuit, precisely in their elusiveness. Berlant (2011) asked how differently positioned people navigated the material and subjective crises of neoliberal capitalism, taking note of the growing difficulty of ascribing genre or narrative coherence to life at a time of widespread precarity and economic and political improvisation. Assessing the devastation wrought by neoliberalism, Berlant lovingly critiqued Eve Kosofky Sedgwick's (2003) concept of 'reparative reading', which Sedgwick proposed as a complement to 'paranoid' Left critical vigilance that constantly anticipates injustice. Where Sedgwick (1990), among Berlant's key comrades in inaugurating queer studies and the affective turn, took pleasure in the deceptively simple axiom that 'people are different from each other' (1990: 22), Berlant counselled that people are also different from themselves – incoherent in both politically-enabling and self-defeating ways.

Alongside their solo-authored works, Berlant collaboratively wrote and edited several influential volumes, including *Sex, or the Unbearable* (Berlant and Edelman, 2014), which theorised the incoherent, awkward, self-shattering, and relational possibilities of sex as a challenge to sovereign subjectivity. *The Hundreds* (Berlant and Stewart, 2019) curated a series of experimental vignettes exploring the relationships between politics, affect, and everyday life that were bound by the formal constraint that they must be exactly 100 (or 200, or 300, etc.) words long. Berlant's notable editorial work included three decades as co-editor of the journal *Critical Inquiry*, special issues of that journal on intimacy (Berlant, 1998), case study methods (2007), and comedy (Berlant and Ngai, 2017), and an edited volume on the cultural politics of the Monica Lewinsky affair (Berlant and Duggan, 2001).

Berlant died in Chicago on June 28, 2021, of complications from a rare form of cancer. Yet Berlant themselves insisted (2005) that events remain open and ongoing as long as new interpretations of them are being generated – a claim that clearly applies to their own highly-salient work. Berlant remains one of the most influential (as well as one of the most enigmatic) recent Anglophone theorists of affect, neoliberalism, citizenship, gender, and sexuality. Berlant's posthumously-published *On the Inconvenience of Other People* (2022) picked up where *Cruel Optimism* left off, examining experiments in collective survival that improvise exit routes from cruel-optimistic attachments while avoiding the sentimental idealisation that characterises some Leftist visions of community. Berlant was in the midst of projects on flatness and humourlessness at the time of their passing, and both are anticipated in this book. Their collected papers are housed at the Feminist Theory Archive in the Pembroke Archives at Brown University Library.

Spatial Contributions

Berlant's work evinced their lifelong fascination with geography, and they once remarked that they would have majored in the field had Oberlin College offered it as a course of study in the 1970s (Anderson et al., 2023a: 135). Berlant corresponded with noted critical geographers including **Neil Smith** and **Ruth Wilson Gilmore**, gave an interview to *Society and Space* (Seitz and Berlant, 2013) and delivered the 2015 *Society and Space* Lecture at the American Association of Geographers Annual Meeting (published as Berlant, 2016), and was a founding member of the editorial board of *Emotion, Space, and Society* (see Berlant, 2008b). While Berlant's geographical legacy will continue to be debated for some time (see Anderson et al., 2023a, 2023b; McLean and Seitz, 2022), at least four key contributions are worth highlighting.

First, Berlant offered a robustly spatial and material account of the production of gender and sexuality. In 'Sex in Public', they argued with Michael Warner that heteronormativity is a publicly subsidised and ideologically articulated utopia, one that is:

> supported and extended by acts less commonly recognized as part of sexual culture: paying taxes, being disgusted, philandering, celebrating a holiday, investing for the future, teaching, disposing of a corpse, carrying wallet photos, buying economy size, being nepotistic, running for president, or owning anything "His" and "Hers" [...] We are describing a constellation of practices that everywhere disperses heterosexual privilege as a tacit but central organizing index of social membership. (Berlant and Warner, 1998b: 555)

Berlant's (2008a) account of white women's sentimental culture likewise emphasised that it is through the public, generic conventions of the personal that participants confide, confess, complain, cope, and shore up their resolve to live with the toll of social relations of heteropatriarchy.

Second, and relatedly, Berlant (2008a) insisted on the non-oppositional relationships between the particular and the general, the case study and its circulation. Ever mindful of the fact that 'anecdote is argument' (Berlant, 1994: 128), their work on case study methods (Berlant, 2007: 672) scrutinised the case as a contested site where the 'cool', detached quality of professional-class epistemologies meets the 'hot' instability of a seemingly singular event. Berlant's aim was not to trivialise case study methods, but to create 'an opening within realism, suggesting where it might travel' (Berlant, 2007: 669) by inviting more 'methodological reflection [*from scholars*] on why the work took *this* narrative shape given *that* archive' (Berlant, 2007: 671; original emphasis).

Third, Berlant's work on citizenship brought psychoanalysis and affect theory to bear on the nation-state, beginning from the deceptively simple premise that 'nations provoke fantasy' (Berlant, 1991: 1). Berlant's (1991: 5) model of citizen-subjectivity both allows for incoherence in the citizen and for geographical contingency in the production of national ideology, demonstrating how 'national culture becomes local – through the images, narratives, monuments, and sites that circulate through personal/collective consciousness'. Rather than a singular logic, they argued, national form was organised around 'many simultaneously "literal" and "metaphorical" meanings, meanings stated and unstated' (Berlant, 1991: 5). Without offering any easy exits from citizenship's double-binds, Berlant sought to slow down the reification of the nation-state by examining the aesthetic, material, and affective stitches that suture subject to citizen and nation to state.

Finally, Berlant's theorisations of affect invited Marxist critiques of ideology to better account for the vicissitudes of consciousness while rescaling psychoanalysis beyond the Oedipal family (see Nast, 2000), such that 'what looked like a conflict between institutional attachment to the world and intimate models of attachment are not [...] in conflict at all but are a part of the problem of imagining and living attachments to lifeworlds' (Seitz and Berlant, 2013: n.p.). Berlant (2011) insisted that it was by attending to shifts in *affect* – a deep, structured, shared sense of the present that might give rise to any number of conscious emotions and bodily states – that critics might make sense of how capitalist ideology works in an ongoing moment of political-economic impasse. Berlant's prescient diagnosis of liberals' cruel-optimistic idealisation of the Obama administration despite its manifest disappointments (Berlant, 2011) ultimately led them to correctly, and grimly, anticipate 'the Trumping of politics' (Berlant, 2012). Their sustained attention to affective double-binds enabled them to both champion efforts to break out cruel-optimistic relations, and to caution that such experiments need to go beyond sentimentality to flourish (Berlant, 2016).

Key Advances and Controversies

The notion of cruel optimism has inspired considerable engagement – and reformulation – from geographers of neoliberal capitalism (see Anderson et al., 2023b). Although Berlant was always modest about the potential for their work to travel beyond the US and Western European contexts that informed it (Berlant, 2011; Seitz and Berlant, 2013), their theorization of the present as a situation without a coherent genre has reverberated in case studies of economic and environmental precarity in a wide range of geographical contexts, both North and South (e.g., Cockayne, 2016; Salas Landa, 2016; Petit, 2020; Linz, 2021). Berlant has likewise been a foundational thinker for geographers of sexuality and gender, who have mapped the ambivalent articulation of feminist, multicultural, and queer utopias *with* neoliberal global capitalist ideologies (e.g., McGlotten, 2014; Luo, 2021).

Arguably, the richest geographical engagements with Berlant come from those who push back at their concepts (see Anderson et al., 2023a). Carter and Acker (2020: 1099) seek an exit from cruel-optimistic attachments to capitalism via the work of US speculative fiction novelist Kim Stanley Robinson, which imagines the formation of revolutionary infrastructures and an 'angry optimism' in the face of geopolitical and ecological disaster. Analysing anti-eviction organising in working-class Black Chicago, Roy (2017: A4) revises cruel optimism from the perspective of 'those who cannot claim possession, not even possession of personhood', shifting from cruel-optimistic anticipation of loss to a defiant attachment to home 'in the *lived* process of loss' (original emphasis). Such pushback has not been limited to studies of the Global North. Interviewing the creators of radical political tours of Hebron, Palestine, Griffiths (2017: 631) insists that 'hope, not hopelessness, remains a potential even amid a supposedly debilitating sensorium'. Thajib's (2017: 132) interviews with queer and trans Muslims in Aceh, Indonesia, point to 'hopeful attachments towards the unworldly' that facilitate 'different possibilities of engaging with the limiting material conditions that constitute ordinary life'.

Berlant's work on citizenship and sovereignty, with its emphasis on the incoherent materialisation of the nation in everyday life, has likewise found an enthusiastic but also critical audience from geographers (e.g., Cowen, 2008; Smith, 2020). Seitz (2017) identifies a resonance between Berlant's approach to citizenship and that of Isin (2002), whose genealogies remind contemporary audiences that citizenship's history is predominantly urban, rather than nationally scaled, and that local experiments in citizenship continue to proliferate, both within and beyond the nation-state. Much of Berlant's work on the breakdown of the sovereign subject in the space of sexual encounter focused on the scale of the individual (Berlant and Edelman, 2015). Yet *On the Inconvenience of Other People* (Berlant, 2022) begins to acknowledge the scalar limitations of that work, responding to scholars of Indigenous decolonization who emphasise the emancipatory possibilities of alternative forms of sovereignty at other scales (see Simpson, 2014; Grant, 2022). As Berlant's dialogues with geographers continue posthumously, it seems that the discipline's encounter with Berlant's thought is only just beginning.

Berlant's Key Works

Berlant, L. (1991) *The Anatomy of National Fantasy: Hawthorne, Utopia, and Everyday Life*. Chicago, IL: University of Chicago Press.

Berlant, L. (1994) '68, or something'. *Critical Inquiry*, 21: 124–55.

Berlant, L. (1997a) 'Feminism and the institutions of intimacy', in E.A. Kaplan and G. Levine (eds), *The Politics of Research*. New Brunswick: Rutgers University Press. pp. 143–61.

Berlant, L. (1997b) *The Queen of America Goes to Washington City: Essays on Sex and Citizenship*. Durham, NC: Duke University Press.

Berlant, L. (1999) 'The subject of true feeling: Pain, privacy, and politics', in S. Ahmed, J. Kilby, C. Lury, M. McNeil and B. Skeggs (eds), *Transformations: Thinking through Feminism*. London: Routledge. pp. 33–47.

Berlant, L. (2005) 'Unfeeling Kerry', *Theory & Event*, 8 (2). doi.org/10.1353/tae.2005.0021

Berlant, L. (2007) 'On the case', *Critical Inquiry*, 33: 663–72.

Berlant, L. (2008a) *The Female Complaint: The Unfinished Business of Sentimentality in America*. Durham, NC: Duke University Press.

Berlant, L. (2011) *Cruel Optimism*. Durham, NC: Duke University Press.

Berlant, L. (2022) *On the Inconvenience of Other People*. Durham, NC: Duke University Press.

Berlant, L. and Edelman, L. (2014) *Sex, or the Unbearable*. Durham, NC: Duke University Press.

Berlant, L. and Stewart, K. (2019) *The Hundreds*. Durham, NC: Duke University Press.

Berlant, L. and Warner, M. (1998) 'Sex in public', *Critical Inquiry*, 24: 547–66.

Secondary Sources and References

Anderson, B., Aitken, S., Bacevic, J., Callard, F., Chung, K.D.M., Coleman, K.S. [...] and Wilkinson, E. (2023a) 'Encountering Berlant part one: Concepts otherwise', *The Geographical Journal*, 189: 117–42.

Anderson, B., Awal, A., Cockayne, D., Greenhough, B., Linz, J., Mazumdar [...] and Williams, A. (2023b) 'Encountering Berlant part two: Cruel and other optimisms', *The Geographical Journal*, 189: 143–60.

Berlant, L. (1998) 'Intimacy: A special issue', *Critical Inquiry*, 24: 281–8.

Berlant, L. (2008b) 'Thinking about feeling historical', *Emotion, Space and Society*, 1: 4–9.

Berlant, L. (2012) 'The Trumping of politics', *Supervalent Thought*, 2 September, https://web.archive.org/web/20220125224409/https://supervalentthought.com/2012/09/02/the-trumping-of-politics/ (accessed October 27, 2023).

Berlant, L. (2016) 'The commons: Infrastructures for troubling times', *Environment and Planning D: Society and Space*, 34: 393–419.

Berlant, L. (2019) 'Why chasing the good life is holding us back, with Lauren Berlant', *Big Brains*, 4 November. https://news.uchicago.edu/podcasts/big-brains/why-chasing-good-life-holding-us-back-lauren-berlant (accessed October 27, 2023).

Berlant, L. (n.d.) 'Lauren Berlant FAQ', University of Chicago. https://english.uchicago.edu/faq-lauren-berlant (accessed October 27, 2023).

Berlant, L. and Duggan, L. (eds) (2001) *Our Monica, Ourselves: The Clinton Affair and the National Interest*. New York, NY: New York University Press.

Berlant, L. and Ngai, S. (2017) 'Comedy has issues', *Critical Inquiry*, 43: 233–49.

Carter, D. and Acker, A. (2020) 'To oblivion and beyond: Imagining infrastructure after collapse', *Environment and Planning D: Society and Space*, 38: 1084–100.

Cockayne, D.G. (2016) 'Entrepreneurial affect: Attachment to work practice in San Francisco's digital media sector', *Environment and Planning D: Society and Space*, 34: 456–73.

Cowen, D. (2008) *Military Workfare: The Soldier and Social Citizenship in Canada*. Toronto: University of Toronto Press.

Grant, S. (2022) 'Impasse, infrastructure, jurisdiction: Thinking with Lauren Berlant', *Antipode Online*, 31 October. https://antipodeonline.org/wp-content/uploads/2022/10/Berlant-forum_Soni-Grant.pdf (accessed October 27, 2023).

Griffiths, M. (2017) 'Hope in Hebron: The political affects of activism in a strangled city', *Antipode*, 49: 617–35.

Isin, E.F. (2002) *Being Political: Genealogies of Citizenship*. Minneapolis, MN: University of Minnesota Press.

Linz, J. (2021) 'Where crises converge: The affective register of gentrification and displacement in post-earthquake Mexico City', *Cultural Geographies*, 28: 285–300.

Luo, M. (2021) 'Circular, transitory, permanent: State and migration pathways among the intranational migrant gay men in China', *Gender, Place and Culture*, 28: 326–46.

McGlotten, S. (2014) 'A brief and improper geography of queerspaces and sexpublics in Austin, Texas', *Gender, Place and Culture,* 21: 471–88.

McLean, H. and Seitz, D.K. (2022) 'Symposium: Fractured kinship and failed utopias: Remembering the brilliant Lauren Berlant', *Antipode Online*, 31 October. https://antipodeonline.org/2022/10/31/remembering-lauren-berlant/

Nast, H.J. (2000) 'Mapping the "unconscious": Racism and the Oedipal family', *Annals of the Association of American Geographers*, 90: 21555.

Pettit, H. (2020) 'Uncomfortable ethnography: Navigating friendship and "cruel hope"with Egypt's disconnected middle-class', *Emotion, Space and Society*, 36: 100714.

Roy, A. (2017) 'Dis/possessive collectivism: Property and personhood at city's end', *Geoforum*, 80: A1–A11.

Salas Landa, M. (2016) 'Crude residues: The workings of failing oil infrastructure in Poza Rica, Veracruz, Mexico', *Environment and Planning A,* 48: 718–35.

Sedgwick, E.K. (1990) *Epistemology of the Closet*. Berkeley, CA: University of California Press.

Sedgwick, E.K. (2003) *Touching Feeling: Affect, Pedagogy, Performativity*. Durham, NC: Duke University Press.

Seitz, D.K. (2017) *A House of Prayer for All People: Contesting Citizenship in a Queer Church*. Minneapolis, MN: University of Minnesota Press.

Seitz, D.K. and Berlant, L. (2013) 'On citizenship and optimism: Lauren Berlant, interviewed by David Seitz'. *Society and Space,* 23 March. https://www.societyandspace.org/articles/on-citizenship-and-optimism (accessed October 27, 2023).

Simpson, A. (2014) *Mohawk Interruptus: Political Life Across the Borders of Settler States*. Durham, NC: Duke University Press.

Smith, S. (2020) *Political Geography: A Critical Introduction*. Hoboken, NJ: Wiley-Blackwell.

Thajib, F. (2017) 'Kaleidoscopic feelings: Faith narratives among Indonesian Muslim queers', *Emotion, Space and Society*, 25: 127–35.

8
Nicholas Blomley

Laam Hae, York University
Reecia Orzeck, Illinois State University

Biographical Details and Theoretical Context

Nick Blomley was born in 1962 in a village near Reading, Berkshire, in the UK. Blomley's interest in law and space began with his PhD work at Bristol University in the mid-1980s – on the role of the *Shops Act* (1950) and the legalisation of Sunday trading in England and Wales. It is no overstatement to say that since then, Blomley, along with a handful of other scholars, has brought the sub-discipline of legal geography into existence, foregrounding the role of law in the making of space and place, and the role of ideas about space and place in the shaping of laws and legal decisions. Nor is it hyperbole to say that Blomley has, through his regular contributions to the sub-discipline, played an outsized role in maintaining its liveliness. While developing his interventions about law, space, and geography, Blomley has also made significant contributions on the themes of property and territory, thereby also contributing to political geography, urban geography, and critical legal studies.

Blomley has, since the start of his career, focused his attention on a wide range of legal topics, from street-level mobilities of the homeless through to questions of gentrification and urban development. His monographs include *Law, Space, and the Geographies of Power* (1994), *Unsettling the City* (2004a), *Rights of Passage* (2011), and *Territory* (2022). Despite this thematic variety, however, a few overarching concerns can be identified in most of his writings. First is a concern with understanding the ways in which law, space, and society are mutually constitutive of one another. Second is an interest in demonstrating the ways in which the law or legal relationships are, despite their appearances to the contrary, fluid, contingent, and unstable. For Blomley, the importance of scholarship that brings this to light is not merely academic, as an understanding of how the law is brought to life through interpretation allows us to see how it can be challenged when necessary. A final pair of concerns central to Blomley's scholarship has been the coercive and disenfranchising ways that law and legal relations have been wielded by powerful social groups, as well as the ways that these have been challenged, unsettled, or reworked by marginalised groups.

These concerns may well be traceable to Blomley's biography. Indeed, Blomley writes eloquently in his preface to *Unsettling the City*

(2004a: xiii) about growing up in a small village in England that had been 'forged through complex and often violent political struggles and legal remakings' despite its 'old' appearance, and which was then being remade by gentrification processes that had deep implications for long-standing residents. And certainly, many of the issues to which Blomley has turned his attention to over the years are relevant to or manifested in the particular places he has lived – the England of his birth and Vancouver, Canada, to which he moved in 1988 to take up an initially temporary lecturing position in Geography at Simon Fraser University.

But if biography has played a role in the trajectory of Blomley's interests, so too has the academic context within which he began and has continued his career. As several scholars have noted, legal geography's approach to theorising space at the time of its emergence was aligned with approaches that were on the rise in the discipline of geography (Chouinard, 1994; Forest, 2000; Blomley, 2008; Orzeck and Hae, 2020). Here, as elsewhere in critical social science, the emergence of a post-structuralist challenge to Marxism resulted in scholarship that, among other things, took language and discourse to be important nodes through which power moved and was contested, rejecting too-muscular accounts of structures in favour of accounts that preserved a place for human agents as effective combatants against structural injustices (e.g., see the entry on **J.K. Gibson-Graham**).

Writing at this juncture, early proponents of critical legal geography, Blomley among them, were critical of what they saw as Marxist scholars' tendency to reduce law to a simple outcome or reflection of structural forces. They were likewise critical of the tendency of mainstream depictions of the law to present it as closed off from society. Self-consciously informed by then-current socio-cultural theories, especially those concerned with meaning and interpretation and those that disavowed foundationalism and reductionism (e.g., see entries on **bell hooks**, **Stuart Hall, Michel Foucault**), legal geographers writing in the late 1980s and early 1990s emphasised the multi-directional ways in which law, space, and society shaped one another, and brought to light the highly contextual and unfinished nature of law. These and other early ideas have endured as central themes within the sub-discipline, even as legal geography has evolved to include many more participants and topics (Orzeck and Hae, 2020).

Spatial Contributions

Blomley has made important contributions to geographers' and legal scholars' way of understanding what he and others have called the 'law-space nexus' – the intersection of law and space. The most important, perhaps, is the notion that law, space, and society are mutually constitutive. Blomley and Clark (1990) argued that this was an important corrective to both perceived Marxist reductionism and mainstream claims of what was often referred to as 'legal closure' – the presentation of law as if it were removed from society. This idea was also advanced in Blomley's 1994 monograph *Law, Space, and the Geographies of Power*. In that text, Blomley (1994: 28) identified two relevant literatures, one 'concerned with mapping the regional diversity of law' which tended to see 'space [...] as a vital structuring agent of law', and one that tended to see 'law as an external variable that affects spatial configurations'. In contrast to these, Blomley (1989) noted an emerging body of scholarship attempting to 'transcend the causality' of earlier accounts, demonstrating how the legal, the spatial, and the social are mutually imbricated (see also Blomley and Clark, 1990; Delaney, 2004). This argument has become a key and enduring leitmotif of legal geography (see Bartel et al., 2013; Bennett and Layard, 2015; Delaney, 2015). To help scholars to 'think beyond binary categories such as "space" and "law"' (Blomley, 2003a: 29–30), Blomley later developed the idea of 'splices' to capture the conjunction of legal and spatial categories and representations.

A related argument is that there is a disconnect between how law and legal relationships appear to us – as settled, static, authoritative – and their reality on the ground. Blomley has demonstrated this in many ways throughout his career, most enduringly through the example of property, and more particularly through what he and others call the 'ownership model' of property. According to Blomley, while the formal view of a category like property insists on its coherence and stability, it is more correct to understand it as a kind of performance, with no existence 'outside its performances and enactments' (Blomley, 2015: 168), one whose meanings and parameters are more aspirational than absolute. The ownership model of property, in which one way of doing property has come to stand in for all possibilities, is similarly more ephemeral than it first seems. Whereas the ownership model makes it seem as though property must always be either private or public, and that private property must always be individualised, unambiguous, and clearly-bounded, Blomley (2016a: 96) unearths a world of exceptions that prove that property is 'a far more commodious category' than the ownership model suggests (see also Blomley, 2004b).

Blomley has added to this argument by tracing the importance of territory to the historical emergence of the ownership model of property (Blomley, 2016b; 2019; 2022). Much like property, he insists territory is not natural or given, but an outcome of social relations and 'a *means* by which such relations are enforced, naturalized and contested' (2022: xiii, original emphasis). In his work on territory, Blomley has brought to light the consequences of the ownership model of property for indigenous peoples, noting its role in the advancement and entrenchment of settler colonialism. Blomley has explored, for example, the roles that cadastral surveys, maps, and grids have played in conquering and commodifying indigenous lands (2003b; 2004a; 2019). He has also written on land treaty negotiations between Indigenous communities and the provincial (British Columbian) and federal (Canadian) state. In these negotiations, the state attempted to transform Indigenous treaty settlement lands into a form of 'fee simple' ownership – the principal form of property relations under Canadian common law (Blomley, 2014; 2015). Here, Blomley paid close attention not only to the rationalisations of the state, but also to the response of the First Nations negotiators who attempted to destabilise and open up the reified and 'bracketed' version of 'fee simple' advanced by their state counterparts and to advocate instead for a version of fee simple more compatible with Indigenous interests.

This example brings us to the final two major arguments found in Blomley's oeuvre. One is that the law, along with legal relationships like property, is often an instrument of disenfranchisement and marginalisation. Throughout his career, Blomley has explored and brought to light many examples of this, focusing on the impact of the law on, among others, homeless and precariously housed people (Blomley, 2007a; 2009; 2011; Blomley et al., 2020), showing how mobility is circumscribed for these populations through their relationship to the court system or the police (Blomley, 2012; Sylvestre et al., 2019). Likewise, he has focused on those dispossessed of property or community resources through gentrification or outright expropriation (Blomley, 2004a; 2007b; 2016a; 2020; Stanger-Ross et al., 2017). Among the insights that Blomley and some of his co-authors have advanced in this context is that marginalised people are not outside of the law, but are both constituted, and made vulnerable, by it (e.g., Blomley, 2009; Blomley and the Right to Remain Collective, 2021). Thus, for example, and in an argument that has been important in studies of gentrification, the status of being dispossessed through gentrification does not result from an absence of law but precisely from the enactment of property law.

Finally, a key argument of Blomley's that unites some of his earliest writings with some of his most recent is that there is value in scholarship that exposes the contingency and the performative nature of the law and legal relationships because these realisations allow us to see opportunities for agency that we previously could not. He suggests that, despite the aspirational nature of the law's self-presentation as closed and absolute, we can 'rework' narrow understandings of property or other legal relationships by '[drawing] upon suppressed meanings and practices' (Blomley, 2004a: 22) and engaging in alternative interpretations and performances. And indeed, Blomley's writings offer both subtle and dramatic examples of people unsettling and challenging legal relationships through such alternative interpretations. This is especially the case in his work on Vancouver's Downtown Eastside and its residents' fight for the collective ownership of their neighbourhood. For Blomley, residents' practical and discursive 'enactments' of space – including 'protest, anti-gentrification walking tours, poetry, occupations and squats, and everyday acts of presence, reiterated use, and engrained inhabitation' (Blomley, 2016a: 98–9) – are a means of communing property through performance. These performances, Blomley (2016a: 99) has argued, 'produce the "effect" that is common property'.

Key Advances and Controversies

As several scholars have noted, legal geography is a field with only a handful of regular contributors, Blomley chief among them (Blomley, 2008; Delaney, 2010; Braverman et al., 2014). Yet legal geographers have examined topics as varied as the geopolitics of outer space (Collis, 2009), legal pluralism at the beach (Hubbard, 2020) and the agency of non-human actants (Ojalammi and Blomley, 2015; Bartel, 2018). Several scholars have argued that legal geography would benefit from utilising a wider repertoire of methodologies, including ethnography (Braverman, 2014; Cuomo and Brickell, 2019). Others have suggested that legal geography would benefit from paying closer attention to legal practitioners such as lawyers and formal legal spaces such as courtrooms (Martin et al., 2010; Jeffrey, 2019) or to the work and concerns of the administrators and engineers who design particular spaces (Blomley, 2011). Still others have suggested that legal geography might be fruitfully integrated with other subdisciplines in Geography, such as political ecology (Andrews and McCarthy, 2014; Kay, 2016), global economic geography (Barkan, 2011), and feminist geography (Brickell and Cuomo, 2019; Cuomo and Brickell, 2019).

More fundamental critiques of legal geography – and Blomley's role in providing the field with its conceptual framing – have also circulated. Chouinard, for example, took issue with the inaccurate way that some legal geography scholars described Marxist approaches to law, suggesting that legal geographers' unflattering representations of Marxist scholarship were derivative of those that had populated the discipline of geography since the early 1980s when 'the representation of Marxism as a static "dead end" was compelling, if not mesmerizing' (Chouinard, 1994: 422). Yet Chouinard applauded legal geography for, among other things, emphasising the ongoing and open-ended nature of 'law-making and implementation' and the actual distance between abstract principles of justice and the local contexts in which they are always rolled out. Chouinard raised concerns, however, about how well a post-structuralist approach to law and geography could serve the radical project of securing real change, given the discursive way in which that approach seemed to imagine both law itself and challenges to it. In Chouinard's estimation, legal geography's tendency to see law as something essentially discursive limited its effectiveness as a tool of radical change.

While legal geography has become far less disparaging of Marxist approaches to law and Marxist concerns, and while scholars involved in the field of legal geography have arguably devoted significant attention to the deleterious impacts of capitalism, scholars have continued to identify a problem, or perhaps an absence, within legal geography when it comes to political-economic or materialist approaches to understanding the social world (Orzeck and Hae, 2020). For example, in an article making a case for the usefulness of **Henri Lefebvre's** writings, which allow us to conceptualise law 'as a tool of the state in its strategies of spatial production' *and* as 'a contested body of spatial representations', Butler (2009: 322–4) warns legal geographers against their 'tendencies to reduce space to a linguistic model and conceive of it as a metaphorical source of indeterminacy and social contingency'. In a similar spirit, Jepson (2012) has critiqued legal geographical scholarship for fetishising legal discourse in a way that misses their actual dynamism, and for not sufficiently grappling with the political-economic interests behind particular legal narratives and claims. In our own critique of legal geography we argued that in spite of the ways in which legal geography has evolved in the years since its founding, it still has a 'contingency orientation' – a tendency to focus on moments of openness and flexibility in law instead of moments of its fixity, and a tendency to focus on the agency of legal actors as opposed to the ways in which that agency is constrained (Orzeck and Hae, 2020). The risk of this contingency orientation is that it may lead us to mis-assess the power of discursive and performative efforts to rework or unsettle law and legal relationships.

Blomley cannot, however, be accused of failing to consider the role of law as a tool of power exercised against marginalised peoples. Nor can he be accused of failing to consider the nefarious impacts of racism, settler colonialism, and capitalism. The critiques of legal geography that have emerged seem rather to eddy around legal geographers' choice of conclusions to draw from their empirical research. Despite the bleakness of the injustices that they often illuminate, scholarly conclusions are often relatively sanguine, highlighting as they do people's agency in the face of significant obstacles. Clarifying what *types* of agency and what types of actions are effective in what contexts would arguably deepen the sub-discipline's contribution to understanding how spaces and legal relations are made and how they might most effectively be remade. With more than thirty years of scholarship behind him and an entire sub-discipline created and devotedly maintained, this task is not one for Blomley, who has already done more than any single scholar's share. Indeed, it is because of Blomley's path-breaking work that there is now a dedicated place in the discipline of Geography for research and conversations of all types about the relations between law, society and space.

Blomley's Key Works

Blomley, N. (1994) *Law, Space, and the Geographies of Power*. New York, NY: The Guilford Press.

Blomley, N. (2004a) *Unsettling the City: Urban Land and the Politics of Property*. New York, NY: Routledge.

Blomley, N. (2011) *Rights of Passage: Sidewalks and the Regulation of Public Flow*. New York, NY: Routledge.

Blomley, N. (2014) 'Making space for property', *Annals of the Association of American Geographers*, 104: 1291–306.

Blomley, N. (2022) *Territory: New Trajectories in Law*. New York, NY: Routledge.

Blomley, N. and Clark, G. (1990) 'Law, theory and geography', *Urban Geography*, 11: 433–46.

Secondary Sources and References

Andrews, E. and McCarthy, J. (2014) 'Scale, shale, and the state: political ecologies and legal geographies of shale gas development in Pennsylvania', *Journal of Environmental Studies and Sciences*, 4: 7–16.

Barkan, J. (2011) 'Law and the geographic analysis of economic globalization', *Progress in Human Geography*, 35: 589–607.

Bartel, R. (2018) 'Place-speaking: Attending to the relational, material and governance messages of Silent Spring', *The Geographical Journal*, 184: 64–74.

Bartel, R., Graham, N., Jackson, S., Jason, Prior, J.H., Robinson, D.F., Sherval, M. and Williams, S. (2013) 'Legal geography: An Australian perspective', *Geographical Research*, 5: 339–53.

Bennett, L. and Layard, A. (2015) 'Legal geography: Becoming spatial detectives', *Geography Compass*, 9: 406–22.

Blomley, N. (1989) 'Text and context: Rethinking the law-space nexus', *Progress in Human Geography* 13: 512–34.

Blomley, N. (2003a) 'From "what?" to "so what?": Law and geography in retrospect', in J. Holder and C. Harrison (eds), *Law and Geography*. Oxford: Oxford University Press. pp. 17–33.

Blomley, N. (2003b) 'Law, property, and the geography of violence: The frontier, the survey, and the grid', *Annals of the Association of American Geographers*, 93: 121–41.

Blomley, N. (2004b) 'Un-real estate: Proprietary space and public gardening', *Antipode*, 36: 614–41.

Blomley, N. (2007a) 'How to turn a beggar into a bus stop: Law, traffic and the "function of the place"', *Urban Studies*, 44: 1697–712.

Blomley, N. (2007b) 'Legal geographies – *Kelo*, contradiction, and capitalism', *Urban Geography*, 28: 198–205.

Blomley, N. (2008) 'Making space for law', in K.R. Cox, M. Low and J. Robinson (eds), *The SAGE Handbook of Political Geography*. London: Sage. pp. 155–67.

Blomley, N. (2009) 'Homelessness, rights, and the delusions of property', *Urban Geography*, 30: 577–90.

Blomley, N. (2012) 'Colored rabbits, dangerous trees, and public sitting: Sidewalks, police and the city', *Urban Geography*, 33: 917–35.

Blomley, N. (2015) 'The ties that blind: Making fee simple in the British Columbia treaty process', *Transactions of the Institute of British Geographers*, 40: 168–79.

Blomley, N. (2016a) 'The right to not be excluded: Common property and the struggle to stay put', in A. Amin and P. Howell (eds), *Releasing the Commons: Rethinking the Futures of the Commons*. New York: Routledge. pp. 89–106.

Blomley, N. (2016b) 'The territory of property', *Progress in Human Geography*, 40: 593–609.

Blomley, N. (2019) 'The territorialization of property in land: Space, power, and practice', *Territory, Politics, Governance*, 72: 233–49.

Blomley, N. (2020) 'Urban commoning and the right not to be excluded', in D. Ozkan and G.B. Buyuksarac (eds), *Commoning the City*. New York, NY: Routledge. pp. 89–103.

Blomley, N., Flynn, A. and Sylvestre, M.-E. (2020) 'Governing the belongings of the precariously housed: A critical legal geography', *Annual Review of Law and Social Science*, 16: 165–81.

Blomley, N. and Right to Remain Collective (2021) 'Making property outlaws: Law and Relegation', *International Journal of Urban and Regional Research*, 45: 911–29.

Braverman, I. (2014) 'Who's afraid of methodology? Advocating a methodological turn in legal Geography', in I. Braverman, N. Blomley, D. Delaney and A. Kedar (eds), *The Expanding Spaces of Law: A Timely Legal Geography*. Stanford, CA: Stanford University Press. pp. 120–41.

Braverman, I., Blomley, N., Delaney, D. and Kedar, A. (2014) 'Introduction: Expanding the spaces of law', in I. Braverman, N. Blomley, D. Delaney and A. Kedar (eds), *The Expanding Spaces of Law: A Timely Legal Geography*. Stanford, CA: Stanford University Press. pp. 1–29.

Brickell, K. and Cuomo, D. (2019) 'Feminist geolegality', *Progress in Human Geography*, 43: 104–22.

Butler, C. (2009) 'Critical legal studies and the politics of space', *Social & Legal Studies*, 18: 313–32.

Chouinard, V. (1994) 'Geography, law, and legal struggles: Which ways ahead?' *Progress in Human Geography*, 18: 415–40.

Collis, C. (2009) 'The geostationary orbit: A critical legal geography of space's most valuable real estate', *The Sociological Review*, 57: 47–65.

Cuomo, D. and Brickell, K. (2019) 'Feminist legal geographies', *Environment and Planning A*, 51: 1043–9.

Delaney, D. (2004) 'Tracing displacements: Or evictions in the nomosphere', *Environment and Planning D: Society and Space*, 22: 847–60.

Delaney, D. (2010) *The Spatial, the Legal and the Pragmatics of World-Making: Nomospheric Investigations*. New York, NY: Routledge.

Delaney, D. (2015) 'Legal geography I: Constitutivities, complexities, and contingencies', *Progress in Human Geography*, 39: 96–102.

Forest, B. (2000) 'Placing the law in geography', *Historical Geography*, 28: 5–12.

Hubbard, P. (2020) 'Legal pluralism at the beach: Public access, land use, and the struggle for the "coastal commons"', *Area*, 52 (2): 420–8.

Jeffrey, A. (2019) 'Legal geography 1: Court materiality', *Progress in Human Geography*, 43: 565–73.

Jepson, W. (2012) 'Claiming space, claiming water: Contested legal geographies of water in South Texas', *Annals of the Association of American Geographers*, 102: 614–31.

Kay, K. (2016) 'Breaking the bundle of rights: Conservation easements and the legal geographies of individuating nature', *Environment and Planning A*, 48: 504–522.

Martin, D.G., Scherr, A.W. and City, C. (2010) 'Making law, making place: Lawyers and the production of space', *Progress in Human Geography*, 34: 175–92.

Ojalammi, S. and Blomley, N. (2015) 'Dancing with wolves: Making legal territory in a more-than-human world', *Geoforum*, 62: 51–60.

Orzeck, R. and Hae, L. (2020) 'Restructuring legal geography', *Progress in Human Geography*, 44: 832–51.

Stanger-Ross, J., Blomley, N. and the Landscapes of Injustice Research Collective (2017) '"My land is worth a million dollars": How Japanese Canadians contested their dispossession in the 1940s', *Law and History Review*, 35: 711–51.

Sylvestre, M.-E., Blomley, N. and Bellot, C. (2019) *Red Zones: Criminal Law and the Territorial Governance of Marginalized People*. Cambridge: Cambridge University Press.

9
Pierre Bourdieu

Gary Bridge, Cardiff University

Biographical Details and Theoretical Context

Pierre Bourdieu was born in 1930 in a small village in south-west France. When he finished high school (as a star pupil and rugby player) he moved to Paris where he studied philosophy at the elite École Normale Supérieure. After graduation, he worked as a teacher for a year before being drafted into the army to fight in the Algerian War between 1956 and 1958. His experiences in Algeria (a conflict over French colonialism that he subsequently described as 'an appalling war') turned him away from philosophy and towards sociology and anthropology: he remained in Algeria after the conflict to teach at the University of Algiers while conducting fieldwork among the Kabyle people of the North-East region. In 1960 he returned to France and held posts at the University of Paris-Sorbonne and the University of Lille, becoming Director of Studies at the École des Hautes Études en Sciences Sociales in 1964. In 1968 he launched the Centre for European Sociology and later established the journal *Actes de la Recherche en Sciences Sociales*, devoted to understanding cultural reproduction and class domination. Bourdieu was elected to a Chair at the prestigious College de France in 1981. He died in 2002.

Over his distinguished career Bourdieu published over forty books and more than 400 articles. These ranged from anthropological texts – such as *The Algerians* (1962) – and related theoretical contributions – most notably, *Outline of a Theory of Practice* (1977a) – through contributions to the understanding of class reproduction (*Distinction: A Social Critique of the Judgement of Taste*, 1984), the sociology of education (*Reproduction in Education, Society and Culture*, 1977b; *The Inheritors: French Students and Their Relation to Culture*, 1979) and action in everyday life (*Practical Reason*, 1998). Bourdieu also critiqued the politics of globalisation and neoliberalism through the voices of the dispossessed (*The Weight of the World*, 1999a), while *The Social Structures of the Economy* (2005) stressed the significance of local circumstances against the discourse of a 'universal' market.

Bourdieu's leading theoretical claim was that his work transcended the dualism between explanations that attributed social change to over-arching structures, and theorisations privileging individual subjective intention or experience. He drew on existentialism to emphasise the importance of practical action and existence (*praxis*) whilst simultaneously rejecting the subjectivist elements of Sartre's existentialism depicting authentic existence as an act of self-creation.

Here, Bourdieu's work shared affinities with US pragmatism (especially George Herbert Mead and John Dewey) in his social conception of agency and emphasis on the body as a site of socially-instilled habit (see Aboulafia, 1999; Bridge, 2005). This response to the dualism of objectivism/subjectivism constituted Bourdieu's 'theory of practice', an idea that ongoing, practical everyday activities – articulated through bodies – contain prior cultural dispositions that weigh upon, but do not define, ongoing actions. He reasoned these ways of acting encompassed deployments of different forms of capital. Here, Bourdieu's innovation was to argue that forms of capital include cultural assets: e.g., the ability to 'consume' rare objects, such as works of art, might demand skills of taste and appreciation derived from family background and/or via institutionalised education (hence his interest in the sociology of education).

'Cultural capital', as defined by Bourdieu, involves embodied dispositions and resources, including competences incorporated via education, as well as symbolic power, which is the power of a certain class to legitimise their 'natural' tastes and so dominate the social order. Cultural and symbolic capital has relations with social capital (based on social networks) and economic capital. Different combinations of capital operate in different 'fields' (economic, political or artistic). If social life is thought of as a 'game' (an analogy Bourdieu used often) then fields are the playing board. Embodied dispositions provide the 'feel for the game;, which cannot be explained wholly by the game's rules. Crucially, Bourdieu argued that the feel for the game varies over social space (a term he used to suggest a 'force field' of conflicting social power). Moreover, he suggested strategies tend to coalesce in relatively enduring combinations and embodied dispositions termed *habitus*. A crucial concept at the heart of Bourdieu's sociology, habitus is an array of inherited dispositions that condition bodily movement, tastes and judgements, according to class position (Bourdieu, 1984).

Spatial Contributions

While there was a paucity of explicit references to space or place in Bourdieu's work, the implicitly spatial aspects of his work as well as the use of Bourdieu's work in explicitly spatial contexts opened a rich vein of research. Beyond the direct influence of his early anthropology of the Kabyle, and later theorisations of practice relating to 'social' time and space, there is a more diffuse, but no less pervasive, set of influences rippling out from the idea of the body as strategy of capital accumulation (for example, the idea of the body as the locus of class reproduction).

However, as Painter (2000) pointed out, the take-up of Bourdieu's ideas in geography has been patchy. That is not to say that he has not been frequently cited, but sustained and direct engagement of geographers with Bourdieu's work has proved surprisingly rare. There were some early skirmishes, most notably Parkes and Thrift's (1980) use of the idea of 'social time' from Bourdieu's study of the Kabyle people in Algeria, a term Bourdieu used to describe how the cultural and economic rhythms of Kabyle life constituted time in that society. Bourdieu's analysis of the times and spaces of the Kabyle house also provided **David Harvey** (1989) with an example of the structuring of space/time that was non-capitalist. For Harvey, the Kabyle example shows time and space can be orientated socially around the body in ways that have deep cultural resonance. Harvey subsequently questioned the degree to which this socio-cultural time/space could survive the increasingly global reach of capitalism – an ordering of time and space that relies on the abstraction of place in accordance with the logic of capital accumulation.

This vague definition of habitus as some kind of authentic, culturally-rich, 'place', resisting what **Henri Lefebvre** called the 'abstract space' of capitalism, accounts for many citations of Bourdieu that make no direct engagement with his ideas. Against this, there has also been explicit exploration of the links between habitus and place which, although not located centrally in geography, involves geographers and urbanists. For example, the edited collection by Jean Hillier and Emma Rooksby (2002) – *Habitus: A Sense of Place* – asked if the idea of habitus remained relevant in a rapidly-globalising world. The book tackled this question at different scales, from national and sub-national, exploring their relationship to symbolic capital and domination through discussions of post-colonialism and habitus as resistance. Specifically geographical contributions included Painter's drawing together of the idea of habitus and **Michel Foucault's** notion of governmentality in the discursive construction of place (Painter, 2002), and Pile's exploration of the unpredictable effects of the *uncanny* in the formation of habitus as sense of place through ideas of ghostliness and places being haunted (Pile, 2002).

Recognising another important departure, Painter (2000) argues that **Nigel Thrift** took some inspiration from Bourdieu's definition of practice as a means of 'going on in the world' but pursued it though performative modes of knowing. In Thrift's (2007) non-representational geography, practice, practical sense and 'a feel for the game' can never be fully captured and represented. If that is acknowledged, then the spatialities of practice raise the possibilities of other ways of being in the world. Spatialities of practice were taken in another direction by **Tim Cresswell** who shows that the geographical 'facts of life' (place, territory, landscape) are constitutive of social life because they relate the body to space through habitus. Cresswell (1996) demonstrates the power of these geographical ordering devices by exploring reactions to their transgression (for instance, by New York graffiti artists, Greenham Common peace protestors and new age travellers). Emphasising geographies of mobility, Cresswell (2002) describes Bourdieu's influence on his work in a glowing appreciation of his legacy.

The concept of habitus has, in more recent times, been most evident in the analysis of urban space and gentrification, where a range of studies have used Bourdieu's work to explain socio-spatial changes in the city. For example, Butler and Robson (2003) argue for a distinct 'metropolitan habitus' of middle-class gentrifiers in London, noting how distinct neighbourhoods house different middle-class fractions who defend their class habitus (e.g., via schooling). Other work stresses the importance of aesthetics in 'loft living' (Podmore, 1998), artistic practice in gentrification aesthetics (Ley, 2003), constructions of cosmopolitanism (May, 1996; Hage, 2000), the aesthetics of middle-class gentrifiers (Wynne and O'Connor, 1998), the impacts of regeneration on working-class habitus (Allen, 2008), and the relationship between cultural capital and various components of cultural capital (especially education and aesthetics) in housing (see Bridge, 2001a, b; 2003; 2006a, b).

The relationship between space, social trajectories and identity was also investigated comprehensively from a Bourdieusian perspective in a large-scale research programme at the UK Centre for Research on Socio-Cultural Change (CRESC) which conducted a comprehensive neighbourhood 'habitus mapping' to investigate the changing nature of class relations, culture and identity in Britain (Bennett et al., 2009). As well as recording the contemporary relevance of different forms of capital in a geographical context this developed new ideas of habitus, such as the significance of 'elective belonging' as part of the social/spatial trajectories of the middle-class (Savage et al., 2005). Work on middle-class habitus continued in a large comparative study of Paris and London looking at the relations of social, cultural and economic capital in a range of different middle-class neighbourhoods from the urban core to peri-urban locations (Bacqué et al., 2015).

Bourdieu's ideas also gained currency in social and cultural geography as part of the rising concern with embodiment in the 1990s and beyond. Here, Foucault's work on the body as the site of the operation of disciplinary power, Merleau-Ponty's (1962) figuring of the body as locus of primary sociality, and Lefebvre's conceptualisation of the body as the locus of expressive 'lived space' were absorbed within Bourdieu's thinking as part of wider discourses about the body as a class/cultural process (Dowling, 1998). For example, Holt (2008) productively combined Bourdieu's ideas of embodied social capital with **Judith Butler's** performativity theory to investigate the role of social capital in young persons' experiences of disability, while O'Brien (2021) used Bourdieu, alongside Lefebvre, to understand deaf spatial experience, and **Gill Valentine** and Joanna Sadgrove (2012) drew on Bourdieu to theorise prejudice against marginalised social groups.

Key Controversies and Advances

There are long-standing and ongoing debates about the degree to which Bourdieu's canon transcends the dualism of objectivism/subjectivism (see Honneth, 1986; Jenkins, 1992). Equally, his idea of habitus has come under fire for being too 1960s, too limited to the French social structure and too static. Bourdieu also had a complicated, and often controversial relationship with feminist scholarship, but his ideas on the importance of embodiment have been developed by feminist scholars, taking the idea of habitus beyond class to explore gendered relations and embodied experience (e.g., Adkins and Skeggs, 2005).

Such questions of embodied inequalities and injustices are also a part of a long-standing concern in human geography with spatial justice. Israel and Frenkel (2018), for example, combined Amartya Sen's capability approach to justice with Bourdieu's understanding of field, capital and habitus, providing a conceptual framework for understanding injustices. Addressing injustices was very much the spirit, as well as the conceptual letter, of much of Bourdieu's work, evident in the passionate descriptions of disadvantage, marginalisation and dispossession in *The Weight of the World* (1999a).

In political geography, Bourdieu's idea of the bureaucratic field has influenced thinking about transnational bureaucracies and the growing interest in 'spatiality beyond the state territory' with a focus on geopolitical practices in the formulation and implementation of transnational policies (Kuus, 2015: 443). The idea of 'field' is seen as especially helpful here because of its open-ended spatial frame, bringing an understanding of how practices are bound together in relations at various scales.

Economic geography has long-acknowledged the role of culture in economic processes (Amin and Thrift, 2007). Spigel (2017) sees Bourdieu's ideas on cultural reproduction, used alongside contemporary relational and performative perspectives in economic geography, as a way of overcoming individual rational choice or macro-structural approaches. In his study of the role of culture in entrepreneurship and innovation, Spigel uses Bourdieu's idea of the relationship between social fields and individual's interpretations of those fields through habitus as a way of understanding entrepreneurial activity within multiple fields at different scales to build a new understanding of regional economies.

Over the last decade or so there has been a renewed effort to demonstrate Bourdieu's relevance for understanding the city and conceptions of class more widely. Savage (2011) points to the spatial elements of Bourdieu's early work on the disruptive transformations of rural social relations through the effects of urbanisation in France (Bourdieu, 2008) and in his later work on social suffering (Bourdieu, 1999a) and the social structures of the housing market (Bourdieu, 2005). Savage argues that Bourdieu's concept of 'field' as a structured set of positions can be understood in terms of the relational space that intermediates symbolic, social, and physical space. For Savage, field analysis can connect research on cities of flows with more conventional studies of class and stratification in the city.

Bourdieu's significance as a key thinker on urban space has been argued most forcefully by the sociologist and erstwhile collaborator with Bourdieu, Loïc Wacquant. Wacquant (2023: xi–xii) sees the city as a privileged site for Bourdieusian analysis as it is the 'accumulation, diversification and contestation of capitals and the terrain for the commingling and collision of variegated habituses'. Wacquant (2008; 2009) has traced what he calls the trialectic of symbolic space (topography of cognitive categories), social space (distributions of capital) and physical space, arguing that neoliberal forms of social/symbolic disparagement, involving social and racial dispossession, produce the 'advanced marginality' of certain urban neighbourhoods. These conditions are entrenched through 'territorial stigmatisation' (in which the institutions of the state are complicit), producing a continuity between segregation and incarceration as the prison becomes an urban institution. Two themed issues of the *International Journal of Urban and Regional Research* on 'Bourdieu Comes to Town", facilitated by Wacquant and Savage, drew contributions from urban sociologists applying Bourdieu's ideas to a range of cities in the Global North and South (for introductions see Wacquant, 2018; Savage 2021). These ideas have influenced urban geography in various ways (see, for example, Peck and Theodore, 2008; Slater, 2018; Sisson, 2021).

Savage's, Wacquant's and other recent contributions demonstrate that the full range of Bourdieu's concepts – distinction, field, forms of capital, habitus – still resonate in thinking about space and place. Wacquant (2023) suggests that rather than using Bourdieu's conceptual canon comprehensively and reverently, we use it adaptively, as a toolbox to challenge concepts and categorisations (including Bourdieu's own) in ongoing social critique. This is, in fact, how Bourdieu's ideas have been used in research on space and place to date, and how they will probably continue to be used in the future. The other lesson that Bourdieu leaves us with is the richness and value of combining theoretical advances with comprehensive and careful empirical work, locating the self-critical voice of the academic alongside the voices of the subjects of knowledge. This is fitting legacy for a 'great and original figure of contemporary sociology', as ***Jacques Derrida*** put it (in the French newspaper *Le Monde,* 25/1/02) when mourning the death of one of France's greatest public intellectuals.

Bourdieu's Key Works

Bourdieu, P. (1962) *The Algerians* (trans A. Ross). Boston, MA: Beacon.

Bourdieu, P. (1977a) *Outline of a Theory of Practice.* Cambridge: Cambridge University Press.

Bourdieu, P. (1977b) *Reproduction in Education, Society and Culture* (trans R. Nice). London: Sage.

Bourdieu, P. (1979) *The Inheritors: French Students and Their Relation to Culture* (trans R. Nice). Chicago, IL: Chicago University Press.

Bourdieu, P. (1984) *Distinction: A Social Critique of the Judgement of Taste.* London: Routledge and Kegan Paul.

Bourdieu, P. (1988) *Homo Academicus.* Cambridge: Polity.

Bourdieu, P. (1990) *The Logic of Practice.* Cambridge: Polity.

Bourdieu, P. (1994) *The Rules of Art.* Cambridge: Polity.

Bourdieu, P. (1998) *Practical Reason.* Cambridge: Polity.

Bourdieu, P. (1999a) *The Weight of the World: Social Suffering in Contemporary Society.* Palo Alto, CA: Stanford University Press.

Bourdieu, P. (1999b) *Acts of Resistance: Against the Tyranny of the Market.* New York, NY: New Press.

Bourdieu, P. (2005) *The Social Structures of the Economy.* Cambridge: Polity.

Bourdieu, P. (2008) *The Bachelor's Ball.* Cambridge: Polity.

Secondary Sources and References

Aboulafia, M. (1999) 'A (neo)-American in Paris: Bourdieu, Mead, and pragmatism', in R. Shusterman (ed), *Bourdieu: A Critical Reader.* Oxford: Blackwell. pp. 153–76.

Adkins, L. and Skeggs, B. (eds) (2005) *Feminism After Bourdieu.* Oxford: Blackwell.

Allen, C. (2008) *Housing Market Renewal and Social Class.* London: Routledge.

Amin, A. and Thrift, N. (2007) 'Cultural-economy and cities', *Progress in Human Geography,* 31 (2): 143–61.

Bacqué, M-H., Bridge, G., Benson, M., Butler, T., Charmes, E., Fijalkow, Y., Jackson, E. Launay, L. and Vermeersch, S. (2015) *The Middle Classes and the City: A Study of Paris and London.* London: Palgrave.

Bennett, T., Savage, M., Silva, E., Warde, A., Gayo-Cal, M. and Wright, D. (2009) *Culture, Class, Distinction.* London: Routledge.

Bridge, G. (2001a) 'Estate agents as interpreters of economic and cultural capital: The "gentrification premium" in the Sydney housing market', *International Journal of Urban and Regional Research,* 25: 81–101.

Bridge, G. (2001b) 'Bourdieu, rational action and the time-space strategy of gentrification', *Transactions of the Institute of British Geographers,* NS 26: 205–16.

Bridge, G. (2003) 'Time-space trajectories in provincial gentrification', *Urban Studies,* 40 (12): 2545–56.

Bridge, G. (2005) *Reason in the City of Difference: Pragmatism, Communicative Action and Contemporary Urbanism.* London: Routledge.

Bridge, G. (2006a) 'It's not just a question of taste: Gentrification, the neighbourhood and cultural capital', *Environment and Planning A,* 38: 1965–78.

Bridge, G. (2006b) 'Perspectives on the neighbourhood and cultural capital', *Urban Studies,* 43: 719–30.

Butler, T. and Robson, G. (2003) *London Calling: The Middle Class and the Remaking of Inner London.* London: Berg.

Cresswell, T. (1996) *In Place/Out of Place: Geography, Ideology and Transgression.*

Minneapolis, MN: University of Minnesota Press.
Cresswell, T. (2002) 'Bourdieu's geographies: In memoriam', Guest editorial, *Environment and Planning D*, 20: 379–82.
Derrida, J. (2002) 'Jacques Derrida's reaction', *Le Monde*, 25 January, https://www.lemonde.fr/archives/article/2002/01/25/la-reaction-de-jacques-derrida_4213396_1819218.html (accessed 20 December 2023).
Hage, G. (2000) *White Nation: Fantasies of White Supremacy in a Multicultural Nation*, London: Routledge.
Harvey, D. (1989) *The Condition of Postmodernity.* Oxford: Blackwell.
Hillier, J. and Rooksby, E. (eds) (2002) *Habitus: A Sense of Place*. Aldershot: Ashgate.
Holt, L. (2008) 'Embodied social capital and geographic perspectives: Performing the habitus', *Progress in Human Geography*, 32: 227–46.
Honneth, A. (1986) 'The fragmented world of symbolic forms: Reflections on Pierre Bourdieu's sociology of culture', *Theory, Culture & Society*, 3: 55–66.
Israel, E. and Frenkel, A. (2018) 'Social justice and spatial inequality: Toward a conceptual framework', *Progress in Human Geography*, 42 (5): 647–65.
Jenkins, R. (1992) *Pierre Bourdieu.* London: Routledge.
Kuus, M. (2015) 'Transnational bureaucracies: How do we know what they know?', *Progress in Human Geography*, 39 (4): 432–48.
Ley, D. (2003) 'Artists, aestheticisation and the field of gentrification', *Urban Studies*, 40: 2527–44.
May, J. (1996) 'Globalisation and the politics of place and identity in an inner London neighbourhood', *Transactions of the Institute of British Geographers*, NS 21: 194–215.
Merleau-Ponty, M. (1962) *Phenomenology of Perception* (C. Smith, trans). London: Routledge.
O'Brien, D. (2021) 'Theorising the deaf body: Using Lefebvre and Bourdieu to understand deaf spatial experience', *Cultural Geographies*, 28 (4): 645–60.
Painter, J. (2000) 'Pierre Bourdieu', in M. Crang and N. Thrift (eds), *Thinking Space*. London: Routledge. pp. 240–55.
Painter, J. (2002) 'Governmentality and regional economic strategies', in J. Hillier and E. Rooksby (eds), *Habitus: A Sense of Place*. Aldershot: Ashgate. pp. 131–256.
Parkes, D. and Thrift, N. (1980) *Times, Spaces and Places: A Chronogeographic Perspective.* Chichester: Wiley.
Peck, J. and Theodore, N. (2008) 'Carceral Chicago: Making the ex-offender employability crisis', *International Journal of Urban and Regional Research*, 32: 251–81.
Pile, S. (2002) 'Spectral cities: Where the repressed returns and other short stories', in J. Hillier and E. Rooksby (eds), *Habitus: A Sense of Place*. Aldershot: Ashgate. pp. 235–57.
Podmore, J. (1998) '(Re)reading the "loft living" habitus in Montreal's inner city', *International Journal of Urban and Regional Research*, 22: 283–302
Savage, M. (2011) 'The lost urban sociology of Pierre Bourdieu', in G. Bridge and S. Watson (eds), *The New Blackwell Companion to the City*. Oxford: Wiley-Blackwell. pp. 511–20.
Savage, M. (2021) 'Bourdieu comes to town Part II', *International Journal of Urban and Regional Research*, 45 (1): 150–3.
Savage, M., Bagnall, G. and Longhurst, B. (2005) *Globalisation and Belonging.* London: Sage.
Sisson, A. (2021) 'Territory and territorial stigmatisation: On the production, consequences and contestation of spatial disrepute', *Progress in Human Geography*, 45 (4): 659–81.
Slater, T. (2018) 'The invention of the "sink estate": Consequential categorisation and

the UK housing crisis', *The Sociological Review*, 66 (4): 877–97.

Spigel, B. (2017) 'Bourdieu, culture, and the economic geography of practice: Entrepreneurial mentorship in Ottawa and Waterloo, Canada', *Journal of Economic Geography*, 17 (2): 287–310.

Thrift, N. (2007) *Non-representational Theory: Space, Politics, Affect*. London: Routledge.

Valentine, G. and Sadgrove, J. (2012) 'Lived difference: A narrative account of spatiotemporal processes of social differentiation', *Environment and Planning A*, 44 (9): 2049–63.

Wacquant, L. (2008) *Urban Outcasts: A Comparative Study of Advanced Marginality*. Cambridge: Polity.

Wacquant, L. (2009) *Prisons of Poverty*. Minneapolis, MN: University of Minnesota Press.

Wacquant, L. (2018) 'Bourdieu comes to town: Pertinence, principles, application', *International Journal of Urban and Regional Research*, 42 (1): 90–105.

Wacquant, L. (2023) *Bourdieu in the City: Challenging Urban Theory*. Cambridge: Polity.

Wynne, D. and O'Connor, J. (1998) 'Consumption and the postmodern city', *Urban Studies*, 5–6: 841–64.

10
Judith Butler

Eden Kinkaid, University of Arizona
Kelsey Emard, Oregon State University

Biographical Details and Theoretical Context

Judith Butler (they/them; born February 24, 1956) is an American philosopher best known for their work in feminist theory, queer theory, and philosophy. Butler obtained their bachelor's and doctoral degrees from Yale University, and is currently Distinguished Professor in the Graduate School and formerly the Maxine Elliot Chair in the Department of Comparative Literature and the Program of Critical Theory at the University of California, Berkeley. Across their career, Butler has engaged and brought together various traditions of thought including, most prominently, psychoanalysis, phenomenology, queer theory, political theory, and feminist theory to ask wide-ranging ethical and political questions. Butler's abiding concerns have centred on theorising the (gendered) subject; developing a social constructivist account of gender (i.e., performativity); analysing the uneven distribution of precarity and vulnerability in collective life; understanding and activating the power of political speech and assembly; and examining the ethical questions that relate and bring together these themes. They have approached these topics in dialogue with the work of thinkers like Lacan, Freud, **Foucault**, de Beauvoir, Derrida, Levinas, Irigaray, Wittig, Kristeva, Adorno, Nietzsche, Merleau-Ponty, Arendt, Austin, and others, and made major contributions to debates and concepts in philosophy and cultural theory.

Butler came to prominence as a feminist and queer philosopher through the publication of their early work on gender. Butler's 1990 book *Gender Trouble* draws on psychoanalytic perspectives to theorise gender. It is here and in their 1993 book *Bodies that Matter* that Butler conceptualises and refines their most well-known concept, the 'performativity' of gender. In the early 2000s, Butler's attention turned toward questions of vulnerability and precarity and examined how bodies become differentially precarious, in part through the uneven ascription of value to bodies and subjects. These two major areas of engagement have, over the course of their career, weaved and blended in various ways (see, for example, Butler, 2009b), often returning to a core set of ethical questions about the self and collective political life. In Butler's more recent work, they return to their early work in performativity and their work on precarity and vulnerability to think through the implications of their theories of subjectivity for political life, public space, collective action, and ethics (2015a; 2015b; 2020; 2022).

Spatial Contributions

While questions of space and geography do not figure explicitly or prominently in Butler's work (though geopolitical themes are evident and explicit questions of space have emerged more recently, see 2015a), geographers have taken up Butler's work to advance geographic debates, including those about identity, the performative nature of space, and political geography, among other topics.

Butler's work first entered geography in the 1990s when geographers began to engage the concept of performativity, which was developed in *Gender Trouble* (1990) and *Bodies that Matter* (1993). Butler's theory of performativity posits that there is no essence to gender; it is not some stable truth to be found in or read from the body. Rather, the illusion of gender's truth is produced through a string of citations of gendered meanings, representations, and behaviours. It is through this reiteration that gender is produced. By virtue of being born into a social system and language that exceeds them, gendered subjects are constrained into these meanings: reiterating them becomes necessary to emerge as a legible subject. However, slippage of these norms is also possible, through which gender norms become queered and reworked.

When Butler's work on performativity entered geography, it came into contact with an intellectual discourse shaped by theories of space – for example, the work of **Henri Lefebvre** and **Doreen Massey** – that argue space is a social product, a terrain of social action and political struggle. Read into these existing geographic debates, performativity became a concept useful not only for thinking about the construction of social identities like gender (as in Butler's original formulation), but also a tool for thinking about the production of space as a historically contingent and power-laden process that is necessarily connected to bodies and social identities.

Performativity and its geographic potential first caught the attention of feminist and queer geographers exploring the imbrication of gender and space (e.g., see various chapters in David Bell and **Gill Valentine**'s 1995 edited volume *Mapping Desire* and Duncan's 1996 edited volume *BodySpace*). A key work that translated Butler's concept of performativity into geography was Nicky Gregson and **Gillian Rose**'s (2000) article, 'Taking Butler elsewhere', in which the authors sought to bring Butler's concepts into direct contact with critical human geography. Gregson and Rose urged geographers to approach space itself as performative and to examine how spatial meanings and practices shape the meanings and practices of identity and vice versa (a focus which is not pronounced in Butler's own approach). Other early applications of Butler's work focused on the gendering and sexualisation of space: how space becomes coded as heterosexual through the reproduction of certain practices and the exclusion of others (e.g., Bell et al., 1994; McDowell, 1995). Building on these early engagements, geographers have since engaged performativity to shed light on the social construction of multiple embodied social identities, including race, gender, and sexuality, in and through spatial practice (e.g., Mahtani, 2002; Johnston, 2008; Rose-Redwood and Rose-Redwood, 2017; Preser, 2018; Sullivan, 2019). The idea that spaces and embodied identities are *actively produced* as hegemonic – and that the power relations and hierarchies underwriting them are not 'natural' or 'given' – has contributed to critical geographic projects imagining space and identity otherwise (Rose-Redwood and Glass, 2015: 16).

This potential of performativity for thinking about the production of space has also been realised in political geography. As Rose-Redwood and Glass suggest, performativity 'applies just as much to assertions of territorial sovereignty, the surveying of private property, or the naming of a city's streets as it does to the embodiment of gendered subjectivities' (2015: 2). Political geographers have applied the idea of the performativity of space to understand the (re)production and politics of geopolitical boundaries, national identities, cartographic practices, and migration, among other themes (see Glass and Rose-Redwood's 2015 edited volume *Performativity, Politics, and the Production of Social Space* for an overview of this work). Like gender identity, geopolitical identities and spaces are seen as the product of repeated and hegemonic performances, ones that are inscribed in geographic representations and imaginaries.

In addition to areas of engagement within geographies of social identity and political geography, performativity has also influenced geographic work on urban space (Rose-Redwood, 2008; Cook and Whowell, 2011; Denis and Pontille, 2014) and the economy (Gibson-Graham, 2006) – though many other applications of the term 'performativity' and 'performance' in economic geography often derive from sources other than Butler such as the sociology of science and actor network theory – as well as work employing a posthuman and multispecies lens (Hovorka, 2015; Schuurman and Franklin, 2015).

Butler has also had an influence in geography (particularly within political geography) beyond the concept of performativity through their work in the 2000s on precarity and (non)violence. In their book *Precarious Life* (2004a), Butler explores ideas of precarity, violence, and mourning through an examination of how some lives were deemed worthy of grief and others undeserving of grief as the United States violently responded to the 9/11 attacks. Through their engagement with war and precarity, Butler conceptualises an ethics of nonviolence as a political struggle to challenge the perpetuation of violence against those whose lives are deemed 'ungrievable' (see *Frames of War* [2009a] and *The Force of Nonviolence* [2020]). Geographers concerned with geopolitics have widely engaged Butler's conceptualizations of precarity in their analyses of the spatial dimensions of violence and nonviolence (Woon, 2014), the spatial relations and practices involved in reducing or extending precarity (Harker, 2012; Green and Estes, 2019), the globally uneven and selective recognition of suffering, and the resulting geographies of warfare, violence, aid, partnership, and support (Walsh, 2011; Jones, 2015; Joronen, 2016; Kaufmann and Cote, 2021).

Key Advances and Controversies

As noted, Butler's work has made its way into geography and become part of the landscape of critical human geographic theory. However, the adoption of Butler's work has not been without tension or critique; indeed, at times it has prompted substantial debate. Be they celebratory or critical, engagements with Butler's work have refashioned and translated their concepts into important ideas in the landscape of geographic thinking and contributed to key advancements in the discipline.

As performativity made its way into geography, it collided with ongoing debates about the subject and questions of structure and agency that were central theoretical concerns in the 1990s. In this context, some geographers posed critical questions of Butler's conception of performativity. Some found that Butler's focus on the constitution of subjectivity through discourse was overly deterministic and abstract; human agency seemed trapped within the terms of dominant discourse (e.g., McNay, 1999; Nelson, 1999). In an early critique, Nelson argues that 'performativity ontologically assumes an abstracted subject (i.e. abstracted as a subject position in a given discourse) and thus provides no space for conscious reflexivity, negotiation or agency in the doing of identity' (Nelson, 1999: 332). She also argues that the subject of Butler's analysis is abstracted from space, place, and time which makes it difficult for geographers to operationalise. Yet at the same time, some were critical that geographers were using performativity in a manner that reinstated and overemphasised the possibility of choice and agency in the 'doing' of identity (Mahtani, 2011). As such, the reception, interpretation, and uptake of Butler's ideas within geography were caught in a tension between overly deterministic accounts of subjectivity as constituted through discourse, on the one hand, and more voluntaristic ideas of agency, self-consciousness, and the 'performance' of identity, on the other.

Others wondered about the role of space in theories of performativity (Lloyd, 1999: 210), urging us to situate performativity in concrete spaces, and arguing that space itself can be seen as performative. As Gregson and Rose (2000: 434) suggested, 'space too needs to be thought of as brought into being through performances and as a performative articulation of power'. In their critical engagement with Butler's work, Gregson and Rose (2000: 434) argue for an uptake of Butler's work that can help elucidate 'the construction of social identity, social difference, social power relations, and the way space might articulate all of these'. They thus urge geographers to not take space for granted and to consider how space is produced in similar ways as gendered subjectivity: through repeated discursive and embodied performances. Embodied performance can also reshape space: as Doan (2010: 638) describes, 'the performance of gender in space not only shifts with each performance, but in a very real way each performance also changes the space in which it is performed.' Through these articulations, geographers have brought an explicitly geographic perspective to bear on Butler's theory of performativity.

While, for some, Butler's early work presents a limited discussion of embodiment, space, and agency in relation to performativity, their later work has fleshed out these concepts more fully without majorly departing from their original formulations. A deeper engagement with the phenomenological threads of Butler's work, which come across strongly in *Senses of the Subject* (2015b), for example, can help explain how time, space, relationality, agency, and selfhood are all integral to Butler's ontology of the subject. Further, in *Notes Toward a Performative Theory of Assembly*, Butler (2015a) more explicitly engages with space, showing how space matters in performativity and contestations of precarity. Specifically, they focus on how bodies assemble in public space and how acts of political assembly are both a testament to and a contestation of embodied precarity. In theorising acts of emergent and collective political speech and embodiment in public space, they revise and complicate their concept of performativity toward a more politically- and ethically-engaged notion of political collectivity.

Most of the critiques and concerns with Butler's work discussed so far seem to be largely resolved; as Johnston (2020: 3) writes: 'By the end of the 1990s it became well established that bodies and places are mutually constitutive and performative'. The controversial conception of subjectivity that Butler introduced in the 1990s has become more or less commonplace as later waves of poststructuralist thinking have moved beyond and redefined the structure/agency debates of the 1990s. Despite the relatively limited theorisations of space in Butler's work, geographers and other spatial thinkers have had no trouble extending performativity and Butler's other concepts like precarity to explicitly think about space and place, be it in the realm of geopolitics, human–environment systems, cartographic practices, or the spatialisation of social identities (see Rose-Redwood and Glass, 2015). All in all, Butler's work has contributed to key advances in thinking about the relations between space, subjectivity, identity, and the political and ethical questions that emerge at these intersections. When brought into contact with a rich history of geographic thinking on these themes, Butler's work promises to continue to be an intellectual resource for critical human geographers.

Butler's Key Works

Butler, J. (1990) *Gender Trouble*. London: Routledge.

Butler, J. (1993) *Bodies That Matter*. London: Routledge.

Butler, J. (1997a) *Excitable Speech*. London: Routledge.

Butler, J. (1997b) *Psychic Life of Power: Theories in Subjection*. Stanford, CA: Stanford University Press.

Butler, J. (2004a) *Precarious Life: The Power of Mourning and Violence*. London: Verso.

Butler, J. (2004b) *Undoing Gender*. London: Routledge.

Butler, J. (2005) *Giving an Account of Oneself*. New York, NY: Fordham University Press.

Butler, J. (2009a) *Frames of War: When is Life Grievable?* London: Verso.

Butler, J. (2015a) *Notes Toward a Performative Theory of Assembly*. Cambridge, MA: Harvard University Press.

Butler, J. (2015b) *Senses of the Subject*. New York, NY: Fordham University Press.

Butler, J. (2020) *The Force of Nonviolence*. London: Verso.

Butler, J. (2022) *What World Is This? A Pandemic Phenomenology*. New York, NY: Columbia University Press.

Secondary Sources and References

Bell, D. and Valentine, G. (1995) *Mapping Desire: Geographies of Sexualities*. New York, NY: Routledge.

Bell, D., Binnie, J., Cream, J. and Valentine, G. (1994) 'All hyped up and no place to go', *Gender, Place and Culture*, 1 (1): 31–47.

Butler, J. (2009b) 'Performativity, precarity and sexual politics', *Revista de Antropología Iberoamericana*, 4 (3): i–xiii.

Cook, I.R. and Whowell, M. (2011) 'Visibility and the policing of public space', *Geography Compass*, 5 (8): 610–22.

Denis, J. and Pontille, D. (2014) 'Maintenance work and the performativity of urban inscriptions: The case of Paris subway signs', *Environment and Planning D: Society & Space*, 32 (3): 404–16.

Doan, P.L. (2010) 'The tyranny of gendered spaces – reflections from beyond the gender dichotomy', *Gender, Place & Culture*, 17 (5): 635–54.

Duncan, N. (1996) *BodySpace: Destabilizing Geographies of Gender and Sexuality*. New York, NY: Routledge.

Gibson-Graham, J.K. (2006) *The End of Capitalism (As We Knew It)*. Minneapolis, MN: University of Minnesota Press.

Glass, M.R. and Rose-Redwood, R. (eds) (2015) *Performativity, Politics, and the Production of Social Space*. New York, NY: Routledge.

Green, W.N. and Estes, J. (2018) 'Precarious debt: Microfinance subjects and intergenerational dependency in Cambodia', *Antipode,* 51 (1): 129–47.

Gregson, N. and Rose, G. (2000) 'Taking Butler elsewhere: Performativities, spatialities and subjectivities', *Environment and Planning D: Society and Space,* 18 (4): 433-52.

Harker, C. (2012) 'Precariousness, precarity, and family: Notes from Palestine', *Environment and Planning A: Economy and Space,* 44 (4): 849–65.

Hovorka, A.J. (2015) 'Feminism and animals: Exploring interspecies relations through intersectionality, performativity and standpoint', *Gender, Place & Culture,* 22 (1): 1–19.

Johnston, L. (2008) 'Mobilizing pride/shame: Lesbians, tourism and parades', *Social & Cultural Geography,* 8 (1): 29–45.

Johnston, L. (2020) *Transforming Gender, Sex, and Place: Gender Variant Geographies.* London: Routledge.

Jones, C.A. (2015) 'Frames of law: Targeting advice and operational law in the Israeli military', *Environment and Planning D: Society and Space,* 33 (4): 676–96.

Joronen, M. (2016) 'Politics of precarious childhood: Ill treatment of Palestinian children under the Israeli military order', *Geopolitics,* 21 (1): 92–114.

Kaufmann, C. and Cote, M. (2021) 'Frames of extractivism: Small-scale goldmining formalization and state violence in Colombia', *Political Geography,* 91: article #102496.

Lloyd, M. (1999) 'Performativity, parody, politics', *Theory, Culture & Society,* 16 (2): 195–213.

Mahtani, M. (2002) 'Tricking the border guards: Performing race', *Environment and Planning D: Society and Space,* 20 (4): 425–40.

Mahtani, M. (2011) 'Judith Butler', in P. Hubbard and R. Kitchin (eds), *Key Thinkers on Space and Place* (2nd edition). London: Sage. pp. 82–90.

McDowell, L. (2003) 'Body work: Heterosexual gender performances in city workplaces', in D. Bell and G. Valentine (eds), *Mapping Desire: Geographies of Sexualities.* New York, NY: Routledge. pp. 67–85.

McNay, L. (1999) 'Subject, psyche and agency: The work of Judith Butler', *Theory, Culture & Society,* 16 (2): 175–93.

Nelson, L. (1999) 'Bodies (and spaces) do matter: The limits of performativity', *Gender, Place & Culture,* 6 (4): 331–53.

Preser, R. (2021) 'What therefore discourse has joined together: Sexuality, performativity and everyday use of public spaces', *Social & Cultural Geography,* 22 (1): 58–75.

Rose-Redwood, R.S. (2008) '"Sixth Avenue is now a memory": Regimes of spatial inscription and the performative limits of the official city-text', *Political Geography,* 27 (8): 875–94.

Rose-Redwood, R.S. and Glass, M. (2015) 'Introduction: Geographies of performativity,' in M. Glass and R. Rose-Redwood (eds), *Performativity, Politics, and the Production of Social Space.* New York, NY: Routledge. pp. 1–36.

Rose-Redwood, C. and Rose-Redwood, R. (2017) '"It definitely felt very white": Race, gender, and the performative politics of assembly at the Women's March in Victoria, British Columbia', *Gender, Place & Culture,* 24 (5): 645–54.

Schuurman, N. and Franklin, A. (2015) 'Performing expertise in human–animal relationships: Performative instability and the role of counter-performance', *Environment and Planning D: Society and Space,* 33 (1): 20–34.

Sullivan, T. (2012) '"I want to be all I can Irish": The role of performance and performativity in the construction of ethnicity', *Social & Cultural Geography,* 13 (5): 429–43.

Walsh, R. (2011) 'What stories we tell when we talk about torture: Mapping the geopolitics of compassion and the post-Abu-Ghraib national family in 24: Redemption and rendition', *Environment and Planning D: Society and Space*, 29 (1): 150–68.

Woon, C.Y. (2014) 'Precarious geopolitics and the possibilities of nonviolence', *Progress in Human Geography*, 38 (5): 654–70.

11
Denis Cosgrove

Keith Lilley, Queen's University Belfast

Biographical Details and Theoretical Context

During an academic career spanning more than three decades, Denis Cosgrove did perhaps more than any other Anglophone academic geographer to ensure that 'landscape' continues to occupy a central place on the disciplinary 'map'. His work on the idea of landscape, which came to the fore in the 1980s especially, profoundly altered the course of landscape study in human geography through a series of theoretically-informed yet empirically-grounded books and articles (Cosgrove, 1998a [1984]; 1985; 1993; Cosgrove and Daniels, 1988). In 1993 he co-founded a new journal – *Ecumene: A Geographical Journal of Environment, Culture and Meaning* (later *cultural geographies*) – providing a forum for research within the humanities tradition of geographical scholarship. Later, in the 1990s, Cosgrove's interests moved towards broader issues of 'mapping' and representation, but his work continued to be characterised by its close attention to empirical – and particularly historical – detail, whilst continuing to engage with theoretical ideas from disciplines such as cultural studies, landscape architecture, cartographic history and creative art (Cosgrove, 1996; 1999a; 2001a; 2001b; 2008). This blending of empirical and theoretical material is a hallmark of Cosgrove's work, and it is this that has ensured that his writing has had considerable impact on contemporary geographical thought and enquiry despite the focus of much of his work on the historical geographies of Renaissance and Enlightenment Europe.

Born in 1948, and educated in Liverpool at St Francis Xavier College and then St Catherine's College, Oxford, Cosgrove established his reputation as a leading practitioner of the 'new' cultural geography while in the Geography Division of Oxford Polytechnic (now Oxford Brookes University), where he served as Chairman (*sic*) from 1976 to 1980, and while at the Department of Geography of Loughborough University (from 1980 to 1993). Alongside those humanistic geographers in the UK and US who critiqued the positivist and quantitative basis of much of human geography, Cosgrove sought to recognise the subjectivity of geographical enquiry, and situate geography within a broader humanities tradition that acknowledged both charisma and context (Cosgrove, 1989; Freytag and Jöns, 2006; Vallerani, 2018). This concern with 'authorship' and 'authority' in geographical writing arguably informed Cosgrove's subsequent scholarship (Cosgrove and Domosh, 1993), and while few contemporary geographers might refer to themselves as 'humanistic', the contributions made by Cosgrove to theoretical debates in the early 1980s about the writing and representation of space have endured (see Cosgrove, 2001b). One important reason for this was Cosgrove's own enthusiasm to engage with post-modern and post-structural thought, marking him out as one of the most important sources of inspiration for those geographers seeking to interrogate the spatial relations of knowledge and power.

Cosgrove's intellectual importance really flourished during his time as Professor of Human Geography at Royal Holloway, University of London, between 1994 and 1999, where he established the Landscape Surgery, inspiring a new generation of cultural geographers whose work in this field owes much to this energetic and expansive period of his academic career. From 1999 Cosgrove was resident in the US as Alexander von Humboldt Professor of Geography at the University of California at Los Angeles (UCLA), where he continued in his role as a mediator between geography, the humanities and arts, becoming Getty Distinguished Scholar for 2008–9, but at the age of 59 he died in March 2008 following a prolonged illness (see Lilley, 2009). His empirical and theoretical insights in cultural and historical geography remain a rich intellectual legacy that have posthumously influenced human geography widely.

Spatial Contributions

One of Cosgrove's key contributions was to engage with key geographical concepts through the motif of *landscape*. Within the context of cultural geography in the UK, Cosgrove's engagement with ideas of 'writing' and 'reading' landscape drew upon the ideas of North American geographers such as Carl Sauer, J.B. Jackson and **Yi-Fu Tuan**, among others (see Cosgrove, 1998b). Taking one of the concepts that was central to the North American geographical imagination, Cosgrove nonetheless departed from North American 'Berkeley' cultural tradition by exploring not just the cultural processes that shaped landscape but also the constitutive role that landscape plays in shaping the lives of those who engage with that landscape, or landscapes. In the early 1980s this was a new and radical way of viewing landscape, for up to that point UK historical geographers were broadly concerned with 'reconstructing' past landscapes from maps, documents and fieldwork in a tradition derived from the work of W.G. Hoskins (as well as M.W. Beresford and M.R.G. Conzen) (see Whitehand, 2001; Wylie, 2007). Cosgrove offered a different approach to landscape, one that recognised that 'landscape is not merely the world we see, it is a construction, a composition of the world', in short, 'an ideological concept' (Cosgrove, 1984: 13, 15). Cosgrove set out this thesis in his first book, *Social Formation and Symbolic Landscape*, introducing the idea that 'landscape is a way of seeing' – a means by which 'Europeans have represented to themselves and to others the world about them and their relationships with it' (Cosgrove, 1984: 1). Looking back on this work in the introduction to the book's second edition, Cosgrove reflects on its 'prime intention [...] to press landscape studies, especially in Geography, towards what seemed to be specific new directions: to locate landscape interpretation within a critical historiography, to theorise the idea of landscape within a broadly Marxian understanding of culture and society, and thus to extend the treatment of landscape beyond what seemed [...] a prevailing narrow focus on design and taste' (Cosgrove, 1998b: xiii). The book certainly did this, and at its heart was Cosgrove's interest in Renaissance Italy (Cosgrove, 1993). Here he found the roots of the 'landscape idea' in Western culture, which was (like the book itself) a humanistic enterprise concerned with new ways of imagining and representing landscape, of 'seeing' the land.

Cosgrove's landscape 'as a way of seeing' idea became more influential in geographical discourse following the publication in 1985 of his paper, 'Prospect, perspective and the evolution of the landscape idea' (Cosgrove, 1985), and with the publication of a collection of papers under the title *The Iconography of Landscape*, edited by Cosgrove and Daniels (1988). In both, Cosgrove reinforced the message of *Social Formation*, that landscape was 'a "way of seeing" that was bourgeois, individualist and related to the exercise of power' (Cosgrove, 1985: 45). Simultaneously, he underlined that this was connected with Renaissance Humanism and scientific advances in mapping, projection, surveying and cosmology. As a 'way of seeing', landscape is thus 'a composition and structuring of the world' that 'may be appropriated by a detached, individual spectator to whom an illusion of order and control is offered through the composition of space according to the certainties of geometry' (Cosgrove, 1985: 55). 'Landscape' was property, owned by those beholding it; capturing and controlling the land through representations of it as landscape in maps and in paintings – and through fashioning landscapes on the ground using design and architecture. The landscape, then, far from being neutral and inert, has social and cultural meanings, a symbolism – an 'iconography'. It was this 'iconography' of landscape that Cosgrove (1998a [1984]) sought to elucidate, and this is perhaps the most enduring and celebrated aspect of Cosgrove's work.

In their editorial introduction to the *Iconography of Landscape* Cosgrove and Daniels began by referring to the landscape as a 'cultural image, a pictorial way of representing, structuring or symbolising surroundings' (Daniels and Cosgrove, 1988: 1). The idea was that in order to understand the material landscape it was necessary not just to make use of 'images and verbal representations of it' for the purposes of illustration, but also to recognise that such images were constitutive of 'landscape' and 'its meaning or meanings' (Daniels and Cosgrove, 1988: 1). To explore this idea, the editors capitalised on the ideas of iconography which 'sought to probe meaning in a work of art by setting it in its historical context and [...] analyse the ideas implicated in its imagery' (Daniels and Cosgrove, 1988: 2) – exactly what Cosgrove had been doing with his work on the emergence of landscape as a 'way of seeing' in late-medieval and Renaissance Italy. Conceptually, this approach owed much to art history, and especially the work of Ernst Cassirer and Erwin Panofsky on Christian iconography, while at the same time the anthropologist Clifford Geertz was drawn on to provide some sense of method with his 'thick description' and 'conceptualisation of culture as a "text"' (Daniels and Cosgrove, 1988: 4). There was little advice apart from this on how to interpret landscapes iconographically. What was instead significant about this book was that it opened up new avenues for understanding 'landscape' – including both the 'imagined' landscapes of maps, literature and art, as well as the 'material' landscape (Cosgrove and Daniels, 1988). Indeed, running throughout much of Cosgrove's work was his concern with 'representation' – especially visual culture, from art and architecture in fifteenth- and sixteenth-century Italy, for example, through to cartographic images of the earth and cosmos in studies of Western and Renaissance cosmography (Cosgrove, 1988; 1992; 2001a; 2001b; 2006; 2007; Cosgrove et al., 1996; Atkinson and Cosgrove, 1998).

Key Advances and Controversies

Certainly, Cosgrove made a key contribution to spatial thought by offering a distinctive take on the interpretation and understanding of landscape. In the light of his work, 'landscape' is no longer considered an unproblematic geographical concept (Nash, 1999; Wylie, 2007). Equally, geographers cannot now refer to landscape without referring to Cosgrove – the two becoming synonymous. Yet it would be wrong to cast Cosgrove's contributions to the thinking of space and place simply in these terms, for his approach to landscape provided a means by which geographers broached broader conceptual issues – especially in matters of subjectivity, representation, power and authority (see Cosgrove, 1988; Freytag and Jöns, 2006). These had of course been present in Cosgrove's earlier work on landscape, but in the 1980s and 1990s they were under closer scrutiny under the rubric of a 'new' cultural geography informed by critical social and cultural theory (particularly feminist and postcolonial critiques). Cosgrove addressed these issues of authority and authorship writing with Mona Domosh, noting the 'crisis of geographical representation' and stressing that 'the very structures of representation are implicated in moral and political discourse' (Cosgrove and Domosh, 1993: 30). This 'crisis' was one in which geographers began to realise that by writing about the world in the ways that they did, they were also reinforcing a particular view of the world; an institutionalised, bourgeois, male-dominated, and sometimes imperialist view. Indeed, others at this time were also becoming conscious of geography's history and duplicity in presenting and shaping the world (see Livingstone, 1992; Driver, 2001). Hence, together with the work of *Peter Jackson* (1989) and Susan Smith (1989) on the scripting of race and racism, James and Nancy Duncan on landscape as text (Duncan and Duncan, 1988; Duncan, 1990), and ***Derek Gregory's*** analysis of the production of geographical knowledge (Gregory, 1994), Cosgrove's idea of landscape as 'way of seeing' (and an ideologically-infused mode of representation) emerged as a key reference point in the formation of a putative 'new' cultural geography.

Though Cosgrove has been identified as a key practitioner of the 'new' cultural geography, it is important to stress that not all those implicated in the making of this 'new' cultural geography agreed on its definition (Dear, 1988). Discussions in the early 1990s flowed back and forth about its newness and also concerns over its tendency to deal more with abstract theory than with the empirical world. There were those, too, who sought to argue that this 'new' cultural geography, with its concern with text, image and metaphor, was making 'landscape' more unstable and slippery (see Muir, 1999). Indeed, there seemed to be a polarisation in the way that historical and cultural geographers were treating landscape – some continuing with their empirically-grounded, largely atheoretical work on landscape history; others more inclined to see landscapes as imaginings, in poetry, art and literature (and accused of detaching these representations from their material contexts) (see Whyte, 2002). This polarisation was in part a conceptual parting of the ways, between, on the one hand, those geographers with more positivist leanings, and on the other, those concerned with taking onboard critical theory. It was an empirical and historical divide too, with the focus of the latter being especially – indeed almost exclusively – on the 'modern' world, particularly the eighteenth, nineteenth and twentieth centuries. It was here that historical and cultural geographers led the field during the

1990s (Driver, 1992; Daniels, 1993; Ogborn, 1998; Driver and Gilbert, 1999; Graham and Nash, 2000), and subsequently in postcolonial debates on landscape and imperialism (Morrissey, 2013).

Since his death in 2008, Geography's – and Cosgrove's own – fetishisation of the visual in landscape has become challenged through 'non-representational' and performative approaches to landscape (e.g., see Howard et al., 2019). Indeed, the landscape of cultural geography has itself been reshaped over subsequent years, but in ways that still owe something to Cosgrove's work and influences, particularly in imagined and imaginary geographies, and in intersections between geography and architecture (see Cosgrove and Della Dora, 2009; Della Dora, 2022). Seen in this context, Cosgrove's particular cultural approach to landscape was not only in and of itself highly original and novel two decades ago, but remains influential in geographers' approach to space and place more generally (e.g., Davies, 2021), catalysing disciplinary changes in cultural and historical geography notably in what has emerged more definably as a field of geohumanities (Vallerani, 2018). The Royal Holloway Centre for the GeoHumanities, launched in 2016, honours Cosgrove's legacy and continued importance in spatial thinking through an annual lecture given in his name. Cosgrove's engagements with environmental humanities, and particularly the connections between water, engineering and landscape (Cosgrove et al., 1996), have developed significantly in recent years as cultural geographers have become important interlocutors between humanities and science discourses on climate change and sustainability (Lekan, 2014).

In Cosgrove's later geographical work, while landscape remained an important element of his writing (e.g., Cosgrove, 1993), his focus increasingly shifted towards issues of visual representation, especially mapping and aerial photography (Cosgrove, 1992; 1999b; 2006; 2008; Cosgrove and Fox, 2009). With a firm historical basis to this work, theoretical insight remained key, as seen in his editorial introduction to *Mappings*. Here Cosgrove (1999b: 9) discussed 'mapping' in its various guises – both literal and metaphorical – probing the 'complex accretion of cultural engagements with the world that surround and underpin the authoring of a map', questioning how maps 'may be regarded as a hinge around which pivot whole systems of meaning'. This concern with mapping and its meanings connected with a growing querying of the 'map' in geographical discourse, largely inspired by ***Brian Harley's*** work in the late 1980s on 'deconstructing the map' (Harley, 2001), and theoretical challenges laid at cartography's door during the 1990s. As with landscape, so with maps and mapping; 'critical cartography' has taken a more performative direction, less concerned with maps as visual media but, instead, more as unfolding objects, where 'maps are constantly in a state of becoming; constantly being remade by those who use them and engagement with them, a process of co-constitutive production between inscription, individual and world' (Kitchin and Dodge, 2007: 335). While theoretical shifts are constantly reworking these core spatial concepts of landscapes and mappings, Cosgrove's sustained focus on Renaissance and Enlightenment geographical knowledges, particularly the representation of the world and globe in cosmology and cosmography, endures (Cosgrove, 2001a; 2001b; 2007). So while for some Cosgrove's subject matter might appear esoteric, his approach was catholic, and from it he always spun a larger and forceful argument that has provided fuel for debate within and beyond geography – a true Renaissance project.

Cosgrove's Key Works

Cosgrove, D.E. (1993) *The Palladian Landscape: Geographical Change and its Cultural Representations in Sixteenth Century Italy*. Leicester: Leicester University Press.

Cosgrove, D.E. (1984/1998a) *Social Formation and Symbolic Landscape*. London: Croom Helm (second edition with additional introductory chapter, Madison: University of Wisconsin Press).

Cosgrove, D.E. (ed) (1999a) *Mappings*. London: Reaktion.

Cosgrove, D.E. (2001a) *Apollo's Eye: A Cartographic Genealogy of the Earth in the Western Imagination*. Baltimore, MA: Johns Hopkins University Press.

Cosgrove, D.E. (2008) *Geography and Vision. Seeing, Imagining and Representing the World*. London: I B Taurus.

Cosgrove, D.E. and Daniels, S. (eds) (1988) *The Iconography of Landscape: Essays on the Symbolic Representation, Design and Use of Past Environments*. Cambridge: Cambridge University Press.

Cosgrove, D.E. and Della Dora, V. (eds) (2009) *High Places: Cultural Geographies of Mountains, Ice and Science*. London: I B Tauris.

Secondary Sources and References

Atkinson, D. and Cosgrove, D.E. (1998) 'Urban rhetoric and embodied identities: City, nation and empire at the Vittorio Emanuele II monument in Rome 1870–1945', *Annals, Association of American Geographers*, 88: 28–49.

Cosgrove, D.E. (1985) 'Prospect perspective and the evolution of the landscape idea', *Transactions of the Institute of British Geographers,* NS 10: 45–62.

Cosgrove, D.E. (1988) 'The geometry of landscape: Practical and speculative arts in 16th century Venice', in D. Cosgrove and S. Daniels (eds), *The Iconography of Landscape*. Cambridge: Cambridge University Press. pp. 254–76.

Cosgrove, D.E. (1989) 'Geography is everywhere: Culture and symbolism in human landscapes', in D. Gregory and R. Walford (eds), *Horizons in Human Geography*. London: Palgrave. pp. 118–35.

Cosgrove, D.E. (1992) 'Mapping new worlds: Culture and cartography in sixteenth-century Venice', *Imago Mundi*, 44: 1–25.

Cosgrove, D.E. (1996) 'The measures of America', in J. Corner and A. MacLean (eds), *Taking Measures Across the American Landscape*. New Haven, CT: Yale University Press. pp. 3–13.

Cosgrove, D.E. (1998b) 'Introductory essay', in D.E. Cosgrove, *Social Formation and Symbolic Landscape*, 2nd edition. Madison, WI: Wisconsin University Press. pp. xi–xxxv.

Cosgrove, D.E. (1999b) 'Mapping meanings', in D.E. Cosgrove (ed), *Mappings*. London: Reaktion. pp. 1–23.

Cosgrove, D.E. (2001b) 'Geography's cosmos: The dream and the whole round earth', in K. Till, S. Hoelscher and P. Adams (eds), *Textures of Place*. Minneapolis, MN: University of Minnesota Press. pp. 326–39.

Cosgrove, D.E. (2006) *Geographical Imagination and the Authority of Images*. Stuttgart: Franz Steiner.

Cosgrove, D.E. (2007) 'Renaissance cosmography 1450–1650', in D. Woodward (ed), *The History of Cartography, Vol.3 'Renaissance cartography'*. Chicago, IL: University of Chicago Press. pp. 55–97.

Cosgrove, D.E. and Domosh, M. (1993) 'Author and authority: Writing the new cultural geography', in J. Duncan. and D. Ley (eds), *Place/culture/representation*. London: Routledge. pp. 25–38.

Cosgrove, D.E. and Fox, W. (2009) *Photography and Flight*. London: Reaktion.

Cosgrove, D.E., Roscoe, B. and Rycroft, S. (1996) 'Landscape and identity at Ladybower Reservoir and Rutland Water', *Transactions of the Institute of British Geographers,* NS 21: 534–51.

Daniels, S. (1993) *Fields of Vision: Landscape Imagery and National Identity in England and the United States.* Cambridge: Polity.

Daniels, S. and Cosgrove, D.E. (1988) 'Introduction: Iconography and landscape', in D.E. Cosgrove and S. Daniels (eds), *Iconography of Landscape.* Cambridge: Cambridge University Press. pp. 1–10.

Davies, A. (2021) 'Landscape semaphore: Seeing mud and mangroves in the Brazilian Northeast', *Transactions of the Institute of British Geographers,* 46: 626–41.

Dear, M. (1988) 'The post-modern challenge: Reconstructing human geography', *Transactions of the Institute of British Geographers,* NS 13: 262–74.

Della Dora, V. (2022) *Where Light in Darkness Lies: The Story of the Lighthouse.* London: Reaktion.

Driver, F. (1992) 'Geography's empire: Histories of geographical knowledge', *Environment and Planning D: Society and Space,* 10: 23–40.

Driver, F. (2001) *Geography Militant: Cultures of Exploration and Empire.* Oxford: Blackwell.

Driver, F. and Gilbert, D. (eds) (1999) *Imperial Cities: Landscape, Display and Identity.* Manchester: Manchester University Press.

Duncan, J. (1990) *City as Text: The Politics of Landscape Interpretation in the Kandyan Kingdom.* Cambridge: Cambridge University Press.

Duncan, J. and Duncan, N. (1988) '(Re)reading the landscape', *Environment and Planning D: Society and Space,* 6: 117–26.

Freytag, T. and Jöns, H. (2006) 'Vision and the cultural in human geography: A biographical interview with Denis Cosgrove', in D.E. Cosgrove (ed), *Geographical Imagination and the Authority of Images.* Stuttgart: Franz Steiner. pp. 79–91.

Graham, B. and Nash, C. (eds) (2000) *Modern Historical Geographies.* Harlow: Pearson.

Gregory, D. (1994) *Geographical Imaginations.* Oxford: Blackwell.

Harley, J.B (2001) *The New Nature of Maps: Essays in the History of Cartography.* Baltimore, MA: Johns Hopkins University Press.

Howard, P., Thompson, I., Waterton, E. and Atha, M. (eds) (2019) *The Routledge Companion to Landscape Studies.* London: Routledge.

Jackson, P. (1989) *Maps of Meaning: An Introduction to Cultural Geography.* London: Unwin Hyman.

Kitchin, R. and Dodge, M. (2007) 'Rethinking maps', *Progress in Human Geography,* 31: 1–14.

Lekan, T.M. (2014) 'Fractal Earth: Visualizing the global environment in the Anthropocene', *Environmental Humanities,* 5 (1): 171–201.

Lilley, K.D. (2009) 'Obituary: Denis E. Cosgrove, 1948–2008', *Social & Cultural Geography,* 10: 19–24.

Livingstone, D.N. (1992) *The Geographical Tradition.* Oxford: Blackwell.

Morrissey, J. (2013) 'The imperial present: Geography, imperialism, and its continued effects', in N.C. Johnson, R.H. Schein and J. Winders (eds), *The Wiley-Blackwell Companion to Cultural Geography.* Chichester: Wiley-Blackwell. pp. 494–507.

Muir, D. (1999) *Approaches to Landscape.* Basingstoke: Macmillan.

Nash, C. (1999) 'Landscapes', in P. Cloke, P. Crang and M. Goodwin (eds), *Introducing Human Geographies.* London: Arnold. pp. 217–25.

Ogborn, M. (1998) *Spaces of Modernity: London's Geographies, 1680–1780.* New York, NY: Guilford.

Smith, S. (1989) *The Politics of 'Race' and Residence: Citizenship, Segregation and White Supremacy in Britain.* Cambridge: Polity.

Vallerani, F. (ed) (2018) *Everyday Geographies and Hidden Memories: Remembering Denis Cosgrove.* Centre for the GeoHumanities Royal Holloway, University of London.

Whitehand, J.W.R. (2001) 'British urban morphology: The Conzenian tradition', *Urban Morphology*, 5: 103–9.

Whyte, I.D. (2002) *Landscape and History since 1500.* London: Reaktion.

Wylie, J. (2007) *Landscape.* London: Routledge.

12
Tim Cresswell

Peter Merriman, Aberystwyth University

Biographical Details and Theoretical Context

Tim Cresswell is a geographer and poet whose academic writing has made major contributions to cultural geography, mobility studies and the geohumanities. He was born in Cambridge, England in 1965 and completed a geography degree at University College London (UCL) in 1986, before moving to the US to complete a Master's degree (1988) and PhD (1992) in the Department of Geography at the University of Wisconsin, Madison. After a brief spell in Connecticut, Cresswell spent six years lecturing at the University of Wales, Lampeter; an important stopping place for many cultural geographers in Britain in the late 1980s and early 1990s (see Philo, 2002). He then moved to the University of Wales, Aberystwyth (1999–2006), Royal Holloway, University of London (2006–13) and Northeastern University, Boston (2013–16), before taking up a leadership role as Dean and Vice-President for Academic Affairs at Trinity College, Connecticut (2016–19). Since 2019 he has been Ogilvie Professor of Human Geography at the University of Edinburgh.

Cresswell's principal works have focused on the concepts of mobility and place, both together and separately. In previous reflections he has stated that this probably results from being an 'airforce kid' who moved around a lot, whose 'being is mobile', and who has 'never lived in a house for more than six or seven years' (Cresswell and Adey, 2022: 156). As an undergraduate at UCL Cresswell was particularly drawn to classes on humanistic geography and the emerging 'new cultural geography' led by ***Peter Jackson*** and Jacqueline Burgess, and he completed an undergraduate dissertation on 'Metaphors of mobility in the lyrics of Bob Dylan' (Cresswell, 2006). In Wisconsin he worked under the supervision of the influential humanistic geographer **Yi-Fu Tuan**, completing a Master's thesis on 'Go-egraphy: Mobility as resistance in modern American popular culture' (1988) and a PhD thesis titled 'In place/out of place: Geography, ideology and transgression' (1992) that became his first book in 1996. In this early work, Cresswell's empirical research focused on examples of mobility that were construed as transgressive, deviant and involving practices of resistance; including representations of mobility in Jack Kerouac's novels (Cresswell, 1993) and the actions of New Age Travellers, graffiti artists and members of the Greenham Common women's peace camp (Cresswell, 1996a). A subsequent project and second book looked at *The Tramp in America* (Cresswell, 2001a).

Mobility lies at the heart of many of these writings in the early and mid-1990s. In 1999, Cresswell co-organised a colloquium that brought together influential colleagues working on 'mobilities' in different fields of the humanities and social sciences. At the same time, **John Urry** (initially on his own, and later with Mimi Sheller) was delineating a new sociology of mobilities, establishing a Centre for Mobilities Research at Lancaster University in the UK in 2003 and a journal of *Mobilities* in 2006, around which a new multi-disciplinary field would emerge. In the same year as *Mobilities* launched and Sheller and Urry's (2006) important 'new mobilities paradigm' article was published, Cresswell's landmark volume *On the Move* examined the social, cultural and political 'production' of mobilities in the modern Western world, echoing **Henri Lefebvre**'s approach to the 'production of space' (1991) by outlining how movements become meaningful, enmeshed in power relations, and shape places (Cresswell, 2006). *On the Move* provided an important and influential critical analysis of different theoretical approaches to movement and mobility. It included seven in-depth studies of how mobility had been represented, regulated, practised, performed, experienced, and produced in the twentieth century. These focused on the photography of Eadweard Muybridge and Etienne-Jules Marey, the time and motion studies of Taylor and Gilbreth, the regulation of ballroom dancing in the UK, Chinese immigration in nineteenth- and twentieth-century America, the motor tours of suffragettes, the production of mobility in Schiphol Airport in Amsterdam, and mobility, rights and citizenship in the US. It remains the most important research monograph on mobility that spans the humanities and social sciences, and its theoretical contributions resonate through Cresswell's subsequent books, articles and co-edited collections, including volumes on *Gendered Mobilities* (Uteng and Cresswell, 2008), *Geographies of Mobilities* (Cresswell and Merriman, 2011), *Moving towards Transition* (Adey et al., 2021), and *Muybridge and Mobility* (Cresswell and Ott, 2022).

A second key theme of his work is place. Since the early 1990s, Cresswell has challenged static and closed-off conceptions of place to examine how movements shape places, how cultural and political struggles constitute places, and how certain kinds of practices and behaviours are seen as 'out of place' (Cresswell, 1996a). He has authored encyclopaedia and handbook entries on place, as well as a leading textbook – *Place* – which is now in its second edition (Cresswell, 2004; 2015a). His work has advanced theoretical understandings of place, as well as telling important and engaging empirical 'stories' that are underpinned by archival and library research.

Following his arrival at Royal Holloway, Cresswell undertook a part-time practice-based creative writing PhD in poetry supervised by Jo Shapcott, taking his thinking on place in more creative and experimental directions (Cresswell, 2015b). His second PhD, titled 'Topo-poetics: Poetry and place', combined a portfolio of fifty of his own poems with a series of critical essays on poetry and place (2015b). Cresswell approaches poems 'as places and spaces' and characterises writing poetry as 'a little form of place creation' (Cresswell, 2017: 319, 324). While Cresswell initially pursued his poetry and academic writing in parallel, as largely separate ventures, his recent scholarly publications explore the relations between poetry and geography (Cresswell, 2014a; 2015b; 2017; 2022), and experiment with more creative writing styles to understand the geographies of places such as Maxwell Street Market in Chicago (Cresswell, 2019). His poetry, which frequently addresses geographical themes, has been published across a broad range of poetry magazines and academic journals, as well as in three single-authored volumes – *Soil* (2013b); *Fence* (2015c); and *Plastiglomerate* (2020).

Cresswell's theoretical influences, thematic foci and methods are wide-ranging. His early work drew influence from **Pierre Bourdieu**'s theories of practice, British cultural studies, humanistic geography, and the emerging 'new cultural geography'. In later work on both mobility and place, Cresswell engaged with traditions of postmodern and post-structuralist theory, phenomenology, and most recently processual philosophy and assemblage thinking. There is a strong sense of social justice, spatial justice and mobility justice in his critical readings of past and present geographies, and he has undertaken detailed engagements with feminist theory and critical race theory that are woven through a number of projects. Beyond his writing, he has been an influential journal editor: of *cultural geographies* between 2007 and 2015, and as inaugural co-editor of *Geohumanities* from 2015.

Spatial Contributions

Place has remained a key preoccupation throughout Cresswell's career. He is drawn to its everydayness, which he contrasts with the seemingly more abstract concept of space and the rarefied debates and cultural contexts which were often entangled with Western traditions of landscape (Cresswell, 1996a, 2003). He shares this concern for place with his former supervisor **Yi-Fu Tuan**, whose two prominent books on *Topophilia* (1974) and *Space and Place* (1977) focused on the meaningful attachments people develop to places (Cresswell and Adey, 2022). However, Tuan tended to ignore the social and political processes shaping places and constraining human actions in places, reinforcing a binary where space is seen as abstract and place as particular, static, and the locus of dwelling (Tuan, 1977; Merriman, 2022). In contrast, Cresswell has advanced more open, multi-dimensional, vibrant and processual understandings of place, and focused on the politics of place and questions of power, resistance and ideology (Cresswell, 1996a, 2002). This approach is informed by a wide range of perspectives, including the writings of 'practice' theorists **Pierre Bourdieu** and ***Michel de Certeau***, post-structuralist geographers such as **Nigel Thrift**, post-structural feminists such as **Judith Butler**, and critical thinkers such as **Doreen Massey**, **David Harvey**, **Henri Lefebvre** and ***Edward Soja*** (see Cresswell, 2002).

In his wide-ranging essay 'Theorizing place', Cresswell seeks to steer a course between 'sedentary' imaginations of authentic, static and rooted places *and* 'nomadic' figurations that emphasise incessant flux and movement. What emerges is a theory of place that emphasises bodily practices and encounters, and an insistence that 'places are never complete, finished or bounded but are always becoming – in process' (Cresswell, 2002: 20). Places, like bodies in **Judith Butler**'s work, are seen to 'operate through constant and reiterative practice', being 'performed on an everyday basis' (2002: 23). In later writings Cresswell also emphasises the materiality of places, describing them as 'particular constellations of material things' (Cresswell, 2008: 135) and as 'constantly being made through gathering/weaving/assembling and constantly being pulled apart' (Cresswell, 2017: 321). In his writings on place and poetry, Cresswell draws upon the ideas of existentialists and phenomenologists such as Martin Heidegger, Maurice Merleau-Ponty, Jeff Malpas and Ed Casey (2015b, 2017, 2022), as well as the assemblage theorist Manuel DeLanda (Cresswell, 2017, 2019).

Cresswell's evolving and multi-faceted theorisation of place is best reflected in *Maxwell Street: Writing and Thinking Place* (2019). Rather than opening the book with a grand, abstract theoretical premise about what place is, he starts with a discussion of different strategies for 'writing place'. This is followed by a lengthy 'exercise in local theory' that provides a montaged history of Maxwell Street Market, and then Cresswell concludes by outlining a mesotheoretical account of 'place that might, at least in part, be applicable elsewhere, in other places' (Cresswell, 2019: 165). His 'mesotheory' brings insights from a vast array of critical theoretical traditions, building upon earlier work on places as formed by a 'gathering/assembling/weaving' of 'things, meanings, and practices' (Cresswell, 2019: 169). As he puts it, 'places both gather and disperse' (Cresswell, 2019: 175), and they change – possessing a temporality (Cresswell, 2019: 187). Places are lived and performed, and they are related to other places.

In his writings on poetry and place, Cresswell uses the term *topo-poetics* to refer to creative practices of 'place-making', tracing the etymology of 'topos' and 'poiesis' to their Greek roots, and coining a phrase which is reminiscent of the earlier work of Gaston Bachelard (1969 [1958]) on 'topo-analysis' and 'topophilia' (as well as **Yi-Fu Tuan**'s related writings on *Topophilia* (1974)). Cresswell also draws influence from Georges Perec's (2008 [1974]) essay on 'Species of spaces' and Heidegger's thoughts on dwelling to suggest that poems can be seen both as 'a kind of place' and an 'act of dwelling' (Cresswell, 2017: 319–26).

Cresswell does not claim to be a big 'T' or grand theorist who projects a single, over-arching theory of how places are formed, structured or experienced. Rather, his mesotheoretical approach is presented as a more grounded, open, and experimental guide to place-making, accompanied by important creative and empirical stories about how places are experienced and practised. This is a more 'qualified' theory of place that responds to several generations of critiques of universal theories by feminists, post-colonial thinkers, post-structuralists, queer theorists and critical race theorists.

Key Advances and Controversies

Cresswell's work has had a major influence on several generations of scholars and students working on mobility and place in cultural geography, mobility studies and the geohumanities. His early writings on the politics of place reflected the theoretical influences that were shaping much of the emerging politically-attuned cultural geography of the late 1980s and early 1990s. His textbooks on *Place* (2004, 2015a) and *Geographic Thought* (2013a) have become go-to resources for undergraduates, postgraduates and academics seeking an authoritative introduction to these fields. But perhaps more than anything, his writings on the cultural and political significance of mobilities have inspired a generation or two of cultural geographers, social scientists and humanities scholars associated with the so-called 'new mobilities paradigm'. In his third major book, *On the Move* (2006), Cresswell indicated his desire to approach mobility in a similar way to how **Lefebvre** treated space in *The Production of Space* (1991 [1974]), namely as something that is socially and politically produced, meaningful and caught up in intense power relations (Cresswell, 2001b, 2006, 2010). It was here that his influential, non-metric 'formula' for approaching mobility was loosely codified as 'mobility = movement + meaning + power' (Adey, 2022: 146).

Cresswell's writings have proposed conceptual shifts in how academics approach place, mobility, writing, landscape and more, but they appear to have sparked very little controversy. While Cresswell addresses big questions, his somewhat modest mesotheoretical conclusions, and the grounding of his arguments in rich stories and in-depth empirical case studies lends them an authority and realism which often appears incontrovertible. His ambition has always been to write in a way that is understandable and accessible, imagining that he is 'writing for [his] mum and dad' (Cresswell and Adey, 2022: 155). This accessibility and clarity were encouraged by Tuan, but while his supervisor did not always consider how others might experience the spaces and places that he described, much of Cresswell's writing has explored minoritarian and marginalised experiences of place. An early example which received some criticism was his study of Jack Kerouac's beat generation novel *On the Road,* in which the mobility of the male characters is presented as a form of rebellion and resistance against the stasis of the domestic sphere and the family home (Cresswell, 1993). In a response to the paper titled 'Off the road: Alternative views of rebellion, resistance and "the beats"', the feminist geographer **Linda McDowell** undertook a critical reading of Kerouac's *On the Road* and Cresswell's paper alongside Carolyn Cassady's memoir *Off the Road,* to suggest that Cresswell had presented 'too simple a dualism between masculinity and femininity, and their respective associations with the road and the home' (McDowell, 1996: 413). What emerged was an important exchange in *Transactions* which highlighted the role of different texts and contexts in both the production and interpretation of cultural representations, including the importance of considering the positionality of readers (and writers) (see Cresswell, 1996b; McDowell, 1996; Rycroft, 1996).

Cresswell's 'everyday' approach to mobility and place has led him to be critical of some processual, post-structuralist and non-representational approaches to the world which emphasise incessant flux, change, movement, contingency and relationality at the expense of an attention to stability, fixity, and agency (see Merriman, 2023). This critique has been directed in two related ways: towards universalising celebrations of flux and movement in cultural studies, mobility studies, and the humanities and social sciences; and towards certain tendencies evident within non-representational approaches to geography. First, in a series of theoretical interventions during the late 1990s and early 2000s, Cresswell was critical of accounts of the world that were grounded in a 'nomadic metaphysics' which puts 'mobility first, has little time for notions of attachment to place, and revels in notions of flow, flux, and dynamism' (Cresswell, 2006: 26; also 1997, 2001b). Scholars who invoked **Gilles Deleuze** and Félix Guattari's figure of the 'nomad' as a somewhat romanticised figure of resistance were a key target, and Cresswell echoed the concerns of postcolonial critics and feminist scholars for whom the nomad appeared to be a 'remarkably unsocial being – unmarked by traces of class, gender, ethnicity, sexuality, and geography' (Cresswell, 1997: 377). As non-representational and processual approaches to movement and mobility have gained increasing purchase in mobility studies, Cresswell has reiterated his concern that such approaches can flatten mobilities and overlook the very real, concrete materialities which halt or regulate movements of different kinds, such as the physical movements which are restricted by the Israel separation wall (Cresswell, 2014b). This understanding has been challenged by those who point out that 'to say that the world is in constant flux and movement' is not to argue that movements are 'uniform, linear and uncontrolled' or to deny that 'movements have different speeds, magnitudes, qualities and affective force and resonance' (Merriman, 2016: 559). What is more, in some of his co-authored research, Cresswell has himself drawn upon processual thinking on 'turbulent mobilities' in order to cut across the mobility/stasis binary, although these conceptual discussions remain closely tied to a single case study of the beached container ship, the MSC Napoli (Cresswell and Martin, 2012: 521).

A second and related focus of Cresswell's critical attention has been on 'non-representational theory', a term coined by **Nigel Thrift** to refer to a broad spectrum of anti-essentialist, contextual theoretical approaches to practice which seek to apprehend movement, embodiment, process, relations, subjectivity, affect, performance, materiality, and the unfolding of events (Thrift, 1996, 2008). Thrift's initial target was the 'new cultural geography' and its model of the world as text. Though Cresswell was one of many scholars who deployed the kinds of approaches that Thrift was critiquing, he also demonstrated a deep and profound engagement with embodied practices in his own work. Additionally, Cresswell – like Thrift – has drawn inspiration from the writings of **Pierre Bourdieu** and Maurice Merleau-Ponty. As an 'interested sceptic' (Lorimer, 2005; Cresswell, 2012), Cresswell disagreed with the way in which Thrift approached dance as a non-representational embodied practice, 'insist[ing] on the continuing importance of seeing bodily mobility within larger social, cultural, and geographical worlds' (Cresswell, 2006: 128). When invited to review the collection *Taking-Place: Non-Representational Theories and Geography*, edited by Ben Anderson and Paul Harrison (2010), Cresswell voiced his broader concerns in a series of personal reflections titled 'non-representational theory and me' (Cresswell, 2012). He suggested that the critique of geographers who treat 'representations' as static was something of 'a straw person' argument (2012: 100). He outlined the failure of many non-representational theorists to move beyond conventional texts as either sources of analysis or as the prime media through which they themselves communicate. Cresswell also demonstrated the continuities between non-representational theoretical concerns and those of earlier traditions of geographic thought, and pointed out that non-representational theories 'level agency too much' (2012: 101) and focus on unmarked and undifferentiated subjects (2012: 102). Further remarks focused on the problems associated with relational approaches and the question of how non-representational approaches address stasis, obduracy and structure. Cresswell's well-informed review is one of very few to provide a balanced critical evaluation of non-representational theories, neither dismissing them quickly, nor uncritically embracing a new orthodoxy.

Cresswell's Key Works

Cresswell, T. (1993) 'Mobility as resistance: A geographical reading of Kerouac's "On the Road"', *Transactions of the Institute of British Geographers*, 18 (2): 249–62.

Cresswell, T. (1996a) *In Place/Out of Place: Geography, Ideology and Transgression*. London: University of Minnesota Press.

Cresswell, T. (2001a) *The Tramp in America*. London: Reaktion.

Cresswell, T. (2002) 'Introduction: Theorizing place', in G. Verstraete and T. Cresswell (eds), *Mobilizing Place, Placing Mobility*. Amsterdam: Rodopi. pp. 11–32.

Cresswell, T. (2003) 'Landscape and the obliteration of practice', in K. Anderson, M. Domosh, S. Pile and N. Thrift (eds), *Handbook of Cultural Geography*. London: Sage. pp. 269–81.

Cresswell, T. (2004) *Place: A Short Introduction*. Oxford: Blackwell.

Cresswell, T. (2006) *On the Move: Mobility in the Modern Western World*. London: Routledge.

Cresswell, T. (2010) 'Towards a politics of mobility', *Environment and Planning D: Society and Space*, 28 (1): 17–31.

Cresswell, T. (2013a) *Geographic Thought: A Critical Introduction*. Chichester: John Wiley & Sons.

Cresswell, T. (2015a) *Place: An Introduction (second edition)*. Chichester: John Wiley & Sons.

Cresswell, T. (2015b) *Topo-Poetics: Poetry and Place*, doctoral dissertation, Royal Holloway, University of London. https://pure.royalholloway.ac.uk/ws/portalfiles/portal/25313757/Complete_poems.2015.final_signed.pdf (accessed October 31, 2023).

Cresswell, T. (2019) *Maxwell Street: Writing and Thinking Place*. Chicago, IL: University of Chicago Press.

Cresswell, T. and Merriman, P. (eds) (2011) *Geographies of Mobilities: Practices, Spaces, Subjects*. Farnham: Ashgate.

Secondary Sources and References

Adey, P. (2022) 'Introduction: Connecting, capturing and clarifying mobility', *Mobility Humanities*, 1 (2): 141–7.

Adey, P., Cresswell, T., Lee, J.Y., Nikolaeva, A., Nóvoa, A. and Temenos, C. (2021) *Moving Towards Transition: Commoning Mobility for a Low-carbon Future*. London: Bloomsbury Publishing.

Anderson, B. and Harrison, P. (eds) (2010) *Taking-Place: Non-Representational Theories and Geography*. Farnham: Ashgate.

Bachelard, G. (1969 [1958]) *The Poetics of Space*. Boston, MA: Beacon Press.

Cresswell, T. (1996b) 'Writing, reading and the problem of resistance: A reply to McDowell', *Transactions of the Institute of British Geographers*, 21 (2): 420–4.

Cresswell, T. (1997) 'Imagining the nomad: Mobility and the postmodern primitive', in G. Benko and U. Strohmayer (eds), *Space and Social Theory: Interpreting Modernity and Postmodernity*. Oxford: Blackwell. pp. 360–79.

Cresswell, T. (2001b) 'The production of mobilities', *New Formations*, 43: 11–25.

Cresswell, T. (2008) 'Place: Encountering geography as philosophy', *Geography*, 93 (3): 132–9.

Cresswell, T. (2012) 'Nonrepresentational theory and me: Notes of an interested sceptic', *Environment and Planning D: Society and Space*, 30 (1): 96–105.

Cresswell, T. (2013b) *Soil*. London: Penned in the Margins.

Cresswell, T. (2014a) 'Geographies of poetry/poetries of geography', *Cultural Geographies*, 21 (1): 141–6.

Cresswell, T. (2014b) 'Mobilities III: Moving on', *Progress in Human Geography*, 38 (5): 712–21.

Cresswell, T. (2015c) *Fence*. London: Penned in the Margins.

Cresswell, T. (2017) 'Towards topopoetics: Space, place and the poem', in B.B. Janz (ed), *Place, Space and Hermeneutics*. Cham: Springer. pp. 319–31.

Cresswell, T. (2020) *Plastiglomerate*. London: Penned in the Margins.

Cresswell, T. (2022) 'Writing (new) worlds: Poetry and place in a time of emergency', *Geografiska Annaler: Series B, Human Geography*, 104 (4): 374–89.

Cresswell, T. and Adey, P. (2022) 'An interview with Tim Cresswell', *Mobility Humanities*, 1 (2): 148–58.

Cresswell, T. and Martin, C. (2012) 'On turbulence: Entanglements of disorder and order on a Devon beach', *Tijdschrift voor Economische en Sociale Geografie*, 103 (5): 516–29.

Cresswell, T. and Ott, J. (2022) *Muybridge and Mobility*. London: University of California Press.

Lefebvre, H. (1991 [1974]) *The Production of Space*. Oxford: Blackwell.

Lorimer, H. (2005) 'Cultural geography: The busyness of being more-than-representational', *Progress in Human Geography*, 29 (1): 83–94.

McDowell, L. (1996) 'Off the road: Alternative views of rebellion, resistance, and "the beats"', *Transactions of the Institute of British Geographers*, 21 (2): 412–19.

Merriman, P. (2016) 'Mobilities II: Cruising', *Progress in Human Geography*, 40 (4): 555–64.

Merriman, P. (2022) *Space*. London: Routledge.

Merriman, P. (2023) 'Mobility/fixity: Rethinking binaries in mobility studies', *Mobility Humanities*, 2 (1): 60–75.

Perec, G. (2008 [1974]) 'Species of spaces/*Espèces d'Espaces*', in G. Perec, *Species of Spaces and Other Pieces*. London: Penguin Books. pp. 1–96.

Philo, C. (2002) '"Our journey through Lampeter": A marginal space of geographical knowledge', unpublished conference paper, available at: https://eprints.gla.ac.uk/116548/1/116548.pdf (accessed October 31, 2023).

Rycroft, S. (1996) 'Changing lanes: Textuality off and on the road', *Transactions of the Institute of British Geographers*, 21 (2): 425–8.

Sheller, M. and Urry, J. (2006) 'The new mobilities paradigm', *Environment and Planning A*, 38 (2): 207–26.

Thrift, N. (1996) *Spatial Formations*. London: Sage.

Thrift, N. (2008) *Non-Representational Theory*. London: Routledge.

Tuan, Y.-F. (1974) *Topophilia: A Study of Environmental Perception, Attitudes, and Values*. Englewood Cliffs, NJ: Prentice-Hall.

Tuan, Y.-F. (1977) *Space and Place*. London: University of Minnesota Press.

Uteng, T.P. and Cresswell, T. (eds) (2008) *Gendered Mobilities*. Aldershot: Ashgate.

13
Gilles Deleuze

Marcus A. Doel, University of Swansea
David B. Clarke, University of Swansea

Biographical Details and Theoretical Context

Gilles Deleuze was born into a bourgeois and conservative Parisian family in 1925. He was made famous by his catalytic role in the uprising of French students and workers in May 1968, for his promotion of a 'desiring revolution' that would destabilise the subjugation of the individual to the social formations that feed off it, and for his part in the emergence of what became known as poststructuralism, particularly through his collaboration with the militant psychoanalyst Félix Guattari (Dosse, 2010). Deleuze died in 1995, having served most of his career as Professor of Philosophy at Université de Paris VIII, Vincennes. He was a well-regarded teacher, and his classes were reputed to have been notable events. During his lifetime Deleuze rarely left France, not least because he disliked the conditions under which intellectuals travelled: flitting across the globe to exchange pleasantries and chatter. This is a good example of his indifference to the academic crowd and disdain for intellectual fashion. Deleuze invariably preferred the margins to the mainstream, employing neglected thinkers and heterodox ideas to wage a thought-provoking insurgency against the combined forces of State Philosophy and the social order it dogmatically sustains, not least through the ostensibly 'good' sense and 'common' sense that logic dictates and reason demands. Like so many other thinkers of his generation, Deleuze espoused a 'philosophy for militants' (Badiou, 2012) which continues to inspire an astonishingly diverse range of radical thinkers.

While Deleuze was not entirely averse to conversation – which may lead to who knows where (Deleuze and Parnet, 2007; 2011) – interviews and debates were anathema to him. 'We do not lack communication', wrote Deleuze and Guattari (1994: 108) in *What is Philosophy?* 'On the contrary, we have too much of it. We lack creation. *We lack resistance to the present*' (original emphasis). When Deleuze collaborated with others, it was not to communicate or exchange, as if he were at a swap meet. He considered collaboration as transformatory; an act of creative depersonalisation that opens a 'line of flight' along which one becomes a stranger to oneself. 'We are no longer ourselves', announced Deleuze and Guattari (1987: 3) at the outset of *A Thousand Plateaus: Capitalism and Schizophrenia*. With Félix Guattari, who was a radical psychoanalyst, schizoanalyst, psychogeographer, and eco-philosopher (Guattari, 1996; 2000; 2013; 2016), Deleuze undertook a long collaboration of convergence that spanned three decades. With Alain Badiou, another militant French philosopher, Deleuze belatedly entered into a 'collaboration of divergence' over their respective struggles to subtract 'One' from 'multiplicity' (Badiou, 1999; Crockett, 2013; Roffe, 2012).

Although Deleuze's interests were legion, he always considered himself to be a philosopher. According to Deleuze, however, there is nothing privileged about philosophy in relation to other activities, each of which is perfectly capable of thinking for itself, as he demonstrated so powerfully in his two books on cinema (Deleuze, 1986; 1989). There is nothing 'essential' or 'fundamental' about philosophy. For Deleuze, it was not the ground of truth, the pinnacle of reason or the font of knowledge. Such claims are at best laughable and at worst despotic. Nobody needs a philosopher to think *for* them, but somebody may benefit from a philosopher thinking *alongside* them: a form of collaboration rather than domination (e.g., Deleuze, 1997; 2000; 2003). Nevertheless, philosophy, like every other academic discipline, has not been averse to serving sectarian interests (nations, states, capital, sexism, racism, etc.). Indeed, philosophy has often been conducted as if it were in the service of a totalitarian police state: an endless succession of interrogations, inquisitions, and tribunals presided over by investigators, legislators, and judges. Deleuze was always attentive to the social and political context of philosophy, and especially how philosophy could be repressive rather than emancipatory. This is why he preferred posing interesting problems and seeking novel solutions to asking probing questions and demanding correct answers. So, rather than answering to a higher authority (and thereby submitting oneself to reason, logic, etc.), Deleuze's radicalism amounted to problematising the world, and thereby opening the potential for something genuinely new to happen (Hallward, 2006; Zourabichvili, 2012; Hardt, 2013; Savransky, 2021).

For Deleuze, then, philosophy is neither communication nor contemplation 'because no one needs philosophy to reflect on anything' (Deleuze and Guattari, 1994: 6). Like any other activity, philosophy is creative, and hence positive and joyful, even when it seems laborious and laboured. 'It is the same in philosophy as in a film or a song: no correct ideas, just ideas' (Deleuze and Parnet, 2007: 9). Needless to say, this constructivist and non-representational attitude 'beyond good and evil' put Deleuze out of step with most other philosophers, intellectuals, and militants, so many of whom tended to succumb to resentment, nihilism, and sadness. Deleuze had no time for doubt, reflection, and criticism or the 'terrible labour of the negative' (the paranoiac disposition of modern philosophy), and even less time for lack, scarcity, and limitation (the repressive myths of social science). Deleuze (1977: 112) stated that what he 'detested more than anything else was Hegelianism and the Dialectic', which misconstrued *difference* as opposition and contradiction, and *repetition* as reconciliation and sublation.

For Deleuze, philosophy is not law-enforcing and judgemental; it is creative and experimental. 'Let us abandon, like an old corpse, the idea that the natural form of thought is judgement. And above all, do not judge: that is a good axiom for thought', as Badiou (2009: 117) put it in his eulogy to Deleuze. Deleuze's philosophy created *passwords* that let slip, rather than *order words* that must be obeyed. This is what makes it radical and potentially earth-shattering. Echoing Nietzsche, who professed to 'philosophize with a hammer' (Deleuze, 1983), Deleuze and Guattari philosophised with 'thinking machines', manufacturing new concepts for a new people and a new earth (Deleuze and Guattari, 1984; 1994).

Of all the concepts Deleuze and Guattari assembled in their factory of geo-philosophy, the one that has been laboured over most often in geography is their concept of 'assemblage' [*agencement*], which is a dynamic articulation, machinic arrangement, and working diagram of disparate components that is both creative and expressive: assemblages make up the world; they make-believe; they make visible and intelligible, etc. This concept has spawned a cottage industry in 'Assemblage Thinking' (Anderson and McFarlane, 2011; Anderson et al., 2012; Buchanan, 2020; DeLanda, 2016). Moreover, what philosophers produce is fabricated for both philosophers and non-philosophers, just as music is produced for both musicians and music-lovers. Such was Deleuze's democratisation of philosophy, which was based on friendship rather than ownership, passion rather than possession, and radicalism rather than criticism. Indeed, Deleuze's willingness to share his thinking with non-humans, including animals, rhizomes, rocks, viruses, and machines, has inspired much of the work branded 'more-than-human', 'post-human', 'new materialist', and 'multinaturalist' (Roffe and Stark, 2015; Viveiros de Castro, 2017).

Eschewing the intellectual snobbery that glorifies thinking by letting it lord over existence and practice, philosophy for Deleuze was simply one 'flow among others; it enjoys no special privilege and enters into relationships of current and counter-current, of back-wash with other flows – the flows of shit, sperm, speech, action, eroticism, money, politics, etc.' (Deleuze, 1977: 114). Such is Deleuze's equalitarianism and anti-Platonism, which returns every hierarchy (e.g., the privileging of mind over matter, essence over appearance, identity over difference, and original over copy) to the superficial abyss whence it came. Additionally, while the vertical axis of hierarchy gives the illusion that identity, stability, and conservation are the rule rather than the exception, the horizontal axis of the *simulacrum* is always open to differentiation, instability, and transformation. Reading Deleuze is unsettling, then, because nothing is fixed in place. 'On the level of style, the reader cannot fail to be struck by the fact that Deleuze makes systematic use of division', noted Jean-Jacques Lecercle (1985: 97), and 'yet the text grows and multiplies in an extremely disquieting manner'.

Spatial Contributions

Even though philosophy is without privilege, Deleuze insisted that it is not without specificity, which brings it into the orbit of geography. On Deleuze's account, philosophers are responsible for creating concepts: not concepts in general (Ideas, Truths, Universals, etc.), 'globalizing' concepts, 'concepts as big as hollow teeth, THE law, THE master, THE rebel' (Deleuze, in Boundas, 1993: 254); but situated concepts, contextual concepts, contingent concepts; 'localizing' concepts that only have meaning, value, and efficacy in relation to some milieu. Concepts must be fabricated, articulated, and earthed. Deleuze treated concepts as territories and events; and philosophy as necessarily geo-philosophical and world-historical (Gasché, 2014; Lambert, 2021). 'Deleuze the thinker is the thinker of the event and always of this event in particular', wrote Jacques Derrida (1998: 3) in his eulogy to Deleuze. What Deleuze offers us, then, is not a first philosophy (which would be privileged, fundamental, foundational, essential, etc.), but a philosophy *for* the event – philosophy *as* an event. His philosophy responds to problems and encounters that *forced* him to think. This is a remarkably modest disposition that chimes with the spirit of most forms of materialism, process philosophy, and the philosophy of praxis, which always find themselves cast adrift in the midst of things: aleatory, errant, vagrant, ex-centric, and nomadic. Yet this is the only way for theoretical practice to become truly earth-shattering; for *thought itself* to actually *change* the world.

By attending to events, Deleuze provided a profound re-conceptualisation of space and time (Deleuze, 1988a; 1994). The Deleuzian 'event' is far removed from the commonsensical notion of something that is simply present-to-hand and fully given (what Derrida calls the 'metaphysics of presence'). Rather than a self-contained Now that happens once and for all, an event is an untimely Meanwhile or Meantime: a becoming-multiple (*and, but*) rather than a being-one (*is*). So, space–time does not consist of points, but of folds (Deleuze, 1988b; 1993) and manifolds (Calamari, 2017; Somers-Hall, 2018; Widder, 2019). However, although an event is never simply present, it is not at all lacking. It is full (i.e., immanent) – but it is fully *distributed*. In Deleuze's version of empiricism, then, an event is as much virtual as it is actual, and both are fully real. An event is not a matter of fact; it is an extraordinary, irruptive, and earth-shattering aberrant movement (Lapoujade, 2017).

Given that modern philosophy has not been especially concerned with events (and still less with spatialisation), one can appreciate why Deleuze – looking back to the Stoics – was often characterised as an outsider, even during his long apprenticeship in the history of philosophy, which gave rise to what he called 'monstrous' caricatures of Henri Bergson (Deleuze, 1988a), David Hume (Deleuze, 1991), Gottfried Leibniz (Deleuze, 1993), Immanuel Kant (Deleuze, 1984), and Baruch Spinoza (Deleuze, 1988c; 1990a), amongst others. These anti-rationalist portraits were not so much bastardisations as 'immaculate conceptions', rendering them *otherwise* through unfaithful reproduction and differential repetition.

As a key thinker of space and place, it is noteworthy that during each 'immaculate conception' Deleuze was eager to disclose the geo-philosophy that subtends the history of philosophy: *multiplicities* in Bergson, *cartography* in **Michel Foucault**, *relations* in Hume, *folds* in Leibniz, *territories* in Kant, *articulations* in Spinoza, and *differences* in Nietzsche. When Deleuze began to write in his own name, in books including *Difference and Repetition* (1994) and *The Logic of Sense* (1990b), this spatial awareness was already acutely developed. When he wrote with Guattari, in books such as *Anti-Oedipus* (1984), *Kafka* (1986), *A Thousand Plateaus* (1987), and *What is Philosophy?* (1994), one has the sense that there is geography, nothing but geography: maps, planes, surfaces, series, lines, strata, spaces, territories, transversals, etc. (Jellis et al., 2019). And although Deleuze's work is often regarded as difficult, he and Guattari were fond of putting ordinary words to extraordinary use: becoming-animal, molecular revolution, Body without Organs, desiring-machines, geology of morals, etc. While this eclectic 'pop philosophy' composed of *deterritorialised* terms – terms estranged from their customary habitats to address problems in other domains – has been appreciated by many, it has also given rise to considerable misunderstanding, criticism, and infringement claims. Indeed, Deleuze and Guattari arguably went out of their way to provoke incredulity when they declared such things as: 'Of course there are werewolves and vampires, we say this with all our hearts' (Deleuze and Guattari, 1987: 275). Their notions of 'schizophrenia as process', 'becoming-woman', 'nomad thought', and 'multiplicity' remain especially contentious.

Since thought is occasioned by events, '[t]here is no heaven for concepts. They must be invented, fabricated, or rather created and would be nothing without their creator's signature' (Deleuze and Guattari, 1994: 5). To put it bluntly, Deleuze felt that communication, interpretation, explanation, representation, reflection, and criticism are more often than not a waste of time, diverting us from the 'true voyage', which is a pursuance of creative encounters. Little wonder, then, that Deleuze should have been called a 'radically horizontal' thinker: of difference rather than identity; seriality rather than hierarchy; and becoming (*and ... and ... and ...*) rather than Being and Nothingness (*is, is not*). As Deleuze and Guattari famously declared at the outset of their delirious by-passing of the Marxian and Freudian impasse in *Anti-Oedipus* (1984: 2), '[a] schizophrenic out for a walk is a better model than a neurotic lying on the analyst's couch.' Rather than enlist academics in the service of Inquisitions, critical tribunals, and show trials (State Philosophy), Deleuze would prefer to take them out into the world for a stroll (nomad thought). Consequently, he was always on the lookout for creative encounters with texts, films, novels, etc. He was always on the lookout for something 'overwhelming' – an event – that would *force* him to think otherwise. Such is his fondness for becoming, transformation, and shape-shifting. Life always unfolds on the edge, on the horizon, and from the midst of things. 'Even if it is for the hundredth time, you must encounter each thing as if you have never known it before', insisted Paul Auster (1989: 7). 'No matter how many times, it must always be the first time. This is next to impossible, I realize, but it is an absolute rule.'

Key Advances and Controversies

During the 1990s, Deleuze's ex-centric empiricism, non-dialectical conception of difference, and immanent materialism began to make ripples in human geography (e.g., Jacobs and Morris, 1996; Doel, 1999). Like salacious gossip, word slowly spread – especially amongst poststructuralist, post-Marxist, and other so-called 'critical' human geographers – that Deleuze created an extraordinary form of *geo*-philosophy that proffered events for everyone (e.g., Bonta and Protevi, 2004; Rawes, 2008; Saldanha, 2017; Keating and Williams, 2022). Not only does this non-representational geo-philosophy provide an innovative basis for rethinking the nature of space and place, it is also a fully fledged 'thinking space' in its own right: not an abstract space *for* thought (a space of consciousness, representation, reflection, theory, etc. that would claim to be removed from the play of the world), but a concrete space *of* thought (a portion, region or milieu within the play of the world) – an honest-to-goodness *thinking* space. Quite simply, Deleuze had a remarkable ability to enliven even the most barren and derelict of spaces. He showed not only that every kind of space teems with life, but that space *itself* is alive. 'He is like a cork floating on a tempestuous ocean: he no longer moves, but is in an element that moves' (Deleuze, 1997: 26). In short, Deleuze's 'thinking space' can help us to become worthy of the specificity of geography: enabling the event of space to take place (Buchanan and Lambert, 2005; Anderson and Harrison, 2010; Frichot and Loo, 2013). And would-be geo-philosophers can now enjoy a vast secondary literature inspired by Deleuze and Guattari that spans practically every discipline, from Architecture to Zoology, including several series of books (e.g., Edinburgh University Press's *Deleuze Connections* and *Plateaus – New Directions in Deleuze Studies* series, and Bloomsbury's *Deleuze Encounters* series, which boast almost 100 volumes between them), and a dedicated *Deleuze and Guattari Studies* journal (also Edinburgh University Press).

We should end by noting that Deleuze was horrified by the prospect of becoming a 'master' thinker with followers and disciples. Given his fondness for encounters, estrangement, and transformation, Deleuze always favoured what was ex-centric (out of place and untimely), impersonal (collective and transhuman), and imperceptible (the becoming of the event). In turning to Deleuze, then, we should not turn him into a shining star that would reign over us: bright, eternal, and heavenly. Rather, we should turn to him as if we were turning towards the event horizon of a black hole: 'I dream not of being invisible, but imperceptible' (Deleuze, 1977: 112).

Deleuze's Key Works

Deleuze, G. (1983 [1962]) *Nietzsche and Philosophy*. New York, NY: Columbia University Press.

Deleuze, G. (1988a [1966]) *Bergsonism*. New York, NY: Zone.

Deleuze, G. (1990a [1968]) *Expressionism in Philosophy: Spinoza*. New York, NY: Zone.

Deleuze, G. (1990b [1969]) *The Logic of Sense*. New York, NY: Columbia University Press.

Deleuze, G. (1993 [1988]) *The Fold: Leibniz and the Baroque*. Minneapolis, MN: University of Minnesota Press.

Deleuze, G. (1994 [1968]) *Difference and Repetition*. London: Athlone.

Deleuze, G. and Guattari, F. (1984 [1972]) *Anti-Oedipus: Capitalism and Schizophrenia*. London: Athlone.

Deleuze, G. and Guattari, F. (1987 [1980]) *A Thousand Plateaus: Capitalism and Schizophrenia*. Minneapolis, MN: University of Minnesota Press.

Deleuze, G. and Guattari, F. (1994 [1991]) *What is Philosophy?* London: Verso.

Secondary Sources and References

Anderson, B. and Harrison, P. (eds) (2010) *Taking-Place: Non-Representational Theories and Geography*. Aldershot: Ashgate.

Anderson, B. and McFarlane, C. (eds) (2011) Special issue: Assemblage and Geography. *Area*, 43 (2).

Anderson, B., Kearnes, M., McFarlane, C. and Swanton, D. (2012) Special issue: On Assemblages and geography. *Dialogues in Human Geography*, 2 (2).

Auster, P. (1989) *In the Country of Last Things*. London: Faber & Faber.

Badiou, A. (1999) *Deleuze: The Clamour of Being*. Minneapolis, MN: University of Minnesota Press.

Badiou, A. (2009) *Pocket Pantheon*. London: Verso.

Badiou, A. (2012) *Philosophy for Militants*. London: Verso.

Bonta, M. and Protevi, J. (2004) *Deleuze and Geophilosophy*. Edinburgh: Edinburgh University Press.

Boundas, C.V. (ed) (1993) *The Deleuze Reader*. New York, NY: Columbia University Press.

Buchanan, I. (2020) *Assemblage Theory and Method*. London: Bloomsbury.

Buchanan, I. and Lambert, G. (eds) (2005) *Deleuze and Space*. Edinburgh: Edinburgh University Press.

Calamari, M. (2017) 'The mathematical model of smooth and striated spaces and the connectivity problem', *Deleuze Studies*, 11 (3): 328–55.

Crockett, C. (2013) *Deleuze Beyond Badiou: Ontology, Multiplicity, and Event*. New York, NY: Columbia University Press.

DeLanda, M. (2016) *Assemblage Theory*. Edinburgh: Edinburgh University Press.

Deleuze, G. (1977 [1973]) 'I have nothing to admit', *Semiotext(e)*, 2 (3): 111–16.

Deleuze, G. (1984 [1983]) *Kant's Critical Philosophy*. London: Athlone.

Deleuze, G. (1986 [1983]) *Cinema 1: The Movement-Image*. Minneapolis, MN: University of Minnesota Press.

Deleuze, G. (1988b [1986]) *Foucault*. Minneapolis, MN: University of Minnesota Press.

Deleuze, G. (1988c [1970]) *Spinoza: Practical Philosophy*. San Francisco, CA: City Lights.

Deleuze, G. (1989 [1985]) *Cinema 2: The Time-Image*. Minneapolis, MN: University of Minnesota Press.

Deleuze, G. (1991 [1953]) *Empiricism and Subjectivity: An Essay on Hume's Theory of Human Nature*. New York, NY: Columbia University Press.

Deleuze, G. (1997 [1993]) *Essays Critical and Clinical*. Minneapolis, MN: University of Minnesota Press.

Deleuze, G. (2000 [1964]) *Proust and Signs*. London: Athlone.

Deleuze, G. (2003 [1981]) *Francis Bacon: The Logic of Sensation*. London: Continuum.

Deleuze, G. and Guattari, F. (1986 [1975]) *Kafka: Toward a Minor Literature*. Minneapolis, MN: University of Minnesota Press.

Deleuze, G. and Parnet, C. (2007 [1977]) *Dialogues II*. New York, NY: Columbia University Press.

Deleuze, G. and Parnet, C. (2011 [1988/1989]) *Gilles Deleuze from A to Z* [DVD]. Los Angeles: Semiotext(e). https://www.youtube.com/playlist?list=PLiR8NqajHNPbaX2rBoA2z6IPGpU0IPlS2 (accessed November 17, 2023).

Derrida, J. (1998) 'I'm going to have to wander all alone', *Philosophy Today*, 42 (1): 3–5.

Doel, M.A. (1999) *Poststructuralist Geographies: The Diabolical Art of Spatial Science*. Edinburgh: Edinburgh University Press.

Dosse, F. (2010) *Gilles Deleuze and Félix Guattari: Intersecting Lives*. New York, NY: Columbia University Press.

Frichot, H. and Loo, S. (eds) (2013) *Deleuze and Architecture*. Edinburgh: Edinburgh University Press.

Gasché, R. (2014) *Geophilosophy: On Gilles Deleuze and Felix Guattari's* What Is Philosophy? Evanston, IL: Northwestern University Press.

Guattari, F. (1996) *Chaosophy: Soft Subversions*. Los Angeles, CA: Semiotext(e).

Guattari, F. (2000) *The Three Ecologies*. London: Athlone.

Guattari, F. (2013) *Schizoanalytic Cartographies*. London: Bloomsbury.

Guattari, F. (2016) *Lines of Flight: For Another World of Possibilities*. London: Bloomsbury.

Hallward, P. (2006) *Out of This World: Deleuze and the Philosophy of Creation*. London: Verso.

Hardt, M. (1993) *Gilles Deleuze: An Apprenticeship in Philosophy*. Minneapolis, MN: University of Minnesota Press.

Jacobs, J.M. and Morris, M. (eds) (1996) Special issue: Deleuze and Geography. *Environment and Planning D*, 14 (4).

Jellis, T., Gerlach, J. and Dewsbury, J.D. (eds) (2019) *Why Guattari? A Liberation of Cartographies, Ecologies and Politics*. Abingdon: Routledge.

Keating, T.P. and Williams, N. (eds) (2022) Special issue: Geophilosophy. *Subjectivity*, 15 (3).

Lambert, G. (2021) *Towards a Geopolitical Image of Thought*. Edinburgh: Edinburgh University Press.

Lapoujade, D. (2017) *Aberrant Movements: The Philosophy of Gilles Deleuze*. South Pasadena, CA: Semiotext(e).
Lecercle, J.-J. (1985) *Philosophy Through the Looking-Glass: Language, Nonsense, Desire*. La Salle, IL: Open Court.
Rawes, P. (2008) *Space, Geometry and Aesthetics: Through Kant and Towards Deleuze*. London: Palgrave Macmillan.
Roffe, J. (2012) *Badiou's Deleuze*. Durham, NC: Acumen.
Roffe, J. and Stark, H. (eds) (2015) *Deleuze and the Non/Human*. London: Palgrave Macmillan.
Saldanha, A. (2017) *Space After Deleuze*. London: Bloomsbury.
Savransky, M. (ed) (2021) Special issue: Problematizing the Problematic, *Theory, Culture & Society*, 38 (2).
Somers-Hall, H. (2018) 'The smooth and the striated', in H. Somers-Hall, J.A. Bell and J. Williams (eds), *A Thousand Plateaus and Philosophy*. Edinburgh: Edinburgh University Press. pp. 242–59.
Viveiros de Castro, E. (2017) *Cannibal Metaphysics*. Minneapolis, MN: University of Minnesota Press.
Widder, N. (2019) 'The mathematics of continuous multiplicities: The role of Riemann in Deleuze's reading of Bergson', *Deleuze and Guattari Studies*, 13 (3): 331–54.
Zourabichvili, F. (2012) *Deleuze: A Philosophy of the Event, and The Vocabulary of Deleuze*. Edinburgh: Edinburgh University Press.

14
Stuart Elden

Kimberley Peters, HIFMB, AWI and the Carl von Ossietzky Universität Oldenburg
Rachael Squire, Royal Holloway, University of London

Biographical Details and Theoretical Context

Stuart Elden's academic career began at Brunel University with a BSc (Hons) in Political and Modern History (1994), followed by a PhD in Political Theory from the same institution (1998). His first position post-PhD was as a Lecturer in the Department of Political and International Studies at Warwick University, UK (PAIS, one of the most renowned schools of political theory in the UK), where he remained for three years before moving to the Department of Geography at Durham University, UK, for 11 years. He returned to PAIS in 2013, the same year he was awarded a Doctor of Letters (DLitt) for post PhD-publications. He is a Fellow of the British Academy, and has held other important cross-disciplinary roles, such as Editor of *Environment and Planning D: Society and Space* (between 2006 and 2015), the premium journal for theory-led contributions to spatial thinking. At PAIS he currently holds the position of Professor of Political Theory and Geography, the title of his Chair reflecting the unique assemblage of ideas defining his work.

Elden's work traverses the disciplines of Political Science, History and Geography, and his unique interdisciplinary contributions can be divided into three main literatures. The first is a detailed socio-spatial analysis of the thinking of (predominantly) critical continental philosophers and philosophy (Elden, 2023a). The second focuses on history, historicisation of philosophical thinking (Elden, 2010a; 2014a) and themes emerging from such thinking: for example – and of particular interest to geographers – concepts such as 'space' (Elden, 2001a), 'geometry' (Elden, 2001b), and 'territory' (Elden, 2013a). The third is Elden's extensive and agenda-setting work on the notion of territory, not as an inert 'explanation' for the world as it is ordered and organised, but as a term that 'needs explaining' and 'understanding' (Elden, 2011: 306). For Elden, concepts such as territory are not given, they have their own (spatial) histories. Examining these histories reveals more clearly the concepts as they are known and used, politically, as well as academically.

On Elden's own website, the rich and regularly-updated *Progressive Geographies*, he delineates his work somewhat similarly as focusing on social/spatial theory, territory, Foucault studies, the work of Shakespeare and other projects, but notes there is 'some overlap between them' (Elden, 2023a: n.p.).

Indeed, Elden's vast back catalogue of work reveals a regular looping of ideas combining deep philosophical engagement with figures that are themselves 'key thinkers' on space and place, including continental philosophers **Michel Foucault**, Martin Heidegger, Friedrich Nietzsche and **Henri Lefebvre** (the first three the focus of Elden's PhD), and also a broader set of scholars, such as Peter Sloterdijk, Immanuel Kant and, more recently, Julia Kristeva. In each case, Elden explores how their work can help geographers to better understand the concepts crucial to their analysis. For Elden, reading through, and from, philosophers (and their readings, sometimes, of each other), using deep historical and archival analysis, has led to manifold spatio-historical contributions that have framed him too as a 'key thinker'.

His spatialised approach – and line of reckoning – became evident in Elden's (2001a) first book, *Mapping the Present: Heidegger, Foucault and the Project of a Spatial History*. Through a reading of Foucault's engagement with Nietzsche, and of the more pervasive influence of Heidegger upon Foucault, Elden demonstrates that concepts essential to the work of both Geography and geographers can be better understood by tracing the history of space as a term (see also Elden, 2001b) and treating history itself *as* necessarily spatial. As Elden writes:

> questions of spatiality have often been thought to be the preserve of geographers and urbanists, and there have been recently a number of attempts in these fields to critically theorize space [...] However practically valuable, useful and interesting these recent studies may have been, they are often worryingly conceptually weak [...] we need to both historicize space and spatialize history. In other words, rather than solely providing an analysis of how the meaning and use of the word 'space' has changed over time – a useful analysis to be sure – we need to recognize how space, place and location are crucial determining factors in any historical study. (Elden, 2001a: 2–3)

Elden here develops a thread that runs throughout his work: that philosophy, history and geography have a meaningful relationship and exploring those connections is necessary to make sense of the phenomena at the heart of Geography as a discipline. Indeed, in an early paper focused on Heidegger, Elden (2001b) historicises the term 'space' stretching back to Aristotle and early understandings of geometry and mathematical reasoning. In doing so, he highlights what the separation of space and place means for modern deployments of calculable space. This work on space led to perhaps Elden's greatest contribution to geography and geographers: his corpus of work on territory.

In later work, he has demonstrated (perhaps surprisingly) how the work of Shakespeare may be an essential tool in making better sense of territory (Elden, 2018; 2019). Through such readings, Elden reveals the specific ways that 'modern' workings of space emerged into mapped territories of control. This is a further important part of Elden's work: his reflections on spatial histories/histories of space are often also read through contemporary concerns. Punctuating Elden's core intellectual projects that feature in monograph form (e.g., see Elden, 2001a; 2009; 2013a; 2018) are papers that use the theoretical openings of this work to reflect upon the role of the European Union (EU) (Bialasiewicz et al., 2005a; Bialasiewicz et al., 2005b), Boko Haram (Elden, 2014b), the War on Terror (Elden, 2005; 2009) and role of the United Nations (Elden, 2006).

This diversity is reflected too on *Progressive Geographies* where Elden curates reading lists and resources for the academic community on key and pressing concerns, as was the case for the Ebola outbreak of 2013–2016 (see Elden, 2023b).

Spatial Contributions

Elden has been a key interlocuter introducing and interpreting the work of continental philosophers for geographical audiences. For example, he has provided notable overviews of the life and works of Lefebvre (Elden, 2004) and ensured that much of this work unpublished in English has eventually reached Anglophone audiences, including significant work on planetary urbanisation and rural land relations (editing compilations of key essays and lectures e.g., Lefebvre, 2022).

But, it is *The Birth of Territory* (Elden, 2013a) which arguably provides Elden's most significant spatial contribution. Following earlier work centred on the import of historical knowledge, this book *historicises* territory, providing a radically different conceptualisation of territory from other work that, as Kearns (2014, n.p.) notes, simply 'reassembles' modern assumptions that territory is 'a bounded space' and 'the object of rule'. However, Elden notes his interest in territory long preceded his 2013 publication:

> While I was certainly thinking about [*territory*] ... both in terms in teaching and research from around 2001, it really became the focus only when I finished the Heidegger book in late 2004. The historical work culminated in *The Birth of Territory*, submitted in 2011 and published in 2013. But while I began that book first, I put it aside for a while to write *Terror and Territory*, submitted in late 2007 and published in 2009. Some of the work from *The Birth of Territory* led directly to *Shakespearean Territories*, which I began in 2011–12 but which I didn't finally complete until 2017. (Elden, 2023a: n.p.)

In *The Birth of Territory* Elden displays, as he does in all of his work, detailed, multilingual, and meticulous historical research, tracing territory from antiquity to the 1800s and revealing, in the process, the 'complexities [*that*] are hidden behind the seemingly straightforward' (Elden, 2013a: 3). He also develops a fundamental way of understanding territory: not as a 'thing' or object of study for the discipline, but as a 'process' (Elden, 2013a: 17) and a 'political technology' (Elden, 2010b: 801).

Here, Elden develops a notion of *territoriality* that is different to prior thinking (whereby territoriality was understood as a means to approach territory). In doing so, Elden contends that '[t]erritory must be approached in itself rather than *through* territoriality, and in relation to land and terrain' (Elden, 2010b: 804, emphasis added). Here, two further themes in Elden's work become clear: first, his conception of territory as a political technology associated with *land* (2010b; 2013a), and, second, the importance of *terrain* and the materiality of territory (Elden, 2017a; 2021a).

Elden's conceptualisation of territory has inspired many. Notably, the edited book *Territory Beyond Terra* (Peters et al., 2018) extends Elden's work on grounded territory and moves it to geophysical spaces of sky and sea, and to slippery territories of mudflats and ice. Here Elden's influence of thinking territory as process is extrapolated to other 'earthly' spaces. This work pays heed to thinking with the very substance of territory, its matter, which Elden (2021a, 170) conceives also as 'political'. This call has been taken seriously in a raft of work focusing on territory's material agencies (e.g., see Thornton, 2015; Squire, 2016a, 2016b; Gordillo, 2023; Jamieson, 2023).

Leading from Elden's work on territory has been his agenda-setting work on volume as a quality through which the political technology of territory operates (Elden, 2013b). In his paper in *Political Geography*, Elden draws from the work of Israeli architect **Eyal Weizman**, German philosopher Peter Sloterjidk, and French theorist ***Paul Virilio***, each with their focus on opening up planes beyond the horizontal, to contend that territory is understood primarily as flat. As he notes, territory is a political technology via which 'areas' are secured, when, in fact, it is *volumes* through which power, control and governmentality, operate (see also **Stephen Graham** on *verticality*). Indeed, as Elden argues:

> Security goes up and down; space is volumetric [...] Most fundamentally, thinking merely straight up and down may blind us to different angles of approach, and the function of the oblique. Only by thinking through all of these aspects can we reflect more profoundly on the politics, metrics and power of volume. (Elden, 2013b: 15)

Like his work on territory, his development of the concept of volume has offered Geography and geographers much food for thought. With many hundreds of citations, the concept has gained firm traction as a key way to think about the politics of space and the workings of (state) power. 'Secure the volume' (Elden, 2013b) laid the foundations for swathes of further work thinking with volume, applying this frame of understanding to examples as varied as vertical urbanism (Harris, 2015), the operation of aerial drones in securing conservation territories (Millner, 2020), subterranean exploration (Pérez, 2015), and (deep) ocean spaces (Steinberg and Peters, 2015). This work has also spurred thinking about terrain in three-dimensional terms (Thornton, 2015; Squire, 2016b; 2021; Gordillo, 2016). Further work has aimed to expand Elden's theorisations, for example, thinking of volume in terms of capacity (Peters and Turner, 2018), and taking seriously the methods that may be required to research worlds of volume (Jackman and Squire, 2021).

Key Advances and Controversies

Elden's key advances on notions of territory and volume have garnered some constructive critique. This has included the absence of non-human life in his articulation of territory and a lack of attention to the ways the world is already inhabited before humans (see Gibbs, 2018; Squire, 2021). Perhaps the strongest critiques, however, originate from feminist and decolonial geographers who draw attention to the fact that 'the history of territory as a practice and an idea is often told as a history of (predominantly 'western') imperial, military, and economic struggles' that is highly gendered (Jackman et al., 2020: 2) (see also Bier, 2022). Halvorsen's (2019: 790) work is key here, arguing that territory had 'been little examined beyond the context of the modern, Eurocentric State'. Engaging with the context of Latin America, Halvorsen (2019: 790) calls for more open 'epistemological dialogue' with 'strategies to decolonise territory in thought and practice, orientated around the themes of land, terrain, and the state'. This involves dismantling colonial regimes of power that are reproduced in 'anglophone scholarship and sustained through dominant imaginations and practice of territory' (Halvorsen, 2019: 790). Halvorsen not only exposes the limits of 'western' thinking on territory, but sheds light on the ways that territory 'is resisted, (re) appropriated and (re)defined in the course of diverse grassroots struggles' (Halvorsen, 2019: 803). Indigenous scholars have also called for greater attention to be paid to the diverse ways that sovereignty is articulated in contested territory (see Correia, 2019; Daigle, 2019). Within this line of critique, Satizábal and Melo Zurita (2021: 267) call for 'plural understandings of territories, bodies, and Earth', emphasising their relational nature whilst thinking through the gendering of territory through practices of colonial and patriarchal violence. As a result they challenge the 'separation of territories, bodies, and disasters', thus opening space to 'counter hegemonic territorial propositions' (Satizábal and Melo Zurita, 2021: 267).

Jackman et al. (2020: 1, emphasis added) pick up on elements of these critiques to call for 'a *feminist* historiography of territory and terrain' that accounts for the gendered evolution of the concepts, and their ongoing reproduction in conceptual debates'. In doing so, they highlight diverse 'embodied accounts and accountings of these concepts' (Jackman et al., 2020: 1). Indeed, this work has paid attention to the body, and also the spiritual, in the process of territory. Paying attention to these personal, subjective, lived and lively elements that are essential, but often overlooked, in understanding territory, Jackman et al. (2020: 10) 'offer an important counterpoint to calculative, technology driven apprehensions of the earth in all its forms'. Such work, building on that which came before it, challenges the 'all too neat and tidy take' on territory, to 'facilitate the creation of more equitable accounts of political geography's messy, muddy, multiple, and lively territories and terrains' (Jackman et al., 2020: 10).

In surveying Elden's work, it is possible to reveal 'multiple' spatial contributions that are also, perhaps controversially, more overlooked. Foucault has long been at the centre of Elden's scholarship – as a window to thinking space, governmentality, control (Crampton and Elden, 2007) – and he has, significantly, written perhaps the most authoritative accounts of Foucault's thinking across four substantial tomes (Elden, 2016; 2017b; 2021b; 2023c). These books, though not published in the linear temporal order of Foucault's life, use extensive archival research to piece together spatial histories of Foucault's influence. Although not exclusively for a geographical audience, these should, as they too gain traction, provide sources of reflection for Geography in thinking about key concerns and the power-space-history nexus.

Elden's work on Foucault also represents a unique feature of his approach – his extensive reading, but also keen and constant engagement with often under-explored and examined ideas in and beyond the discipline. Running throughout Elden's work is an attention to that which fails to garner attention, and a call to the importance of history (Elden, 2014a). But Elden is also a scholar who looks forwards. Notably, he is attentive to interdisciplinary work across career stages, including that of early career scholars. Moreover, in *Progressive Geographies*, Elden offers not only regular reflections on the core interests of his work but also uses this space to share his expertise widely, for example, offering detailed and pointed advice on where academics can best approach working with Sloterjidk or Lefebvre when they are new to such dense theory, much of it only available via translation (see 2023b, n.p.). It is vital to note such attributes, not always evident in the counting of citations, awards or accolades: Elden is a thinker who thinks *of*, as well as *with* and *about*, some of the core tenets of geography: space and territory.

Elden's Key Works

Elden, S. (2001a) *Mapping the Present: Heidegger, Foucault and the Project of a Spatial History*. London: Continuum.

Elden, S. (2009) *Terror and Territory: The Spatial Extent of Sovereignty*. Minneapolis, MN: University of Minnesota Press.

Elden, S. (2010b) 'Land, terrain, territory', *Progress in Human Geography*, 34 (6): 799–817.

Elden, S. (2013a) *The Birth of Territory*. Chicago, IL: University of Chicago Press.

Elden, S. (2013b) 'Secure the volume: Vertical geopolitics and the depth of power', *Political Geography*, 34: 35–51.

Elden, S. (2018) *Shakespearean Territories*. Chicago, IL: University of Chicago Press.

Elden, S. (2021a) 'Terrain, politics, history', *Dialogues in Human Geography*, 11 (2): 170–89.

Secondary Sources and References

Bialasiewicz, L., Elden, S. and Painter, J. (2005a) 'The constitution of EU territory', *Comparative European Politics*, 3: 333–63.

Bialasiewicz, L., Elden, S. and Painter, J. (2005b) '"The best defence of our security lies in the spread of our values": Europe, America, and the question of values', *Environment and Planning D: Society and Space*, 23 (2):159–64.

Bier, J. (2022) 'Depth beyond 3D: The decolonial dimensions of volume', *Progress in Human Geography*, 46 (2): 672–88.

Correia, J.E. (2019) 'Unsettling territory', *Journal of Latin American Geography*, 18 (1): 11–37.

Crampton, J.W. and Elden, S. (eds) (2007) *Space, Knowledge and Power: Foucault and Geography*. Farnham: Ashgate.

Daigle, M. (2019) 'Tracing the terrain of Indigenous food sovereignties', *The Journal of Peasant Studies*, 46 (2): 297–315.

Elden, S. (2001b) 'The place of geometry: Heidegger's mathematical excursus on Aristotle', *The Heythrop Journal*, 42 (3): 311–28.

Elden, S. (2004) *Understanding Henri Lefebvre*. London: Bloomsbury.

Elden, S. (2005) 'Territorial integrity and the war on terror', *Environment and Planning A*, 37 (12): 2083–2104.

Elden, S. (2006) 'Contingent sovereignty, territorial integrity and the sanctity of borders', *SAIS Review of International Affairs*, 26: 11–24.

Elden, S. (2010a) 'Thinking territory historically', *Geopolitics*, 15 (4): 757–61.

Elden, S. (2011) 'What's shifting?', *Dialogues in Human Geography*, 1 (3): 304–7.

Elden, S. (2014a) 'Necessary, but not sufficient: Geography, territory and the history of ideas', *Dialogues in Human Geography*, 4 (3): 320–23.

Elden, S. (2014b) 'The geopolitics of Boko Haram and Nigeria's "war on terror"', *The Geographical Journal*, 180 (4): 414–25.

Elden, S. (2016) *Foucault's Last Decade*. Cambridge: Polity Press.

Elden, S. (2017a) 'Legal terrain – the political materiality of territory', *London Review of International Law*, 5 (2): 199–224.

Elden, S. (2017b) *Foucault: The Birth of Power*. Cambridge: Polity.

Elden, S. (2019) 'Why should people interested in territory read Shakespeare?', *Territory, Politics, Governance*, 7 (3): 289–96.

Elden, S. (2021b) *The Early Foucault*. Cambridge: Polity.

Elden, S. (2023a) 'Main periods of research', *Progressive Geographies*. Available at: https://progressivegeographies.com/about/main-periods-of-research/ (accessed November 1, 2023).

Elden, S. (2023b) 'Resources', *Progressive Geographies*. Available at: https://progressivegeographies.com/resources/ (accessed November 1, 2023).

Elden, S. (2023c) *The Archaeology of Foucault*. Cambridge: Polity.

Gibbs, L. (2018) 'Shores: Sharks, nets and more-than-human territory in Eastern Australia', in K. Peters, P. Steinberg and E. Stratford (eds), *Territory Beyond Terra*. London: Rowman & Littlefield. pp. 203–19.

Gordillo, G. (2023) 'Hostile terrain: On the spatial and affective conditions for revolution', *Territory, Politics, Governance*. doi.org/10.1080/21622671.2023.2172450

Halvorsen, S. (2019) 'Decolonising territory: Dialogues with Latin American knowledges and grassroots strategies', *Progress in Human Geography*, 43 (5): 790–814.

Harris, A. (2015) 'Vertical urbanisms: Opening up geographies of the three-dimensional city', *Progress in Human Geography*, 39 (5): 601–20.

Jackman, A. and Squire, R. (2021) 'Forging volumetric methods', *Area*, 53 (3): 492–500.

Jackman, A., Squire, R., Bruun, J. and Thornton, P. (2020) 'Unearthing feminist territories and terrains', *Political Geography*, 80, 102180.

Jamieson, W. (2023) 'All lines flow in: Excavating the geophilosophical relations of Singapore's infrastructure through SEA STATE', *Cultural Geographies*, 30 (1): 51–69.

Kearns, G. (2014) 'The birth of territory by Stuart Elden', *Society + Space*, June 2. https://www.societyandspace.org/articles/the-birth-of-territory-by-stuart-elden (accessed November 1, 2023).

Lefebvre, H. (2022) *On the Rural: Economy, Sociology, Geography*. (Ed. S. Elden and A. Morton. Trans. R. Bononno). Minneapolis, MN: University of Minnesota Press.

Millner, N. (2020) 'As the drone flies: Configuring a vertical politics of contestation within forest conservation', *Political Geography*, 80: article #102163.

Pérez, M.A. (2015) 'Exploring the vertical: Science and sociality in the field among cavers in Venezuela', *Social & Cultural Geography*, 16 (2): 226–47.

Peters, K. and Turner, J. (2018) 'Unlock the volume: Towards a politics of capacity', *Antipode*, 50 (4): 1037–56.

Peters, K., Stratford, E. and Steinberg, P. (eds) (2018) *Territory Beyond Terra*. London: Rowman & Littlefield.

Satizábal, P. and Melo Zurita, M.d.L. (2021) 'Bodies-holding-bodies: The trembling of women's territorio-cuerpo-tierra and the feminist responses to the earthquakes in Mexico City', *Third World Thematics: A TWQ Journal*, 6 (4–6): 267–89.

Squire, R. (2016a) 'Immersive terrain: The US Navy, Sealab and Cold War undersea geopolitics', *Area*, 48 (3): 332–8.

Squire, R. (2016b) 'Rock, water, air and fire: Foregrounding the elements in the Gibraltar–Spain dispute', *Environment and Planning D: Society and Space*, 34 (3): 545–63.

Squire, R. (2021) *Undersea Geopolitics: Sealab, Science, and the Cold War*. London: Rowman & Littlefield.

Steinberg, P. and Peters, K. (2015) 'Wet ontologies, fluid spaces: Giving depth to volume through oceanic thinking', *Environment and Planning D: Society and Space*, 33 (2): 247–64.

Thornton, P. (2015) 'The meaning of light: Seeing and being on the battlefield', *Cultural Geographies*, 22 (4): 567–83.

15
Sarah Elwood

Matthew W. Wilson, University of Kentucky

Biographical Details and Theoretical Context

Born in Rhinelander, Wisconsin in 1973, Sarah Elwood is a leading figure in critical human geography, especially with respect to its engagement with technological methods and the uses of geographic information systems (GIS). Indeed, Elwood's path-defining scholarship links critical social theory with the techniques and technologies of GIScience and has directly contributed to the development of participatory and public participation GIS, critical GIS, qualitative GIS, as well as community geographies and digital geographies, especially as these are informed by studies of structural inequalities and theorised from radical, feminist, queer, and Black studies.

In her first semester as an undergraduate at Macalester College in Saint Paul, Minnesota, Elwood's writing professor (Charles Norman) advised students to take two classes they had either always wanted to take or that they knew little about. Elwood chose geography and anthropology, as neither were taught in the schools she had attended. Geography became her academic home. Completing her BA in Geography in 1994, Elwood entered graduate school at the University of Minnesota, graduating with an MA in Geography in 1996 and a PhD in Geography in 2000. Her deep connections to the educational spaces of the discipline are only perhaps rivalled by her lasting commitments to the places of these institutions and the people that made up her community. Elwood has described how having a twin sister, Nora, 'wires you for a complexity: twinship is the holding of sameness and difference, where your first pronoun is *we*' (pers. comm). Descended from church-going people from the midwestern US, Elwood's style is working hard through adversity, rarely complaining and always speaking up.

In Minneapolis, Elwood participated in local politics around LGBTQIA+ rights and in bringing attention to gender violence and homelessness. Elwood's coursework and academic training involved urban geography, computer cartographies, and early developments in GIS. She was inspired by feminist political theory, queer studies, social movements, liberation theologies, and Black and Latinx feminisms. Minneapolis became the site for Elwood to better understand how community-based organisations made use of mapping and GIS to advocate for their concerns and interests. It was perhaps where Elwood's unique socio-technical politics of place emerged. It was also here, in Minneapolis, that Elwood met her partner Lisa.

Elwood's scholarly foundations charted a course that drew together radical moorings in critical human geography with the techniques and processes of GIScience. Her first sole-author publications underlined the breadth of her disciplinary contributions: 'Lesbian living space: Multiple meanings of home', published in 2000 in *The Journal of Lesbian Studies*, and 'GIS use in community planning: A multi-dimensional analysis of empowerment', published in 2002 in *Environment and Planning A*. While the former piece provides evidence of Elwood's long-held interest in the intersectional dynamics of place-based experiences, the latter piece traces these important considerations into the realm of the technological: 'The question is less whether GIS is empowering or disempowering, but in what ways does it foster empowerment and disempowerment, and *for whom*?' (Elwood, 2002: 907–908, emphasis added).

Elwood and her partner would make the move to Chicago in 2000 for Elwood's first faculty position at DePaul University, a position she held until 2004. Bringing her experience of working with a neighbourhood organisation in Minneapolis, Elwood wasted no time in making important connections with non-profits and community organisations in Chicago, thereby building a research program in critical, place-based, participatory GIS. Her research entered debates in urban geography on the role of participatory uses of technology in the contexts of neoliberal policies at multiple scales. She connected this work to the classroom developing a model of experiential learning with GIS that continues in various forms and places today. While at DePaul, Elwood was awarded a prestigious CAREER grant from the National Science Foundation, which funded empirical, participatory research from 2003 through 2010 into how community-based organisations might utilise GIS to transform local urban governance.

Their move in 2004 to Tucson and the University of Arizona would provide an important, if brief, launch of Elwood's research at a major doctoral program. The NSF CAREER grant continued to generate work on urban politics and the co-implication of GIS-based narratives, and her publications at this time also captured a national disciplinary conversation on the digital, felt across the social sciences and humanities. In 2006, the opportunity to relocate to Seattle would prove too tempting. With the relief of being able to move to a place where same-sex partners had greater recognition and support, Elwood would also be able to return to the US northwest, a region where she had history, or more aptly, a geography, having attended high school in Oregon.

Her arrival at the University of Washington (UW) in 2006 signalled a second generation of GIS scholarship. The Department of Geography at UW had been an important centre of the so-called quantitative revolution in the discipline of Geography, and had built the department's scholarship on GIScience upon that foundation. Elwood's arrival marked a turn to work with and on GIScience that was avowedly critical, participatory, queer, and radical. Her early years at UW saw Elwood continue her thinking through the case of Humboldt Park in Chicago and the role of neighbourhood groups as stakeholders in local decision-making and ethnographic methods. This work in critical GIS found outlets in new human geography and GIScience collaborations around youth mapping, volunteered geographic information and spatial media, and relational poverty. As the collection of technologies under the umbrella of GIScience continued to expand, she was at the forefront of emerging digital geographies. Elwood became a name synonymous with mentorship, actively fostering the next generation of critical digital geographic and critical GIS scholarship. From 2018, in her position as department chair, Elwood deepened her longstanding commitments to expand the diversity of perspectives in critical geography and create a GIScience, discipline, and academy that is more inclusive.

Spatial Contributions

Elwood's contributions to spatial theory and practice are evident in four programs of study, investigating: (1) the epistemological shifts that emerge within currents of participatory, collaborative, critical and community-based deployments of GIS, (2) the expansion of spatial media, its attendant practices and data, and the impacts on critical mapping/GIS, (3) the dynamics of power and privilege, oppression and resistance, in everyday digital geographies, and (4) the relational politics of poverty amid increasing inequities and precarities. Whether in projects that develop new GIS capacities for community-based nonprofits or in work that examines the implications of digital societies or in comparative analysis of classed place-making, Elwood's concept of space is one of hopeful potential. However, this disposition toward space is not born of naive optimism, but of a long analysis of people utilising their knowledge to better articulate what is urgently necessary. While being resolutely nondeterministic, Elwood's space is where the action is and, therefore, is also formed of an awareness of the partiality and incompleteness of its representation.

For Elwood, this peculiar cartographic crisis of representation, in which an analysis of mapmaking and GIS reveals violent and dispossessing power geometries, requires that critical cartographers and GIS practitioners take up the responsibilities of these techniques to interrupt and change these geometries. To develop this spatial practice, Elwood's (2000) dissertation research in the Powderhorn Park neighbourhood of Minneapolis constituted a form of urban geographic and GIScience research that was quite novel: to study the use of maps and GIS by community-based organisations, to both improve capacities while also analysing their social situations and implications. This research involved interviews, participant observation of organisation, neighbourhood, and city government processes, as well as analysis of maps, spatial data, and systems for spatial analysis. At issue in the Minneapolis work were questions of empowerment, legitimacy, and access, especially as these might be addressed differently across scales. The neighbourhood organisation's capacity to use GIS, albeit through an instrumentalised rationality, also reinforced the state and its hegemonic approach to planning, in Elwood's nuanced and complex analysis. In so doing, she sought to provide an analysis of the shifting relationship between state and civil society *and* the technical affordances and limitations of community-based GIS practices.

Elwood's long history of developing classroom experiences that address complex university–community relations and centre a diversity of experiences and voices is a mainstay of her commitment to remaking the university in service of vulnerable peoples and places (see Elwood, 2009; Elwood and Wilson, 2017). At DePaul, Elwood developed field experiences in a course called 'Explore Chicago'. Here, students took up topics of gentrification, public transportation access, and affordable housing. The NSF CAREER-funded project in the Humboldt Park neighbourhood of Chicago deepened these inquiries as a collaboration that drew undergraduate students in Elwood's courses into questions that shaped everyday experiences in a neighbourhood beset with the issues of disinvestment.

In the Chicago project, Elwood's (2006a; 2006b; 2006c) research examined systematically the forms of politics that emerged in participatory GIS settings, challenging more traditional GIS analysis that predominantly thought that, under the right socio-technical conditions, power and politics could be controlled for and mitigated. For Elwood, politics were imbued, always-already co-constitutive within the spaces of community-based GIS use. Indeed, her publications highlighted the contradictions of participatory GIS, not as an attempt to resolve these, but to recognise them as part of the complexity of engaged work. Her data and field notes included moments where community groups were articulating concerns and directions for development using maps and other outputs from GIS. Elwood describes these as *spatial narratives*, drawing upon ***Michel de Certeau*** (1984), defining these as 'flexible frameworks that produce and reproduce the neighbourhoods in different ways for different audiences, projects, or arguments' (Elwood, 2006a: 332). Importantly, there were also lessons for GIScience, that issues of spatial data handling, access, and interoperability take on new urgency as a greater diversity of organisations make use of GIS to advance their concerns. Despite the contradictions, GIS is an important instrument in urban politics. Elwood writes, 'GIS is tremendously important because it is such a powerful mediator of spatial knowledge, social and political power, and intellectual practice in geography. In short, the answer [to the question, "Why use GIS?"] is the same today as it was ten years ago: Because the stakes are very high' (Elwood 2006b: 693).

While the project in Humboldt continued throughout Elwood's move to Arizona and then to Washington, a new direction in research was emerging around the role of new spatial media (see Elwood, 2008a). Rapid changes in internet-based technologies around what was called 'web 2.0' (enabling user-produced content) was creating new opportunities for geographers to both examine spatial phenomena and develop and share web-based maps. Here again, Elwood (2008b) maintains space as a potentiality and her publications on 'volunteered geographic information' (VGI) encouraged GIScientists to centre questions of power, positionality, and identity as they shifted their research agendas toward these new sources of data and computational capacity. As the first woman to be invited to write the reports on GIScience for *Progress in Human Geography*, Elwood carefully connected these new Internet-based technologies around the geospatial web (or geoweb; web-based GIS and new networked, interactive, participatory spatial technologies, and georeferenced information organised and delivered through the internet) to questions motivating critical GIS research. Indeed, for Elwood (and her co-author and first PhD student at Washington, Agnieszka Leszczynski; see Elwood and Leszczynski, 2011, 2013, 2018), the geoweb also represented a new area of privacy concerns and of new closures on the potentiality of space in the wake of digital industries. Drawing upon feminist theory, Latinx and Black studies, this work contributed to an emerging field of digital geography.

In some ways, Elwood's interest in space as a hopeful potential provides a direct connection from these more GIS-centric and digital geographic research projects to her collaborative work developing a relational poverty network. In writing with colleague Vicky Lawson and others in their network, Elwood et al. (2015) examine sites where politics around poverty emerge, whether in the comparative analysis of middle-class residential areas or in the visual forms of resistance to homelessness and impoverishment. Elwood remains committed to both advancing a spatial analytics of vulnerability and providing a visual spatial narrative bringing critical attention to such precarities.

Key Advances and Controversies

Elwood has advanced the GIS and Society agenda, begun in the 1990s, and has expanded the concerns of this field to incorporate questions of inclusivity and difference. For her, GIS and digital tools remain too potent to remain in the hands of the powerful, and her scholarship continues to centre the political in areas where concern about the technical dominates. In doing so, Elwood has championed a new generation of critical GIS scholars and digital geographers, and her reworking of university classrooms to be spaces where critical mapping and study can occur with community partners has positively shaped the academy. Indeed, current work in community geographies benefits from these earlier struggles around participatory and qualitative GIS, and especially Elwood's work, to create an ethics where the ability to utilise GIS and digital tools come with great responsibility.

Importantly, it is Elwood's approach to scholarship as perpetual bridge-building that perhaps enables the work to steer clear of unproductive controversy. Of course, this is not to say that her scholarship avoids complexity, but that her approach is to listen and amplify, to utilise her position in service of those voices that are absent from the decision-making spaces of state power or the discourses of the academy. That these forms of scholarship may take the shape of a map or a podcast or a journal article underlines her steadfast work to carefully introduce trouble where the stakes are high and the issues are urgent.

Her scholarship demonstrates an important link between spatial technologies and spatial knowledge. While this link is technically supported by scholarship in mainstream GIScience, the political and, more precisely, epistemological dimensions of Elwood's scholarship remain to be taken up. To put it coarsely, if GIScience is concerned with refining the methods and models that represent the *truth* of spatiality (albeit perhaps multiple and plural), then Elwood's critical GIScience recognises that this concern is always-already impacted by the non-neutrality of the technology, its attendant practices, and its situation. Her co-authored publications on volunteered geographic information with Mike Goodchild and Dan Sui (e.g. Elwood et al., 2012), have broad use across research in citizen science and crowdsourcing (Sieber and Haklay, 2015), social network analysis (Jiang et al., 2015), environmental monitoring (Ghermandi and Sinclair, 2019), disaster management and humanitarian response coordination (Haworth and Bruce, 2015) and smart city systems (Sagl et al., 2015) among others. Her work with Goodchild and Sui provided an intellectual framework to better understand *and* further the proliferation of tools, 'neo-geographic' practices, and new spatial data infrastructures within the geoweb. They write, 'geographers no longer have exclusive claims on the production of either geographic information or geographic knowledge' (Elwood et al., 2012: 585), and this seems particularly directed toward scholars in GIScience.

As Elwood's work in critical GIS, qualitative GIS, and participatory GIS has demonstrated, claims to information and knowledge were only ever partial and always imbricated in power geometries. Her work is required reading for those interested in spatial media, the geoweb, and digital geographies – especially as these interests intersect questions around power, politics, visuality, privacy, and resistance. For instance, Shelton et al. (2014) drew upon Twitter data around Hurricane Sandy to highlight the importance of mixed methods, and specifically *smaller* big data, in the rush of VGI. Elwood's scholarship has proven instructive here, from her path-defining contributions in public participation GIS to the experimental and radical work in qualitative GIS (see her edited collection with Meghan Cope in 2009). Her contributions continue to widen its focus, beyond the important but limiting knowledge politics that point to new sources of data by new participants, toward being 'epistemologically open to seeing how, against all odds, illegible subjects deploy digitality toward thriving and re-inventing' (Elwood, 2022: 439; see also Elwood, 2021).

While mainstream GIScience remains fixated on the affordances of artificial intelligence and automation despite Elwood's push to examine the knowledge politics of spatial technologies, her work has greatly impacted the development of digital geographies (and the study of AI and automation as objects and not only as methods/techniques). Ash, Kitchin, and Leszczynski (2018: 27) described this turn toward the digital as creating 'geographies produced *through*, produced *by*, and *of* the digital' (their emphasis), all of which are evident in Elwood's scholarship. Indeed, on Elwood's watch, the field loosely organised as digital geographies has enveloped concerns in critical GIS while widening its focus on digitality. Here, work by Richardson (2018), Wilmott (2019), and Gieseking (2018) expand Elwood's commitment to feminist and queer approaches to the study of spatial media, and her former students' publications reflect her mentorship toward a more inclusive digital geographies in the study of urban platformisation (Leszczynski, 2020), community groups and smart city initiatives (Burns and Welker, 2023), digital colonialisms (Young, 2019), poverty discourses (Gordon, 2013), digital inequalities (Slager, 2018), and digital Indigenous epistemologies (Rivera, 2023).

Elwood's Key Works

Cope, M. and Elwood, S.A. (eds) (2009) *Qualitative GIS: A Mixed Methods Approach*. London: Sage.

Elwood, S.A. (2006a) 'Beyond cooptation or resistance: Urban spatial politics, community organizations, and GIS-based spatial narratives', *Annals of the Association of American Geographers*, 96 (2): 323–41.

Elwood, S.A. (2006b) 'Critical issues in participatory GIS: Deconstructions, reconstructions, and new research directions', *Transactions in GIS*, 10 (5): 693–708.

Elwood, S.A. (2006c) 'Negotiating knowledge production: The everyday inclusions, exclusions, and contradictions of participatory GIS research', *The Professional Geographer*, 58 (2): 197–208.

Elwood, S.A. (2008a) 'Grassroots groups as stakeholders in spatial data infrastructures: Challenges and opportunities for local data development and sharing', *International Journal of Geographical Information Science*, 22 (1): 71–90.

Elwood, S.A. (2008b) 'Volunteered geographic information: Future research directions motivated by critical, participatory, and feminist GIS', *GeoJournal*, 72: 173–83.

Elwood, S.A. (2009) 'Integrating participatory action research and GIS education: Negotiating methodologies, politics and technologies', *Journal of Geography in Higher Education*, 33 (1): 51–65.

Elwood, S.A. (2021) 'Digital geographies, feminist relationality, Black and queer code studies: Thriving otherwise', *Progress in Human Geography*, 45 (2): 209–28.

Elwood, S.A., Goodchild, M.F. and Sui, D.Z. (2012) 'Researching volunteered geographic information: Spatial data, geographic research, and new social practice', *Annals of the Association of American Geographers*, 102 (3): 571–90.

Elwood, S.A., Lawson, V. and Nowak, S. (2015) 'Negotiating poverty and privilege: Middle class place-making and poverty politics', *Annals of the Association of American Geographers*, 105 (1): 123–43.

Elwood, S.A. and Leszczynski, A. (2013) 'New spatial media, new knowledge politics', *Transactions of the IBG*, 38 (4): 544–59.

Elwood, S.A. and Leszczynski, A. (2018) 'Feminist digital geographies', *Gender, Place & Culture*, 25 (5): 629–44.

Elwood, S.A. and Wilson, M.W. (2017) 'Critical GIS pedagogies beyond "Week 10: Ethics"', *International Journal of Geographical Information Science*, 31 (10): 2098–116.

Secondary Sources and References

Ash, J., Kitchin, R. and Leszczynski, A. (2018) 'Digital turn, digital geographies?', *Progress in Human Geography*, 42 (1): 25–43.

Burns, R. and Welker, P. (2023) 'Interstitiality in the smart city: More than top-down and bottom-up smartness', *Urban Studies*, 60 (2): 308–24.

de Certeau, M. (1984) *The Practice of Everyday Life*. Berkeley, CA: University of California Press.

Elwood, S.A. (2000) *Information for Change: The Social and Political Impacts of Geographic Information Technologies*. PhD dissertation, Geography, University of Minnesota.

Elwood, S.A. (2000) 'Lesbian living spaces: Multiple meanings of home', in G. Valentine (ed), *From Nowhere to Everywhere: Lesbian Geographies*. New York, NY: Harrington Park. pp. 11–28.

Elwood, S.A. (2002) 'GIS use in community planning: A multidimensional analysis of empowerment', *Environment and Planning A*, 34: 905–22.

Elwood, S.A. (2022) 'Toward a fourth generation of critical GIS: Extraordinary politics', *ACME: An International Journal for Critical Geographies*, 21 (4): 436–47.

Elwood, S.A. and Leszczynski, A. (2011) 'Privacy, reconsidered: New representations, data practices, and the geoweb', *Geoforum*, 42: 6–15.

Ghermandi, A. and Sinclair, M. (2019) 'Passive crowdsourcing of social media in environmental research: A systematic map', *Global Environmental Change*, 55: 36–47.

Gieseking, J. (2018) 'Operating anew: Queering GIS with good enough software', *The Canadian Geographer*, 62 (1): 55–66.

Gordon, E. (2013) 'Under-served and un-deserving: Youth empowerment programs, poverty discourses and subject formation', *Geoforum*, 50: 107–16.

Haworth, B. and Bruce, E. (2015) 'A review of volunteered geographic information for disaster management', *Geography Compass*, 9 (5): 237–50.

Jiang, S., Alves, A., Rodrigues, F., Ferreira, Jr., J. and Pereira, F.C. (2015) 'Mining point-of-interest data from social networks for urban land use classification and disaggregation', *Computers, Environment and Urban Systems*, 53: 36–46.

Leszczynski, A. (2020) 'Glitchy vignettes of platform urbanism', *Environment &*

Planning D: Society and Space, 38 (2): 189–208.

Richardson, L. (2018) 'Feminist geographies of digital work', *Progress in Human Geography*, 42 (2): 244–63.

Rivera, I. (2023) 'Undoing settler imaginaries: (Re)imagining digital knowledge politics', *Progress in Human Geography*, 47 (2). doi.org/10.1177/03091325231154873

Sagl, G., Resch, B. and Blaschke, T. (2015) 'Contextual sensing: Integrating contextual information with human and technical geo-sensor information for smart cities', *Sensors*, 15 (7): 17013–35.

Shelton, T., Poorthuis, A., Graham, M. and Zook, M.A. (2014) 'Mapping the data shadows of Hurricane Sandy: Uncovering the sociospatial dimensions of "Big Data"', *Geoforum*, 52: 167–79.

Sieber, R.E. and Haklay, M. (2015) 'The epistemology(s) of volunteered geographic information: A critique', *Geo: Geography and Environment*, 2 (2): 107–63.

Slager, E.J. (2018) 'Infrastructures of survival: Digital justice and black poetics in community Internet provision'. PhD dissertation, Geography, University of Washington. Available from http://hdl.handle.net/1773/43394 (accessed November 1, 2023).

Wilmott, C. (2019) '"Mapping-with": The politics of (counter-)classification in OpenStreetMap', *Cartographic Perspectives*, 92: 43–57.

Young, J.C. (2019) 'The new knowledge politics of digital colonialism', *Environment and Planning A: Economy and Space*, 51 (7): 1424–41.

16
Arturo Escobar

Simon P.J. Batterbury, University of Melbourne
Jude L. Fernando, Clark University

Biographical Details and Theoretical Context

Arturo Escobar was born in Manizales, Colombia in 1952 and trained first as a chemical engineer, graduating from the Universidad del Valle in Cali in 1975. After a year of studying biochemistry, he relocated to the US, completing a Masters in Food Science and International Nutrition at Cornell University in 1978, before enrolling in a PhD program to study nutrition at University of California, Berkeley. His interests – spurred on by the presence of **Michel Foucault** on campus – shifted towards the social sciences and questions of power, international development and planning. Doctoral work (completed in 1987) saw early expression in a brilliant article published in the Indian journal *Alternatives,* where he applied Foucault's notions of power to the study of international development (Escobar, 1984). Having completed three years as a lecturer in Latin American Studies at the University of California at Santa Cruz in 1989, Escobar moved into the discipline of anthropology, beginning as assistant professor at Smith College, and then the University of Massachusetts, where he taught for five years. In 2000 Escobar became the Kenan Distinguished Professor of Anthropology, University of North Carolina, and over the years also spent short periods teaching in Colombia, the UK, and Spain. He retired in 2018. Professionally active beyond retirement, he has continued to work in Colombia, working with activists to imagine alternative futures for the Cauca River Valley.

Few twentieth-century ideas sparked as much controversy as international and economic development. Here, Escobar's theoretical argument – that development should be seen as a *discourse* of power and control – was new and challenging, and by the early 1990s Escobar was established as a leading thinker among a strong group of 'post-development' theorists including Ashis Nandy, Wolfgang Sachs and James Ferguson (Escobar, 1992; Rahnema with Bawtree, 1997). Escobar's first major monograph in 1995, *Encountering Development* (2012 [1995]), elevated him to the status of post-development icon. Later, his ideas – including a rephrasing as alternatives *to* development (Escobar, 2016; 2020) became commonplace across the social sciences, but at the time were more revolutionary. Escobar, and others, were highlighting the impasse of development thinking in the 1980s, with 'post-development' ideas exposing the pervasiveness of its cultural discourse, with profound consequences for the production of social reality in the so-called 'Third World' (Escobar, 2000: 11).

In subsequent work, Escobar began to look more concretely at forms of social change led by new social movements, activists, and progressive non-governmental organisations: 'narratives of transition' (Alvarez and Escobar, 1992; Alvarez et al., 1998). He conducted fieldwork on the Pacific coast of Colombia in 1993 and 1994, the first of several periods during which he worked with Afro-Colombians (descendants from African slaves) and their activist organisations (Escobar and Pedrosa, 1996). The coast is a hot-spot of biological diversity, and is also home to social movements with a strong sense of place and territory (Satizábal and Batterbury, 2018). Escobar's research continued to emphasise his belief in the power of these place-based social movements as alternatives to national and Western-style modernity. He combined this analysis with an interest in nature, which he describes as a 'constructed' category immersed in discursive and material struggles (over the meaning of biodiversity and sustainability, for example). Some of his work in Colombia has been framed in the language of political ecology, with geographical concepts of place and territory also critical to his analysis (Escobar, 1996; 1998; 2001; 2008).

A later project was to explore 'pluriversal' concepts, challenging the stability of fixed ontologies and asking his audience to 'imagine possibility differently' at a time of planetary crisis (Escobar, 2020: x). Escobar writes that 'it is easier to imagine the end of the world than the end of modernity' (Escobar, 2018: 209). From his perspective, people too often take the dominant world for granted (in terms of markets, individualism, productivity, competitiveness, the need for economic growth, etc.), which has only served to cement the universals of modernity and their attendant capitalist institutions with techniques of dominance, control, violence, and war. These are, he argues, both irrelevant and destructive to pluralistic politics. More complex features of his argument revolved around the transformation of socio-cultural configurations and the use of 'ontological design' (Escobar, 2018), and he has stressed the importance of imagination and reconstruction in 'relational worlds' which challenge modernist discourses, institutions and policies.

Spatial Contributions

Escobar (2012 [1995]: 14) has been 'imagining alternatives' for almost half a century, and his work has been stimulating and provocative. Aside from a wide-ranging conversation that has developed around his post-development critique, some geographers have used his analysis as a point of departure for studying social movements and development alternatives. Escobar himself has engaged with the discipline, for example co-editing a book series with Dianne Rocheleau (*New Ecologies for the Twenty-First Century*, Duke University Press) and frequently attending events and conferences. In *Encountering Development*, he details the spatial outcomes of the hegemonic 'development' discourse since colonial times. Western 'development', particularly during the Cold War, was behind the construction of almost all aspects of social reality in the Global South in such a pervasive way 'that even its opponents were obliged to phrase their critiques in development terms: another development, participatory development, socialist development, and so on' (Peet and Hartwick, 1999: 145). Discourse, Escobar argued, has the power to influence reality, following **Edward Said**'s theory of Orientalism (Said, 1978: 3).

Escobar's early critiques of modernity and development were purposely harsh, suggesting an instrumental aim lying behind a universal (Western) paradigm in which many people, including Western geographers, anthropologists, and other social scientists, were more or less complicit. The very idea of development was framed by the Western geopolitical imagination that seeks to 'subordinate, contain and assimilate the Third World as Other. Such Western imagination is a violation of the rights of the other societies' (Slater, 1993: 421). Escobar argues that in Latin America, from 1998 to 2015, there was progress as progressive movements and several governments shifted to the left, only to reverse course later, resulting once more in authoritarian regimes and neoliberalism. Experiments in 'commoning' have been crushed by right-wing regimes with neoliberal and patron-clientelist policies. Latin America still relies far too much on the commodification and exploitation of natural resources, linked to displacement and dispossession.

In *Designs for the Pluriverse: Radical Interdependence, Autonomy, and the Making of Worlds* (2018) Escobar follows Ezio Manzini's work on design: 'a culture and a practice about how spaces in terms of their design are vital to actualize desired meanings, functions, and goals' (Manzini, 2015: 13). Design is integral to fostering mutually beneficial relationships between physical and social environments (Escobar, 2018: xvii). He not only provides a critique of design, but also reimagines it in line with his thoughts on resisting a current world order controlled by capitalist forces, including neoliberal globalisation, individualism, sexism, etc. (García, 2019). Escobar argues that design as a practice should shift away from objects and technological change, anchored in individualism, 'expert' knowledge and markets, to a process that articulates people's everyday lives and social and ecological relations that govern them in their respective places and cultures. Design is therefore a praxis not only limited to producing physical places, but also focusing on the subjectivities embodied in them (2018: 110), demonstrating how people experience and engage with rearticulating the realities they face.

With the slow dismantling of Western aid flows in the twenty-first century, more attention is now paid to alternatives, and locally constructed agendas. 'Development geography' has also changed direction to recognise some pluriversal concepts. Escobar's contributions have been recognised by scholars who listen to or work with Indigenous/autonomous social movements, and localised strategies to manage modernity (Matthews, 2010). Place-based social movements need 'territory', in which their livelihoods or life projects are largely conducted, but they also operate in a 'region-territory', which is a 'political construction for the defense of the territories and their sustainability' (Escobar, 2001: 162). This can include global support networks, including those now offered online; his examples include the U'wa Indigenous group in Colombia who mobilised against the Colombian government, Occidental Petroleum, and later Ecopetrol to oppose oil and gas exploration (Escobar and Harcourt, 2002). Escobar follows many scholars of Indigenous rights in stressing the attachment such groups feel to particular places. As he puts it, 'The struggle for territory is thus a cultural struggle for autonomy and self-determination. This explains why for many people of the Pacific the loss of territory would amount to a return to slavery or, worse perhaps, to becoming "common citizens"' (Escobar, 2001: 162). Social movements are therefore intimately linked to local geographies; culture resides in places.

The commoning of culturally-specific localities that comprise the 'pluriverse' (Escobar, 2018; 2020) means 'creating another world' as an alternative to predatory capitalism. Community design practices, conceptualised as political ontologies, allow a transition from understanding the world as a single universal entity to a relational pluriverse, a concept based on the agency of autonomous places driving change against a dominant system. The pluriverse serves as an 'autonomous design', allowing social movements that strive for the defence of their territories to effect change themselves in their social and ecological struggles (2018: 172).

Escobar has also argued that nature is deeply imbricated with technology and social relations which are 'hybrid and multiform' (Escobar, 1999: 1; 1998; 2008). Fields like political ecology should be anti-essentialist (2016): his definition of the field is 'the contingent study of the manifold articulations of history and biology and the cultural mediations through which such articulations are necessarily established' (1999: 3). He suggests we should be interested in the material transformations of the natural world by human actions, while realising nature is a historically- and socially-constructed category (Escobar, 2008). *Territories of Difference* (2008) linked nature to social movements in an exhaustive account anchored in the Colombian Pacific.

Key Advances and Controversies

Since Escobar's major argument thirty years ago was to challenge the language, discourse, and project of mainstream development as a failed modernist project and he has not wavered from this, a critical response to these ideas was inevitable. It has come from several quarters. For instance, a group of critics argued that his attack on development missed the target, since 'the problem is not so much with development, even less so with modernity, than with capitalism' (Escobar, 2000: 12). Richard Peet and ***Michael Watts*** (2004) and most of their contributors to their book, *Liberation Ecologies*, wished to balance the attention given to powerful discourses with analysis of the impact of material transformations on lives, while Peet and Hartwick (1999) and Kiely (1999) were more blunt, suggesting capitalist material relations have penetrated all corners of the globe. Thus, development discourse arose first and foremost from capitalism in which autonomous spaces are firmly embedded, intentionally or involuntarily. This critique sees post-development, and postcapitalist and alternative pluriversal imaginations (Escobar, 2020: 138) as requiring the transformation of the social relations that produced and that sustain this discourse. This is a question of active intervention against capitalism (i.e., a new and revolutionary form of development itself), with global, not just local, solidarity.

A second concern is that development – including the actions of the state – has, of course, long contained progressive critical voices and practices, and is far from monolithic in its silencing of local and marginal voices (Gardner and Lewis, 1996; Lehmann, 1997). Lumping together progressive aid agencies or the actions of radical NGOs, with the worst technocratic and domineering aid projects practising their craft on their 'clients', and then collectively condemning the totality as an instrument of Western power (as passages in *Encountering Development* came close to doing), denies these differences.

It ignores initiatives that are genuinely involved in radical praxis. Some anthropologists also tried to create *better* development or state policies, taking a more active role (Little and Painter, 1995; Gardner and Lewis, 1996). Escobar's early critiques of modernity and development may, therefore, throw together contested positions, simply because they 'share the same discursive space'. He may, therefore, have been practising a form of essentialism after all (Kiely, 1999; Fernando and Kamat, 2000).

A third question concerns the power of place-based social movements. Local activist struggles cannot be romanticised (Matthews, 2010). Some of their actors are concerned with gaining access to resources, some of these being modern or Western (e.g., capital, paid labour, formal education, health) (Li, 2014). Again, capitalist forces permeate almost everywhere, including to distant social movements and their members. This point was made in a comradely critique of post-development in general, and of Escobar's work in particular, by geographer Tony Bebbington over 20 years ago (Bebbington, 2000). Bebbington wanted to resolve the debate between post-development, and the neoliberal interpretations informed by neoclassical economics (pro-market and modernising), to call for a notion of development that is alternative and critical, but practicable. Criticisms of development's failures in Latin America must recognise cases in the Andes where symbols of failed development – high levels of out-migration, increased consumption of Western commodities, and imported knowledge and technologies – have been accompanied by 'increased Indigenous control of everything from municipal government, to regional textile markets, to bus companies' and 'assertive and ever more ethnically self-conscious social organizations' (Bebbington, 2000: 496). Migrant remittances from the West are partly responsible, as Tania Li (2014) explores in her work in highland Indonesia. Post-development and pluriversal thinking can also overlook the actions of the state (particularly its support of land reform), NGOs, and churches. For sure, capitalist modernity is reproduced, and its internal crises are resolved by incorporating, reconfiguring, and disciplining the pluriverse; in the process, capital may potentially produce local hierarchies and power inequalities detrimental to the interests of local communities, while enabling some capital and knowledge transfers. We need to foreground 'problems of livelihood and production as much as problems of politics and power' and 'emphasize negotiation and accommodation as much as resistance' (Bebbington, 2000: 449).

Escobar (2020: 36) sees 'inevitable bits of modernism clinging to historical materialism', while encouraging critical debate across Latin America. He disputes the strong arguments of his critics that assert 'the primacy of the material over the discursive' (Escobar, 2000: 12; 2008; 2020). He counters that the 'heteropatriarchal capitalist modern/colonial world system' (Escobar 2020: xii) is in turn challenged by 'pluriversal' networks of power that sometimes require political coalitions. Social struggles, like those of the U'wa, form networks, particularly around gender responsibilities and rights via 'theoretical and practical action – that is, in the production of alternative discourses' (Escobar and Harcourt, 2002: 4). Celebrating embodied multiplicity in these struggles, however, differs widely from a universal and fundamental notion of a cosmopolitan world, as **Amartya Sen** described it. Universal abstractions and categorical dualities, according to Escobar, are difficult and ineffective for progressive social movements. Autonomy is internally and externally dynamic in its drive to sustain the pluriverse.

A persistent criticism of Escobar's work is that no political and economic programme for change is presented. He says this is precisely the normative goal of most development and modernist thinking, and it is something he and those that he works with are trying to escape from. The future is up to a wide range of movements, and they must elaborate their own paths in that 'pluriverse' (Escobar, 2000; 2008; 2018). These critiques do not diminish the power of Escobar's key political message – local cultures, places, and territories matter (2008). His works contrast with sweeping accounts of social change, and are usually localised – for him, 'grand theorizing is part of the problem' (Escobar and Harcourt, 2002: 4).

Escobar has also been instrumental in promoting 'world anthropologies' for their 'richer cosmopolitics', levelling the terrain of anthropological work and recognising non-Western scholarship (Ribeiro and Escobar, 2005). Nonetheless, Marxist and post-structuralist critics of Escobar's work suggest that whatever one's particular ideology, change cannot be left entirely open. Movements sit in political economies. One place to start is the recognition that capital, leading the overarching economic system, is directly responsible for social and human injustices. This also means that we look critically at social movements themselves – they cannot be romanticised given many are nationalist, fundamentalist, or insensitive to human rights and gender difference (Kiely, 1999). Escobar is, of course, well aware of this (Escobar and Harcourt, 2002).

In a recent article, Escobar and colleagues respond to critics. They claim their 'critique of "modernity"' does not imply a 'rejection of all that is "modern", nor an uncritical acceptance of all that is "traditional"' (Demaria et al., 2023: 62). But on what basis can inhabitants of the pluriverse cope with the hegemonising powers of the modern, or actually confront modernity? This is unclear, although Demaria et al. expect: 'Alternatives to development practices and worldviews' will 're-politicize the debate on much-needed socio-ecological transformation, affirming dissidence with the current world representations of sustainable development and searching for alternative ones' (2023: 67). This is reasonable, but it is a global project, like capitalism itself. By acknowledging the interplay between different regions, we can work towards alternative worldviews, prioritising socio-ecological transformation.

Escobar's thoughts on political ecology and the study of nature have attracted some commentary. David Cleveland (1999: 18) says that in a form of organic nature where Escobar identifies that 'nature is not manipulated', premodern peoples did exploit natural capital to an advanced degree (see Escobar, 1998, who partially addressed this). One of the founders of the field, Piers Blaikie (2000) took on the implicit populism and support of the underdog in post-development thinking more generally, arguing analysis was lacking on development processes and policy.

To conclude, Escobar's 'journey of the imagination' has taken him a long way towards 'reconceiving and reconstructing the world from the perspective of, and along with, those subaltern groups that continue to enact a cultural politics of difference as they struggle to defend their places, ecologies, and cultures' (Escobar, 2000: 15). The sentiment is laudable, and his work has helped to foreground the importance of local culture within thinking on development hegemonies and practices.

Escobar's Key Works

Escobar, A. (1984) 'Discourse and power in development: Michel Foucault and the relevance of his work to the Third World', *Alternatives,* 10 (3): 377–400.

Escobar, A. (1999) 'After nature: Steps to an anti-essentialist political ecology', *Current Anthropology,* 40 (1): 1–30.

Escobar, A. (2008) *Territories of Difference: Place, Movements, Life, Redes.* Durham, NC: Duke University Press.

Escobar, A. (2012 [1995]) *Encountering Development: The Making and Unmaking of the Third World.* Princeton, NJ: Princeton University Press.

Escobar, A. (2018) *Designs for the Pluriverse: Radical Interdependence, Autonomy, and the Making of Worlds.* Durham, NC: Duke University Press.

Escobar, A. (2020) *Pluriversal Politics: The Real and the Possible.* Durham, NC: Duke University Press.

Ribeiro, G.L. and Escobar, A. (eds) (2005) *World Anthropologies: Disciplinary Transformations within Systems of Power.* Oxford: Berg.

Secondary Sources and References

Alvarez, S. and Escobar, A. (eds) (1992) *The Making of Social Movements in Latin America: Identity, Strategy, and Democracy.* Boulder, CO: Westview.

Alvarez, S., Dagnino, E. and Escobar, A. (eds) (1998) *Cultures of Politics/Politics of Cultures: Revisioning Latin American Social Movements.* Boulder, CO: Westview.

Bebbington, A.J. (2000) 'Re-encountering development: Livelihood transitions and place transformations in the Andes', *Annals of the Association of American Geographers,* 90 (3): 495–520.

Cleveland, D. (1999) 'Comments on Escobar', *Current Anthropology,* 40 (1): 17–18.

Demaria, F., Kothari, A., Salleh, A., Escobar, A. and Acosta, A. (2023) 'Post-development: From the critique of development to a pluriverse of alternatives', in S. Villamayor-Tomas and R. Muradian (eds), *The Barcelona School of Ecological Economics and Political Ecology.* Cham: Springer. pp. 5–59.

Escobar, A. (1992) 'Planning', in W. Sachs (ed), *The Development Dictionary: A Guide to Knowledge as Power.* London: Zed. pp. 132–45.

Escobar, A. (1996) 'Constructing nature: Elements for a poststructuralist political ecology', in R. Peet and M.J. Watts (eds), *Liberation Ecologies: Environment, Development, Social Movements.* London: Routledge. pp. 46–68.

Escobar, A. (1998) 'Whose knowledge, whose nature? Biodiversity, conservation, and the political ecology of social movements', *Journal of Political Ecology,* 5 (1): 53–82.

Escobar, A. (2000) 'Beyond the search for a paradigm? Post-development and beyond', *Development,* 43 (4): 11–14.

Escobar, A. (2001) 'Culture sits in places. Reflections on globalism and subaltern strategies of globalization', *Political Geography,* 20: 139–74.

Escobar, A. (2016) 'Thinking-feeling with the earth: Territorial struggles and the ontological dimension of the epistemologies of the South', *AIBR Revista de Antropologia Iberoamericana,* 11 (1): 11–32.

Escobar, A. and Harcourt, W. (2002) 'Women and the politics of place', *Development,* 45 (1): 7–14.

Escobar, A. and Pedrosa, A. (eds) (1996) *Pacífico, Desarrollo o Diversidad? Estado, Capital y Movimientos Sociales en el Pacífico Colombiano.* Bogotá: CEREC/Ecofondo.

García, N.A. (2019) 'Contemporary world as a massive design failure: A way out? Review', *Journal of Extreme Anthropology,* 3 (1): 174–7.

Gardner, K. and Lewis, D. (1996) *Anthropology, Development and the Post-Modern Challenge.* London: Pluto.

Kiely, R. (1999) 'The last refuge of the noble savage? A critical assessment of post-development theory', *The European Journal of Development Research,* 11 (1): 30–55.

Lehmann, D. (1997) 'An opportunity lost: Escobar's deconstruction of development', *Journal of Development Studies,* 33 (4): 568–78.

Li, T.M. (2014) *Land's End: Capitalist Relations on an Indigenous Frontier*. Durham, NC: Duke University Press.

Little, P. and Painter, M. (1995) 'Discourse, politics, and the development process: Reflections on Escobar's anthropology and the development encounter', *American Ethnologist,* 22 (3): 602–16.

Manzini, E. (2015) *Design, When Everybody Designs: An Introduction to Design for Social Innovation*. Cambridge, MA: MIT Press.

Matthews, S.J. (2010) 'Post-development theory', in S.J. Matthews (ed), *Oxford Research Encyclopedia of International Studies*. Oxford: Oxford University Press.

Peet, R. and Hartwick, E. (1999) *Theories of Development*. New York, NY: Guilford.

Peet, R. and Watts, M.J. (eds) (2004) *Liberation Ecologies: Environment, Development, Social Movements*. London: Routledge.

Rahnema, M. with Bawtree, V. (1997) *The Post-Development Reader*. London: Zed.

Said, E. (1978) *Orientalism*. London: Penguin.

Satizábal, P. and Batterbury, S.P.J. (2018) 'Fluid geographies: Marine territorialisation and the scaling up of local aquatic epistemologies on the Pacific coast of Colombia', *Transactions of the Institute of British Geographers,* 43: 61–78.

Slater, D. (1993) 'The geopolitical imagination and the enframing of development theory', *Transactions of the Institute of British Geographers,* 18: 419–37.

17
Michel Foucault

Chris Philo, Glasgow University

Biographical Details and Theoretical Context

Born in Poitiers, France (1926), Michel Foucault's education started inauspiciously, but, spurred on by the promise that learning philosophy would reveal 'the secret of secrets' (Sheridan, 1980: 2), he did well enough to pass his school-level *baccalauréat* with credit. Securing a place at the highly prestigious École Normale Supérieure (ENS), Paris, he took his *licence de philosophie* in 1948, but quickly became disillusioned with philosophy, turning instead to psychology, taking his *licence de psychologie* in 1952 and commencing research on psycho-pathology.

The 1950s–1960s were a heady period in French intellectual and political circles, with challenges to Marxism and structuralism arising from various strains of existentialism and phenomenology, and with the first stirrings of the broad post-structuralist movement with which Foucault is commonly aligned. Reconfiguring his approach to philosophy, psychology and, indeed, science, while wandering between jobs on the fringes of academia, Foucault eventually completed a doctoral thesis in 1959 and returned to the academy. Two years later his first major book, *Histoire de la Folie*, appeared. In the early 1960s he returned to the ENS as Professor of Systems of Thought, a position that he held for many years alongside visiting Professorships to institutions elsewhere. He was progressively 'globalised' as his ideas began to reach many different audiences and destinations, bequeathing a 'Foucauldian' (or 'Foucaultian') approach to social inquiry that ultimately led him to be feted as one of the leading intellectuals of the twentieth century. He died in 1984 of AIDS, aged 57, having done enough to be acknowledged as one of the most influential intellectuals of the twentieth century, prompting one commentator to propose, not entirely flippantly, that '[i]n many ways, we are all Foucauldians now – in the ways we think about gender, normalisation, psychiatry, confinement, surveillance' (Behrent, 2021: n.p.).

Foucault's reputation grew from innovative theoretical interventions that problematised the production of knowledge and, crucially for the present age when it has become so ragged and disputed, the constitution of 'truth'. Foucault himself remains elusive, the 'truth' of his own life and practices subject to increasingly troubling claims and counterclaims. Similarly, the 'truth' of his academic work defies easy characterisation, while his many writings – with shifting subject-matters and conceptual shape-shifting – resist simple, unitary depictions.

He never acceded to expectations about the theories and positions that he should take, and in an introduction to the English edition of *The Archaeology of Knowledge* he insisted: 'Do not ask me who I am and do not ask me to remain the same', adding 'Leave it to our bureaucrats and our police to see that our papers are in order' (Foucault, 1972b: 17).

Foucault's first four major texts are usually cast as his *archaeologies*, excavating the 'discourses' (or organised bodies of knowledge) that emerged within European history as the foundations for both intellectual orthodoxy and practical endeavour. *Histoire de la folie* (1961; translated in truncated form as *Madness and Civilization*, 1965, and then in full as *History of Madness*, 2007) and *Naissance de la clinic* (1963; translated as *The Birth of the Clinic*, 1973) probed the discourses present within prevailing understandings of mental ill-health ('madness') and physical illness, revealing how these gave rise to the 'invention' of both the mental hospital ('asylum') and the modern hospital. *Les mots et les choses* (1966; translated as *The Order of Things*, 1970) interrogated the wider discursive formations (or *épistèmes*) present within European conceptions of language, economics and nature, laying bare subtle transitions in the scaffolding of what Europeans have taken as the root 'order' of the world (the supposed links between 'words' and 'things'). *L'archéologie du savoir* (1969; translated as *The Archaeology of Knowledge*, 1972b) reflected still more broadly on the making of knowledge – the text was more a topical investigation of 'what is knowledge?' than a methodological treatise on 'how to produce knowledge' – and in so doing examined the conjoint temporality and spatiality of statements, discourses and their ordering in the 'archive'. Such texts betrayed the influence of structuralism, notably in the sense that human subjects appear to be 'spoken' by discourses rather than possessing their own autonomous agency, and that objects of knowledge only seem properly explicable within discourse rather than through any reference to external reality.

Foucault's next four major texts are usually cast as his *genealogies*, wherein he skipped from the realm of discourse to addressing more directly the mechanics of power suturing knowledges to objects and practices. *Surveiller et punir* (1975; translated as *Discipline and Punish*, 1976) ostensibly traced the spread of prisons and reformatories throughout later-eighteenth and nineteenth-century Europe, but also interrogated the transition from an older regime of violent 'sovereign power' (whereby monarchies terrorised their populaces into obedience through the bloody spectacle of the scaffold) to a modern calculus of 'disciplinary power' as less an absolute possession and more a subtle – but ultimately more effective – relational play of forces between the state and its subjects. Foucault argued that the occupants of identifiable spaces (whether closed institutions or national territories) were quietly disciplined as 'docile minds and bodies' compliant with the demands of capital accumulation and civic responsibility.

The next three books in the 'sexuality' series – translated as *The History of Sexuality Volume One: An Introduction* (1978), *The History of Sexuality Volume Two: The Uses of Pleasure* (1985) and *The History of Sexuality Volume Three: The Care of the Self* (1986a) – furnished 'chapters' within a projected larger survey of how Europe has 'produced' notions of sexuality, of sexual conduct both accepted and shunned, from ancient times through to the present. Foucault demonstrated that these notions have never been fixed, but rather have differed according to the status, class, gender, age and place of the peoples concerned, and have been converted into the objects of discourse (in everything from self-help manuals to confessional whisperings) wherein the possibilities for sexual expression have been curtailed on many occasions but enlarged on others. If *Discipline and Punish* emphasised the shaping of human subjects from without, through anonymous forces inserting individuals into disciplinary apparatuses of one kind or another, *The History of Sexuality* mingled this focus with a sense of how individuals could be more knowingly, wittingly even, enlisted into their own self-fashioning not just as sexual beings but as agents cognisant of the wider socio-cultural rules governing 'the conduct of conduct' (Foucault, 1994: 237).

Numerous texts have continued to emerge under Foucault's name after his death in 1984, notably in the form of transcribed and translated Collège de France lecture series (1970–1984). There are too many to cite, but of especial importance is how those from the mid-1970s saw Foucault adjusting his focus from an 'anatomo-politics' of 'disciplinary power' – latching on to the body of the prisoner, patient or other seemingly troublesome individual – in favour of emphasising a 'biopolitics' of 'biopower' that addressed overall populations, confined in institutions or spread across countrysides, cities and regions, in terms of such vital life matters as birth, death, health, illness, capacity and competency. The three lecture series, translated as *'Society Must Be Defended'* (Foucault, 2003), *Security, Territory, Population* (Foucault, 2007) and *The Birth of Biopolitics* (Foucault, 2008), elaborated this trajectory in Foucault's thinking, also clarifying Foucault's theorisations of 'race' (meaning all those marked out as 'biologically' different from the majority), 'governmentality' (ideas-practices behind governing collectivities), the state (and its links to political economy) and liberalism. The latter three angles brought Foucault into debates about neo-liberalism and the kinds of 'truths' it calculates and diffuses, and also the forms of active human 'subject-citizen' seemingly demanded by contingent collusions between state and capital.

Spatial Contributions

It has been persuasively argued that Foucault extends a thorough-going geographical provocation to critical social thought, demanding sustained alertness to questions of space, place, environment and landscape in a manner rarely encountered from someone not a professional geographer (Crampton and Elden, 2007). **Stuart Elden** characterises Foucault as a practitioner of 'spatial history':

> Foucault's historical studies are spatial through and through, and that this is a fundamental legacy of his work to those interested in the question of space [...] Understanding how space is fundamental to the use of power and to historical research into the exercise of power allows us to recast Foucault's work not just as a history of the present but as a mapping of the present. (Elden, 2001: 152)

It is necessary to stress that Foucault used historical research – converting it into what Dean (1994) calls 'critical and effective histories' – to understand how 'we' have arrived at where 'we' are today (in short, how modernity has been shaped, complete with all its tangled inequalities). By teasing out how space 'works' in history, tracing the spatial configurations that expose how power and knowledge operate in countless (mal)treatments of 'the unloved', he was seemingly able to throw into relief many of the more questionable contours of the present.

Most abstractly, Foucault advanced a fierce critique of what he called the project of *total history*: 'one that seeks to reconstitute the overall form of a civilization, the principle – material or spiritual – of a society, the significance common to all the phenomena of a period, the law that accounts for their cohesion – what is called metaphorically the "face" of a period' (Foucault, 1972b: 9). Such a project was anathema to Foucault, since its ambitions stood squarely against his own belief that 'nothing is fundamental: this is what is interesting in the analysis of society' (Foucault, 1972b: 16). He advocated instead a *general history*, a 'bellicose history' (Lemert and Gillan, 1982: 39), which militated against the tidying up of the past to give neat patterns, steadfastly resisting the rush from the small details of lived struggle to the grander pronouncements of historians (particularly those of social scientists dabbling in the practice of history). Foucault thereby drew this distinction: 'a total description draws all phenomena around a single centre – a principle, a meaning, a spirit, a world-view, an overall shape; a general history, on the other hand, would deploy the space of a dispersion' (Foucault, 1972b: 10). Comments about 'spaces of dispersion' offered a spatialised ontology of the social world: a vision of a space or plane across which the events and phenomena relevant to the substantive inquiry are 'dispersed' (Philo, 1992).

Foucault's own histories were spatial not solely in the philosophical sense of talking about 'dispersions' *vs.* 'totalisations', nor because they charted changing conceptions of space (although a history of space in this guise does flicker through his 1967 lecture 'On other spaces': Foucault, 1986b). As Elden (2001: 118) argues, though, Foucault's 'histories are not merely ones in which space is yet another area analysed but have space as a central part of the approach itself', meaning that, 'rather than merely writing histories of space, Foucault is *[indeed]* writing spatial histories.' In one regard, this was simply because of the insistence on bringing details of past phenomena to the fore, as he acknowledged when borrowing from Nietzsche's notion of 'genealogy' as 'grey, meticulous and patiently documentary':

> [It] requires patience and a knowledge of *details*, and it depends upon a vast accumulation of source material. Its 'cyclopean monuments' are constructed from 'discreet and apparently insignificant truths and according to a rigorous method'; they cannot be the product of 'large and well-meaning errors'. (Foucault, 1986c: 76–77, emphasis added)

Hutcheon identifies Foucault's 'assault on all the centralising forces of unity and continuity', and on his requirement – using terms that should immediately arrest the geographer – that 'the particular, the local and the specific' be pursued in place of 'the general, the universal and the eternal' (Hutcheon, 1988: 120). Whilst no straightforward empiricist, it remains the case that taking particularity, specificity and locality seriously was fundamental to Foucault's notion of 'spaces of dispersion', with his spatial histories furnishing systematic insights into the play of spatial relations in the historical record.

Key Advances and Controversies

Foucault the person may have departed but Foucault the intellectual – and Foucauldian scholarship (as in the dedicated journal

Foucault Studies) – remains alive and definitely kicking. **Stuart Elden** (2016; 2017; 2021; 2023) has also completed a remarkable four-volume exposition of Foucault's intellectual manoeuvres across four decades. Meanwhile, Foucault's writings are still eagerly consulted and deployed across the breadth of academic philosophy, social sciences, cultural studies and beyond. His ideas are routinely inspected and dissected, sometimes fiercely critiqued, and challenging questions are asked about what his work can bring to – and whether he should himself be critiqued within – contemporary debates around feminism, anti-racism, postcolonialism, queer theory, radical environmentalism, and more. It is impossible to do justice to the richness, complexity and electricity of these debates, so the following are brief gestural observations concerning these 'spatial contributions'.

In *The Order of Things*, Foucault suggested an opposition between 'the Same' and 'the Other' (Philo, 1986) that framed many of his major historical studies. Firstly, he identified his inquiries into discourse and knowledge as reconstructions of what it is that a given society considers the Same, incorporating both the leading statements of 'experts' (academics, politicians, moralists) and the taken-for-granted assumptions figuring in the everyday lives of the populace. Secondly, he identified his social histories of 'the mad, the sad and the bad' as reconstructions of what it is that a given society regards as the Other, as the unacceptable mass of activities, people and places beyond the boundaries of what is deemed as 'normal' and which thereby necessitate some response of policing, removal or even eradication:

> Ostensibly, [*Foucault's*] project is to describe the mechanisms of order and exclusion that have operated within European society since the sixteenth century, and above all since the late-eighteenth century. In a motif that recalls Bataille's reflections on Hegel, Foucault sees a conflict in history against which it can define itself, just as every 'master' needs a 'slave'. When such an 'Other' is absent, it must be invented. (Megill, 1985: 192)

For Foucault, madness, sickness and criminality, as well as sexual dissidence, were hence traced historically in their Otherness to European norms, laying out the shifting bases for their constitution as oft-feared moments of alterity.

In this respect *Madness and Civilisation* examined how 'Reason' (or the Same) has progressively identified, named, stigmatised and sought to exclude 'Unreason' (or the Other). It is telling to repeat the assessment that Serres offers of this text, since he 'interprets Foucault's categories of inclusion and exclusion in terms of spatial relationships, and [...] views Foucault's concept of Unreason as a "geometry of negativities"' (in Major-Poetzl, 1983: 120). Such a geometry embraces projections in which society imaginatively positions itself over and against those phenomena – especially peoples reckoned less-than-human in their madness, sickness, criminality and so on – that are supposed to transgress the limits of the sanctioned. Yet, beyond these projections, and paralleling the spatialised vocabulary deployed when charting what occurred at 'the level of the imaginary', it is also true that Foucault's spatial sensibility transferred to 'the level of the real' (Elden, 2001: 93). Thus, he was clearly interested in 'the physical divide of segregation and exclusion' distancing the Other from the Same, and for this reason he ended up 'conceiv[ing] of madness and reason, sickness and health in spatial terms, and then examin[ing] the groups that inhabit these liminal areas' (Elden, 2001: 94–5). He thereby paid repeated attention to specific 'liminal areas', notably the bricks-and-mortar solutions of institutions such as asylums, hospitals and prisons (see also Philo, 2000; 2002; 2003) designed to confine, 'reform' and, where appropriate, to 'cure' those displaying signs of such difference.

Cross-cutting these spatial histories are different conceptualisations of power, themselves the basis for a profound theorisation of space and power (or power/knowledge: Gordon, 1980). Most abstractly, Foucault (1976: 215) insisted that power be understood through its 'micro-physics' – 'its techniques, procedures, levels of applications, targets' – and hence in a thoroughly relational fashion that subsequent theorists have readily elaborated in terms of 'capillaries', 'transmissions' and 'relays' of power through specific spatial fields (Driver, 1985, 1993; Hannah, 1997). More empirically, Foucault (1976: 141–9) analysed 'the art of distributions' underlying a host of nineteenth-century disciplinary mechanisms, tracing the enactment of spatial innovations across all manner of institutions from Bentham's design for an ideal prison, the high-walled 'Panopticon', to the example of an unwalled reformatory at Mettray. There is a danger that many readings of *Discipline and Punish* reduce Foucault's claims to the figure of the Panopticon, neglecting the significance of his arguments about Mettray (Driver, 1990), but it remains the case that the Panopticon has now become a dramatic spatial provocation for social theorists of power. With its internal spatial arrangements allowing a constant (threat of) inspection, a surveillance that captures inmates in an overall field of visibility while prompting them to convert the external eye of the inspection tower into the internal eye of conscience (Bender, 1987), it is unsurprising that many have found here keys to unlock the broader workings of disciplinary power throughout modernity.

The mid-1970s lecture series introduced further spatial inflections to Foucault's inquiries into power and its workings (Philo, 2012), particularly as he began to emphasise the significance of both territory (i.e., states' lands and bounded jurisdictions) and environment (whether 'natural' situations, such as swampy ground, or cities as laid out and buildings as constructed) for the workings of biopower. As he claimed, '[b]iopolitics' last domain is [...] control over relations between the human race, or human beings insofar as they are a species, insofar as they are living beings, and the environment, the *milieu*, in which they live' (Foucault, 2003: 244–5). Together with the 'sexuality' books, these contributions solidified an interest – as already noted – in 'governmentality' (Foucault, 1979), encompassing the government of both populations, wherein states and religions intervene in the processes of life, birth and death, and individuals, especially their everyday sexual and reproductive conduct. While he said less about space in this latter regard, a spatial sensitivity still recurred in what he said about the countless specific sites – from the confessional to the late-Roman city-state (Sharp et al., 2000: 16–19) – implicated in the persuading of people (or, rather, in people persuading themselves) to take seriously 'the relationship that one ought to have with one's status, one's functions, one's activities, and one's obligations' (Foucault, 1986a: 84).

The 'late Foucault' turned increasingly back to the latter, to different iterations of 'pastoral power' and spiritual-philosophical guidance from Medieval Europe and Ancient Greece. But he also returned to the broader problematic of 'truth' – what constitutes 'truth-telling', how truth is tested, forms of 'courageous truth-speaking' (*parrhesia*) against the wishes of authority – in a manner recalling his earlier concerns for 'otherness' and the 'other truths' sometimes uttered by those cast as 'mad' or otherwise marginalised. In the final written note for his last ever lecture, Foucault concluded: 'There is no establishment of the truth without an essential position of otherness; the truth is never the same; there can be truth only in the form of the other world and the other life' (Foucault, 2008: 340).

Foucault's Key Works

Foucault, M. (1965 [1961]) *Madness and Civilization: A History of Insanity in the Age of Reason* (*abridged text*; trans. Richard Howard). New York, NY: Random House.

Foucault, M. (1970 [1966]) *The Order of Things: An Archaeology of the Human Sciences*. London: Tavistock.

Foucault, M. (1972b [1969]) *The Archaeology of Knowledge*. London: Tavistock.

Foucault, M. (1973 [1963]) *The Birth of the Clinic: An Archaeology of Medical Perception*. London: Tavistock Publications.

Foucault, M. (1976) *Discipline and Punish: The Birth of the Prison*. London: Allen Lane.

Foucault, M. (1978) *The History of Sexuality, Volume 1: An Introduction*. New York, NY: Random House.

Foucault, M. (1985) *The History of Sexuality, Volume 2: The Uses of Pleasure*. Harmondsworth: Penguin.

Foucault, M. (1986a) *The History of Sexuality, Volume 3: The Care of the Self*. Harmondsworth: Penguin.

Foucault, M. (2003) *Society Must be Defended. Lectures at the Collège de France, 1975–1976*. London: Allen Lane.

Foucault, M. (2005) *The Hermeneutics of the Subject. Lectures at the Collège de France, 1981–1982*. London: Picador.

Foucault, M. (2007) *Security, Territory and Population. Lectures at the Collège de France, 1977–1978*. London: Picador.

Foucault, M. (2008) *The Birth of Biopolitics. Lectures at the Collège de France, 1978–1979*. London: Picador.

Secondary Sources and References

Bender, J. (1987) *Imagining the Penitentiary: Fiction and the Architecture of Mind in Eighteenth-Century England*. Chicago, IL: University of Chicago Press.

Behrent, M. (2021) 'The true Foucault', *Dissent Magazine*, September 30. https://www.dissentmagazine.org/online_articles/the-true-foucault/ (accessed November 3, 2023).

Crampton, J.W. and Elden, S. (eds) (2007) *Space, Knowledge and Power: Foucault and Geography*. Aldershot: Ashgate.

Dean, M. (1994) *Critical and Effective Histories: Foucault's Methods and Historical Sociology*. London: Routledge.

Driver, F. (1985) 'Power, space and the body: A critical assessment of *Discipline and Punish*', *Environment and Planning D: Society and Space*, 3: 425–46.

Driver, F. (1990) 'Discipline without frontiers? Representations of the Mettray Reformatory Colony in Britain, 1840–1880', *Journal of Historical Sociology*, 3: 272–93.

Driver, F. (1993) 'Bodies in space: Foucault's account of disciplinary power', in C. Jones and R. Porter (eds), *Reassessing Foucault: Power, Medicine and the Body*. London: Routledge. pp. 113–31.

Elden, S. (2001) *Mapping the Present: Heidegger, Foucault and the Project of a Spatial History*. London: Continuum.

Elden, S. (2016) *Foucault's Last Decade*. Cambridge: Polity.

Elden, S. (2017) *Foucault: The Birth of Power*. Cambridge: Polity.

Elden, S. (2021) *The Early Foucault*. Cambridge: Polity.

Elden, S. (2023) *The Archaeology of Foucault*. Cambridge: Polity.

Foucault, M. (1979) 'Governmentality', *Ideology and Consciousness*, 6: 5–21.

Foucault, M. (1986b) 'Of other spaces', *Diacritics*, 16: 22–7.

Foucault, M. (1986c) 'Nietzsche, genealogy, history', in P. Rabinow (ed), *The Foucault Reader*. Harmondsworth: Penguin. pp. 63–77.

Foucault, M. (1994) *Dits et écrits IV*. Paris: Gallimard.

Foucault, M. (2007) *The History of Madness* (Unabridged English re-translation by Jonathan Murphy and Jean Khalfa). London: Routledge.

Gordon, C. (ed) (1980) *Power/Knowledge: Selected Interviews and Other Writings, 1972–1977*. Hemel Hempstead: Harvester.

Hannah, M. (1997) 'Space and the structuring of disciplinary power: An interpretive review', *Geografiska Annaler*, 79B: 171–80.

Hutcheon, L. (1988) *A Poetics of Post-modernism: History, Theory and Fiction*. London: Routledge.

Lemert, C.C. and Gillan, G. (1982) *Michel Foucault: Social Theory as Transgression*. New York, NY: Columbia University Press.

Major-Poetzl, P. (1983) *Michel Foucault's Archaeology of Western Culture: Towards a New Science of History*. Brighton: Harvester.

Megill, A. (1985) *Prophets of Extremity: Nietzsche, Heidegger, Foucault, Derrida*. San Francisco, CA: University of California Press.

Philo, C. (1986) *'The Same and the Other': On geographies, madness and outsiders*, University of Loughborough, Department of Geography, Occasional Paper No. 11. https://eprints.gla.ac.uk/116553/1/116553.pdf (accessed November 3, 2023).

Philo, C. (1992) 'Foucault's geography', *Environment and Planning D: Society and Space*, 10: 137–62.

Philo, C. (2000) '*The Birth of the Clinic*: An unknown work of medical geography', *Area*, 32: 11–19.

Philo, C. (2002) 'Accumulating populations: Bodies, spaces, institutions', *International Journal of Population Geography*, 7: 473–90.

Philo, C. (2003) *The Space Reserved for Insanity: An Historical Geography of the Mad-Business in England and Wales to the 1860s*. Lampeter: Edwin Mellen.

Philo, C. (2012) 'A "new Foucault" with lively implications – or "the crawfish advances sideways"', *Transactions of the Institute of British Geographers*, 37: 496–514.

Sharp, J.P, Routledge, P., Philo, C. and Paddison, R. (2000) 'Entanglements of power: Geographies of domination/resistance', in J.P. Sharp, P. Routledge, C. Philo and R. Paddison (eds), *Entanglements of Power: Geographies of Domination/Resistance*. London: Routledge. pp. 1–42.

Sheridan, A. (1980) *Michel Foucault: The Will to Truth*. London: Tavistock.

18
J.K. Gibson-Graham

Kelly Dombroski, Massey University
Stephen Healy, Western Sydney University
Wendy Larner, Cardiff University
Katharine McKinnon, University of Canberra

Biographical Details and Theoretical Context

J.K. Gibson-Graham is the academic pseudonym of feminist economic geographers Katherine Gibson and Julie Graham. Writing under this authorial identity, the duo established the field of *diverse economies*, a theoretically-sophisticated feminist post-structuralist economic geography that transformed debates within, and well beyond, the discipline (Barnes and Christophers, 2018). Core to their approach was a consistent refusal of totalising discourses of economy, and particularly 'capitalocentric' representations of economy, a commitment to 'reading for difference', and a methodological stance that looks for how the world we live in is always already more-than-capitalist.

Katherine and Julie first met as PhD students in the late-1970s at Clark University in Worcester, Massachusetts. Katherine Gibson moved from her home country Australia to the US for doctoral study following a geography undergraduate degree from the University of Sydney and further postgraduate study in community development at Macquarie University. When Katherine came into the discipline in the 1970s she was developing her skills in participatory research with inner-city Sydney's remnant working-class community, while quantitative and humanistic approaches were pulling geography in different directions. Julie Graham was from the US, with her intellectual grounding including a degree in English literature from Smith College (Massachusetts) followed by several years of freelance writing, editing, and researching and supporting the feminist movement. Katherine has previously spoken about her move to Clark; her intellectual pleasure on discovering the rigours of Marxian political economy, but perhaps even more importantly her meeting, and subsequent friendship, with Julie Graham in her 'beat-up yellow Volkswagen' and its lasting professional and personal impact (Gibson, 1999).

The intellectual camaraderie and friendship gave rise to intellectually influential collaborations. Both Katherine and Julie wrote doctoral theses that involved rigorous theoretical engagements with Marxian political economy and substantive empirical engagements with processes of economic and industrial restructuring in Australia and the US respectively. The single writing persona of Gibson-Graham was born in a dormitory room at a feminist conference at Rutgers University in 1992 (Gibson-Graham, 1996: xi) and first appeared in print with the publication of 'Waiting for the revolution, or how to smash capitalism while working at home in your spare time', published in *Rethinking Marxism* (Gibson-Graham, 1993).

Despite the substantial physical distance between the two scholars, their ongoing engagement with political economy, increasingly adventurous theoretical experimentation, and robust intellectual engagements with each other continued unabated over the next three decades. As they themselves explain; 'From the theory sluts of 1996 to the self-help junkies of 2006 we have navigated a personal path that ever enriches as new challenges of relating and thinking/writing together arise' (Gibson-Graham, 2006: xi). Their collaboration spanned the Pacific Ocean and many time zones; at a time when maintaining this level of commitment required expensive international phone calls and infrequent intercontinental visits.

In early publications Gibson and Graham developed their individual and collective interests in Marxian political economy. The first co-authored Gibson and Graham (as opposed to Gibson-Graham) article used Marxian theory to include new forms of international contract labour migration in understandings of economic restructuring (Gibson and Graham, 1986). Further signs of wider intellectual engagements came when Julie published an early article on postmodernism and Marxism (Graham, 1988) now widely considered a disciplinary classic. Similarly, Katherine (Gibson, 1991) began to explicitly engage with debates around what we would now call 'positionality' by stressing the intellectual distinctiveness of 'northern' and southern' Marxist geography and accounting for the empirical specificity of the Australian social formation when theorising economic change. A series of high-profile articles followed, leading to their first, path-breaking book *The End of Capitalism (as We Knew It): A Feminist Critique of Political Economy* (Gibson-Graham, 1996), subsequently published in multiple languages across two editions.

Both carried these research concerns forward as they took up academic positions in economic geography in their own countries. Katherine took a series of prestigious positions in the Australian academy, eventually assuming a professorial fellowship at the Institute for Culture and Society at Western Sydney University. Julie, meanwhile, moved to the University of Massachusetts, Amherst, subsequently becoming Professor and Associate Head of the Department of Geosciences (until her death in 2010 at the age of 65).

Following her appointment at the University of Massachusetts, Julie attended the seminars of anti-essentialist Marxist economists Stephen Resnick and Richard Wolff, who argued for alternative theorisations of class and capitalism. Inspired by her experiences in Resnick and Wolff's seminars Julie began co-organising seminars with student cohorts on industrial geography, economic subjectivity, regional development, and post-capitalist politics. This intellectually-generative environment helped shape the publication of the edited collections *Class and its Others* (Gibson-Graham et al., 2000) and *Re-presenting Class* (Gibson-Graham et al., 2001), both co-edited with Resnick and Wolff, and featuring the work of students they co-mentored. Further, these writing collaborations coincided with major research projects, focused on two post-industrial settings, the Pioneer Valley in the US and the Latrobe Valley in Australia. These informed *A*

Postcapitalist Politics (Gibson-Graham, 2006). Through this work the Gibson-Graham authorial presence reflected a relationship of shared thinking and co-writing, in which neither distinguished whose contribution was whose. Since Julie's death in 2010, Katherine has honoured this shared thinking by using J.K. Gibson-Graham as an authorial presence in works that continue to build directly on their shared thinking in many articles, the popular action book *Take Back the Economy* (Gibson-Graham et al., 2013), and the edited academic volumes *Making Other Worlds Possible* (Roelvink et al., 2015) and *The Handbook of Diverse Economies* (Gibson-Graham and Dombroski, 2020).

Spatial Contributions

One major contribution of Gibson-Graham's work was to represent economic space as a space of heterogeneity and difference. Whereas both the proponents and critics of mainstream economics tended to accept the vision of a capitalist totality, Gibson-Graham argued that economic space in fact encompasses many different economic practices. This recasting of 'the economy' as an always more-than-capitalist composite of many different economies is a key contribution of the subdiscipline of diverse economies and continues to reshape economic geography and many cognate fields (Barnes and Christophers, 2018).

The diverse economies project has implications for space on a local and global scale, but its viability as a shared project depended firstly upon a reconceptualization of authorial space. The Gibson-Graham persona was founded in a recognition of scholarship as a site of shared thinking and creative exchange, rather than a heroic, individual endeavour. This feminist rethink of the site of scholarship is consistent with the injunction to 'start where you are' which constitutes a feminist ethical thinking-strategy, as well as an approach to empirical research, theorising, pedagogy, and engagement. Put simply, starting where you are signals an acceptance that you can never step outside your present context and the limitations that this inevitably places on radical thought and action. The response required then is a theoretical and methodological commitment to reading for difference, entailing a careful refusal to *a priori* allocate agency and power and a generative commitment to experimentation as the basis for generating new thought (Sharpe, 2014).

The intellectual space that Gibson-Graham managed to create is no dour factory but a promiscuous workshop of creation. Sharpe (2014) describes this thinking-space as one that proceeds through a process of disjunctive synthesis, or a willingness to put in apposition seemingly disparate ideas in order to see what happens. For example, what might embodied feminist theories of performance say about the power and potential of a politics of collective action? What might queer theory contribute to an analysis of global finance? This disjunctive thinking-practice is one that requires a sort of intellectual patience, and an extension of goodwill, and a sense of faith that there *is* a connection to be made. Julie was asked in the 2010 Janice Monk lecture, what happens when you disagree? Her answer was to say, 'we never disagree in our writings, there are just spaces where we are not yet being clear with one another' (pers. comm., Healy, 2010). 'Starting where you are' also means committing to a process of ongoing conversation and dialogue – a willingness to work on intellectual collaboration as a relationship. What the pen-name Gibson-Graham makes explicit is how the means of thinking, much like other means of production, is always socialised (even if the fruits of intellectual labour are not).

One of the innovative and empowering spatial arguments J.K. Gibson-Graham brought to geography was their approach to economic transformation as a global politics that can be punctiform and ubiquitous. Drawing inspiration from the feminist movement, which managed to effect social change in a widespread manner without a central organising committee or hierarchy, Gibson-Graham argued that capitalism might be overcome through similarly decentralised and seemingly disorganised politics (Gibson-Graham, 1993). In the same way the neglect of the feminist movement and women's political activism reproduced phallocentric theory, they argued that a capitalocentric logic meant theorists overlooked the anti-capitalist movement that was already all around them (Gibson-Graham, 1996). This argument advanced thinking on the spatiality of global politics and social transformation via considering scale not as a set of nested hierarchies, but a complex assemblage of relationships across space (and through time). While others have contributed to this thinking (Marston et al., 2005; Massey, 2005), Gibson-Graham's use of 'weak theory', their commitment to both place-based action research and radical transformation of a global capitalist economy, and their experience of feminist political activism were singular. One example of this line of thought is the politics of ubiquity captured by the argument that 'if women are everywhere, a woman is always somewhere, and those somewheres are what the project is about: places being created, strengthened, defended, augmented, or transformed by women' (Gibson-Graham, 2016: 131). Thus, the place 'where you are' becomes the only space in which it is possible to think and to act, even with the intention that the repercussions of those actions might spread beyond your reach.

Their particular combination of serious Marxist and feminist thinking and action, poststructural analysis and explorations of Buddhist philosophy influenced Gibson-Graham's commitment to a stance of 'not knowing' or 'beginner's mind'. This stance shaped the three-pronged politics laid out in *A Postcapitalist Politics* (2006): a politics of collective action, a politics of language and a politics of the subject. This understanding of social and economic transformation embraces a kind of 'both/and' approach to thinking about change, where the power of discourse, the power of collectives and tangible action, and the power of self-transformation are all recognised. If the economy is 'overdetermined' by a range of overlapping histories, trajectories, power relations, and social norms, so too is 'the revolution' – there are multiple spaces of political action which we are all able to access. A decentralised understanding of global transformation is used to both explain and prompt transformation, as was made explicit in the 2013 popular publication *Take Back the Economy*.

Key Advances and Controversies

Gibson-Graham's critique of the capitalocentric nature of much geographic theorising was timely and transformative. Their intervention into rethinking the economy in the 1990s was important not just in terms of the economy in economic geography, but the way geographers used theory in general

(Barnes, 2009). Theory was not about a realist, rationalist account of how a 'real' object of study (such as the economy) worked, but also became conceived of as 'a vocabulary to achieve new ends, rather than mirroring the object of investigation' (Barnes, 2009: 319). For Gibson-Graham, 'promiscuity' in theory was thus valuable, as they drew together various theoretical influences from a range of writers outside geography, including **Foucault**, **Deleuze** and Guattari, Eve Sedgwick, **Judith Butler**, George Demartino, **Arturo Escobar** and more. Theory was intentionally adopted to *perform* different kinds of possibilities and even realities. This playful engagement with theory involved a different kind of affect than the seriousness of Marxist materialist theorists and thinkers – and indeed the quantitative economic geographers who had dominated prior to them (Barnes, 2009). Yet playfulness did not mean a lack of commitment to serious topics that required intellectual and activist attention, for the ethics of both feminism and Marxism involved commitment to real and grounded change.

The form of action research taken up by J.K. Gibson-Graham and their collaborators thus built on both an attention to affect and performativity. The approach requires research participants to *feel* and *act* differently in order to understand economy as a space of affect and potentiality. Committing to an ethic of action and hope meant that action researchers worked alongside communities, often with community researchers, to *intervene* in their economies through activities that encouraged different kinds of subjectivities, collective action and language. This means that despite the apparent playfulness of their theoretical approach, very often their work has impact on the ground as the research itself unfolds, as demonstrated in partnerships with various groups in the Philippines, the Pacific, rural Australia and US, and many more (Community Economies Collective, 2001; Gibson-Graham, 2005; Healy and Graham, 2008; McKinnon et al., 2016; Dombroski et al., 2022).

An approach that highlights action, affect and engagement enabled Gibson-Graham to embrace an even more expansive collective from the mid-2000s. As the degradation of the planet and the climate crisis became more publicly visible, the authorial persona of Gibson-Graham partnered with environmental humanities thinkers and others to start thinking about diverse and community economies with reference to the more-than-human. Given the many compounding crises of climate, environment and economy, this work has sought to perform different kinds of economic relationships by highlighting the possibilities of building on what is already done, always in collaboration in more-than-human collectives. In later publications, the role of a butterfly hovering at ecofeminist Val Plumwood's funeral, the presence of the Georges River in thinking about ecology and economy, the ongoing collaboration of J.K. Gibson-Graham beyond Julie's death, and the co-authorship with a wider group through the 'Community Economies Collective' all point to the importance of a collaborative authorship in their intellectual work (e.g., Gibson-Graham and Miller, 2015; Gibson-Graham et al., 2021).

One consequence of this shift in the disposition of scholarship is that the work of J.K. Gibson-Graham, and this wider collective, has been critiqued for not offering a complete answer to the problem of capitalism, or for being overly optimistic in its reading of the economy as a site of possibility. As Soaita (2022) identifies, the 'optimism' of Gibson-Graham is not about a dismissal of the failures and inequities of capitalist economies, but a choice of focus (see also Gibson-Graham and Dombroski, 2020). Gibson-Graham and colleagues recognise that, firstly, there is plenty of existing scholarship that describes exploitation, welfare insecurity, challenges of scale and co-option – but not so much that brings to light the times when these things offer hope and possibility for 'making other worlds'. Secondly, the critiques often stem from an ontological elision between the diverse economies project of performing and bringing into being new economies, and the realist project of describing what is 'actually' out there. These are fundamental differences in understanding the purpose of research and mean that the answers to critique rarely satisfy.

Despite these concerns, the influence of Gibson-Graham's work has continued to grow, and the generative potential of their vision for a radically rethought economy and a decentralised process of global transformation continues to have a growing influence on scholars and activists well beyond the boundaries of geography as a discipline. The Community Economies Research Network (CERN), established to connect scholars, artists, and activists from around the world who are engaged with Gibson-Graham's work on diverse and community economies, is evidence of this. By 2022, the Community Economies Research Network had over 350 members worldwide and, supported by the not-for-profit Community Economies Institute, members are enacting change grounded in Gibson-Graham's thinking around place-based yet global vision:

> Rather than 'waiting for the revolution' to transform a global economy and governance system at the world scale, we [engage] with others to transform local economies *here* and *now*, in an everyday ethical and political practice of constructing 'community economies' in the face of globalization. (Gibson-Graham, 2016: 132)

Gibson-Graham's Key Works

Gibson-Graham, J.K. (1993) 'Waiting for the revolution, or how to smash capitalism while working at home in your spare time', *Rethinking Marxism*, 6: 10–24.

Gibson-Graham, J.K. (1996) *The End of Capitalism (As We Knew It). A Feminist Critique of Political Economy*. Minneapolis, MN: University of Minnesota Press.

Gibson-Graham, J.K. (2006) *A Postcapitalist Politics*. Minneapolis, MN: University of Minnesota Press.

Gibson-Graham, J.K., Cameron, J. and Healy, S. (2013) *Take Back the Economy: An Ethical Guide for Transforming our Communities*. Minneapolis, MN: University of Minnesota Press.

Gibson-Graham, J.K. and Dombroski, K. (2020) *The Handbook of Diverse Economies*. Cheltenham: Edward Elgar.

Secondary Sources and References

Barnes, T.J. (2009) 'Economic geography', in R. Kitchin and N. Thrift (eds), *International Encyclopedia of Human Geography*. Oxford: Elsevier. pp. 315–27.

Barnes, T.J. and Christophers, B. (2018) *Economic Geography: A Critical Introduction*. London: John Wiley & Sons.

Community Economies Collective (2001) 'Imagining and enacting noncapitalist futures', *Socialist Review*, 28 (3/4): 93–135.

Dombroski, K., Duojie, C. and McKinnon, K. (2022) 'Surviving well: From diverse economies to community economies in Asia-Pacific', *Asia Pacific Viewpoint,* 63: 5–11.

Gibson, K. (1991) 'Considerations on northern Marxist geography: A review from the Antipodes', *The Australian Geographer,* 22: 75–81.

Gibson, K. (1999) 'Community economies: Economic politics outside the binary frame', Paper presented at the Rethinking Economy Conference, ANU, Australia, August. www.communityeconomies.org/publications/conference-papers/community-economies-economic-politics-outside-binary-frame (accessed 2 Jan 2024).

Gibson, K. and Graham, J. (1986) 'Situating migrants in theory: The case of Filipino migrant contract construction workers', *Capital & Class,* 10: 130–49.

Gibson-Graham, J.K. (2005) 'Surplus possibilities: Postdevelopment and community economies', *Singapore Journal of Tropical Geography,* 26 (1): 145–62.

Gibson-Graham, J.K. (2016) 'Building community economies: Women and the politics of place', in W. Harcourt (ed), *The Palgrave Handbook of Gender and Development.* London: Palgrave Macmillan. pp. 287–311.

Gibson-Graham, J.K. and Miller, E. (2015) 'Economy as ecological livelihood', *Manifesto for Living in the Anthropocene*: 7–16.

Gibson-Graham, J.K., Cameron, J., Dombroski, K., Healy, S. and Miller, E. (2021) 'Cultivating community economies: Tools for building a liveable world', in J.G. Speth and K. Courrier (eds), *The New Systems Reader Alternatives to a Failed Economy.* London: Routledge. pp. 410–32.

Gibson-Graham, J.K., Resnick, S. and Wolff, R. (2000) *Class and its Others.* Minneapolis, MN: University of Minnesota Press.

Gibson-Graham, J.K., Resnick, S. and Wolff, R. (2001) *Re/presenting Class: Essays in Postmodern Marxism.* Durham, NC: Duke University Press.

Graham, J. (1988) 'Post-modernism and Marxism', *Antipode,* 20: 60–6.

Healy, S. and Graham, J. (2008) 'Building community economies: A postcapitalist project of sustainable development', in D. Ruccio (ed), *Economic Representations.* London: Routledge. pp. 291–314.

Marston, S.A., Jones, J.P. III and Woodward, K. (2005) 'Human geography without scale', *Transactions of the Institute of British Geographers,* 30: 416–32.

Massey, D. (2005) *For Space.* Thousand Oaks, CA: Sage.

McKinnon, K., Carnegie, M., Gibson, K. and Rowland, C. (2016) 'Gender equality and economic empowerment in the Solomon Islands and Fiji: A place-based approach', *Gender, Place & Culture,* 23 (10): 1376–91.

Roelvink, G. and Gibson-Graham, J.K. (2009) 'A postcapitalist politics of dwelling: Ecological humanities and community economies in conversation', *Australian Humanities Review,* 46: 145–58.

Roelvink, G., St. Martin, K. and Gibson-Graham, J.K. (2015) *Making Other Worlds Possible: Performing Diverse Economies.* Minneapolis, MN: University of Minnesota Press.

Sharpe, S. (2014) 'Potentiality and impotentiality in J.K. Gibson-Graham', *Rethinking Marxism,* 26: 27–43.

Soaita, A.M. (2022) 'The handbook of diverse economies', *Housing, Theory and Society,* 39: 118–20.

19
Ruth Wilson Gilmore

Unusah Aziz, Lauren Cape-Davenhill, and Deirdre Conlon, University of Leeds

Biographical Details and Theoretical Context

Ruth Wilson Gilmore was born in 1950 in New Haven, Connecticut. The daughter of a machinist, union organiser, and activist against racism, Gilmore grew up at the peak of the civil rights movement. From an early age she learned that activism 'was completely normal, that no matter what one did, what one should be doing at the same time is changing the world' (Johnson, 2006: n.p.). It is no coincidence, then, that she has become a renowned scholar-activist, whose critical work on prisons, racial capitalism, inequalities and injustice, and abolition is encapsulated with the statement 'abolition requires that we change one thing: everything' (Gilmore, forthcoming). In a 2006 interview, Gilmore remarked 'I've had many lives, many, many lives' (Johnson, 2006: n.p.); she worked as a mechanic, a higher education consultant, and studied to be an actor. She completed undergraduate and master's degrees in Dramatic Literature and Criticism at Yale University, Connecticut, where she was among the first co-educational cohorts to complete an undergraduate degree at the university. Prior to Yale, Gilmore attended Swarthmore College for a brief period; there she and a group of other Black students, including Angela Davis's sister, Fania, were involved in campus activism aimed at expanding college enrolment for Black students (Kushner, 2019). Between 1994 and 1998, Gilmore completed her doctoral degree at Rutgers University, New Jersey. Initially, she intended to study urban planning but switched to the geography programme under the supervision of Marxist geographer **Neil Smith**. Gilmore's thesis examined the transition from military Keynesianism to post-Keynesian militarism and elaborated the crisis of global capitalism in the US context as characterised by 'surpluses of finance capital, land, labor, and state capacity' (Gilmore, 1999a: 174) that were catalysts for the massive growth and expansion of prisons in the US, specifically California, from the 1980s onwards.

After completing her PhD Gilmore worked as Assistant Professor at UC Berkeley, California. Subsequently, she joined the faculty at the Graduate Center, City University of New York, where she is currently professor of Earth and Environmental Sciences and American Studies, director of the Center for Place, Culture, and Politics, and a member of the Executive Committee of the Institute for Research on the African Diaspora in the Americas and the Caribbean. Gilmore's long-standing activities with grassroots organising include co-founding Critical Resistance (1997), the California Prison Moratorium Project (1998), and the Central California Environmental Justice Network (1999). Her activism has, thus, always taken place in dialogue with her scholarship.

Gilmore has received numerous awards and honours for both scholarship and activism. In 2021 she was elected to the American Academy of Arts and Sciences. Along with Angela Y. Davis and ***Mike Davis***, she was awarded the Lannan Foundation Lifetime Cultural Freedom Prize in 2020. She received the American Association of Geographers (AAG) Lifetime Achievement Award in 2020 and the AAG's Harold Rose Award for Anti-Racist Research and Practice in 2014. American Studies Association (ASA) honours include the Richard A. Yarborough Mentorship Award in 2017 and the Angela Y. Davis award for Public Scholarship in 2012. In September 2020, *Prospect* magazine identified Gilmore as one of 'the world's top 50 thinkers for the COVID-19 age' (Clark, 2020) and a coalition of youth activists, artists, and educators in New Haven, Connecticut is supporting a mural honoring her life and 'the work she creates at the intersection of abolition feminism, transformative justice, Black liberation, coalition-building and climate justice' (Intergenerational Abolition Project, 2023).

Gilmore's work and interests articulate and interrogate the deep-rooted and intersecting structures of racism and inequality; resource extraction, organised abandonment, and organised violence as expressions of state power under capitalism; and the carceral geography and infrastructure of the prison industrial complex as a key spatial fix for surpluses and for social control in contemporary society. Gilmore's oeuvre includes enormously influential books *Golden Gulag: Prisons, Surplus, Crisis, and Opposition in Globalizing California* (2007) and *Abolition Geographies: Essays Towards Liberation* (2022), dozens of articles and contributions to book publications, a film produced by the Antipode Foundation (Gilmore, 2020b), as well as countless contributions to organising resistance. Her work critically examines 'the complexities and contradictions of globalisation' (Gilmore, 1999a: 174) through a rigorously grounded and deliberately expansive analysis of how the political economy of state and society mutually reinforce one another producing 'group-differentiated vulnerabilities' (Gilmore, 2022: 107). Refusing the 'anti-state state' (C. Gilmore and R.W. Gilmore, 2008) and steadfastly guided by her indelible commitment to 'changing the world', Gilmore's work lives and breathes a politics of 'abolition as presence' (see Kushner, 2019: n.p.). Here collectivities and connectedness are immanent with injustices and inequalities, as well as the realisation of the possibility to change everything.

Spatial Contributions

Gilmore is possibly best known in geography for challenging mass incarceration in the US, and for her thinking and praxis in abolition geography. Indeed, Kushner (2019: n.p.) observes that Gilmore's work delineating carceral geography is 'a category of scholarship she more or less single-handedly invented'. The US has the highest incarceration rate in the world: possessing less than five percent of the world's population, it has 20 percent of its prisoners. Gilmore unpacks how incarceration at this unprecedented scale has come about, showing how prison and policing have become *the* dominant responses to what, following **David Harvey**, she calls 'organized abandonment' (Harvey, 1989: 303; Gilmore, 2008: 31) and 'organized violence' (R.W. Gilmore and C. Gilmore, 2016: 83). These terms refer to austerity under neoliberal government and how it materialises as, for instance, criminalisation, prison, detention, and deportation. Here, she rejects the hegemonic rationale of 'law and order' ideology, drawing together analysis of political economy, space, and race to theorise mass incarceration as a 'spatial fix'. This fix is in response to surpluses of rural land underused due to debt and drought; surplus labour because of deindustrialisation; surplus finance capital seeking investment; and surplus state capacity (Gilmore, 1999b; 2007). Bringing together space, race, and capital, Gilmore elucidates capitalism as characterised by uneven access to state support and protection. In this context, working-class, racialised, and minoritised groups are blamed for social issues that ensue with this 'organized abandonment' and are subjected to surveillance, criminalisation, and mass incarceration. The 'prison fix', therefore, serves as a 'project of state-building' (Gilmore, 2002: 16) where state activities do not shrink with austerity but instead are redirected to 'organized violence'.

This analysis of what Gilmore describes as the 'anti-state state' also reveals a key contemporary contradiction where the state is both shrinking *and* growing as welfare and state provisions are rolled back while incarceration and organised violence expand. With a rigorous, persuasive socio-spatial analysis, Gilmore connects often-discrete sites and scales and demonstrates the systemic and relational aspects of prison as an accumulation strategy. This analysis of mass incarceration is framed by the concept of 'racial capitalism' which centres race in understanding how capitalism operates. Building on Cedric Robinson's work on the Black Radical Tradition (see Robinson, 1983), Gilmore observes how, through history, the distinction between the owners of the means of production and those whose labour is exploited has always been articulated in racial terms, noting that 'capitalism requires inequality, and racism enshrines it' (Gilmore, 2017: 240). Gilmore insists on the interdependence of capitalism and race in the Black Marxist tradition, in contrast to the racist and developmentalist assumptions of Marx around the historical progression of European societies. She develops analyses of racial capitalism further by focusing attention on its imbrications with different spaces: rural and urban communities facing struggles about exploitative resource extraction; dispossessed neighbourhoods; and prison buildings themselves. Gilmore's analysis enables an understanding of how struggles against racism and capitalism are intimately linked. As she notes: 'We can't undo racism without undoing capitalism' (Gilmore, 2020a: n.p.).

Gilmore's (2007: 28) understanding of racism, defined as 'the state-sanctioned or extra-legal production and exploitation of group differentiated vulnerability to premature death' is then significant. This is a deliberately expansive framing that draws together historical materialism and racial capitalism, grounded in contexts that trace and engage with the histories and strategies of a range of social movements and grassroots groups. These include Gilmore's work with Black women in Los Angeles fighting to free their children from excessively long prison sentences (Gilmore, 1999b) and the development of coalitions between rural and urban activists to challenge both the environmental and social harms of prisons (Gilmore, 2007). From this conceptualisation of racism, Gilmore illuminates the connected differentiation of dispossession and abandonment and the possibilities that emerge for 'solidarity between unlikely allies' (Gilmore, 2007: 249), as well as providing important insights into the histories and strategies of an array of labour, anti-racist, anti-carceral, and environmental justice movements. Gilmore's research also points to ways forward in organising around social justice. Through concrete examples of what such organising looks like in practice, her work is indispensable, providing a 'way in' for people seeking to understand or engage with the abolitionist framework and praxis that Gilmore's oeuvre embodies.

Gilmore's development of abolitionist scholarship and praxis is tremendously influential. Her research reveals the enormous damage inflicted on individuals and communities by mass incarceration in the US. As an activist, she makes the case that abolition, rather than reform, is the only possible response to this damage. Gilmore moves abolition beyond merely closing prisons, to a vision of societal transformation and human flourishing for all (Gilmore, 2022). Challenging the view that incarceration is a necessary response to crime, she argues that we need to change the social and environmental conditions that produce harm, with a focus on education, healthcare, sustainability, and developing new forms of justice. Building on Andre Gorz's 1967 concept of reformist versus non-reformist reforms (Gilmore, 2007: 242), her research traces debates in grassroots activism and demonstrates how proposed policy changes would alter the power of the carceral state. This has resulted in specific strategic guidance for abolitionist organising, such as the need to challenge the system rather than making individual or group-based claims around 'innocence' as the basis for decarceration (Loyd, 2012: 43).

Gilmore's abolition geography advances a vision, and a world, that is inclusive and interconnected. She explains 'the abolition I speak of has roots in all radical movements for liberation and particularly in the Black Radical Tradition' (Gilmore, 2011: 258). She also holds that 'abolition has to be green [...] and to be green it has to be red' (Gilmore, 2020c: n.p.) meaning abolition is inseparable from struggles for environmental justice and for the resources to support vulnerable groups everywhere. In short, 'abolition is a totality and it is ontological. It is the context and content of struggle, the site where culture recouples with the political' (Gilmore, 2011: 258). And, while this vision might be taken to suggest an abstracted, totalising, utopia, Gilmore's work always insists that 'freedom is a place' (2017: 227), which means not only is Gilmore's analysis grounded in and inseparable from her political activism, in addition, the geographical dimensions and specific dynamics of struggles for, and organising around, abolition are always in the foreground.

Key Advances and Controversies

Gilmore's influence, like her work, is capacious. It extends into and beyond geography, history, sociology, criminology, cultural studies, race, ethnic, and indigenous studies, critical migration and border studies, and gender and sexuality, to name a few. Within geography, Gilmore's work has formed and continues to shape carceral and abolition geographies, both expansive sub- and interdisciplinary terrains. It has advanced a diversity of scholarship including the study of prisons, incarceration, carceral landscapes, economies, and infrastructures (see Moran et al., 2017; Story, 2019); critical interventions on policing, borders, immigration enforcement, detention, and refugee camps (see, for example, Loyd et al., 2012; Massaro and Boyce, 2021); as well as articulations of abolition from growing numbers of collectives, including disabled, queer, feminist, and indigenous people and groups (see Mei-Singh, 2021; Bierra et al., 2022).

To illustrate the scope, affordances, and array of ways Gilmore's work has been taken up, one example is instructive. Her analysis of prison expansion as 'a geographical solution to socio-economic problems, politically organised by the state which is itself in the process of radical restructuring' (Gilmore, 1999a: 174) refuses simplification and instead advances intricate attention to geographic scales from the intimate to the global. This approach has been hugely generative: first, her insistence on complexity invites a dialectical approach that holds onto and weaves specificities with adroit, radical left theorising. Second, this approach means that ordinarily distinct concerns – prisons and environmental issues, for instance – combine and cohere in vitally important, interdisciplinary ways. Not only is this significant for research and scholarship; in addition, it has advanced strategies and approaches to organising. For example, the analysis of prisons as a form of environmental racism opens new possibilities for activism and coalition-building. Third, and relatedly, this mode of knowledge production and action, centred on experiences and lives, has forged activist scholarship. Gilmore's work insists that researching struggles and social movements cannot simply be for its own sake. Rather, this work enables us, together, collectively, to 'identify and promote multiple routes out of crisis' (Gilmore, 2011: 257). Gilmore's work embodies a moral imperative for academics to actively engage in the struggle for social justice demanding that '[I]f you're going to talk that talk, you've got to walk that walk' (Gilmore, 1993: 71).

Calls for prison abolition and, increasingly, to de-fund the police force in the US, especially, where violence toward Black and minoritised people is endemic, are frequently met with a reactionary question 'but what is to be done about crime and people who commit crimes?' Gilmore, too, has been questioned about her abolitionist position and thinking. In response she has repeatedly noted the following: first, what counts as crime, for whom, and how crime is dealt with, changes depending on hegemonic political, economic, and cultural ideology. Nevertheless, the response is always racialised, differentiating, and harmful. Abolition calls for a different response, one that recognises that 'where life is precious, life is precious' (Kushner, 2019: n.p.). Second, Gilmore reminds us that calling for abolition does not mean 'glossing over the issue of harm that crimes cause' (Gilmore, 2020c: n.p.). Rather, it entails asking and desiring something else, which is 'to see something else in our communities besides harm' (Gilmore, 2020c: n.p.). Gilmore's work continues, extends, and amplifies 'the [...] work of the nineteenth century movement for the abolition of slavery' (Goldstein, 2023: 359) and the contemporary prison abolition movement by framing abolition not as absence or 'getting rid' but as 'presence'. For Gilmore, 'abolition is figuring out how to work with people to make something rather than figuring out how to erase something [...]. Abolition is a theory of change, it's a theory of social life. It's about making things' (Gilmore, 2022: 351). Gilmore's elaboration of abolition is immensely significant, as it borrows from, and resonates with, an array of struggles in the current moment. The deaths of Trayvon Martin, Michael Brown, Eric Garner, Sandra Bland, Freddie Gray, Terence Crutcher, Philando Castile, Ahmaud Arbery, Breonna Taylor, and George Floyd, to name just a few of the many, many Black men and women who have been killed at the hands of police while in police custody, or who are subjected to 'state-sanctioned violence', gave rise to Black Lives Matter (BLM). BLM and other social movements have taken up and continue to advance Gilmore's work to articulate and show why police or prison reforms are ineffective and to advocate defunding, decarceration, and abolition.

For scholars and in universities, Gilmore's rendering of 'abolition as presence' (Kushner, 2019: n.p.) propels critical and vital attention to a tremendous diversity of issues, including the infrastructure of the university itself. It shapes initiatives to decolonise and/or to 'abolish the university' (Boggs et al., 2019), with actions calling for, among other things, 'cops off campus', commitments to anti-racist and anti-colonial politics and practices, and ending student debt, as well as theses calling for the materialisation of the 'undercommons' (Moten and Harney, 2004). This radical proposal adopts Gilmore's alignment of abolition with liberation and calls out universities as places characterised by exploitative labour conditions, problematic professional managerialism, and other questionable principles. The 'undercommons' calls for subversion of the existing university system toward more liberatory ends. The expansive, enlivening understanding of abolition that Gilmore's work invokes also generates visionary research, activity, and praxis that includes a diverse array of issues and spaces including, for instance, scholarship on access to fresh air, clean water, and resources for sustainable lives (Heynen and Ybarra, 2021) and building solidarities across international spaces. Such amplifications of 'the "we" of abolition geography [*generate*] experimentation, practice, and learning in everyday collective care to build new life worlds' (Chavez-Norgaard et al., 2022: n.p.). As such, we are reminded of Gilmore's early years when she learned that activism and changing the world are normal. Across the breadth and depth of her entire body of work to date, through intricately woven, critical, deliberate, rigorously connected, generous, and generative thinking and praxis, Gilmore advances possibilities for liberation that can – and do – change everything.

Gilmore's Key Works

Gilmore, R.W. (1993) 'Public enemies and private intellectuals: Apartheid USA', *Race & Class*, 35 (1): 69–78.

Gilmore, R.W. (1999a) 'Globalisation and US prison growth: From military Keynesianism to post-Keynesian militarism', *Race & Class*, 40 (2–3): 171–88.

Gilmore, R.W. (1999b) 'You have dislodged a boulder: Mothers and prisoners in the post Keynesian California landscape', *Transforming Anthropology*, 8 (1–2): 12–38.

Gilmore, R.W. (2002) 'Fatal couplings of power and difference: Notes on racism and geography', *The Professional Geographer*, 54 (1): 15–24.

Gilmore, R.W. (2007) *Golden Gulag: Prisons, Surplus, Crisis, and Opposition in Globalizing California*. Berkeley, CA: University of California Press.

Gilmore, R.W. (2011) 'What is to be done?', *American Quarterly*, 63 (2): 245–65.

Gilmore, R.W. (2017) 'Abolition geography and the problem of innocence', in G.T. Johnson and A. Lubin (eds), *The Futures of Black Radicalism*. London: Verso. pp. 225–40.

Gilmore, R.W. (2020a) 'The case for prison abolition: Ruth Wilson Gilmore on COVID-19, racial capitalism and decarceration', *Democracy Now!*, May 5. Available from: https://www.democracynow.org/2020/5/5/ruth_wilson_gilmore_abolition_coronavirus (accessed November 4, 2023).

Gilmore, R.W. (2020b) *Geographies of Racial Capitalism with Ruth Wilson Gilmore*. An Antipode Foundation film directed by Kenton Card, BFD Productions. Available at: https://antipodeonline.org/geographies-of-racial-capitalism/ (accessed November 4, 2023).

Gilmore, R.W. (2022) *Abolition Geography: Essays Towards Liberation*. London: Verso.

Gilmore, R.W. (forthcoming) *Change Everything: Racial Capitalism and the Case for Abolition*. Chicago, IL: Haymarket Books.

Secondary Sources and References

Bierra, A., Caruthers, J. and Lober, B. (2022) *Organizing, Survival, and Transformative Practice. Vol. 1; Abolition Feminism.* Chicago, IL: Haymarket.

Boggs, A., Meyerhoff, E., Mitchell, N. and Schwartz-Weinstein, Z. (2019) *Abolitionist University Studies: An Invitation,* April 7. Swarthmore, PA: Swarthmore College. https://abolition.university/invitation/ (accessed November 4, 2023).

Chavez-Norgaard, S., Montange, L., Sayers, J.E. and Ybarra, M. (2022) 'Introduction: making abolition through one million experiments. Making Abolition in Geography forum', *Society and Space,* October 31. https://www.societyandspace.org/forums/making-abolition-in-geography (accessed November 4, 2023).

Clark, T. (2020) 'The world's top 50 thinkers 2020', *Prospect Magazine,* Sept. 2. https://www.prospectmagazine.co.uk/world/40533/the-worlds-top-50-thinkers-2020-the-winner (accessed November 4, 2023).

Gilmore, R.W. (2008) 'Forgotten places and the seeds of grassroots planning', in C. Hale (ed), *Engaging Contradictions: Theory, Politics, and Methods of Activist Scholarship*. Berkeley, CA: University of California Press. pp. 31–61.

Gilmore, R.W. (2020c) 'Ruth Wilson Gilmore makes the case for abolition', hosted by C. Kumanyika. *The Intercept,* June 10. https://theintercept.com/2020/06/10/ruth-wilson-gilmore-makes-the-case-for-abolition/ (accessed November 4, 2023).

Gilmore, C. and Gilmore, R.W. (2008) 'Restating the obvious', in M. Sorkin (ed), *Indefensible Space: The Architecture of the National Insecurity State*. London: Routledge. pp. 141–62.

Gilmore, R.W. and Gilmore, C. (2016) 'Beyond Bratton', in J.T. Camp and C. Heatherton (eds), *Policing the Planet: Why the Policing Crisis Led to Black Lives Matter.* London: Verso. pp. 82–90.

Goldstein, A. (2023) 'Introduction: Abolitionist worldmaking', *American Quarterly*, 75 (2): 359–64.

Harvey, D. (1989) *The Condition of Postmodernity: An Enquiry into the Origins of Cultural Change*. Oxford: Blackwell.

Heynen, N. and Ybarra, M. (2021) 'On abolition ecologies and making "freedom as a place"', *Antipode,* 53 (1): 21–35.

Intergenerational Abolition Project (2023) Mural Honoring Ruth Wilson Gilmore in Her Hometown of New Haven. https://fundraising.fracturedatlas.org/roots-that-reach-toward-the-sky/campaigns/5686 (accessed November 4, 2023).

Johnson, P. (2006) 'Portrait of an activist-academic', *USC Dornsife,* Oct. 1. https://dornsife.usc.edu/news/stories/portrait-of-an-activist-academic/ (accessed November 4, 2023).

Kushner, R. (2019) 'Is prison necessary? Ruth Wilson Gilmore might change your mind', *The New York Times*, April 17. https://www.nytimes.com/2019/04/17/magazine/prison-abolition-ruth-wilson-gilmore.html (accessed November 4, 2023).

Loyd, J. (2012) 'Race, capitalist crisis, and abolitionist organizing: An interview with

Ruth Wilson Gilmore', in J. Loyd, M. Mitchelson and A. Burridge (eds), *Beyond Walls and Cages: Prisons, Borders, and Global Crisis*. Athens, GA: University of Georgia Press. pp. 42–54.

Loyd, J.M., Mitchelson, M. and Burridge, A. (2012) *Beyond Walls and Cages: Prisons, Borders, and Global Crisis*. Athens, GA: University of Georgia Press.

Massaro, V. and Boyce, G. (2021) 'Carceral geographies, police geographies, and the networked continuum of state-sanctioned coercion and control', *ACME: An International Journal for Critical Geographies*, 20 (6): 569–78.

Mei-Singh, L. (2021) 'Accompaniment through carceral geographies: Abolitionist research partnerships with Indigenous communities', *Antipode*, 53 (1): 74–94.

Moran, D., Turner, J. and Schliehe, A.K. (2017) 'Conceptualizing the carceral in carceral geography', *Progress in Human Geography*, 42 (5): 666–86.

Moten, F. and Harney, S. (2004) 'The University and the undercommons: Seven theses', *Social Text*, 79 (22/2): 101–15.

Robinson, C. (1983) *Black Marxism: The Making of the Black Radical Tradition*. Chapel Hill, NC: University of North Carolina Press.

Story, B. (2019) *Prison Land: Mapping Carceral Power across Neoliberal America*. Minneapolis, MN: University of Minnesota Press.

20
Stephen Graham

Jonathan Rutherford, Ecole des Ponts ParisTech

Biographical Details and Theoretical Context

Stephen Graham was born on Tyneside (UK) in 1965. After completing a BSc in Geography at the University of Southampton and an MPhil in Town and Country Planning at the University of Newcastle, he worked as an urban planner for Sheffield City Council from 1989 to 1992. Returning to academia, he completed his doctorate on urban telecommunications projects at the University of Manchester while also working as Lecturer in the Planning department at the University of Newcastle. At Newcastle, Graham established the Centre for Urban Technology (CUT) with his close colleague Simon Marvin, and along with Simon Guy, undertook an intensive critical research agenda on urban infrastructures and technology systems. This research led to two highly influential monographs, *Telecommunications and the City* (Graham and Marvin, 1996) and *Splintering Urbanism* (Graham and Marvin, 2001). After a year as a visiting Professor at MIT, he was made Professor at Newcastle in 2001. He joined Durham University's Geography department in 2005, where he developed a research agenda on the intimate entwining of security logics, militarisation and urban space. He returned to the Global Urban Research Unit (GURU) at the University of Newcastle in 2010. In 2019 he was made a Fellow of the British Academy.

Graham's wide-ranging work over more than thirty years, translated into more than a dozen languages, has had an enormous impact in Geography, Urban Studies and the wider social sciences by constantly seeking to locate, situate and understand evolving processes of technological (re)mediation of urban space and society in terms of their fundamental intrinsic political context and outcomes. Through an engagement with critical social theory and leveraging a capacity to synthesise often kaleidoscopic and wide-ranging materials and resources, his work has centred a set of emerging contemporary technological and political trends, logics and processes, associated with ever deepening neoliberalisation, that have profound, often worrying, implications for cities and society as a whole, but that are also open to being contested and co-shaped otherwise. This work can be divided into three broad and overlapping themes: telecommunications, infrastructure and digital technology; surveillance, security and militarisation of urban space; and vertical and atmospheric perspectives on the city and society.

Graham's early work around telecommunications networks debunked the dazzling hype and ageographical oversell of increasingly ubiquitous Information and Communication Technologies (ICTs). It is all too easy, many years later, to forget the influence in some quarters of simplistic 'death of distance' arguments propounded at the time as an inevitable consequence of global telecommunication. Graham was resolute that geography continued to matter, often more so in a world of ostensibly fluid exchange. For all the focus on new virtual content and logics (e.g., see Mitchell, 1995), materiality continued to be constitutive of urban technological relations as '[m]aterial space and electronic space are increasingly being produced together' (Graham, 1998a: 174). Graham thus sought 'to advance understanding of the complex *co-evolutionary* processes linking new information technologies and space, place and human territoriality' through what he called their 'recursive interaction' (Graham, 1998a: 171, 174; also Graham, 2004a).

A further substantive contribution was to demonstrate how a newly deregulated, liberalised and globalising context in telecommunications was increasingly exacerbating socio-spatial polarisation in the provision of telecommunications services, with companies competing for the custom of lucrative users, whilst leaving more marginal groups and areas often completely disconnected. In *Splintering Urbanism* (2001) the interlinking of political economic context and highly unequal socio-spatial outcomes was extended from telecommunications to multiple infrastructure sectors (water, energy, transport) and from a predominantly UK/European focus to a global overview. In a 'network society', technological advances and trends toward privatisation and liberalisation of infrastructure provision lead to increasing selectivity and differentiation between valued and less valued groups. As such, the book placed infrastructure networks at the heart of the social and political make-up of cities worldwide.

To study and understand these polarising tendencies, Graham's work provided a fully relational understanding of cities and their socio-technologies, as cutting across scales and spaces and bringing into juxtaposition a diversity of groups (with differing capacities), components and processes (see Amin and Graham, 1997; Graham and Healey, 1999). He also applied this relational perspective to the study of digital surveillance and the practices of socio-spatial sorting of individuals and groups (Graham, 1998b, 2005a; Graham and Wood, 2003).

Though Graham has never abandoned the study of urban infrastructure (see Graham, 2009; Graham and McFarlane, 2015; Graham and Thrift, 2007), his attention widened over time to encompass strategic military and security concerns. This broadening of focus was partly the result of attending what he thought was a social science conference in Israel in 2002 which turned out to be a post-9/11 military event organised by US and Israeli armed forces about optimising military operations in hostile urban areas (Graham, 2005b). Uncovering a partially hidden world of urban technoscience for direct military application, Graham spent much of the next decade bringing to light and critically analysing how the city was becoming the 'target' site for a vast array of applications and techniques that would enable various military-industrial complexes/groups to dominate, conquer, vanquish urban space and the constraints, perceived 'threats' and resistances of recalcitrant social groups and populations (Graham, 2004b). In *Cities Under Siege* (2010) he detailed the ostensibly mundane, unexceptional and stealthy workings of 'military urbanism' that constitute nonetheless 'dramatic attempts to translate longstanding military dreams of high-tech omniscience and rationality into the governance of urban civil society' (Graham 2010, xii). He traced how interlinked policing, surveillance, security techniques, data mining, bordering and border control operations, conservative ideology and popular culture are reaching deeper and deeper into urban space and life to divide, contain, discipline and format cities and citizens according to Orwellian government-corporate requirements.

Graham's most recent work has focused on the volumetric dimensions of cities and urban life. His book *Vertical* (2016) moved beyond a flat, two dimensional geographic view of territory to a fully three-dimensional perspective attentive to the worlds above and below us and how they increasingly constitute 'geographies of inequality, politics and identity'. His analysis moves layer by layer from the 'crucial but neglected manufactured environment of satellites and space junk' (Graham, 2016: 25) at the edge of the earth's atmosphere to the ever deeper subterranean bunkers, tunnels and mines that offer sanctuary or resource-extraction possibilities. In so doing, *Vertical* charted the multiple strata and volumes that organise and constitute urban life and through which power, wealth and status are increasingly inscribed, thus reinforcing existing socio-spatial inequalities.

Spatial Contributions

Graham's work explores how, in an era of globalisation and technological development, cities and urban life are being reshaped by all manner of increasing flows, circulations and mobilities, including through the widespread use of digital technologies, and how the resulting tensions between fixity and motion work through and intensify socio-spatial inequalities. One of his main contributions then has been to uncover and critically analyse geographies of socio-technical change. In this respect he is part of a lineage that goes back to Lewis Mumford and subsequent urban technology and Science and Technology Studies scholarship, but also builds on **Doreen Massey**'s relational geographical thinking. By reflecting on and studying a variety of urban networks, systems and worlds, Graham has constantly drawn attention to, and expanded our understanding of, the uneven, unequal ways in which technology is deployed and experienced, sustaining 'socio-technical geometries of power' (Graham and Marvin, 2001: 11).

Furthering the analysis of how reticular connectivity largely supplants immediate geographical contiguity in the construction of functional territories, in Graham's work, urban space is thus materially constituted through combinations of 'local' and 'global', 'near' and 'far' relations and processes in which social interests, values and practices (and their spatial inscriptions) are indissociably entwined with complex technological systems. The affordances, constraints and possibilities of technology are at once shaped by, and shaping, the (place-based) people, groups and institutions that mobilise and appropriate them. This is a firm rejection of technological determinism and social construction in favour of a co-evolutionary perspective on urban spatial change that, he argued, was more adroit in pinpointing the 'contingency with bias' of systems/networks (Graham and Marvin, 1996: 381). Graham argues that close attention needs to be paid to the 'recursive interactions' between social relations and interests and the particular materiality and technicity of components that together produce and sustain how a technological configuration works in differential, uneven ways. In this sense, his socio-technical perspective has been, like many others, attuned to finding a fertile middle ground between critical urban political economy and a poststructuralist Latourian sensitivity to hybrid relationality (see **Bruno Latour**).

His analysis of the modalities and outcomes of rising digital surveillance and security produced pioneering identification of how 'computerised code continually orchestrates inequalities through technological systems embedded within urban environments' (Graham, 2005a: 562). Ever finer differentiation and separation of marginalised and privileged groups and spaces was becoming possible, allowing authorities and companies in the domains of, for example, street-level CCTV, GIS-based geodemographic systems, and transport and mobility to 'sort' individuals, households and social groups according to whether they were desirable or undesirable as customers or as occupants of public space. These 'software sorted geographies' have become more widespread and mainstream as such systems have been optimised and deepened, and we see the results every day in terms of perpetuating highly differential access to increasingly digital-only platforms and the algorithmic filtering of people's mobilities and movements, and access to banking, insurance and housing markets (see e.g., Kitchin and Dodge, 2011; M. Graham et al., 2013).

His relational view and conceptualisation of space and place is also a key trope running through all his work, pushing beyond a Euclidean flat view of geographic space and urban places as predominantly bounded, static areas. **Doreen Massey**'s (1993) influential understanding of the inherent relationality of places and the importance of their wide-ranging connections to elsewheres in constructing the meaning and experience of those places was important here, as was, inevitably, the objects of Graham's research. Telecommunications and other infrastructure systems lent themselves neatly to the development of multi- or cross-scalar understandings of urban connection and disconnection, but the power of Graham's work is not just in showing these, but also in drawing out the crucial socio-spatial workings and consequences of flows and connections – and how access to these for some can be directly excluding and disempowering, even deadly, for many others.

His later work has added an extra dimension to this horizontal politics of material relationality by exploring the logics, forms and socio-political implications of vertical relationality. This work has been important in shifting recent geographical thinking onto a more volumetric and three-dimensional plane (see **Stuart Elden**, 2013). Urban space is shown to be constituted also through its links to above and below, and as such can be volumetrically enclosed to create 'bubbles' of privileged existence and inhabitation protected from problematic outsides. His political ecology of air (Graham, 2015), for example, demonstrates the capacity for three-dimensional social differentiation that now exists to separate clean elite atmospheres from the impure polluted air of the rest of the city. This kind of trend and development powerfully underscores that there is a spatial politics to even the most fundamental and basic human life support system.

Much of Graham's work focuses on dark and worrying trends and logics in urban socio-technical worlds, and as such is unapologetically critical. It is, however, at the same time ultimately optimistic in its belief in the 'richness of cities' and the potential for transformative social interactions and democratic politics. Graham ends *Cities Under Siege* (2010) and *Vertical* (2016), for example, by seeking out alternative possibilities and ways of seeing/doing in grassroots or social movements whose counter-geographic tactics might be a means of revealing or subverting elite verticalities, military urbanism and dominant security logics. His critique of socio-spatial trends and inequalities always lead therefore to a potential progressive geography through which urban society and urban life might be reshaped along more egalitarian and democratic paths.

Key Advances and Controversies

Graham's research has been highly influential in urban geography, urban studies and the wider social sciences. He has demonstrated an acute eye and aptitude for identifying early on emerging spatial tendencies and logics in the socio-technical organisation and functioning of urban society and life that have profound social and political implications and consequences. His work, including in actively supporting and mentoring younger colleagues, has set the stage for and inspired scholars to follow up and extend his path-shaping analyses. As such, he contributed to advancing cyberspace and urban place debates in the mid-late 1990s, leading more and more researchers to identify that the complex and intertwined relations between ICTs and cities/geography actually reinforced the latter. His work foregrounding the rise of increasingly sophisticated surveillance technologies in securing urban space, and, more broadly, uncovering a variety of hidden practices and techno-mediated logics that undergird the production of cities and make urban space inherently political has been taken up by scholars keen on tracing further the complex, often murky webs through which urban life is becoming ever more subject to the control and influence of particular economic and political forces and interests.

Splintering Urbanism (Graham and Marvin, 2001) in particular foresaw and helped to shape the 'infrastructure turn' in urban studies and the social sciences more widely. As a result, it has become quite commonplace to use exploration of the materiality of the urban fabric as an entry point into urban controversies, conflicts and politics (Wiig et al., 2022). Indeed, 'splintering urbanism' has become a more widely-used notion, a shorthand for the broader fragmentary nature of contemporary urban and spatial change. Its extraordinary influence has not been devoid of critique, however this is somewhat inevitable given the book's big theoretical ambitions and global perspective. For example, Coutard's (2002) exchange with Graham (2002) over the forms and outcomes of 'premium network spaces' set the stage for further debate about the extent to which the book's overarching argument and its core concepts were applicable to all urban contexts.

Coutard (2008) subsequently oversaw a programme of empirical work applying elements of the splintering urbanism thesis to varying urban contexts around the world where infrastructure provision was/is not always shaped by logics of bypass, exclusion and secession. He concluded by foregrounding the diverse, highly contextualised and situated configurations and experiences of access to essential urban services like water and energy, and thus rejected any idea of a generalised tendency toward 'splintering urbanism'. Urban scholars working in the Global South have also highlighted some of the limits to the theory for understanding and explaining urban infrastructural change in and across these contexts. These critiques contest the assumption that Southern cities are waiting or aiming for similar universal infrastructure ideals as rolled out in Europe, that fragmented contexts of infrastructure provision are the result of political economic reforms rather than the already existing social norm, and the lack of focus on ostensibly 'informal' forms of access to basic services which, without fetishising these, show that in many cases heterogeneous configurations constitute a malleable and, in some cases, practical 'solution' adapted to local contextual specificities (see e.g., Furlong, 2014). Thus, while the global applicability of the thesis has been pushed back on, splintering urbanism continues to offer a fecund, critical framework for the study of infrastructure developments in and across a diversity of settings. As such, the book speaks to a longstanding debate within Geography regarding the interplay between the search for general theories and the need for attention to particular idiographic situations.

Graham's work has become arguably more critically, even politically, engaged with time. While never the detached neutral observer in his early work, the themes and contexts he has studied in the twenty-first century (security, militarisation, elite spaces) clearly do not easily lend themselves to sitting on the fence. His work has thus provoked some critical discussion, primarily around its methodological limitations. Fine-grained accounts of urban security techniques and the militarisation of urban space in *Cities Under Siege* (2010) do not usually extend to include the viewpoints and everyday experiences of people living in 'targeted' cities (Giglioli, 2012). As such, his analysis is said to be 'broader than it is deep' (Vitale, 2012), privileging an extensive range of powerful vignettes over in-depth study of specific place-based configurations. Harris (2015) also argues that there is a wealth of additional conceptual and methodological resources for engaging with the vertical and volumetric dimensions of urban life as well as dominant concerns for militarised security and control technologies and urban division. Graham does recognise himself the limitations of an approach that seeks 'to demystify and make visible the invisible technologies of control, tracking, and surveillance which now thoroughly permeate everyday objects, architectures, environments and infrastructures' (Graham, 2010: 363).

While generally praising *Cities Under Siege*, Jennifer Light (2012) took issue with its focus on change over continuity. Light argued that her own work on the topic (Light, 2003) showed the importance of prevailing trends and logics from the Cold War period when the US military-industrial-research complex was already actively trialling technological solutions in civilian contexts and cities were already deploying quite advanced surveillance applications. As a historian of technology, she argued that '[c]laims about the newness of technologies, especially the novelty of their social effects and implications, typically awaken my scepticism'. This is something that Graham himself might have written in the 1990s in relation to ICTs. Like many of those studying technology with a critical lens, he is only too aware of the tension and tricky balance that has to be found between making visible the complex modalities and implications of technological configurations, while not assuming that these configurations are necessarily all-powerful and unlimited in their reach or even that they function as claimed all the time (Graham, 2010: 364). His work on urban repair and maintenance (Graham and Thrift, 2007) and infrastructure disruption (Graham, 2009) are a notable reminder of the fragility of complex systems that we can be fascinated by, and critique, without falling into the trap of technological fetishism or determinism.

Graham's Key Works

Graham, S. (1998a) 'The end of geography or the explosion of place? Conceptualizing space, place and information technology', *Progress in Human Geography*, 22 (2): 165–85.

Graham, S. (ed) (2004a) *The Cybercities Reader*. London: Routledge.

Graham, S. (ed) (2004b) *Cities, War and Terrorism: Towards an Urban Geopolitics*. Oxford: Blackwell.

Graham, S. (2005a) 'Software-sorted geographies', *Progress in Human Geography*, 29 (5): 562–80.

Graham, S. (ed) (2009) *Disrupted Cities: When Infrastructure Fails*. New York, NY: Routledge.

Graham, S. (2010) *Cities Under Siege: The New Military Urbanism*. London: Verso.

Graham, S. (2016) *Vertical: The City from Satellites to Bunkers*. London: Verso.

Graham, S. and Marvin, S. (1996) *Telecommunications and the City: Electronic Spaces, Urban Places*. London: Routledge.

Graham, S. and Marvin, S. (2001) *Splintering Urbanism: Networked Infrastructures, Technological Mobilities and the Urban Condition*. London: Routledge.

Graham, S. and McFarlane, C. (eds) (2015) *Infrastructural Lives: Urban Infrastructure in Context*. London: Routledge.

Graham, S. and Thrift, N. (2007) 'Out of order: Understanding maintenance and repair', *Theory, Culture & Society*, 24 (3): 1–25.

Secondary Sources and References

Amin, A. and Graham, S. (1997) 'The ordinary city', *Transactions of the Institute of British Geographers*, 22 (4): 411–29.

Coutard, O. (2002) '"Premium network spaces": A comment', *International Journal of Urban and Regional Research*, 26 (1): 166–74.

Coutard, O. (2008) 'Placing splintering urbanism: Introduction', *Geoforum*, 39 (6): 1815–20.

Elden, S. (2013) 'Secure the volume: Vertical geopolitics and the depth of power', *Political Geography*, 34: 35–51.

Furlong, K. (2014) 'STS beyond the "modern infrastructure ideal": Extending theory by engaging with infrastructure challenges in the South', *Technology in Society*, 38: 139–47.

Giglioli, I. (2012) '"Cities Under Siege: The New Military Urbanism" by Stephen Graham', *Berkeley Planning Journal*, 25 (1): 235–7.

Graham, M., Zook, M. and Boulton, A. (2013) 'Augmented reality in urban places: Contested content and the duplicity of code', *Transactions of the Institute of British Geographers*, 38: 464–79.

Graham, S. (1998b) 'Spaces of surveillant-simulation: New technologies, digital representations, and material geographies', *Environment and Planning D: Society and Space*, 16: 483–504.

Graham, S. (2002) 'On technology, infrastructure and the contemporary urban condition: A response to Coutard', *International Journal of Urban and Regional Research*, 26 (1): 175–82.

Graham, S. (2005b) 'Remember Fallujah: Demonising place, constructing atrocity', *Environment and Planning D: Society and Space*, 23 (1): 1–10.

Graham, S. (2015) 'Life support: The political ecology of urban air', *City*, 19 (2–3): 192–215.

Graham, S. and Healey, P. (1999) 'Relational concepts of space and place: Issues for planning theory and practice', *European Planning Studies*, 7 (5): 623–46.

Graham, S. and Wood, D. (2003) 'Digitizing surveillance: Categorization, space, inequality', *Critical Social Policy*, 23 (2): 227–48.

Harris, A. (2015) 'Vertical urbanisms: Opening up geographies of the three-dimensional city', *Progress in Human Geography,* 39 (5): 601–20.

Kitchin, R. and Dodge, M. (2011) *Code/Space: Software and Everyday Life*. London: MIT Press.

Light, J. (2003) *From Warfare to Welfare: Defense Intellectuals and Urban Problems in Cold War America*. Baltimore, MA: Johns Hopkins University Press.

Light, J. (2012) 'Review of Cities Under Siege', *Society and Space,* July 7. https://www.societyandspace.org/book-review-forums/cities-under-siege-by-stephen-graham (accessed November 4, 2023).

Massey, D. (1993) 'Power-geometry and a progressive sense of place', in J. Bird, B. Curtis, T. Putnam, G. Robertson and L. Tickner (eds), *Mapping the Futures: Local Cultures, Global Change*. London: Routledge. pp. 59–69.

Mitchell, W. (1995) *City of Bits: Space, Place and the Infobahn*. Cambridge, MA: MIT Press.

Vitale, P. (2012) 'Cities Under Siege: The New Military Urbanism by Stephen Graham', *cultural geographies*, 19 (2): 271–2.

Wiig, A., Karvonen, A., McFarlane, C. and Rutherford, J. (2022) 'Splintering Urbanism at 20: Mapping trajectories of research on urban infrastructures', *Journal of Urban Technology*, 29 (1): 1–11.

21
Jack (Judith) Halberstam

Sage Brice, Durham University

Biographical Details and Theoretical Context

Jack Halberstam was born in England in 1961 to Heini Halberstam – a rabbi's son, *kindertransport* survivor and mathematician – and his first wife, Heather Peacock – a teacher and opera singer who died in a car accident when Jack was still a child. Growing up in the industrial city of Nottingham, England, he recalls a mundane if somewhat bleak and violent childhood brightened only by the punk, ska, and South Asian music subcultures in which he immersed himself as a teen, and which have proven an enduring influence on his aesthetic and political allegiances as a scholar of subcultural lives.

Previously known as Judith, Halberstam goes by Jack professionally and specifies no preferred pronoun, moving fluidly between she, he, and they. By his own account (in Williams, 2011), Halberstam emigrated to the US after failing university entrance exams in the UK. Initially following his father to the University of Illinois at Urbana-Champaign (where Heini had taken up a position as Head of Mathematics), Jack became bored and drifted to the San Francisco area in search of community, became involved in the dyke bar scene there, and later applied on a whim to the University of California at Berkeley. There he earned a BA in English, followed by an MA and a PhD in English Literature at the University of Minnesota. True to the spirit of his haphazard pathway into academia, Halberstam maintains a cultivated irreverence towards institutionalised hierarchies of the intellect, arguing for the importance of stupidity and failure as well as 'low' culture and 'low theory' as crucial modes of knowledge-production (Halberstam, 2011).

After teaching at the University of California San Diego for over a decade, Halberstam served as Professor of American Studies and Ethnicity, Gender Studies, and Comparative Literature and as Director of The Center for Feminist Research at the University of Southern California, and then as Professor of Gender Studies and English in the Department of English and Comparative Literature and Director of the Institute for Research on Women, Gender, and Sexuality at Columbia University, from 2017.

Halberstam is the author of six monographs of which three stand out as particularly influential contributions. *Female Masculinity*, published in 1998, offered a much-needed account of gender divergence and transgression among girls, women, and (in today's vocabulary) people assigned female at birth – phenomena that were frequently (and to some extent still are) disregarded amid popular obsessions with trans femininity and womanhood. The book aims to reshape the terms of reference through which sex, sexuality, and gender expression are mapped, grounding this project in the assertion that masculinity 'does not belong to men, has not been produced only by men, and does not properly express male heterosexuality' (Halberstam, 1998: 241). Here Halberstam is careful to avoid constructing a monolithic model of female masculinity while also refuting any attempt to draw hard lines between butch, drag, transgender, and transexual expressions. He notes both that individuals find respite in different identities over time, and that many move fluidly between – or simultaneously inhabit – seemingly contradictory roles and subject positions.

In a Queer Time and Place: Transgender Bodies, Subcultural Lives (2005) undertook a wider study of transgender representations in film, fiction, art, and music and in the discourses surrounding their production and dissemination. The first part of the book examines cultural archives attached to the story of Brandon Teena, a young gender-nonconforming person who was murdered, execution-style, alongside two friends in rural Nebraska in 1993. It explores the representational tensions that emerge in narration of minority histories through the lives of extraordinary individuals, highlighting how the various retellings of Teena's story reflect persistent problems of whiteness, urban bias against the rural, and tensions between trans and cis gay/lesbian advocates seeking to claim Teena as one of their own. The second part of the book reviews a selection of queer artistic and subcultural representations, drawing out the contours of a transgender aesthetic which in various ways refuses neat and orderly categorisation, thereby preserving a sense of the ambiguity, non-linearity, and incommensurability of unconventionally gendered bodies, lives, and expressions.

The Queer Art of Failure (2011: 174) sets out 'to take failure apart' by revealing it as a category mobilised by 'the winners' to 'ensure that all future radical ventures will be measured as cost-ineffective'. To counter this totalising move, Halberstam seeks to open out and hold a space for failed and flawed modes of resistance and subversion. The emphasis here is on gendered, sexualised, and raced dimensions of failure as a queer, anti-disciplinary form of knowing. The book draws on an eclectic mix of cultural sources; Toni Morrison's *Beloved* sits alongside Disney Pixar's *Finding Nemo* in an account of radical forgetfulness, while a discussion of Nazi-themed erotica teases out the multifaceted and sometimes tense relationships between fascism and homosexuality. The aim of the book is not to 'recuperate' failure as a necessary step on the road to success. Rather, Halberstam is at pains to celebrate failure on its own terms, as a disruption to the reproductive logics of progress, productivity, and generational transmission which would judge the value (read: 'success') of radical ventures based solely on their heritable legacies of social change.

Spatial Contributions

Jack Halberstam is a scholar of English literature and American cultural studies whose work has nevertheless inflected thinking on space and place through a resolute commitment to celebrating and amplifying non-normative configurations of temporality, location, gender, culture, and everyday life. Halberstam has brought critical queer, trans, and feminist sensibilities to bear on the reproductive organisational logics through which space and time are both structured and theorised, arguing that any geographical critique of naturalised temporalities must consider not only the forces of capital accumulation but also the workings of familial, individual, national, and cultural modes of reproduction.

These modes of reproduction are made visible, for Halberstam, in two ways. The first is through analysis of the cultural practices of evasion, subversion, refusal, and failure adopted by those attempting to make lives for themselves outside conventional relational frameworks of family and state. The second is by identifying the ways in which mainstream cultural accounts of such practices misapprehend or attempt to capture and recuperate them. Taken together, these two lines of analysis inform a deep-reaching critique of the politics of representation that in turn informs geographical research across a wide range of specialisms encompassing rural, urban, queer, and counter-cultural spaces.

Halberstam first cast ripples in geography with the anthology *Posthuman Bodies* (1995), co-edited with Ira Livingston and taken up in the social sciences at a time when anxieties were surfacing about the interface between human and technological forms of embodiment but also, and perhaps more importantly for human geography, when the humanist intellectual tradition was undergoing energetic and critical re-evaluation as a fraught and exclusionary project predicated on a fragile but resolute human exceptionalism. His first two books, *Skin Shows* and *Female Masculinity*, also had some impact in geography, primarily in the study of space, gender and sexuality, but it was his third monograph, *In a Queer Time and Place,* that most noticeably brought Halberstam's work into geographical focus.

The book's opening gambit, *Queer Temporality and Postmodern Geographies*, attempts to reckon with the canonical legacy of geographers such as **David Harvey** and ***Edward Soja*** and literary scholar Frederic Jameson, whose work in Halberstam's eyes 'actively excluded sexuality as a category of analysis', thereby missing out on a crucial dimension of normativity in the social construction of time and space (Halberstam, 2005: 5). This exclusion formed part of the wider entrenchment by postmodern geographers of a hierarchical distinction between the bodily, local, and personal on the one hand, and class, global, and political analyses on the other, the latter set of concerns being understood as the 'proper object' of serious geographical enquiry. For Halberstam, postmodernism instead presents an opportunity to rethink supposedly stable terms of reference for form and meaning, inviting a critical consideration of cultural production and the hierarchies, power dynamics, and modes of resistance or capitulation which that cultural production (re)inscribes.

Halberstam's principal beef in this influential chapter is with **David Harvey's** concept of time/space compression. Harvey (1990) demonstrates that values and meanings associated with particular forms of temporality become naturalised in service to the organising logics of late capitalism. Halberstam applauds both the denaturalising impulse of this account and its specific close attention to how time becomes subordinated to organising logics of capital accumulation. However, he finds in Harvey's theories a missed opportunity to address the *normative* dimensions of temporal naturalisation. Paying attention to the ways in which 'hegemonic constructions of time and space are uniquely gendered and sexualized' (Halberstam, 2005: 8) reveals a further underlying logic in their organisation besides that of capital accumulation: that of reproduction in all its forms. In short, Halberstam joins feminist critiques of a totalising and masculinist Marxist tradition that attempts to subsume anything and everything within the two-dimensional explanatory framework of class and economy.

Feminist geographers such as **Doreen Massey** (1995), **J.K. Gibson-Graham** (1996), and **Cindi Katz** (2004) have sought to complicate and extend Marxist frames of analysis by highlighting the importance of often-unwaged female/feminised labours and spaces of social reproduction in sustaining processes of capital accumulation and capture, and by pushing for a messier, less generalised and finer-grained account of global and social difference as well as variations in the relationship between globalised and localised workings of capital. To this project Halberstam (2005: 5) brings an argument against privileging productivity and use value, and an overarching critique of 'repro-time'. Cultural values attached to stability, longevity, progeny, and heritability continue to govern the naturalisation of normative temporalities in service of a reproductive logic that is not fully accounted for by class- and economy-based readings of politics.

Whereas Harvey, Soja, and Jameson at least pay lip service to racial and gendered dimensions of postmodern space, Halberstam (2005: 10) sees in their silence on sexuality a failure to reckon with the ways in which some lives are lived 'outside of reproductive and familial time as well as on the edges of logics of labor and production'. For Halberstam, the non-normative and non-linear temporalities of queer lives constitute an important take-down and refusal of reproductive time; one which not only deserves attention in its own right, but has consequences for any serious consideration of time and space.

The book's other notable contribution to geographical thought is the concept of metronormativity, which names the erroneous association of queerness with urban spaces and communities, and the concurrent de-prioritisation of rural lives and practices in queer geographic imaginaries.

The Queer Art of Failure (2011), though arguably less influential than its immediate precursor, has provided a useful stimulus to geographers interested in messy, antidisciplinary, and unconventional modes of knowledge-production (e.g., see Harrowell et al., 2018). Halberstam's introduction outlines a proposition for 'low theory' set against the monumental edifice of doctrinal Marxism and its perceived claims to a theoretical 'high ground'. Halberstam's core model for envisioning low theory is the 'bottom-up' histories of opposition to capitalism recounted by Linebaugh and Rediker (2002) in their book *The Many-Headed Hydra*. At stake here is a rejection of masculinist hero-narratives in which politics is construed as the imposition of order on chaos. Instead Linebaugh and Rediker advance an alternative model for understanding alternative political formations without organising them into monolithic or teleological accounts of political and historical progression. In *The Queer Art of Failure*, low theory follows a strategy of reading popular animated movies alongside marginalised contemporary artists and writers to draw out the queer disruptions present even in the most mainstream of corporate cinema production.

Key Advances and Controversies

While Halberstam is widely cited in the geographical literature, much of this literature makes only passing reference to his ideas, taking key concepts such as queer temporality and failure as points of departure or as critical asides rather than drivers of sustained analysis. It is rare that a geographer engages in significant depth with Halberstam's case studies and examples, especially outside of specifically LGBTQIA+ geographies (though see Wilkinson, 2014 and Secor and Linz, 2017 for possible exceptions). This relatively light-touch engagement may be due in part to the limited appeal, in geography, of his particular flavour of Anglo-American literary and cultural studies and its emphasis on sub- and pop-cultural representations.

A further limitation may arise from Halberstam's cultural frames of reference, which through focusing on a critique of liberal assimilationism in a largely white American and Anglo context produces a somewhat narrow conception of the forms in which resistance and normativity can present. Thus while geographers of international or non-Euro-American contexts have valued Halberstam's critical attention to the spatial distortions of metronormativity or to the transgressive potential of queer temporality, the detail of these critiques has not always mapped productively onto other contexts, such as the urban world-making practices of queer Filipinos (Manalansan IV, 2015), or conditions of 'extreme precarity' experienced in South Sudan 'in the absence of the liberal state' (Newhouse, 2017: 505).

To say that geographical engagement with Halberstam's ideas is light touch, however, is not to claim that his influence has been superficial. While the fine detail of his cultural analyses has had less traction in the discipline than his evocative titles, Halberstam's continued visibility within the discipline suggests geographers have nevertheless found value in his provocations and incitements to think transgressively with time and space. Perhaps Halberstam's gift to geography, then, is not so much in his detailed cultural analyses as in his gift of naming phenomena whose very elusiveness and counter-foundational orientations might otherwise have hampered their recognition or sustained uptake in social and cultural geographical research.

The concepts of *female masculinity*, *queer time*, *metronormativity*, and *the art of failure* all offered compelling challenges to established orders of recognition and value, and have served to reinsert contrarian viewpoints into geographical analyses of gender, temporality, urbanism, and success. Halberstam does not necessarily claim these concepts as entirely his own ('queer temporality' derives from Barber and Clarke's (2002) readings of Eve Sedgwick and **Michel Foucault;** 'queer failure' is borrowed from José Esteban Muñoz (2010); and 'low theory' from **Stuart Hall** (1991)); his work nevertheless brought these ideas to the attention of geographers in a way that enabled their wider take-up in the discipline, to significant effect.

Halberstam has been open about numerous moments of learning over the years, from his epiphanous encounter with queer rural audiences (Halberstam, 2005) to the critical feedback of trans men (Halberstam, 1998) and trans women/transfeminine people (Halberstam, 2018) who at various times felt misrepresented or overlooked in his analyses. These efforts to listen and take on board the critical feedback of disenfranchised voices, and to be transparent in the process of doing so, are integral to Halberstam's anti-foundational approach to knowledge production. Nevertheless, his unapologetically adversarial style has frequently left critics frustrated and uninspired (Serano, 2014). In his role as self-appointed defender of all that is anti-programmatic, Halberstam has perhaps sometimes become a victim of his own nostalgic aesthetic attachments to a particular generational past; one unnamed critic memorably dubbed him the 'sports dad of queer theory' for his diatribes against 'the kids today' (Halberstam, 2018: 85).

A second, related criticism of Halberstam's work is that its underlying narrative arc pits the queer transgressive potential of late twentieth century political and cultural movements against a perceived naivety and assimilationism among subsequent generations that is difficult to evidence. In this view, Halberstam's critique is arrayed against a neoliberal 'moment' of transgender respectability which has – by and large – failed to materialise. It is also arranged in defence of a transgender past which, as Gill-Peterson (2018; see also Snorton, 2017) contends, assumes a simplistic generational history and decries the arrival of a new and 'smooth(er) passage from trans childhood to adulthood' which is not fully representative of the complex terrains in which gender plasticity has been (and continues to be) contested across generations.

At times, Halberstam's commitment to a non-programmatic reading of trans has seemed to come at the expense of taking actual trans people seriously (the notion of trans-with-an-asterisk, which Halberstam championed in his book *Trans** (2018), has had a mixed reception among trans scholars; see e.g., Gieseking, 2019). Critics have observed a marked unevenness in sympathy and sensibility between his treatment of masculinities and femininities – one perhaps not fully justified by an acknowledged emphasis on his own lived experience, nor even by the need to redress the genuine and widespread erasure of female and trans masculinities in popular discussions of trans existence. While Halberstam's early work on trans masculinities includes some exquisitely sensitive reflections on the importance of 'counting', 'naming', and 'noticing' non-normative gender (1998: 5) and on the knotty questions of who gets to tell, represent, and embody whose stories (2005), this careful attention is less evident in later treatments of trans femininities and the controversies that beset them. This can be seen, for example, in his heavy reliance, in the introduction to *Trans**, on cisgender male performances of femininity at the expense of existing transfeminine examples. The resulting analysis conflates non-committal flirtations with gender transgression by those who largely uphold structures of transphobia and misogyny with the subject positions of trans people who are targeted by those structures (indeed, many of the examples Halberstam relies on have either faced allegations of bullying and sexual misconduct, or later came out publicly against trans rights and recognition).

Though it takes the AIDS epidemic as its starting point and engages cases such as the tragic murder of trans young person Brandon Teena, Halberstam's project of articulating a queer temporality is by and large a celebratory project. Importantly, also, his account foregrounds choice and refusal as the hallmark of queer resistance and non-normative temporalities and subcultures. This emphasis on choice may have limited traction with those geographers who seek to complicate how we think about agency, volition, and resistance by adopting a more relational, even ecological, approach to spacing the subject.

Finally, what Halberstam does not do is attempt a sustained consideration of the ways in which failure or inability to comply with the demands and constraints of hetero-reproductive time also impinge upon the lives of those cast outside its organising parameters. Here the question of volition seems crucial, since while it is important to celebrate the exuberance and radical potential manifested in queer strategies of refusal, it is equally important to recognise the force of exclusion that hetero-reproductive time exerts on the lives of queer people who ultimately may have (had) little choice but to 'opt out' of an organising logic that refuses to accommodate them.

To date, Halberstam is one of few (out) trans scholars to have been around long enough to make a real impact in geography (some current trans scholars in geography include Jack Gieseking (2020), Eden Kinkaid (2023), Rae Rosenberg (2017), and Matt Smith (Smith et al., 2023)). Other trans thinkers will continue to develop their own critical perspectives in the spaces that he has helped to open for theorising the transgender body as a key site in which late liberalism's greatest contradictions play out, in sometimes enlightening ways.

In conclusion, over the course of a prolific career, Halberstam has issued a series of generative provocations to geographers to take seriously the significance of non-normative and countercultural lives as revealing both the hegemonic power and the ruptures and interstices of repro-normative spatial relations. His account of queer temporalities aims to 'dispel' the persistent idea 'that people are always in relationships in ways that are parental and generational', highlighting the extent to which this supposition underpins conventional frameworks for organising and theorising both time and space. The impact of Halberstam's thinking has been wide-reaching in geography, influencing critical geographic perspectives on urban/rural spatialities, gentrification, queer and trans representation, posthuman bodies, knowledge production, wildness, and failure, to name but a few examples. If Halberstam's primary focus on the representational politics of Euro-American popular and subcultural expressions does not immediately present itself to geographical audiences, his flair for relating their discussion to core geographical concerns has facilitated a noteworthy contribution to geographical thinking on time and space.

Halberstam's Key Works

Halberstam, J. (1995) *Skin Shows: Gothic Horror and the Technology of Monsters*. Durham, NC: Duke University Press.

Halberstam, J. (1998) *Female Masculinity*. Durham, NC: Duke University Press.

Halberstam, J. (2005) *In a Queer Time and Place: Transgender Bodies, Subcultural Lives*. New York, NY: New York University Press.

Halberstam, J. (2011) *The Queer Art of Failure*. Illustrated edition. Durham, NC: Duke University Press Books.

Halberstam, J.J. (2013) *Gaga Feminism: Sex, Gender, and the End of Normal: 7*. Boston, MA: Beacon.

Halberstam, J. (2018) *Trans*: A Quick and Quirky Account of Gender Variability (American Studies Now: Critical Histories of the Present): 3*: Oakland, CA: University of California Press.

Halberstam, J. (2020) *Wild Things: The Disorder of Desire*. Durham, NC: Duke University Press Books.

Halberstam, J. and Livingston, I. (eds) (1995) *Posthuman Bodies*. Bloomington, IN: Indiana University Press.

Halberstam, J. 'Jack' and Volcano, D.L. (1999) *The Drag King Book*. London: Serpent's Tail.

Secondary Sources and References

Barber, S.M. and Clark, D.L. (2002) 'Queer moments: The performative temporalities of Eve Kosofsky Sedgwick', in D.L. Clark and S.M. Barber (eds), *Regarding Sedgwick: Essays on Queer Culture and Critical Theory*. New York, NY: Routledge. pp. 1–53.

Gibson-Graham, J.K. (1996) *The End of Capitalism (As We Knew It): A Feminist Critique of Political Economy*. Oxford: Blackwell.

Gieseking, J.J. (2019) 'Review: Trans*, a quick and quirky account of gender variability', *Antipode Online*, vol. Book Review Symposium-Queer Geographies. https://antipodeonline.org/wp-content/uploads/2019/05/6.-jen-jack-gieseking_ak.pdf (accessed November 4, 2023).

Gill-Peterson, J. (2018) *Histories of the Transgender Child*. Minneapolis, MN: University of Minnesota Press.

Hall, S. (1991) 'Old and new identities, old and new ethnicities', in A.D. King (ed), *Culture, Globalization and the World System*. London: Macmillan. pp. 42–69.

Harrowell, E., Davies, T. and Disney, T. (2018) 'Making space for failure in geographic research', *The Professional Geographer*, 70 (2): 230–38.

Harvey, D. (1990) *The Condition of Postmodernity*. Oxford: Blackwell.

Katz, C. (2004) *Growing Up Global: Economic Restructuring and Children's Everyday Lives*. Minneapolis, MN: University of Minnesota Press.

Linebaugh, P. and Rediker, M. (2002) *The Many-Headed Hydra: The Hidden History of the Revolutionary Atlantic*. London: Verso.

Manalansan IV, M.F. (2015) 'Queer worldings: The messy art of being global in Manila and New York', *Antipode*, 47 (3): 566–79.

Massey, D. (1995) *Spatial Divisions of Labour: Social Structures and the Geography of Production*. Houndmills: Red Globe.

Muñoz, J.E. (2010) *Cruising Utopia: The Then and There of Queer Futurity*. New York, NY: New York University Press.

Newhouse, L.S. (2017) 'Uncertain futures and everyday hedging in a humanitarian city', *Transactions of the Institute of British Geographers*, 42 (4): 503–15.

Rosenberg, R. (2017) 'When temporal orbits collide: Embodied trans temporalities in U.S. prisons', *Somatechnics*, 7 (1): 74–94.

Secor, A. and Linz, J. (2017) 'Becoming minor', *Environment and Planning D: Society and Space*, 35 (4): 568–73.

Serano, J. (2014) 'Regarding "Generation Wars": Some reflections upon reading the recent Jack Halberstam essay', *Whipping Girl*, July 13. https://juliaserano.blogspot.com/2014/07/regarding-generation-wars-some.html (accessed November 4, 2023).

Snorton, C.R. (2017) *Black on Both Sides: A Racial History of Trans Identity*. Minneapolis, MN: University of Minnesota Press.

Wilkinson, E. (2014) 'Single people's geographies of home: Intimacy and friendship beyond "the family"', *Environment and Planning A: Economy and Space*, 46 (10): 2452–68.

Williams, J.J. (2011) 'The drag of masculinity: An interview with Judith "Jack" Halberstam', *symploke*, University of Nebraska Press, 19 (1–2): 361–81.

22
Stuart Hall

Jordan T. Camp, Trinity College, Hartford, Connecticut

Biographical Details and Theoretical Context

Stuart Hall, one of the most important radical intellectuals and theorists of his generation, was born in Kingston, Jamaica in 1932. He studied English at Jamaica College in Kingston and won a Rhodes scholarship to attend Oxford University. He lived and worked in England from 1951 until his passing in 2014. In that time, Hall occupied many professional roles, including Director of the Birmingham Centre for Contemporary Cultural Studies, Professor of Sociology at the Open University, President of the British Sociological Association, and the inaugural editor of the *New Left Review.* A prolific essayist, his articles were published in influential venues and collected volumes. He regularly appeared on television programmes on the BBC and for the Open University. He was author, co-author and co-editor of significant books and essays on politics and culture including *Resistance Through Rituals* (Hall and Jefferson, 1975), *Policing the Crisis* (Hall et al., 2013 [1978]), and *The Hard Road to Renewal* (Hall, 2021 [1988]). Hall made significant contributions to spatial thinking through his work in British cultural studies and New Left politics.

During his time at Oxford, Hall became immersed in a vital political culture shaped by the 'Caribbean Marxism' of black intellectuals such as C.L.R. James, Eric Williams, and others who articulated a decisive critique of imperialism. This tradition was formative to his own intellectual and political project. He became active in the independent socialist movement, a project that was critical of mainstream Labour Party politics as well as the Stalinist politics of the Communist Party (CP) (Rustin, 2017: 337; Gilroy, 2021). Hall participated in his first mass protest in London's Trafalgar Square against the Soviet invasion of Hungary in 1956. These events represented a decisive turning point for Hall. As he put it, the 'New Left was formed in the political space which opened up between the two opposing coordinates', the authoritarianisms of the 'nuclear-armed Soviet empire' and the 'revived aggression of British imperialism' (Hall, 2017: 243). The political impasse sparked his imagination about the need for a new space for politics, leading him to focus his studies on Marxist theory (Mitchell, 2004).

As Hall commented, the events of 1956 were as significant to his generation of the 'first New Left', as '1968 was to become to a later generation' (Hall, 2017: 243). The 'generation of 1956' sought to produce a space for politics that broke from the confines of the Cold War and to lead the global struggle for nuclear disarmament (Hall, 2017: 245). Hall's activity in the New Left, specifically the Campaign for Nuclear Disarmament (CND), was critical to how he came to know Britain 'geographically' by travelling to attend CND meetings and speaking in distinct regional venues across the country (Hall, 2017: 229). Hall also contributed to the production of New Left spaces – such as clubs, libraries, and coffee houses – that were critical to the building of a broader anti-imperialist and democratic socialist movement (Hall, 2017).

In 1957, Hall became a founding editor of the socialist magazine *Universities and Left Review* (ULR) alongside his Oxford friend Raphael Samuel. The journal combined socialist essays with cultural analysis, distinguishing it among left publications (Merrick, 2024: 6). Their editorial insights were grounded in a critical understanding of place and social transformation. Hall, Samuel, and their colleagues established ULR clubs in solidarity with Afro-Caribbean migrants in the wake of the 1958 Notting Hill race riots. They protested the routine exclusion of migrant workers from housing and jobs, highlighted racist state and vigilante violence, and promoted direct-action protest. These experiences would come to inform Hall's spatial analysis. Specifically, they shaped his distinctive analysis of how race and class were lived in particular historical geographies, as well as how racist ideologies sustained and justified repression in domestic spaces and resurgent imperialism around the world (Hall, 2017).

In 1960, Hall co-founded the journal *New Left Review* with John Saville. This new journal was a product of a merger between *Universities and Left Review* and *The New Reasoner*, edited by E.P. Thompson and John Saville, both former members of the Communist Party. Hall and his *NLR* colleagues theorised a distinct path amidst the political 'impasse of Stalinism', the resurgence of English imperialism, and the threat of nuclear war. As Hall put it, 'it was out of these practical activities that an expansive conception of the political also began to take root in me' (Hall, 2017: 228–29).

Hall's work as a movement intellectual shaped his engagement with cultural studies. In 1964, Professor Richard Hoggart invited Hall to join the newly-founded Birmingham Centre for Cultural Studies as a research fellow. The Centre provided a unique space for staff and postgraduate students to engage in research and publishing in cultural studies, an emergent field that focused on cultural struggles over the production of meaning. In 1968, Hall became the Centre's director, a position he held until 1979. As a founding figure in the 'Birmingham School of Cultural Studies', Hall, with his colleagues, sought to transform intellectual work in the classroom and to contribute to urgent scholarly and political debates about geographies of race and class, culture, and politics. The Birmingham School distinguished cultural studies as an intellectual and political project through a commitment to collaborative interdisciplinary research and public intellectual praxis with an emphasis on popular cultural texts and forms. They produced a number of important theoretical interventions in the Centre's journal, *Working Papers in Cultural Studies*. In doing so they aimed to provide an 'intellectual *intervention*' and thus to 'define and occupy a space' (Hall, 1996 [1980]: 2).

In 1979, Hall left Birmingham to accept a position in the Department of Sociology at the Open University. Over the next three and a half decades, he participated in debates about race, class, and law and order. These interventions were motivated by events such as the Black working-class uprising in Brixton in 1981, and Margaret Thatcher's authoritarian state's response to it. At the same time,

Hall's students like Hazel Carby, Paul Gilroy, and others provided generative interventions about interpreting the politics of the African diaspora, expressive culture, migration, the blues, sexual politics, reggae, and the policing of urban space. They developed an interdisciplinary project that linked cultural analysis to socialist theory and political strategy (Centre for Contemporary Cultural Studies, 1982). In this way, Hall's work has inspired a growing body of scholarship in cultural studies and geography on the articulation of race, class, gender, 'common sense', place, space, and politics (Carby, 1999; Watts, 2001; Denning, 2021; Gilroy, 2021; Hart, 2024).

Spatial Contributions

Hall's most significant spatial contributions are tied to his analysis of specific historical conjunctures and geographical contexts. He further developed Antonio Gramsci's 'socio-spatial' method of conjunctural analysis for analysing race and class (Ekers et al., 2013; Hall, 2021 [1988]). By conjuncture, Gramsci and Hall were referring to historical moments when political, economic, military, and social forces take a distinct shape (Hall and Massey, 2010; see also Camp and Greenburg, 2020). Gramsci had developed his conjunctural analysis during Mussolini's rise to power in Italy. Hall extended Gramsci's conjunctural analysis to understand the Thatcher era in Britain. In their respective times and places, both showed that concrete analyses of situations were needed to develop effective interventions. For Hall, conjunctural analysis proceeds through a recognition that politics is 'open-ended', a product of concrete relations of force operating in specific regional contexts (Hall, 2021 [1988]: 171, 169). Geographers and other spatial thinkers have subsequently taken up Gramsci and Hall's methods to better analyse conjunctures in diverse sites and political contexts (Hart, 2024).

In 1978, Hall, and his Birmingham colleagues Charles Critcher, Tony Jefferson, John Clarke, and Brian Roberts published their game-changing collaborative work of Gramscian-Marxist cultural studies, *Policing the Crisis* (2013 [1978]). It emphasised the importance of the method of conjunctural analysis for understanding the crisis of the 1970s. Hall and his colleagues provided a conjunctural analysis attentive to the articulation of race, class, region, and state power. They theorised how a tilt toward state authoritarianism stemmed from a 'crisis of hegemony' for capital and the state, and demonstrated the centrality of a racist and populist common sense in securing consent to an authoritarian solution to crisis. By examining the articulation of racism and policing, this work essentially connects the making of an authoritarian state form to the neoliberal counterrevolution (Hall et al., 2013 [1978]).

In 1979, Hall published the essay, 'The great moving right show', in *Marxism Today* (Hall, 1979). It introduced the concept of 'Thatcherism' to describe the rightwing politics of British prime minister Margaret Thatcher, locating it as an authoritarian response to an 'organic crisis' of the state and capital. Hall mobilised Gramsci to theorise how the 'swing to the Right' was less a 'reflection' of, and more a political '*response to*', an 'organic' crisis that had taken shape after the events of 1968 (Hall, 1979: 15). Hall famously argued that one had had to deal with the specific features of the Right and recognise how it was 'engaged in a struggle for hegemony' (Hall, 1979: 16). In doing so, he noted that this shift toward the right marked a 'move toward "authoritarian populism"' – an 'exceptional form of the capitalist state' that had been able to secure 'an active popular consent' (Hall, 1979: 15).

Hall's engagement with Birmingham's 'Policing the Crisis' collective and the research seminar on 'Populism, discourse, and political ideologies' provided the foundations for his own book, *The Hard Road to Renewal* (Hall, 2021 [1988]). In this timely intervention, Hall engaged with the work of Nicos Poulantzas – in particular, his idea of 'authoritarian statism' – as he sought to theorise the challenges of the authoritarian right during early neoliberalism (Hall, 2021 [1988]: 3). Poulantzas argued that a new phase of capitalism had given rise to a new authoritarian state form, marked by the decline of democratic participation and increasingly centralised authoritarian power. He specified that the state, like capital, is not simply a relationship, but rather a '*material condensation*' of relations between classes, and a 'mechanism for organising social space' (Poulantzas, 2000 [1978]: 128, 100). Hall saw important resonance between Poulantzas's project and *Policing the Crisis*. At the same time, he argued that Poulantzas neglected the '"anti-statist" strategy of the "new right"' (Hall, 2021 [1988]: 152). Hall identified how the far right had secured consent by the incorporation of elements of opposition into their programme and described how consent to 'authoritarian populism' had been rooted in racist and populist discontent (Hall, 2021 [1988]: 153). In this work, Hall anticipated how the rise of the Right would become the central challenge of the neoliberal era. He developed the concept of authoritarian populism in the 'heat of conjunctural analysis' in order to theorise the shift toward the right as it unfolded (Hall, 2021 [1988]: 153). Perhaps most importantly, his conjunctural analysis of challenge of the authoritarian right also provides a strategy for reorganising social space.

Scholars have taken up and extended Hall and his readings of Gramsci and Poulantzas, in historical-geographical contexts such as: the authoritarian state in Britain, post-apartheid South Africa, the contemporary US, and beyond (Centre for Contemporary Cultural Studies, 1982; Gilmore, 2007; Camp, 2016; Hart, 2020). Taken together, these works complicate the most influential narratives in geography about the history of neoliberalism (Harvey, 2005). They demonstrate that neoliberalism is not only a project of class restoration. Instead, they illustrate that it is a political project that has deployed racialised narratives to justify the expansion of policing, prisons, and borders (Kundnani, 2023). The social consequences of this political project have been manifest in nationalist articulations of race and class linked with a racist common sense about crime and law and order that paved the way for the rise of the 'neoliberal carceral state' (Camp, 2016; Hart, 2020; Hall, 2021 [1988]). Racist ideologies and projects facilitated the implementation of neoliberal policies, offering hegemonic strategies that rendered people simultaneously less worthy of the neoliberal state's shrunken largesse and more deserving of its expanded punitive capacities (Camp and Heatherton, 2016).

Critical geographers now understand conjunctures as 'spatio-historical' moments that represent an 'accumulation of contradictions' (Hart, 2024: 138–9). Such conjunctural moments represent ruptures in the relations of force, but they do not necessarily lead to revolutionary transformation. They also represent situations where legitimacy can be restored 'through a modification in the modes of hegemony', which enable understandings of the hegemonic process that enables capitalism's survival (Camp and Greenburg, 2020; Hall et al., 2013 [1978]; Hall, 2021 [1988]: 130). For example, our current moment represents a 'crisis of hegemony' of the state and capitalism on a geographical scale not seen since the 1930s. In this conjuncture, marked by the spectacle of increasing social polarisation, the resurgence of racist and authoritarian nationalisms, climate chaos, and a significant refugee crisis, the prevailing common sense has unravelled (Silver and Payne, 2020). Hall's writings show the importance of a 'spatio-historical' interpretation of the crisis, attentive to a common sense that mediates the crisis at different geographical scales (Hall and Massey, 2010; Hart, 2020).

Key Advances and Controversies

In 1980, Hall published the essential essay, 'Race, articulation, and societies structured in dominance', in a UNESCO anthology on social theories of race and colonialism. It represented a decisive advance within the race *versus* class debates in South Africa during the late 1970s (Hart, 2024). Hall famously mobilised Gramsci to illustrate how 'race' became the 'modality' through 'which class is lived' (Hall, 1980: 341). In doing so, he provided a Gramscian analysis of the 'dialectics of space and time' that was especially powerful politically for liberation movements against South African racial capitalism (Chari, 2017).

During this period, Hall also published a highly influential essay, 'Gramsci's relevance for the study of race and ethnicity' (1986), which was first delivered as a lecture at a UNESCO colloquium. He illuminated the centrality of Gramsci's core concepts and method of conjunctural analysis, particularly his attention to the 'complex dialectic of class and regional factors', for theorising 'racisms' in the late twentieth century. Hall argues that Gramsci's relevance stems from his 'emphasis on historical specificity', and his innovative methodological approach to questions of ideology, common sense, and the state. He then demonstrates the political stakes in theorising the precise forms that racisms take. He further provides a way of recognising that racisms take distinct shape in different historical and geographical conjunctures, across space and over time. For Hall, avoiding analysis of a specific conjuncture lends weight to the spurious idea that because 'racism everywhere is a deeply anti-human and anti-social practice, that therefore it is everywhere *the same*' in its form, structure, and effect (Hall, 1986: 23). Hall's use of Gramscian conjunctural analysis helps to disrupt an imagined fixity of racism, revealing the 'existence of multiple, geographically and historically situated – yet deeply interconnected–racisms' (Hawthorne, 2022: 146).

In 'Gramsci's relevance', Hall elaborates his spatial insights. He notes Gramsci's attention to the uneven geographies between North and South in his unfinished 1926 essay on 'The Southern question'. It was here that Gramsci argued for specific attention to spaces of uneven capitalist development and to the struggles of workers in the North and peasants in the South. As Hall showed, this geographical insight was essential to the method of conjunctural analysis developed in his *Prison Notebooks*. Hall argued that Gramsci provided key methodological criteria for interpreting the articulation of racism and capitalism in different contexts, including: the incorporation of underdeveloped areas of Southern Italy, the unevenly developed economies of Asia and Latin America, slavery and settler colonialism's centrality to the 'primitive accumulation' of capital in North America, the super-exploitation of migrant workers in Europe, and the presence of Bantustans ('homelands') within the capitalist economy of Southern Africa (Hall, 1986: 24–5).

Yet, there has been some controversy around Hall's assertion that Gramsci 'did not write directly about race, ethnicity, or racism' (Hall, 1986: 10). Scholars have demonstrated that Gramsci clearly wrote about race and racism, suggesting that Hall's reading suffered from a lack of attention to his core texts (Green, 2013; Hawthorne, 2022: 146). Yet it is also the case that many of Gramsci's integral texts had been unavailable or only partially translated into English at the time. Later publications, including a volume published in the wake of Hall's death on subaltern social groups and classes, have offered a more complete view (Gramsci, 2021; Hawthorne, 2022; Kipfer and Mallick, 2022: 4, 11).

Hall's readings of Gramsci as well as Louis Althusser have also been the subject of ongoing and productive debates in critical geography and social theory. For example, some scholars suggest his engagement with Althusser's conjunctural analysis represented a shift in his political and theoretical work away from his collaborative research during the Birmingham years, and towards a more abstract, 'idealist hyper-antihumanism' typical of Althusserianism (Mitchell, 2004: 163). Still others have argued that Hall's efforts to 'navigate a path' between his reading of Gramsci and Althusser was a 'perilous' one. This debate is far from over (Hart, 2024: 136). After all, as Hall himself argued, there are no innocent borrowings between one theoretical and methodological approach and another. It is important to confront their 'theoretical effect', Hall writes, before 'concepts can be borrowed and transformed' (Hall, 2000: xiv).

There has been a renewed interest in Hall's writings on race and class, culture, identity, and nationalisms amidst the expansion of antiracist class struggles and the resurgence of the racist far right. Scholars of Black, cultural and feminist geographies have conducted original and generative research on the politics of race, gender, class, migration, postcoloniality, and surveillance. They demonstrate that Hall's methods remain relevant for understanding space and politics in places as varied as Brazil, England, India, Italy, South Africa, the US, and beyond (Jackson, 2016: 368; Hawthorne, 2022). In turn, they illuminate the political stakes in 'translating' his insights to meet the challenges of our time (Hart, 2020).

As Hall taught us, conjunctural analysis involves looking at the relations of forces in order to effectively intervene. From his campaigning in the anti-nuclear movement to his theoretical work in Birmingham to his political essays on race and class, Hall demonstrated how to intervene in precise historical and geographical conjunctures, and in doing so, change the world. Hall's conjunctural analysis continues to provide an indispensable approach for a new generation of spatial thinkers.

Hall's Key Works

Hall, S. (1979) 'The great moving right show', *Marxism Today*, 23 (1): 14–20.

Hall, S. (1980) 'Race, articulation, and societies structured in dominance', in UNESCO *Sociological Theories: Race and Colonialism*. Paris: UNESCO. pp. 305–45.

Hall, S. (1986) 'Gramsci's relevance for the study of race and ethnicity', *Journal of Communications Inquiry*, 10: 5–27.

Hall, S. (1996 [1980]) 'Cultural Studies and the Centre: Some problematics and problems', in S. Hall, D. Hobson, A. Lowe and P. Willis (eds), *Culture, Media, Language: Working Papers in Cultural Studies, 1972–1979*. New York, NY: Routledge. pp. 2–35

Hall, S. (2000) 'Introduction', in N. Poulantzas, *State, Power, Socialism*. London: Verso. pp. xii–xviii.

Hall, S. (with B. Schwarz) (2017) *Familiar Stranger: A Life Between Two Islands*. Durham, NC: Duke University Press.

Hall, S. (2021 [1988]) *The Hard Road to Renewal: Thatcherism and the Crisis of the Left*. London: Verso.

Hall, S. and Jefferson, T. (eds) (1975) *Resistance Through Rituals: Youth Subcultures in Post-War Britain*. London. Unwin Hyman.

Hall, S. and Massey, D. (2010) 'Interpreting the crisis', *Soundings*, 44: 57–71.

Hall, S., Critcher, C., Jefferson, T., Clarke, J. and Roberts, B. (2013 [1978]) *Policing the Crisis: Mugging, the State, and Law and Order*. London: Palgrave.

Secondary Sources and References

Camp, J.T. (2016) *Incarcerating the Crisis: Freedom Struggles and the Rise of the Neoliberal State*. Oakland, CA: University of California Press.

Camp, J.T. and Greenburg, J. (2020) 'Counterinsurgency re-examined: Racism, capitalism, and U.S. military doctrine', *Antipode*, 52 (2): 430–51.

Camp, J.T. and Heatherton, C. (eds) (2016) *Policing the Planet: Why the Policing Crisis Led to Black Lives Matter*. New York, NY: Verso.

Carby, H.V. (1999) *Culture in Babylon: Black Britain and African America*. London: Verso.

Centre for Contemporary Cultural Studies (ed) (1982) *The Empire Strikes Back: Race and Racism in 70s Britain*. London: Hutchinson.

Chari, S. (2017) 'Three moments of Stuart Hall in South Africa: Postcolonial-postsocialist Marxisms of the future', *Critical Sociology*, 43 (6): 831–45.

Denning, M. (2021) 'Everyone a legislator', *New Left Review*, 129: 29–44.

Ekers, M., Hart, G., Kipfer, S. and Loftus, A. (eds) (2013) *Gramsci: Space, Nature, Politics*. Malden, MA: Wiley-Blackwell.

Gilmore, R.W. (2007) *Golden Gulag: Prisons, Surplus, Crisis, and Opposition in Globalizing California*. Berkeley, CA: University of California Press.

Gilroy, P. (2021) 'Introduction', in P. Gilroy and R.W. Gilmore (eds), *Selected Writings on Race and Difference (Stuart Hall)*. Durham, NC: Duke University Press. pp. 1–19.

Gramsci, A. (2021) *Subaltern Social Groups: A Critical Edition of Prison Notebook 25* (J. A. Buttigieg and M.E. Green, eds and trans). New York, NY: Columbia University Press.

Green, M.E. (2013) 'Race, class, and religion: Gramsci's concept of subalternity', in C. Zene (ed), *The Political Philosophies of Antonio Gramsci and B.R. Ambedkar: Itineraries of Dalits and Subalterns*. New York, NY: Routledge. pp. 116–28.

Hart, G. (2020) 'Why did it take so long? Trump-Bannonism in a global conjunctural frame', *Geografiska Annaler: Series B, Human Geography*, 102 (3): 239–66.

Hart, G. (2024) 'Modalities of conjunctural analysis: "Seeing the present differently" through global lenses', *Antipode*, 56 (1): 135–64.

Harvey, D. (2005) *A Brief History of Neoliberalism*. New York, NY: Oxford University Press.

Hawthorne, C. (2022) *Contesting Race and Citizenship: Youth Politics in the Black Mediterranean*. Ithaca, NY: Cornell University Press.

Jackson, P. (2016) 'New directions in cultural geography revisited', *Area*, 48 (3): 367–70.

Kipfer, S.A. and Mallick, A. (2022) '"Stretch" and "translate": Gramscian lineages,

Fanonist convergences in the (post) colony', *Historical Materialism*, 30 (4): 137–73.

Kundnani, A. (2023) *What is Antiracism? And Why It Means Anticapitalism*. London: Verso.

Merrick, J. (2024) 'Introduction: Making of a people's historian', in R. Samuel, *Workshop of the World: Essays in People's History* (J. Merrick, ed). London: Verso. pp. 1–23.

Mitchell, D. (2004) 'Stuart Hall', in P. Hubbard, R. Kitchin and G. Valentine (eds), *Key Contemporary Thinkers on Space and Place*. London: Sage. pp. 160–6.

Poulantzas, N. (2000 [1978]) *State, Power, Socialism*. London: Verso.

Rustin, M. (2017) 'Afterword: Stuart Hall as a political intellectual', in S. Davison, D. Featherstone, M. Rustin and B. Schwartz (eds), *Selected Political Writings: The Great Moving Right Show and Other Essays*. Durham, NC: Duke University Press. pp. 336–53.

Silver, B. and Payne, C. (2020) 'Crises of world hegemony and the speeding up of social history', in P. Dutkiewicz, T. Casier and J.A. Scholte (eds), *Hegemony and World Order: Reimagining Power in World Politics*. New York, NY: Routledge. pp. 17–31.

Watts, M. (2001) '1968 and all that', *Progress in Human Geography*, 25 (2): 157–88.

23
Donna Haraway

Mara Miele, Cardiff University

Biographical Details and Theoretical Context

Donna Haraway is Distinguished Professor Emerita in the History of Consciousness and Feminist Studies departments at the University of California, Santa Cruz. She was born in Denver, Colorado, in 1944. Haraway studied zoology and philosophy at Colorado College (The European Graduate School, 2019), before receiving a Fulbright scholarship to study at the University of Paris 6. She completed her PhD in Biology at Yale in 1972, with a dissertation entitled *The Search for Organizing Relations: An Organismic Paradigm in 20th-Century Developmental Biology*. She had started her doctoral research in experimental biology, but then became interested in the cultures, politics, and languages of science and technology, as her work began to engage with other fields – and departments – of philosophy, and history of science and medicine. This interdisciplinarity became the foundation of her research in the following years. Having taught history of science and women's studies at the University of Hawaii (1971–1974) and at Johns Hopkins University (1974–80), she joined the History of Consciousness program at Santa Cruz in 1980.

Haraway subsequently worked on the history of biology, the place of technology in feminist and socialist debates over objectivity, the politics and ethics of genetic interventions into lifeforms, and the expressions of multispecies companionship, kinship, and flourishing. Her research echoes the concerns of fellow feminist scholars and Science and Technology Studies practitioners, such as Karen Barad (2007), **Bruno Latour** (1993; 2004; 2005), Isabelle Stengers (2005), Vinciane Despret (2004; 2013), and **Anna Tsing** (2013; 2015) among others, and has been key in undoing a number of oppositions endorsed in science, particularly biological sciences: namely the opposition of nature/culture, science/society, human/nonhuman animals, animal/machine, mind/body, male/female, fiction/reality. Haraway's work has been characterised by theoretical innovation with political commitment and has helped to develop realist, pragmatist, feminist, and antiracist debates, always asking unsettling questions about the politics and aesthetics of knowing. Her extensive work has addressed several interlinked themes: metaphors in primate studies, race politics and gender (1989), feminism and technoscience (1997a), human–nonhuman animals' relations (2003 and 2008), mass extinction and environmental devastation (2016), population, intimacy, multispecies kinship, and environmental justice (2018).

In 1999 Haraway received the Ludwik Fleck Prize for her book *Modest_Witness@Second_Millennium. FemaleMan©_Meets_OncoMouse™ Feminism and Technoscience*, an extremely vivid presentation of the physical and semiotic world of technoscience. Herein she asked the question 'What counts as nature, for whom, and at what cost?' (Haraway, 1997a: 75). In September 2000, she was awarded the highest honour given by the Society for Social Studies of Science, the J.D. Bernal Prize, for lifetime contributions to the field and her landmark work, *A Cyborg Manifesto: Science, Technology, and Socialist-Feminism in the Late Twentieth Century*.

Spatial Contributions

In 1985, Haraway published 'A Manifesto for Cyborgs: Science, Technology, and Socialist Feminism in the 1980s' (reproduced in Haraway, 1991). The concept of the cyborg, which draws inspiration from the realm of science fiction, entails the integration of the biological body with technological apparatus (Haraway, 1991). This work was an attempt to construct 'an ironic political parable attentive to feminism, socialism, and materialism' (Haraway, 1991: 7). The image of the cyborg emphasised the mistake of producing universal, totalising theories that fail to account for most of reality. It was foundational for promoting the concept of situated knowledges: that knowing and thinking are inseparable from a multitude of relations that also make possible the worlds we think with (Haraway, 1988). Biotechnology, genetic engineering, cloning, and other related fields highlight the ways in which technoscience increasingly works with biological materials in ways that undermine conventional distinctions between the human and the non-human, the biological and the technical, and nature and culture. The figure of the cyborg is crucial in Haraway's early writing as a means to describe the ways in which humans and other organisms are increasingly entwined in machine-like knots of material, biological and digital technologies, and as a metaphor for forms of political and social identity which do not conform to conventional models of identification, such as gender, race or sexuality. For Haraway, the figure of the cyborg undoes the presented 'naturality' of race, gender, age, sexuality, and their association with specific social roles and cultural representations. Here a *'cyborg politics'* may resonate with marginalised groups' attempt to work towards emancipation and empowerment. For example, **Gillian Rose**, has used a 'cyborg politics' to denounce how women are socially and spatially marginalised. In her book *Feminism and Geography* (1993) Rose addressed questions about the politics of knowledge production in relation to the discipline of geography itself developing a post-structural critique of masculinist geographies centred on what Donna Haraway termed the *'master subject'*: that is, the subject constituted as white, bourgeois, heterosexual and masculine.

Haraway's book *Primate Visions: Gender, Race, and Nature in the World of Modern Science*, published in 1989, examined what geographer Nelson defined as the 'Cold War era primatology and corporate oil-funded media productions made to educate the American public and others about nature, culture, race and coloniality' (2017: 145). Haraway focused on primate research and primatology, and the cultural activity that comprised Dian Fossey, Francine (Penny) Patterson and Jane Goodall's work in Tanzania, in co-producing 'natureculture' narratives (1989: 152). The jungle, or Africa, was the space where Haraway was curious to discover what knowledge (in science, here primatology and ethology, as well as in the communication of science) was 'originated, produced, or developed'. She found that Jane Goodall was working with the financial support and advertising competence of the Gulf Oil Corporation that framed her as white woman and mother figure, as the connection between scientific knowledge in the west (chiefly male), and the wild environment on which it depends (usually labelled female). Goodall, as well as Fossey and Patterson, as scientists and Western white women, personified the maternal figure to the chimpanzees and gorillas, and they forged the association between Africa and the scientific knowledge production that comes from the West: 'In the advertisement sponsored by Gulf, there is a photo of Jane holding hands with a chimpanzee. Here is the connection between male/female, white/black, West/Africa, culture/nature' (Haraway, 2010: 78–9).

Two main concerns animated these books: the *'women question in science'* – i.e., the position of women practising science (see also Keller, 1985) – and the *'science question in feminism'* – the feminist critical approach to science as a practice itself (see Puig de la Bellacasa, 2012: 207). In response, Haraway's work was accused of relativism and criticised for looking at science as a practice, to which she replied in the context of the 'science wars' debates in the 1990s:

> The major ethical and epistemological issue for me, in trying to understand what kinds of undertakings comprise the biological sciences, is that knowledge is always an engaged material practice and never a disembodied set of ideas. Knowledge is embedded in projects; knowledge is always for (in many senses of for) some things and not others, and knowers are always themselves formed by their projects, just as they shape what they can know. Such shapings never occur in some unearthly realm; they always are about the material and meaningful interactions of located humans and nonhumans-machines, organisms, people, land, institutions, money, molecules, and many other kinds of things. It is because scientific knowledge is not 'transcendent' that it can make solid claims about material beings that are not reducible to matters of opinion, even as they are never separate from interpretation. (Haraway, 1997b: 124)

In her following works, Haraway focused on human–nonhuman animal relations. *The Companion Species Manifesto: Dogs, People, and Significant Otherness* (Haraway, 2003) is 'a story of co-habitation, co-evolution, and embodied cross-species sociality', 'a story of biopower and biosociality, as well as of technoscience'. In this text she introduced the concept of 'multispecies justice' (2003: 75) to stress how human existence is inevitably entangled with the lives of other beings. Relatedly, *When Species Meet* (2008) tackles questions of 'intra- and inter-action', asking 'who "we" will become when species meet', emphasising that nonhuman animals are not passive recipients of human agency but, rather, 'socially [...] active partners' (Haraway, 1997a: 8). This work has been particularly relevant for the development of geographies of technologies (Hinchliffe, 1996), geographies of nature (Hinchliffe, 2007; Hinchliffe et al., 2005) and animal geographies (Buller, 2014; Miele and Bear, 2022).

A concern for multispecies co-habitation is also the main theme of *Staying with the Trouble: Making Kin in the Chthulucene* (Haraway, 2016). This book is a critique of the concept of the Anthropocene and the inability of the concept to express the extent to which people are 'situated human beings in complicated histories' that have produced the current situation, not a generic 'human' species. The book is, as its name implies:

> About staying with all this trouble. It is about a becoming-with-each-other in another new term: what I'm calling the chthulucene. This can't be done in the mode of critique, which is never enough. In the chthulucene, critique is one of many practices tempered by others that lead to an opening of what is still possible. Chthulu comes from chthonic – the earthly powers and processes – human too, but much more than human. But this isn't some sort of ancient, destroyed-by-modernity story. It's an ongoing and present story. These chthonic powers, forces, and entities are coupled with the suffix –cene, drawn from kainos, or the thick now, the present, which is not instantaneous but extends into many kinds of time, into presence. Into cultivating response-ability. (Haraway cited in O'Neill-Butler, 2016: n.p.)

A later collection of six essays co-edited with Adele Clarke (2018), *Making Kin not Population: Reconceiving Generations* constituted a small but multifaceted book with chapters from leading Black, Indigenous and feminist science studies scholars. It offered different analyses and contrasting views about 'making kin' and about the concept of population, addressing issues of interspecies intimacy and kinship, as well as environmental justice. Including chapters by Adele E. Clarke, Ruha Benjamin, Donna Haraway, Michelle Murphy, Yu-Ling Huang and Chia-Ling Wu, and Kim TallBear, the book examines the massive impact and complexity of human numbers on the Earth and how meaningful relationships of interdependence are forged in the face of destructive social and political regimes. In so doing, it challenges usual ideas and theories in modern materialist and transnational studies, making bold (and hugely controversial) propositions for a drastic reduction in the number of human beings (via a *no-natality*).

Key Advances and Controversies

Over the last forty years, Haraway's work has profoundly influenced a number of debates concerning knowledge production, gender, race, science and technology, nature and culture. Her work has been influential in many social sciences, including feminist studies, sociology, science and technology studies, anthropology, cultural studies, and geography.

Researchers interested in *more-than-human* geographies, geo-humanities and environmental humanities have drawn increasingly upon the work of Haraway to address multispecies sociability and care. These themes are explicitly addressed in Haraway's *Companion Species Manifesto* (2003) and *When Species Meet* (2008), wherein Haraway proposed that concepts and techniques for attending to the somatic and affective dimensions of human life might enhance understandings of animal behaviour. These two books have brought to the fore the agency of non-human beings (e.g., dogs, chickens, microbes) and have become key references for the emerging field of animal geographies (see Lorimer and Davies, 2010; Gillespie and Collard, 2015; Miele and Bear, 2022).

Haraway's interspecies research has been particularly relevant to advancing recent debates about multispecies companionship and multispecies justice. María Puig de la Bellacasa (2012) has argued that, for Haraway, creating knowledge is thinking with care, it is a relational practice with important consequences in the shaping of possible worlds: 'That knowledge is situated means that knowing and thinking are inconceivable without a multitude of relations that also make possible the worlds we think with' (Puig de la Bellacasa, 2012: 198). Puig de la Bellacasa points out that one of the key aspects of Haraway's thinking-with is the invite:

> To enlarge our ontological and political sense of kinship and alliance, to dare in exercises of category transgression, of boundary redefinition that put to test the scope of humanist care. [...] Kinships and alliances become transformative connections – merging inherited and constructed relations. (Puig de la Bellacasa, 2012: 201)

In the *Companion Species Manifesto* (2003), Haraway explores the 'cobbling together' of caring relationality in human–dog love, the creation of 'significant otherness', and asserts: 'Dogs, in their historical complexity, matter here. Dogs are not an alibi for other themes [...] Dogs are not surrogates for theory; they are not here just to think with. They are here to live with' (Haraway, 2003: 5). Puig de la Bellacasa, quoting this, affirms:

> Although this assertion refers to interspecies love, its acknowledgement of interdependency offers a generic warning against idealized forms of caring. For living with is laborious. Relations of otherness are more than about accommodating 'difference', co-existing or tolerating. Thinking with should always be a living with, aware that relations of significant otherness transform those who relate and the worlds they live in. (Puig de la Bellacasa, 2012: 207)

Haraway's work has been particularly influential for the development of a new genre of ethnographic studies: multispecies ethnographies. This methodological approach was first introduced by Kirksey and Helmreich (2010) to address the strategies for looking beyond human-centred worlds and for overcoming language as means for communication of meaning. It is an approach that emphasises the lives of overlooked subjects (nonhuman animals, plants, microbes, fungi, lava, etc.) and challenges conventional modes of representation. Multispecies ethnographies involve studies of the subjectivity and agency of nonhuman beings that are inseparably interconnected to human worlds. They have been used for enlivening specific aspects of human/nonhuman relations: having fun with companion dogs in agility training (Haraway, 2008), human–dog companionship (Fox et al., 2022), constructing a bond between police dogs and police officers (Smith et al., 2021), extinction (van Dooren, 2021), and a bear's life and death in a national park (Blue, 2016). It is a posthumanist approach that enlivens the lives of 'neglected' subjects and troubles conventional modes of representation.

Such ethnographies have used innovative methods to apprehend animal bodies, including visual and digital technologies (see Blue, 2016). Moreover, the rise of multispecies ethnographies has promoted interdisciplinary collaborations with natural scientists and ethologists (Lorimer et al., 2017). Van Dooren and Bird Rose (2016) proposed the term 'lively ethographies' (eliminating the 'n' to explicitly move away from established approaches to 'ethnography') as a mode of knowing, engaging, and narrating that recognises the meaningful lives of others and that, in so doing, enlivens our capacity to respond. These authors argue that the aim of ethnographies is not to produce objective accounts of the world. Instead, this approach is based on the idea that making others 'fleshy and thick on the page', exposing readers to their lives and deaths, may lead to closeness and ethical entanglement, care and concern. They share with Haraway the idea that storytelling is an ethical practice, based on the belief that stories are powerful tools for shaping our shared world: 'rejecting a simple division between the real and the narrated, we see storytelling as a dynamic act of "storying."'(van Dooren and Bird Rose, 2016: 89).

They explain that ethographic storytelling is about the art of becoming-witness, which includes both attention to others and expression of that experience: to stand *as* witness and actively to *bear* witness. Bearing witness and telling stories with the aim of drawing audiences into others' lives in new and consequential ways, telling stories that foster the capacity for *response*. This focus on response enables a more interesting approach to questions of ethics in multispecies worlds (van Dooren and Bird Rose, 2016: 90).

Haraway's later books (Haraway, 2016; Clarke and Haraway, 2018) have addressed some highly controversial issues regarding the making of kin, care and multispecies justice. As Kirksey and Chao (2022) have pointed out, justice and injustice have always been part of multispecies studies but only recently have gained central focus. These authors ask a range of intersecting questions: 'Who are the subjects of justice in our shared worlds? What is at stake when they are captured by juridical-legal systems and social movements? Who has claimed a monopoly over justice in the past, and in the present, and how might we contest their sense of propriety in the future?' (Kirksey and Chao, 2022: 2). These are difficult questions that echo Haraway's concerns about the growth of human population, the disappearance of other species and her proposal for interspecies kinship. Haraway received strong criticism from feminists (Murphy, 2017, 2018; Dow and Lamoreaux, 2020) for employing the highly contested concept of 'population' in her argument, for its historical origins in racist and eugenic biopolitics in the past centuries. Michelle Murphy has argued that '[p]opulation, as an artifact of a particular way of counting, bundles up bodies into a single tally, creating distance and abstraction for a managerial gaze [...]. It is a formulation that allows the anonymization of lives into deletable data points' (Murphy, 2018: 103).

Dow and Lamoreaux (2020) also criticise Haraway and Clarke's proposal and have argued that 'making kin, especially non-biogenetic kin' is presented as a means for achieving 'multi-species reproductive justice fully integrated with human and non-human environmental justice and savvy environmental action.' This call to make 'oddkin' in (bio)diverse and various ways has been welcomed for successfully connecting issues of reproduction and the environment that are too often kept apart. However, Dow and Lamoreaux argue that the 'non-natalism' proposal and Haraway's call to 'make kin, not babies' (Haraway, 2016: 1) 'potentially places unnecessary and disproportionately distributed constraints on reproductive freedoms and bodily sovereignty as well as the further separation of kin into kinds' (Dow and Lamoreaux, 2020: 476). These critiques of Haraway's use of the term 'population' are a good example of what Haraway means by 'staying with the trouble.' Her provocation is 'to stay with the troubles of persistent capitalist-fuelled systems of oppression and exploitation such as social injustices, racism, environmental crises, and reproductive injustices' (Richter, 2021: 15).

Haraway's Key Works

Haraway, D. (1988) 'Situated knowledges: The science question in feminism and the privilege of partial perspective', *Feminist Studies*, 14 (3): 575–99.

Haraway, D. (1989) *Primate Visions: Gender, Race and Nature in the World of Modern Science*. New York, NY: Routledge.

Haraway, D. (1991) *Simians, Cyborgs and Women: The Reinvention of Nature*. London: Free Association.

Haraway, D. (1997a) *Modest_Witness@Second Millenium: FemaleMan#_Meets_ OncomouseTM: Feminisms and Technoscience*. New York, NY: Routledge.

Haraway, D. (2003) *The Companion Species Manifesto: Dogs, People, and Significant Otherness*. Chicago, IL: Prickly Paradigm.

Haraway, D. (2004) *The Haraway Reader*. New York, NY: Routledge.

Haraway, D. (2008) *When Species Meet*. Minneapolis, MN: University of Minnesota Press.

Haraway D. (2016) *Staying with the Trouble: Making Kin in the Chthulucene*. Durham, NC: Duke University Press.

Haraway, D. (2018) 'Staying with the trouble for multispecies environmental justice', *Dialogues in Human Geography*, 8 (1): 102–5.

Clarke, A. and Haraway, D. (eds) (2018) *Making Kin Not Population: Reconceiving Generations*. Chicago, IL: Prickly Paradigm.

Secondary Sources and References

Barad, K. (2007) *Meeting the Universe Halfway: Quantum Physics and the Entanglement of Matter and Meaning*, Durham, NC: Duke University Press.

Blue, G. (2016) 'Public attunement with more-than-human others: Witnessing the life and death of Bear 71', *GeoHumanities*, 2 (1): 42–57.

Buller, H. (2014) 'Animal geographies I', *Progress in Human Geography*, 38 (2): 308–18.

Despret, V. (2004) 'The body we care for: Figures of anthropo-zoo-genesis', *Body & Society*, 10 (2–3): 111–34.

Despret, V. (2013) 'Responding bodies and partial affinities in human–animal worlds', *Theory, Culture & Society*, 30 (7–8): 51–76.

Dow, K. and Lamoreaux, J. (2020) 'Situated kinmaking and the population "problem"', *Environmental Humanities*, 12 (2): 475–91.

European Graduate School (2019) https://egs.edu/biography/donna-haraway (accessed January 24, 2024).

Fox, R., Charles, N., Smith, H. and Miele, M. (2023) '"Imagine you are a dog": Embodied communication in dog training', *Cultural Geographies*, 30 (3): 429–52.

Gillespie, K. and Collard, R-C. (eds) (2015) *Critical Animal Geographies: Politics, Intersections, and Hierarchies in a Multispecies World*. New York, NY: Routledge.

Haraway, D. (1997b) 'enlightenment@science_wars.com: A personal reflection on love and war', *Social Text*, 50: 123–9.

Haraway, D. (2010) 'Staying with the trouble: Becoming worldly with companion species', paper presented at the Colloquium Series of the Center for Cultural Studies, UC Santa Cruz, 10 October.

Hinchliffe, S. (1996) 'Technology, power, and space – the means and ends of geographies of technology', *Environment and Planning D: Society and Space*, 14 (6): 659–82.

Hinchliffe, S. (2007) *Geographies of Nature: Societies, Environments, Ecologies*. London: Sage.

Hinchliffe, S., Kearnes, M.B., Degen, M. and Whatmore, S. (2005) 'Urban wild things: A cosmopolitical experiment', *Environment and Planning D: Society and Space*, 23 (5): 643–58.

Keller, E.F. (1985) *Reflections on Gender and Science*. New Haven, CT: Yale University Press.

Kirksey, E. and Chao, S. (2022) 'Introduction: Who benefits from multispecies justice?' in S. Chao, K. Bolender and E. Kirksey (eds), *The Promise of Multispecies Justice*. Durham, NC: Duke University Press. pp. 1–21.

Kirksey, S. and Helmreich, S. (2010) 'The emergence of multispecies ethnography', *Cultural Anthropology*, 25 (4): 545–76.

Latour, B. (1993) *We Have Never Been Modern*. Hemel Hempstead: Harvester Wheatsheaf.

Latour, B. (2004) *The Politics of Nature*. Cambridge, MA: Harvard University Press.

Latour, B. (2005) *Reassembling the Social: An Introduction to Actor-Network-Theory*. Milton Keynes: Open University Press.

Lorimer, J. and Davies, G. (2010) 'When species meet', *Environment and Planning D: Society and Space*, 28 (1): 32–3.

Lorimer, J., Hodgetts, T. and Barua, M. (2017) 'Animals' atmospheres', *Progress in Human Geography*, 43 (1): 26–45.

Miele, M. and Bear, C. (2022) 'Geography and posthumanism', in S. Herbrechter, I. Callus, M. Rossini, M. Grech, M. de Bruin-Molé and C.J. Müller (eds), *Palgrave Handbook of Critical Posthumanism*. Cham: Springer. pp. 749–72.

Murphy, M. (2017) *The Economization of Life*. Durham, NC: Duke University Press.

Murphy, M. (2018) 'Against population, towards alterlife', in A.E. Clarke and D. Haraway (eds), *Making Kin Not Population*. Chicago, IL: Prickly Paradigm. pp. 101–24.

Nelson, I.L. (2017) 'Interspecies care and aging in a gorilla 2.0 world', *Geoforum*, 79: 144–52.

O'Neill-Butler, L. (2016) 'Donna J. Haraway speaks about her recent book', *Artforum*, September 6. https://www.artforum.com/interviews/donna-j-haraway-speaks-about-her-latest-book-63147 (accessed November 4, 2023).

Puig de la Bellacasa, M. (2012) '"Nothing comes without its world": Thinking with care', *The Sociological Review*, 60 (2): 197–216.

Richter, A. (2021) 'My Oh My Myoma! An autoethnographical experiment of thinking my personal reproductive struggles with Haraway's planetary ethic of making kin, not babies', https://liu.diva-portal.org/smash/get/diva2:1571479/FULLTEXT01.pdf (accessed November 4, 2023).

Rose, G. (1993) *Feminism & Geography: The Limits of Geographical Knowledge*. Cambridge: Polity.

Smith, H., Miele, M., Charles, N. and Fox, R. (2021) 'Becoming with a police dog: Training technologies for bonding', *Transactions of the Institute of British Geographers*, 46 (2): 478–94.

Stengers, I. (2005) 'The cosmopolitical proposal', in L. Bruno and W. Peter (eds), *Making Things Public*. Cambridge, MA: MIT Press. pp. 994–1003.

Tsing, A. (2013) 'More-than-human sociality: A call for critical description', in K. Hastrup (ed), *Anthropology and Nature*. London: Taylor and Francis. pp. 27–42.

Tsing, A. (2015) *The Mushroom at the End of the World: On the Possibility of Life in Capitalist Ruins*. Princeton, NJ: Princeton University Press.

van Dooren, T. and Bird Rose, D. (2016) 'Lively ethography storying animist worlds', *Environmental Humanities*, 8 (1): 77–94.

24
David Harvey

Noel Castree, Manchester University and University of Technology Sydney

Biographical Details and Theoretical Context

David Harvey, born in England in 1935, and raised in Gillingham, Kent, can reasonably lay claim to being the most influential academic geographer to have ever lived. He completed his BA and PhD in the University of Cambridge's Department of Geography, leaving in 1960. During the 1950s, as was common in British geography departments, Cambridge stressed the importance of regional study – that is, the analysis of geographical particularity – in its degree programmes. Much of this involved looking at regions and regional change through time. Accordingly, Harvey's doctoral thesis focused on patterns of hop production in nineteenth-century Kent, considering its geographies within a wider UK and overseas context. However, no sooner had the results of this research been published (Harvey, 1963), than Harvey moved in an altogether different intellectual direction. Following a post-doctoral year at the University of Uppsala (where he encountered free-thinking geographer **Gunnar Olsson**), Harvey spent the 1960s as Lecturer in Geography at Bristol University.

It was while at Bristol – punctuated by study leave at Pennsylvania State University, a major centre for the new scientific and quantitative geography – that Harvey made his name as a philosopher and theorist of positivism. His landmark book *Explanation in Geography* (1969) gave the profusion of statistical methods and techniques in 1960s human geography a robust intellectual foundation. Therein, Harvey explored the nature of scientific explanation in general; the role of theories, hypotheses, laws and models; the use of mathematics, geometry and probability as languages of geographic description; the nature of observation, data classification and data representation; and different forms of scientific explanation. *Explanation in Geography* served two purposes. First, it gave Harvey's generation of 'spatial scientists' a heavyweight justification and manifesto for their project of identifying spatial patterns, departing from an earlier generation concerned with what German-American geographer Frederic Schaefer (1953) had called 'exceptionalism.' Second, *Explanation in Geography* aligned human geography with natural sciences like physics, boosting the discipline's self-image. This was achieved by Harvey drawing upon the ideas of respected philosophers of science, including Karl Popper.

Explanation in Geography, ironically, marked both the high-point and end-point of Harvey's desire to fashion the foundations of a properly 'scientific' geography. Almost immediately after the book's publication, Harvey took up an Associate Professorship in the Department of Geography and Environmental Engineering at Johns Hopkins University, Baltimore. This was triply significant for Harvey. First, though a port city like Bristol, Baltimore was far poorer and had suffered major post-war economic decline. Second, Harvey's new department was less narrowly disciplinary than Bristol, and this gave him license to roam intellectually. Third, Harvey encountered graduate students and academics who wanted to understand the radical ideas of the nineteenth-century political economist Karl Marx. These ideas had profoundly shaped the communist world after the Russian Revolution of 1917, but had not taken hold in Anglo-American societies.

It was in this context that Harvey wrote *Social Justice and the City* (1973), a very different book to its predecessor, but equally significant. As its title suggests, this focused on how to alleviate urban ills like concentrated poverty. Where *Explanation in Geography* was strangely detached from its wider socio-political context – namely, the attempts of the Labour government to make Britain a fairer society, and more distant civil rights and anti-Vietnam protests in the US – *Social Justice* was thoroughly immersed in it. The book charted Harvey's rapidly-changing intellectual and political trajectory. Part One, 'Liberal formulations', focused on how existing mechanisms of (Western) urban planning could produce a more just urban society. But Part Two, 'Socialist formulations', moved beyond the reformism of Part One to propose a revolution in urban affairs predicated on the ideas of radical political economists (notably, Marx). One of the book's key essays, 'Revolutionary and counter-revolutionary theory in geography and the problem of ghetto formation', proposed nothing less than a wholesale rejection of the way human geographers approached urban social problems through scientific, 'objective' analysis of 'the facts'.

Social Justice established Harvey as a major dissenting voice in urban studies, causing a stir for at least three reasons. First, it chronicled Harvey's reinvention as a 'radical geographer', taking many spatial scientists by surprise. Like Gunnar Olsson's (1975) *Eggs in Bird/Birds in Egg*, *Social Justice* was deliberately schizophrenic, literally a book of two halves. Second, it challenged the prevailing view in scientific geography that 'facts' and 'values' could be kept separate: Harvey called for a geography that could study the world to make it more socially just. For Harvey, analysis, evaluation and practical interventions were interconnected; facts did not precede values. Finally, *Social Justice* was the first major book-length exposition of a politically Left-wing geography, helping consolidate the subsequent development of Marxist geography, of which Harvey was to be a pioneer along with some of his graduate students (see **Neil Smith**).

Harvey's turn to Marxism led to a preoccupation with *capitalism* and its workings. Yet he did not entirely abandon earlier preoccupations. Indeed, *Explanation in Geography* ended with the declaration that 'without theory we cannot hope for [...] consistent, and rational, explanation of events' (Harvey, 1969: 486). This endeavour animated his 1970s writings, which were an attempt to fashion a Marxian theory of how geography is both produced by, and alters, the workings of the capitalist economy. By 'geography' Harvey did not mean the discipline of that name, but the material landscape of towns, cities, farms and transport networks that are the 'arteries' of capitalism. To fashion this 'historical-geographical materialism'

Harvey spent the mid-to-late 1970s reading Marx's major later works, namely *Capital* and *Grundrisse*. This culminated in Harvey's most sophisticated and most theoretical book *The Limits to Capital* (2006 [1982]). This was nothing less than a reconstruction and extension of Marx's theory of capitalism; the extensions were primarily geographical because Marx had paid little attention to capitalism's 'spatial fix'.

In the mid-1980s, with his reputation as the leading Marxist (and, indeed, human) geographer established, Harvey returned to the urban issues central to *Social Justice*. Inquiring into the nature and problems of capitalist cities, his two 'Studies in the History and Consciousness of Urbanisation' essentially took the perspectives of 'structure' and 'agency' respectively. *The Urbanization of Capital* (1985a) unfolded a Marxian theory of how capitalism produces cities with characteristic material properties and contradictions, while *Consciousness and the Urban Experience* (1985b) theorised how people respond to having to live in capitalist cities, especially those who lack economic and political power. Both volumes were combined in abridged form in *The Urban Experience* (1989a).

Shortly after publishing these two books, Harvey returned to England as Halford Mackinder Professor of Geography at the University of Oxford. Leaving Ronald Reagan's America for Margaret Thatcher's Britain, Harvey felt himself confronted with an increasingly conservative, anti-Marxist British academic Left and said so in an outspoken critique of UK urban studies (Harvey, 1987a). While at Oxford, Harvey also wrote his best-selling book *The Condition of Postmodernity* (1989b), a critical analysis of the rise of 'postmodernism' in contemporary cultural life (including urban consumption, architecture and academia). Returning to Hopkins in 1993 for academic and personal reasons, he then revisited the theme of social justice and linked it to environmental issues in his highly-ambitious book *Justice, Nature and the Geography of Difference* (1996). Harvey subsequently moved to New York University in 2001 to become Professor of Anthropology, publishing two collections of essays at the start of the new century, one a 'greatest hits' volume (*Spaces of Capital*, 2001), the other about the continued relevance of Marxism in creating a better socio-environmental future (*Spaces of Hope*, 2000).

As the latter book recounts, Harvey's designation as a 'Marxist' represented something very different by 2000 in the Anglophone world compared with the early 1970s. Marxism was seen as rather 'old-hat' among many radicals, which made Harvey's continued commitment to it all the more remarkable given the number of post- and ex-Marxists that littered the intellectual landscape (such as ***Manuel Castells*** and **Doreen Massey**). This said, Marx's work enjoyed a minor revival in the light of the sub-prime lending crisis in the US in 2007. Harvey played a role in this revival by authoring two books on contemporary global affairs, namely *The New Imperialism* (2003) and *A Brief History of Neoliberalism* (2005). The first analysed America's overseas ventures in Iraq and Afghanistan after the World Trade Center attacks in New York City, the latter the spread of neoliberal ideas and policies from the early 1970s. Both texts used Marx's basic insights about the dynamics of capital accumulation to explain global trends. Simultaneously, *The Limits to Capital* was reissued for a second time (2006) – a reflection of its continuing relevance despite the economic turbulence of the previous twenty years.

Following the global financial crisis of 2007–8, and his participation in the Occupy Movement (2011–12), Harvey published a series of well-received books applying Marxist thought to current urban, political and economic affairs (e.g., *The Enigma of Capital and the Crises of Capitalism*, 2010, *Rebel Cities: From the Right to the City to the Urban Revolution*, 2012; *Seventeen Contradictions and the End of Capitalism*, 2014). Moreover, Harvey made a concerted effort to make Marx's later works widely known in the 21st century via video recordings of his graduate seminars deposited on his personal website (www.davidharvey.org) and curated collections of Marx's writings via the publisher Verso (e.g., A Companion to Marx's *Grundrisse*, 2023). These activities mark a 'public turn' in Harvey's pursuits. At this point, he has spent over 20 years trying to shape the thinking of Left political activists and the wider citizenry. Aside from books written in an accessible, largely non-academic style, he maintains the above-mentioned website, has a podcast series and, prior to the Covid-19 pandemic, was a frequent speaker at Leftist-political events around the world.

Spatial Contributions

Harvey was among the first geographers to insist that terrestrial 'space' is not given and absolute, an empty 'container' into which intrinsically 'non-spatial' things are stuffed. As he put it in *Social Justice* (1973: 14), 'The question "What is space?" [must] [...] be replaced by the question "how is it that distinctive human practices create and make use of distinctive [...] space[s]?"'. For Harvey, social practices and processes create spaces while these spaces, in turn, constrain, enable and alter those practices and processes – what ***Edward Soja*** (1989: 78) once called a 'socio-spatial dialectic'. This means that Harvey has long rejected the popular belief that space has no social effects (common among twentieth-century critical theorists) or that it has effects 'in itself' – what US geographer Sack (1974) called 'spatial separatism'. This latter belief sometimes crept into the work of Anglophone geography's spatial scientists, who oftentimes fixated on spatial pattern rather than the social process creating these. Harvey's mid-way position has entailed arguing that 'space' – the material form that processes assume 'on the ground' as buildings, infrastructure, farms, consumption sites and so on – is *both* cause and effect in/of social life – what he once called an 'active moment' (Harvey, 1985a: 3) in human affairs (see also Harvey, 1997).

Harvey's general belief that space is relative and constructed leads to his reading of capitalist space. For Marx, capitalism is a contradictory economic system based on three 'logics': accumulation for accumulation's sake (i.e., the quest for profit as an end in itself); competition between rival producers fighting for market share; and technological innovation in production processes and products. This trinity leads to internal contradictions in the capitalist system, as Marx's opus *Capital* explains. Periodically, these erupt in major 'crises of over-accumulation' in which pools of unutilised workers and commodities can find no productive use. One reason for these crises is that it is 'rational' in a capitalist system for businesses to replace workers with machinery to make production cheaper or more efficient. The paradox is that if many, or most, businesses do this then a collective problem arises: because workers are also consumers, laying off workers removes a major market for commodities, precipitating an economic crisis.

What has any of this to do with geography? In *The Limits to Capital* (2006 [1982]) Harvey argues splurges of investment in buildings, roads and other infrastructure provide capitalism with a mechanism for crisis displacement. Two of the unique features of built environments is that they are exceedingly expensive to build and the return on investment is strung out over decades. Thus, a new convention centre might cost $80 million, the returns on which will trickle back to investors in the form of repeat conference fees over many years. What this means is that at times of incipient economic crisis, the capitalist system tends to 'switch' investment from current production into long-term, fixed capital investments in the geographical landscape. These investments stave off economic crisis by tying up excess capital and amount to what Harvey termed a 'spatial fix' for the contradictions of capitalism. However, these physical investments can later become barriers to further accumulation because they are so hard to replace once made. Thus, when automobiles challenged railways in the early twentieth century as a principal way of moving people and commodities, the enormous investments in rail networks acted as temporary hindrance to new rounds of fixed capital investment. In sum, for Harvey space is *produced* within capitalism and expresses the system's inner contradictions.

Another important theme in Harvey's writings is the space–place relationship. By 'place' Harvey means the unique conjunction of built environments, cultures, peoples, etc., that distinguish one locality from another. However, unlike the regional geographers of the pre-1960s era, Harvey does not conceptualise places as *singular* – that is, *absolutely* different – because in a capitalist world economy *different* places are linked within a *common* economic framework. For geographers, one of the major intellectual challenges is thus to understand the persistence of place distinctiveness amidst heightened place-interdependence. Harvey's answer to this challenge was disarmingly simple. In *The Urbanisation of Capital* (1985a) he argued that capitalism is a circulation process where money is advanced by businesses to purchase labour, inputs and means of production to produce commodities that are sold for the amount of their production cost plus an increment (profit). This process involves commodity production in different places becoming subject to a logic that is indifferent to the qualitative specificities of these places. Because capitalism brings different places within the same economic universe it becomes possible 'to make universal generalisations about the evident unique particularities of [pl]ace' (Harvey, 1985a: 45; see also Merrifield, 1993). Uneven geographical development is central to capitalism as it seeks to profit by exploiting spatial differences (e.g., in labour costs) over time.

Beyond the 'objective' or material side of the particular general/place space relationship there is its 'subjective' side: the way it is perceived and acted on by people living in real places. To be aware of how lives in one place are affected by the unseen actions of distant strangers elsewhere is to possess what Harvey (1990) termed a 'geographical imagination'. To lack such awareness is to fall into the trap of the inter-place competition and rivalry that are part of the geographical logic of capitalism (Harvey, 1985c). In *Justice, Nature and the Geography of Difference* (1996), Harvey highlighted the dilemmas that beset all attempts to make progressive anti-capitalist links between oppressed peoples in different places. One of these relates to 'militant particularism', which is the elemental fact that any wider movement against global capitalism will necessarily be composed of multiple actions by locally-based people. The problem, Harvey argued, is that what makes sense at the local scale may not make sense at a broader geographical scale. Harvey illustrated this key problem in a book co-edited during his time at Oxford,

The Factory and the City (Harvey and Hayter, 1993), about the fight by Austin Rover car workers in Oxford to keep their jobs in the late 1980s. Therein, Harvey argued that while car production creates much needed local employment, at a broader scale it is hardly a worthy socialist objective to make polluting automobiles for middle-class consumers. The question then becomes at which scale – local or global – should political objectives be set?

Ultimately, the key contribution Harvey has made to contemporary conceptions of space and place is to insist that the two can be *theorised* not just studied empirically. This is not as trite a statement as it at first sight seems. ***Andrew Sayer*** (1985) argued that space and place cannot be meaningfully theorised because the difference they make to social practices and processes is contingent: in other words, the material importance of place and space cannot be specified at the level of theory but only at the level of detailed, evidence-based research. This risks relegating geography as a discipline (and geographical matters more generally) to largely factual concerns. Harvey has long disagreed with this view, showing it is possible to make substantive theoretical statements about the role of space in modern life (see Castree, 2002). More particularly, he has been an exponent of 'grand' or 'meta-theory'. For instance, in *The Condition of Postmodernity* (1989b) Harvey argued the 1980s idea that we are now in a 'postmodern era' where difference, complexity and plurality have undermined the search for grand theory is a fiction. Underlying various diverse forms of postmodern expression – in architecture and consumption, for example – Harvey saw the same old logics of capitalism using the production of difference (real and symbolic) as a means of making money.

Key Advances and Controversies

Harvey's writings have persuaded many Marxists at large that space and place matter to capitalism and are thus integral to any project to overthrow it. Harvey is hence celebrated as a Marxist thinker whose work transcends disciplinary boundaries, but he arguably has made two key contributions to geography. First, he gave Marxist geography the intellectual foundation it needed through the 1970s and '80s, along with other emerging Marxists like **Neil Smith,** Erik Swyngedouw and Richard Walker (three of Harvey's graduate students at Hopkins). Second, he set a precedent for radical geography more generally, emboldening younger human geographers to take up the cause of women, gays and lesbians, people of colour and so on.

Examples of how Harvey's Marxist ideas were developed were widespread by the 1980s. In economic geography, several writers applied ideas of the spatial (and temporal) fix to specific research issues – like the way farmers use credit to finance long-term investments in capitalist agriculture (Henderson, 2001). In urban geography, Harvey's idea of how inter-place competition is a necessary feature of the capitalist mode of production inspired detailed studies of the so-called urban 'growth coalitions' that vie for jobs and inward investment (e.g., Cox and Mair, 1988). Indeed, the literature on 'urban entrepreneurialism' and the 'zero-sum' game of place promotion drew (and still draws) substantially on Harvey's theoretical imagination in its attempt to make sense of the lengths that cities will go to entice capital to 'come to town'.

However, Harvey's work was subject to increasing criticism after *The Condition of Postmodernity* was published, especially from human geographers. In part, this was because the intellectual *zeitgeist* had changed, with Marxism being seen as a rather out-dated critical theory by a younger generation of radical academics (despite signs of a renewed interest since 2000). Three criticisms of Harvey's work have loomed large. First, he has been upbraided for the 'muscular' nature of his theoretical ideas: feminists including Deutsche (1991) and Morris (1992) criticised Harvey's apparent belief that his Marxism offers a privileged perspective on the dismal 'truth' of the world. They argued that his confidence in his theoretical claims bespeaks a distinctively masculine desire for cognitive mastery and control, excluding other legitimate ways of knowing the world. The result, they maintained, was that Harvey remained blind to all manner of important real-world processes and events that do not fit with his Marxist worldview.

This links, second, with the criticism that Harvey's Marxism cannot accommodate valued forms of social difference. Those on the academic Left have oftentimes accentuated social difference as a way of escaping the Marxist obsession with social class as the primary axis of oppression and resistance in modern societies. The differences celebrated include those within and between women, gay men, lesbians, trans people, disabled people and people of colour. Harvey (1996) clarified his position by arguing that non-class differences between people must be linked, but are not reducible, to any project to overthrow capitalism, a position he has maintained ever since.

Third, some (e.g., Dennis, 1987) have worried that Harvey's grand theoretical abstractions are insufficiently grounded in the messy empirical complexities of the real world. Harvey (1987b; 1989a) responded with the argument that theory is not intended to *exhaustively* explain any given situation but rather to open a window onto the key processes at work. Because of these (and some other) criticisms, many geographers developed new arguments in direct reaction or opposition to Harvey's. Examples included Herod's (2001) work on how wage labourers fight back against business in place and across space (contra Harvey's preoccupation with the power of capital) and ***Peter Jackson***'s (1999) work on the consumption of commodities, which gave consumers a more active personal and political role than Harvey was seemingly prepared to.

Harvey's later works have continued to be influential, but not because they necessarily break fresh intellectual ground or respond deeply to such critiques. *The New Imperialism* (2003) and *A Brief History of Neoliberalism* (2005), for example, revisit Marx's basic theory of capital accumulation to explain current global affairs. Though not theoretically original, their insights appear fresh to a younger generation of readers unfamiliar with Marxist ideas. His continued 'big picture' perspective on the world still incites productive debates about key issues of our time, such as the changing nature of 'imperialism' now that China is challenging the global hegemony of the US (e.g., Harvey, 2018). Moreover, Harvey's knack of coining memorable neologisms – such as 'accumulation by dispossession' – has continued to make his work eminently citable. He has recently reflected on his long and distinguished career in print (Harvey, 2022), and his life's work has been extensively – and favourably – surveyed (Castree et al., 2023).

Harvey's Key Works

Harvey, D. (1969) *Explanation in Geography*. London: Edward Arnold.

Harvey, D. (1973) *Social Justice and the City*. London: Arnold.

Harvey, D. (2006 [1982]) *The Limits to Capital* (new edition). London: Verso.
Harvey, D. (1985a) *The Urbanization of Capital*. Oxford: Blackwell.
Harvey, D. (1985b) *Consciousness and the Urban Experience*. Oxford: Blackwell.
Harvey, D. (1989a) *The Urban Experience*. Oxford: Blackwell.
Harvey, D. (1989b) *The Condition of Postmodernity*. Oxford: Blackwell.
Harvey, D. (1996) *Justice, Nature and the Geography of Difference*. Oxford: Blackwell.
Harvey, D. (2000) *Spaces of Hope*. Edinburgh: Edinburgh University Press.
Harvey, D. (2003) *The New Imperialism*. Oxford: Oxford University Press.
Harvey, D. (2005) *A Brief History of Neoliberalism*. Oxford: Oxford University Press.
Harvey, D. (2010) *The Enigma of Capital and the Crises of Capitalism*. New York: Profile Books.
Harvey, D. (2012) *Rebel Cities: From the Right to the City to the Urban Revolution*. London: Verso.
Harvey, D. (2014) *Seventeen Contradictions and the End of Capitalism*. New York, NY: Profile.
Harvey, D. (2017) *Marx, Capital and the Madness of Economic Reason*. New York: Profile.
Harvey, D. (2018) *A Companion to Marx's Capital: The Complete Edition*. London: Verso.
Harvey, D. (2023) *A Companion to Marx's Grundrisse*. London: Verso.

Secondary Sources and References

Castree, N. (2002) 'From spaces of antagonism to spaces of engagement', in A. Brown, S. Fleetwood and J.M. Roberts (eds), *Critical Realism and Marxism*. London: Routledge. pp. 187–214.
Castree, N., Charnock, G. and Christophers, B. (2023) *David Harvey: A Critical Introduction to his Thought*. New York, NY: Routledge.
Cox, K. and Mair, A. (1989) 'Locality and community in the politics of local economic development', *Annals of the Association of American Geographers*, 78: 307–25
Dennis, R. (1987) 'Faith in the city', *Journal of Historical Geography*, 13: 310–16.
Deutsche, R. (1991) 'Boy's town', *Environment and Planning D: Society and Space*, 9: 5–30.
Harvey, D. (1963) 'Locational change in the Kentish hop industry and the analysis of land use patterns', *Transactions of the Institute of British Geographers*, 33: 123–40.
Harvey, D. (1985c) 'The geopolitics of capitalism', in D. Gregory and J. Urry (eds), *Social Relations and Spatial Structures*. London: Macmillan. pp. 128–63.
Harvey, D. (1987a) 'Three myths is search of a reality in urban studies', *Environment and Planning D: Society and Space*, 5: 367–76.
Harvey, D. (1987b) 'The representation of urban life', *Journal of Historical Geography*, 13: 317–21.
Harvey, D. (1990) 'Between space and time: Reflections on the geographical imagination', *Annals of the Association of American Geographers*, 80: 418–34.
Harvey, D. (1997) 'Contested cities: Social process and spatial form', in N. Jewson and S. MacGregor (eds), *Transforming Cities*. London: Routledge. pp. 19–27.
Harvey, D. (2001) *Spaces of Capital*. Edinburgh: Edinburgh University Press.
Harvey, D. (2018) 'Realities on the ground: David Harvey replies to John Smith', *Review of African Political Economy*, February 5. https://web.archive.org/web/20230326025623/http://roape.net/2018/02/05/realities-ground-david-harvey-replies-john-smith/ (accessed November 6, 2023).
Harvey, D. (2022) 'Reflections on an academic life', *Human Geography*, 15: 14–24.
Harvey, D. and Hayter, T. (eds) (1993) *The Factory and the City. Story of the Cowley*

Automobile Workers in Oxford. London: Mansell.

Henderson, G. (1999) *California and the Fictions of Capital*. Oxford: Oxford University Press.

Herod, A. (2001) *Labor Geographies*. New York, NY: Guilford.

Jackson, P. (1999) 'Commodity cultures: The traffic in things', *Transactions of the Institute of British Geographers*, 24: 95–108

Merrifield, A. (1993) 'Space and place: A Lefebvrian reconciliation', *Transactions of the Institute of British Geographers*, 18: 516–31.

Morris, M. (1992) 'The man in the mirror', *Theory, Culture and Society*, 9: 253–79.

Olsson, G. (1975) *Eggs in Bird/Birds in Egg*. London: Pion.

Sack, R. (1974) 'The spatial separatist theme in geography', *Economic Geography*, 50: 1–19.

Sayer, A. (1985) 'The difference that space makes', in D. Gregory and J. Urry (eds), *Social Relations and Spatial Structures*. London: Macmillan. pp. 49–66.

Schaefer, F. (1953) 'Exceptionalism in geography', *Annals of the Association of American Geographers*, 43: 226–9.

Soja, E. (1989) *Postmodern Geographies*. London: Verso.

25
bell hooks

Katherine McKittrick, Queen's University, Kingston, Canada

Biographical Details and Theoretical Context

bell hooks is one of the most prominent and influential feminist philosophers of recent years. Her devotion to black activism and organising, intellectual life, and cultural politics in the black diaspora shaped her commitment to dismantling what she termed 'white supremacist capitalist patriarchy' via an emphasis on the liberatory politics inherent in blackness, community, and healing. It is from the margins – specifically the experiences and theories shared by disenfranchised communities – that she offered meaningful ways to work through oppressive systems that are animated by race, class, gender, sexuality, and location.

bell hooks/Gloria Jean Watkins was born in Hopkinsville, Kentucky, in 1952. She has argued that her familial and household politics forced her, as a young child, to theorise the problems of black patriarchy, sexism, and gender subordination (hooks, 1994a: 59–76). hooks received her BA from Stanford University, and her MA from the University of Wisconsin. In 1983 she received her PhD from the University of California at Santa Cruz, writing her dissertation on the US novelist and academic Toni Morrison. hooks taught at the University of Southern California, Yale University, Oberlin University, and the City College of New York, among other colleges and universities. In 2004, hooks joined Berea College in Kentucky as Distinguished Professor and in 2014 the bell hooks centre was established at Berea College. In addition to holding hooks' papers, the centre is an abolitionist hub, offering programming that centres the worlds of black, feminist, and gender-queer students and community members. In 2021, hooks passed, leaving behind her beautiful archive that provides lessons for imagining and enacting collective liberation.

bell hooks' experiences as an undergraduate and graduate student advanced her earlier feminist thinking and her contributions to black feminist thought. First, hooks recognised the lack of historical and contemporary writings by and about black women; second, she noticed that second-wave feminism – although not formally named as 'white' or 'white supremacist' – unjustly marginalised, and in some cases explicitly dismissed, non-white women. This discursive and experiential racism and sexism led her to produce two texts that 'spoke back' to these interconnected forms of black marginalisation. Prior to receiving her PhD, hooks (1981) completed *Ain't I A Woman? Black Women and Feminism*, a text she had been preparing for several years. *Ain't I A Woman?* maps out US black women's contemporary and historical experiences in a contribution to the long tradition of diasporic black feminist thought (see McKittrick, 2006, for an overview). *Ain't I A Woman?* also calls into question the white-racialised meaning of womanhood, asking the reader to consider alternative ways of conceptualising gender, feminism, and political unity *vis-à-vis* black women's struggles and histories.

hooks' (1984) *Feminist Theory: From Margin to Center* focuses specifically on feminist politics and struggles within and outside black communities. *From Margin to Center* outlines a series of feminist issues (such as parenting, sexual oppression, paid and unpaid labour, and political allegiances) and argues that black women's 'special vantage point' (hooks, 1984: 15) necessarily enriches feminist theory and praxis. *From Margin to Center* also demonstrates that racism, sexism, and exclusion have not prevented black women from feminist theorising in the academy and the everyday.

Following these two texts, hooks produced a series of theoretical texts that strengthened her commentary on exclusion, black feminist thought, and cultural politics. hooks' later texts address pedagogy, representational politics, popular culture, desire, internalised racism, black femininity, stereotypes, and visual art. Her many texts consider black experiences and knowledges in relation to processes of racial, sexual, and economic domination (see especially hooks 1989, 1990, 1992, 1994a, 1994b). hooks' 2016 essay on Beyoncé Knowles' visual album *Lemonade*, for example, skilfully draws attention to how black popular culture continues to be nested in – and beholden to – white supremacist capitalist patriarchy yet offers moments that radically reimagine black femininity (hooks, 2016).

For hooks, pervasive forms of racial, economic, and sexual domination demand that black politics, and activism in general, must be radically open and oppositional. In addition to this, she suggests the Self must be repaired by being radically open and oppositional (hence, 'talking back' to, and looking back at, the intricacies and practitioners of oppression). The underlying messages in many of her texts, most notably *All About Love* (2000a), *Feminism is for Everybody: Passionate Politics* (2000b), and *Salvation: Black People and Love* (2001), are that processes of self-renewal and resistance, and associated practices of love and desire, must be fostered in a landscape which has systematically discouraged loving blackness and social differences. Her arguments centre healing, ethical-love-politics, and caring for the planet, and weave together many of her key ideas by infusing her political agenda with questions of decolonisation, critical thinking, black feminist subjectivity, and the combination of care, knowledge, responsibility, respect, trust and commitment (hooks, 2001).

Spatial Contributions

Like many black feminists and theorists, hooks has contributed to the ways 'identity' is understood in social theory. Identity, for hooks, is formed in relation to white supremacist capitalist patriarchy. This means that a black sense of self is constructed according to broader ideological and material patterns that inscribe 'difference' or 'otherness' on the black psyche, black personhood, and the black community. Identity is shaped by processes of marginalisation, and *it is within* this marginal space that diverse black identities are produced. What hooks gives a sense of is the doubleness of black identities – identities that are produced according to crude forms of race/racism *and* identities that are diversely challenging the meaning of these forms of race/racism. This understanding of identity is nuanced by hooks' commitment to oppositional strategies. Thus, the black subject (re)produces identity and black politics by implicitly or explicitly 'coming to voice [...] where one moves from being object to being subject' (hooks, 1989: 12). hooks' conceptualisation of black identities contributes to the geographical shift away from stable, white Euro-American, and androcentric identity formations (see Kobayashi and Peake, 1994). By positing the possibility of, for example, diverse black feminine subjectivities that are actively producing their identities *vis-à-vis* domination, resistance, and the margin, hooks suggests that the production of space is simultaneously contextual and hegemonic.

In 'Choosing the margin as a space of radical openness' (in hooks, 1990: 145–53), hooks builds on her work in *From Margin to Center* (1984). hooks argues that the margin and the centre are neither antithetical nor an indication of a disconnection between white and non-white communities. Instead, she demonstrates that racial, sexual, economic, and other social differences shape a response to, and therefore a connection with, existing cultural norms. This connection, as it is understood from the margins, is significantly oppositional:

> [...] marginality [*is*] much more than a site of deprivation [...] it is also the site of radical possibility, a space of resistance [...] a central location for the production of counter-hegemonic discourse that is not just found in works but in habits of being and the way one lives. As such, I was not speaking of a marginality which one wants to lose – to give up or surrender as part of moving to the center – but rather a site one stays in [...] It offers to one the possibility of radical perspective from which to see and create, to imagine alternatives, new worlds. (hooks, 1990: 149–50)

hooks thus stresses the interplay between margin-centre relations, arguing that the black subject's positionality informs political alternatives. Furthermore, hooks shows that the margin is a legitimate location from which to produce knowledge and confront pain. Spatialising black identities and knowledges, hooks' work on space, place, and the margin underscore how one verbally and non-verbally negotiates one's political self, which generates a space of radical openness: a space 'which affirms and sustains our subjectivity, which gives us a new location from which to articulate our sense of the world' (hooks, 1990: 153).

In *Yearning: Race, Gender and Cultural Politics* (1990), hooks reframes white feminist approaches to home, the family, and community. Second-wave feminists critiqued the home, drawing attention to how it can be a feminised space where patriarchal norms are reproduced, violence is enacted, and women's unpaid work is de-valued. For bell hooks, homeplace is a subversive and feminist space. Her reconceptualisation of home begins with her childhood journeys through unfamiliar white neighbourhoods. These travels were laden with real or potential white supremacist dangers, which were diminished upon arriving 'home': 'the feeling of safety, of arrival, of homecoming [...] the warmth and comfort of shelter, the feeding of our bodies, the nurturing of our souls' (hooks, 1990: 41). By positing that the homeplace was, and is, a potential (but often unrecognised) site of resistance for black women and black communities, hooks overturns the oppressive nature of domestic space, while also challenging stereotypes such as the overpowering matriarch, the 'natural' house servant, and the nurturer of whiteness. Rather than suggesting that black communities neatly replicate white feminist theories, hooks uncovers the history of black homeplaces: spaces that were most often created by black women and provided an environment in which the black community can learn to love and respect blackness. Although sexist norms are certainly evident in black homes, hooks calls for the reclamation of the kind of place where resistance is possible and these norms are discouraged. This is nicely paired with hooks' interest in housing, architecture, and aesthetics, which is illuminated by her observations that resisting oppression is deeply entwined with transforming how we live, produce, and design infrastructure (hooks, 1995: 145–51). These threads are picked up in *Belonging: A Culture of Place* (2008), where hooks focuses on land ownership, territory, race, gender, economy, and home. Here, hooks uncovers the spatial and racial underpinnings of segregation, labour, and migration, and brings into focus how the black community's sense of the environment has engendered a politics of belonging that is not structured by profit-driven logics, racism, and environmental decline. Indeed, to collectively attend to our fragile environment and to ethically practise belonging involves a refusal to profit from longstanding, commonsense, socio-spatial inequities.

The concepts that lie at the heart of hooks' work – identity, the margin, homeplace, belonging – are necessarily linked to space and place, with the spatialisation of difference being a key motif in hooks' writing. Feminist and human geographers have thus used hooks' writing to clarify the ways space and place not only adversely shape black choices and subjectivities, but also to explore the complexities inherent in spatial production. There are three ways this is done. First, geographers turn to hooks' understandings of the ways whiteness and racism shape identity formation. hooks' work, primarily in *Black Looks* (1992) and *Yearning* (1990), describes the historical and everyday terrors black people encounter in a white, capitalist, patriarchal world (Rose, 1993; Ruddick, 1996)

Second, feminist and human geographers have turned to hooks' politicisation of location. hooks' discussion of black homeplace(s) and the margin as a site of resistance have contributed to how some geographers theorise the mutual flexibility of identity and place. Feminist geographers such as Nancy Duncan (1996), **Doreen Massey** (1994) and **Linda McDowell** (1999) have briefly touched on hooks' analysis of home in order to clarify the contradictory nature of femininity, race, and domestic/private spaces. Duncan, furthermore, suggests that hooks' analysis can speak to other feminist spaces – specifically those produced by, and in relation to, women who have experienced domestic violence and sexual minorities. These spaces, although sites of deep sexualisation and gendering, also hold in them the possibility of politicised resistance (1996: 142–3). Others have used hooks to identify how social differences challenge geographical theory and the production of space. Relatedly, ***Edward Soja*** (1996) unpacked hooks' textual and experiential geographies and suggested that they were indicative of a postmodern way of understanding space and difference.

Third, some feminist and human geographers pair hooks' *experiences in*, and *descriptions of* her spaces and places, with her more conceptual/metaphorical advances – such as the margin. This underscores the ways geography and hooks herself are, together, 'multiple and intersecting, provisional and shifting' (Rose, 1993: 155). hooks' concept of belonging (2008) has been useful to geographers in that she entangles the physical geographies of Kentucky (trees, hills, rural landscapes) with memories and emotional attachments; this illuminates the dynamism between place, affect, community, and liberatory possiblities (Antonsich, 2010; Yarker, 2019).

Key Advances and Controversies

bell hooks' extensive and accessible production of theory and texts about blackness and black femininity, her additional writings in *Vibe*, *The Village Voice*, and *Z Magazine*, her canonisation within introductory women's studies classes (and cognate disciplines), and the breadth of her subject matter – from slavery and film to homeplace and the classroom – made her one of the most-cited black feminist authors within the discipline of geography for decades. Often this kind of citation practice functioned to 'add' race or blackness to an otherwise white discipline while also erasing the work of other non-US black feminists, the work of hooks' US black colleagues, and the work of black geographers in general (see Sanders, 1990; Woods, 1998; Gilmore, 2002; Wilson, 2002). What this does, then, is reduce hooks/black womanhood to an authentic embodiment of blackness who is cited to lend texture to normative theories (see Keith and Pile, 1993). As Valerie Smith (1989) argues, this use of black womanhood risks re-colonising blackness and homogenising all black women's experiences and identities.

bell hooks' output was incredibly productive, particularly in the 1990s and into the 2000s. Apart from **Gillian Rose** (1993), many critical engagements with hooks during this window often only described her experiences; many theorists of geography resisted engaging with her deeper concerns, such as the radical potential of loving blackness or the importance of collective anti-racist organising. hooks' output lessened in about 2010, and since that time critical geographers have, for the most part, shied away from engaging with her ideas extensively. However, both implicitly and explicitly, her work has informed the field of black geographies (see McKittrick, 2006; Browne, 2015; Adeniyi-Ogunyankin, 2022). Especially meaningful are the projects insisting that bell hooks is, in fact, deeply connected to the field of black geographies and invested in abolition struggles (Winston, 2021; Eaves and Purifoy, 2022; Graddy-Lovelace, 2023). hooks is a leading feminist theorist who has provided a grammar to help us better understand geographies of difference more broadly and black diasporic geographies specifically. The kinds of insights she has offered, in terms of her own sense of place and the complex ways white supremacist capitalist patriarchy is geographically expressed, not only calls for a working through of how the production of space is simultaneously material, subjective, empowering and unjust, but also how multiple geographical knowledges, emerging from communities on the margins, open up new spaces to ponder, struggle toward, and enact liberation.

Hooks' Key Works

hooks, b. (1981) *Ain't I A Woman? Black Women and Feminism*. Boston, MA: South End Press.

hooks, b. (1984) *Feminist Theory: From Margin to Center.* Boston, MA: South End.

hooks, b. (1989) *Talking Back: Thinking Feminist, Thinking Black.* Toronto: Between the Lines.

hooks, b. (1990) *Yearning: Race, Gender and Cultural Politics.* Toronto: Between the Lines.

hooks, b. (1992) *Black Looks: Race and Representation.* Toronto: Between the Lines.

hooks, b. (1994a) *Teaching to Transgress: Education as the Practice of Freedom.* New York, NY: Routledge.

hooks, b. (1994b) *Outlaw Culture: Resisting Representations.* New York, NY: Routledge.

hooks, b. (1995) *Art on My Mind.* New York, NY: The New Press.

hooks, b. (2000a) *All About Love: New Visions.* New York, NY: HarperCollins.

hooks, b. (2000b) *Feminism is for Everybody: Passionate Politics.* Boston, MA: South End Press.

hooks, b. (2001) *Salvation: Black People and Love.* New York, NY: HarperCollins.

hooks, b. (2008) *Belonging: A Culture of Place.* New York, NY: Routledge.

hooks, b. (2016) 'Moving Beyond Pain', *bellhooksbooks.* https://bellhooksbooks.com/moving-beyond-pain/ (accessed November 6, 2023).

Secondary Sources and References

Adeniyi-Ogunyankin, G. (2022) 'Grappling with the fragmentations: Black feminisms, African feminisms and the possibilities of Black geographies in Canada', *Topia: A Canadian Journal of Cultural Studies,* 44: 175–84.

Antonsich, M. (2010) 'Searching for belonging: an analytical framework', *Geography Compass,* 4 (6): 644–59.

Browne, S. (2015) *Dark Matters: On the Surveillance of Blackness.* Durham, NC: Duke University Press.

Duncan, N. (1996) 'Renegotiating gender and sexuality in public and private spaces', in N. Duncan (ed), *BodySpace: Destabilizing Geographies of Gender and Sexuality.* New York, NY: Routledge. pp. 127–45.

Eaves, L.E. and Purifoy, D. (2022) '"We must situate race": Bobby Wilson's black Geographies', *Southeastern Geographer,* 62 (3): 182–6.

Gilmore, R.W. (2002) 'Fatal couplings of power and difference: Notes on racism and geography', *The Professional Geographer,* 54: 15–24.

Graddy-Lovelace, G. (2023) 'Towards abolitionist agrarian geographies of Kentucky', *Environment and Planning E: Nature and Space,* 6 (3): 1561–89.

Keith, M. and Pile. S. (1993) 'Introduction: The politics of place', in M. Keith and S. Pile (eds), *Place and the Politics of Identity.* London: Routledge. pp. 1–21.

Kobayashi, A. and Peake, L. (1994) 'Unnatural discourse: "Race" and Gender in geography', *Gender, Place & Culture,* 1: 225–43.

Massey, D. (1994) *Space, Place and Gender.* Minneapolis, MN: University of Minnesota Press.

McKittrick, K. (2006) *Demonic Grounds: Black Women and the Cartographies of Struggle.* Minneapolis, MN: University of Minnesota Press.

McDowell, L. (1999) *Gender, Identity and Place: Understanding Feminist Geographies.* Minneapolis, MN: University of Minnesota Press.

Rose, G. (1993) *Feminism and Geography: The Limits of Geographical Knowledge.* Cambridge: Polity.

Ruddick, S. (1996) 'Constructing difference in public spaces: Race, class and gender as interlocking systems', *Urban Geography,* 17: 132–51.

Sanders, R. (1990) 'Integrating race and ethnicity into geographic gender studies', *The Professional Geographer*, 42: 228–31.

Smith, V. (1989) 'Black feminist theory and the representation of the "Other"', in C. A. Wall (ed), *Changing Our Words*. New Brunswick, NJ: Rutgers University Press. pp. 38–57.

Soja, E. (1996) *Thirdspace: Journeys to Los Angeles and Other Real-and-Imagined Places*. Oxford: Blackwell.

Wilson, B.M. (2002) 'Critically understanding race-connected practices: A reading of W.E.B. Du Bois and Richard Wright', *The Professional Geographer*, 54 (1): 31–41.

Winston, C. (2021) 'Black ground truths and police abolition', *Transactions of the Institute of British Geographers*, 46 (4): 815–17.

Woods, C. (1998) *Development Arrested: The Blues and Plantation Power in the Mississippi Delta*. New York, NY: Verso.

Yarker, S. (2019) 'Reconceptualising comfort as part of local belonging: The use of confidence, commitment and irony', *Social & Cultural Geography*, 20 (4): 534–50.

26
Tim Ingold

Hayden Lorimer, University of Edinburgh

Biographical Details and Theoretical Context

Tim Ingold is a social anthropologist who has developed a particular and individual vision for a discipline he defines as 'philosophy with the people in' (Ingold, 1992: 696). Though anthropology is his academic home, Ingold's thinking and writing has increasingly been directed outwards, generating dialogue and published *Correspondences* (2021) with adjacent disciplines and neighbouring practitioners, specifically in Archaeology, Architecture and Art. While, by dint of its unhelpful first letter, Geography has not found an easy fit in this first-order of 'A-class' subjects, Ingold's writing has significantly shaped human geographers' understanding of ecological processes of place-making and inhabitation as well as the ways social relations encompass, and are constituted by, more-than-human forms of life.

Born in 1948, Ingold grew up in Kent in the UK and studied at the University of Cambridge. Having initially intended to read natural science, he became quickly disillusioned with the institutional structures, professional hierarchies and the military-industrial patronage upon which the modern scientific establishment was dependent. A switch of studies to social anthropology carried different appeal: a freer academic community; a subject which did not yet know itself fully; and precious opportunities to forge experimental links between the arts and humanities on the one hand, and the natural sciences on the other. He graduated with a BA in 1970. Ingold's subsequent doctoral research was based on the findings of ethnographic fieldwork undertaken among the Skolt Sámi of north-eastern Finland. Published in 1976, this study positioned traditional and adaptive practices of reindeer herding centrally in the livelihoods of a re-settled minority community.

Following a year spent at the University of Helsinki, Ingold was appointed to a Lectureship in Social Anthropology at the University of Manchester in 1974. He was promoted to a Chair in 1990, and five years later became Max Gluckman Professor of Social Anthropology. At Manchester, his research on northern circumpolar reindeer hunting and herding cultures continued, alongside comparative analysis of hunter-gatherer and pastoral societies, and a more general concern with human–animal relations, exemplified by the edited volume *What is an Animal?* (1988). In parallel, he pursued work on anthropological theory which sought to establish explanations of human distinctiveness via a critical reconsideration of toolmaking, making connections between technology, language and art across the span of human evolution. By the 1990s, this expansive intellectual project crystalised around the concept of 'skilled practice'. This pushed anthropological studies of material culture away from the idea of objects as the primary locus of creative expression, and instead towards the embodied processes of enskillment necessary for the making of things, whether the building of a wall, the tying of a knot, or the throwing of a pot.

An appreciation of skilled practice is also pivotal to Ingold's anthropology of environmental perception, or put more simply, his ongoing inquiries into how environments are perceived, shaped and understood by humans. Dissatisfied with traditional models of genetic and genealogical transmission, founded upon an alliance of neo-Darwinian biology and cognitive science that sustained the theory of cultural transmission accepted by his anthropological forebears, he sought a more satisfactory alternative for understanding human creativity. The ecological approach offers nothing less than a replacement paradigm, reflecting his preference for the idea of growth embodied in the skills of perception and action existing within particular social and environmental contexts of development. As transformative solutions sometimes do, this appeared in a flash, following long years of puzzling:

> I vividly remember one Saturday morning in April 1988 – an entirely ordinary one for Manchester at that time of year, with grey skies and a little rain – when, on my way to catch a bus, it suddenly dawned on me that the organism and the person could be one and the same. Instead of trying to reconstruct the complete human being from two separate but complementary components, respectively biophysical and sociocultural, held together with a film of psychological cement, it struck me that we should be trying to find a way of talking about human life that eliminates the need to splice it up into these different layers. Everything I have written since has been driven by this agenda. (Ingold, 2000: 3)

This realisation had catalytic effect, resulting in *The Perception of the Environment: Essays on Livelihood, Dwelling and Skill* (2000), a major work comprised of 23 essays. Each can be read as a stand-alone piece, uniting to form a phenomenological appreciation of different 'designs for life' existing *between* the biological and the cultural. Two further major collections have since followed, *Being Alive: Essays on Movement, Knowledge and Description* (2011), and *Imagining for Real: Essays on Creation, Attention and Correspondence* (2022), forming a landmark trilogy.

Ingold departed Manchester after 25 years to take up a Chair of Social Anthropology at the University of Aberdeen in 2000 where he was instrumental in re-establishing the subject as a department and a taught degree. His ecological anthropology continued through comparative theorisations of different skilled practices – writing, drawing, story-telling and walking (see also Ingold, 2004; Ingold and Lee Vergunst, 2008) – as exercises in 'line-making'. *Lines: A Brief History* (2007a) was the cumulative result. A work of intellectual and historical synthesis, it defies easy categorisation, advancing a philosophy of line-making as a creative, generative, sustaining and enriching process in a relational world. Here, his earliest experiences from ethnographic fieldwork were still deployed, noting among the Sámi in Lapland, people and other living beings are understood to live lives along paths. These paths are formed and come to be known and remembered as a result of physical movement through the landscape. Stories might be told according to these pathways, allowing associations to be made between person and place. Such lifelines of existence have significant implications for the formulation of personhood. For Ingold, persons should be thought of not as self-contained objects, or as sovereign things, on the move between point A and point B. Rather, a person is *immanent*: a trajectory that exists as it goes along. Lines are the paths of its own movement, and its relationships, through a world always in formation (Ingold, 2006c; 2007a).

Ingold's efforts to re-fashion theories of life according to an ecological approach can be traced through a history of theoretical ideas from existential phenomenology (Maurice Merleau-Ponty, Gaston Bachelard and Martin Heidegger); process philosophy and vitalism (Alfred North Whitehead, **Gilles Deleuze**); labour and value theory (Karl Marx and E.P. Thompson) alongside non-conformist thinkers in ecological psychology (James Gibson), anthropology (Gregory Bateson), ethology (Jakob von Uexküll) and new materialism (Timothy Morton). From geography, it is the ideas of those who shaped the discipline's humanistic episode that feature most prominently (**Denis Cosgrove**, Stephen Daniels, **Yi-Fu Tuan** and David Lowenthal), though sometimes only as counterpoints allowing for the advancement of a distinctive anthropological perspective. Such intellectual debts acknowledged, Ingold's writing is noticeable for being less studded with the names of other theorists (or doctrinal quotations from their works) than might ordinarily be encountered in academic literature. In part, this reflects personal style and a lightness of touch, favouring the long-form essay where argument can exist unfettered by exhaustive lists of citations. But it represents a political statement and ethical commitment too. For Ingold, philosophical lessons are to be sourced *in life*, that is in the skills and practices by which people perceive and understand their immediate surroundings and so make themselves at home as they go about making the world.

When advancing or substantiating argument, his habit is to turn first to anthropologists' ethnographic observations of non-Western, indigenous lifestyles, and the cosmologies of which these day-to-day practicalities are an informing part. In any detailed description of a task's careful undertaking – such as, say, basket-weaving, animal-naming, wayfinding or tool-making – cross-references with social theory might be footnoted, sometimes surfacing briefly in the main text, but true wisdom is found first in the realities of the lived world, in places where social practice is organic, embodied and emergent. His ontological claims therefore hold to the axiom that 'to be is to know, and to know is to be'. So too do his preferences for teaching practice, set within an educational vision about protecting the founding idea of the university (Ingold, 2018). Attentively learning the embodied skills of practice, both by its rudiments and the intricacies of technique, is a scholastic tradition Ingold has observed through experimental, participative workshops, enlisting the tutelage of skilled practitioners and fellow apprentices comprised of colleagues and students. Retiring in 2018, his prodigious output has not slackened, and the award of a Commander of the British Empire (CBE) for services to anthropology in 2022 was merely one among many international honours.

Spatial Contributions

Care is necessary in any assessment of Ingold's spatial contribution since he has been no great advocate of spatial thinking. Assessing the concept's potential as analytical tool he declares himself a 'flat-earther' who harbours first-principle objections: 'I just cannot get out of my head the idea of space as a void, as non-world, as absence rather than co-presence' (Ingold, 2006b: 892). It seems reasonable to assume, therefore, that he would determine 'Spatial Contributions' as a theoretical formulation that – whilst reflecting accurately the 'spatial turn' taken in much social sciences research of late – is ultimately unhelpful, both in its reasoning and usage (Ingold, 2006b). Though scepticism verges on antagonism, this stance taken *against* space is principled, stemming from a strong dislike for theoretical discourse detached from the lived world. By his account, space is unhelpful since it cannot but connote lifelessness, fixity, finitude and enclosure. Philosophically speaking, Ingold's foundational preference is for time, or more specifically *life*-times, and for inquiries into the rhythms and temporal textures of existence. Thus, where geographers find time-space everywhere abounding (and submit to extended debate concerning nuances of position and interpretation), Ingold observes instead the lifeworlds and lifelines of living beings. He is at pains to point out that this is not simply a case of semantics but also styles of language. Finding virtue in clarity of expression, his prickliest comment is saved for academic expression that alienates non-academic readers. This social theory of 'anti-social theorising' is explained thus:

> [It] [...] could be called the reification of hyperabstraction: start with an abstraction, turn this into a quality of something yet more abstract, and then imagine that this meta-abstraction is concretely and plurally present in the world, instantiated in the very things from which the whole process of abstraction started in the first place. For philosophers the attraction of this strategy lies in its circularity: at no point in the cycle is one actually required to observe or engage directly with the world itself. (Ingold, 2006b: 892)

Abstract riffing on space ('Space-Spatial-Spatiality-Spatialities') is cited as a prime example of the habit, and comparable proliferations around 'materialities', 'temporalities' and 'multiplicities' judged just as problematic. Criticism in this vein might conceivably align Ingold with traditional, conservative voices in geography, given to occasional, jaundiced prognostications about the future and the fate of the discipline. This would be an accidental effect, and a rogue alliance. But his sharp eye for post-structuralist pretension noted, intellectually Ingold has enjoyed a rapport with practising cultural geographers concerned with creativity, expression and embodiment. Here, three identifiable fields of geographical study show Ingold's influence. All are informed by work published in the middle and later phases of his academic career, with Ingold's work coming to the attention of the geographical community in the 1990s primarily through original published versions of material (Ingold, 1993; 1995a) later appearing in *The Perception of the Environment* (2000).

The first key area of influence might usefully be labelled the *push for practice*. Ingold's theory of skilled practice bolstered a forceful argument, voiced variously within human geography, that the prevailing understanding of cultural expression and subjectivity was too narrowly focused on a search for interpretable meaning in textual representations, significations and discursive forms of identity formation (Thrift, 1996). Embodiments, enactments and performances constituted complex worlds of spatial practice, everywhere apparent and yet, paradoxically, left largely untouched during the discipline's 'cultural turn'. The effect of this was a deadening of each geographer's sensorium and their collective sensibilities (Thrift and Dewsbury, 2000). To breathe life back into geographers' encounters with the world, as something apprehensible and experiential, Ingold's advocacy for the concept of skilled practice can therefore be read as an act of strategic conscription. However, here Ingold was but one name among the diverse theoretical reference points – phenomenology, poststructuralist thought and science-and-technology studies – that encouraged *non-representational theory* to emerge as an influential episode in geography's modern intellectual history (Thrift, 2007; Lorimer, 2008) (see also entry on **Nigel Thrift**). Ingold did not offer direct comment on what was often an inward-facing debate, but his work clearly powered geography's shift towards ethnographic studies of the ordinary activities, social relations and day-to-day phenomena by which the world continuously takes place.

Ingold's second signature contribution to geographical thought centres on the 'dwelling perspective' informing approaches to the study of landscape and place – concepts long cherished by geographers, and frequently deployed as synonyms for one other, though each possessing of a distinctive intellectual history (Wylie, 2007). Ingold introduced the dwelling perspective in two essays (Ingold, 1993; 1995b; see also Ingold, 2000). The latter begins by directly engaging with the 'iconographic' view of landscape as presented by Cosgrove and Daniels (1988), originating in geography's humanities tradition. Ingold did not accept their view that landscape exists as 'a cultural image, a pictorial way of representing or symbolizing surroundings' (Cosgrove and Daniels, 1988: 1). To accept such a premise, he argued, was to ignore the material existence of the places where people dwell, and to render nature as something existing apart. The favouring of landscape's iconography thereby cemented a flawed division between the inner and outer worlds of human experience, making distinctions between mind and matter, meaning and substance, where none actually exist (Ingold, 2000: 203).

While accommodating cultural meaning and symbolism, Ingold's orientation was fundamentally different, conceiving of landscape as a lived phenomenon in which people form themselves as active participants, rather than being figured as peripheral spectators. Colour and character are brought to this argument in a memorable description of the sixteenth-century painting by Pieter Bruegel the Elder, *The Harvesters*. Readers are invited to imagine themselves present amid the framed scene, and to become familiar with its local topography shaped through the materially inscriptive worlds of work, the round of seasonal tasks being undertaken (tending, tilling, picking, pruning), and the movements this entails (tellingly, a reproduction of *The Harvesters* is to be found in the entrance lobby of Ingold's family home, hung above the shoe-rack, behind the coat and hat stand). Gathered up and bundled together, he terms these rhythms and intimacies of encounter, the 'taskscape'. The influence of these concepts in geography was seen first in research on the place of trees where observations of arboriculture entangle the natural and the cultural to a point where it is difficult to discern any difference (Cloke and Jones, 1999; 2001; Jones and Cloke, 2002). Subsequent studies have carried the twin concepts of dwelling and taskscape elsewhere, producing narratives about landscapes made meaningful by the meshed lives of animals, humans and other agents in their living surrounds (Lorimer, 2006). Cast in a certain light, Ingold's feeling for dwelling has unexpected effects, giving cause to revisit earlier geographical traditions of studying folk cultures, and observing the customs and habits by which people leave marks or monumentalise their presence in landscapes.

Ingold's third and final contribution is, in part, anticipated in the closing remarks offered about the second: namely, his work has aided geographers' understanding of the lives of non-human animals in human geographies, and, conversely, the placing of the lives of humans in animals' geographies (Philo and Wilbert, 2000) or more-than-human geographies (Braun, 2008). On the one hand, ontological concerns are raised about anthropomorphism, and on the other, anxieties remain about the ultimately unknowable beastly qualities and capacities of being animal. His treatment of animality and humanity as relational expressions of personhood dissolves the idea of an ontological impasse and has inspired a range of research where interspecies relations produce properties of togetherness embedded in daily operations (i.e., Grasseni, 2007; Johnston, 2008).

Key Advances and Controversies

Ingold is not shy of engaging in critical academic debate and has proven himself a formidable foe in print (Ingold, 2005; 2007a; 2007b), though this fierce streak is always belied in person. His treatment of faulty structures of knowledge or lazy argument can be totalising, clearing the ground for fresh thinking or radical re-interpretation. By way of example, witness his analytical deconstruction of 'soundscape' (Ingold, 2008b). Equally so, he has submitted to errors in his own work, making plain the reasons for revisions in thinking: for example, criticism that the dwelling perspective might imply too great an emphasis on cultural stasis and earthy settlement (Hinchliffe, 2003) led to his proposal that processes of 'inhabiting' were a preferable alternative (Ingold, 2008a). Another feature of Ingold's thought worthy of comment is the relation he envisages between theory and method. Readers will search his corpus in vain for direct statements on methodology, or for that matter, the operations of ethnographic fieldwork. What then is Ingold's relation to method? In his version of events, a simple and transferable motif suffices, demanding little elaboration: all lessons are learned by participating.

For some, there will be aspects to Ingold's relational ecology that are problematic given its quiet espousal of a more rustic, idealistic, even neo-romantic existence. Certainly, his writings exhibit an obvious personal preference for textures and rhythms of place-making which are elemental, organic, borne of the open air and habits of clean-living and self-sufficiency (Ingold, 2008a). This makes for a thoroughly wholesome and democratic worldview, but one veering away from activist research on matters of environmental justice or indigenous rights. Ingold (2005c) has countered such critique by pointing to detailed explanations of the foundational wisdoms inherent in non-Western ontologies. In the light of accelerating environmental crisis and widespread ecological breakdown, more widespread appreciation of and deeper attention paid to these exemplary lifeworlds is an urgent project.

For the environmental humanities scholar, Ingold's work speaks compellingly of how place is perceived by persons, human or otherwise, and made real through different modes of living. If this take-home message has now been metabolised into geographical thought, then his 'philosophy with the people in' offers continued and valuable reminders of the benefits to be gained from dealing with the world as it is experienced through first-hand encounter and movement (Ingold, 2006a), and how an understanding of different sorts of environment is formed most thoroughly through the practicalities of living. The same enquiring spirit impels published essays (Ingold, 2005c; 2006c; 2007d; 2008a; 2021) where his attentions tilt upwards to the skies, turns 360 degrees to take in forests, coastlines and oceans, or the mycelial underworlds thriving beneath our feet. In this fashion Ingold is able to explore the visual and haptic perception of 'weather-worlds' and elemental phenomena such as light, sound and wind. It is through these kinds of planetary attunement – to atmospheric, terrestrial and aquatic environments – that he reckons the world can remain a live-able and sustainable place, as we devise adaptive designs for life in response to anthropogenic climate change.

Ingold's Key Works

Ingold, T. (1976) *The Skolt Lapps Today*. Cambridge: Cambridge University Press.

Ingold, T. (1988) (ed) *What Is An Animal?* London: Unwin Hyman.

Ingold, T. (1993) 'The temporality of the landscape', *World Archaeology*, 25: 152–74.

Ingold, T. (1995b) 'Building, dwelling, living: How animals and people make themselves at home in the world', in M. Strathern (ed), *Shifting Contexts*. London: Routledge. pp. 57–80.

Ingold, T. (2000) *The Perception of the Environment: Essays on Livelihood, Dwelling and Skill*. London: Routledge.

Ingold, T. (2011) *Being Alive: Essays on Movement, Knowledge and Description*. London: Routledge.

Ingold, T. (2007a) *Lines: A Brief History*. London: Routledge.

Ingold, T. (2018) *Anthropology and/as Education*. London: Routledge.

Ingold. T. (2021) *Correspondences*. Cambridge: Polity.

Ingold, T. (2022) *Imagining for Real: Essays on Creation, Attention and Correspondence*. London: Routledge.

Secondary Sources and References

Braun, B. (2008) 'Environmental issues: Inventive life', *Progress in Human Geography*, 32: 667–79.

Cloke, P. and Jones, O. (1999) 'From wasteland to woodland to "Little Switzerland": Environmental and recreational management in place, culture and time', in J. Tribe and X. Font (eds), *Tourism and Recreation: Case Studies in Environmental Management*. Oxford: CABI Publishing. pp. 161–82.

Cloke, P. and Jones, O. (2001) 'Dwelling, place, and landscape: An orchard in Somerset', *Environment and Planning A*, 33: 649–66.

Cosgrove, D. and Daniels, S. (1998) (eds) *The Iconography of Landscape*. Cambridge: Cambridge University Press.

Grasseni, C. (2007) *Skilled Visions: Between Apprenticeship and Standards*. Oxford: Berghahn.

Hinchliffe, S. (2003) 'Inhabiting: Landscapes and natures', in K. Anderson, M. Domosh, S. Pile, and N. Thrift (eds), *Handbook of Cultural Geography*. London: Sage. pp. 207–25.

Ingold, T. (1992) 'Editorial', *Man*, 27 (4): 693–6.

Ingold, T. (1995a) 'Work, time and industry', *Time and Society*, 4: 5–28.

Ingold, T. (2004) 'Culture on the ground: The world perceived through the feet', *Journal of Material Culture*, 9 (3): 315–40.

Ingold, T. (2005) 'Comments on Christopher Tilley: *Materiality of Stone: Explorations in Landscape Phenomenology* (Oxford: Berg)', *Norwegian Archaeological Review*, 38 (2): 122–9.

Ingold, T. (2006a) 'Up, across and along', *Place and Location: Studies in Environmental Aesthetics and Semiotics*, 5: 21–36.

Ingold, T. (2006b) 'Book review of: Massey, D. *For Space* (London: Sage)', *Journal of Historical Geography*, 32: 891–3.

Ingold, T. (2006c) 'Rethinking the animate, re-animating thought', *Ethnos*, 71: 9–20.

Ingold, T. (2007b) 'Materials against materiality', *Archaeological Dialogues*, 14: 1–16.

Ingold, T. (2007c) 'Writing texts, reading materials: A response to my critics', *Archaeological Dialogues*, 14: 31–8.

Ingold, T. (2007d) 'Earth, sky, wind and weather', *Journal of the Royal Anthropological Institute* 13, (s1): S19–S38.

Ingold, T. (2008a) 'Bindings against boundaries: Entanglements of life in an open world', *Environment and Planning A*, 40 (8): 1796–810.

Ingold, T. (2008b) 'Against soundscape', in A. Carlyle (ed), *Autumn Leaves: Sound and the Environment in Artistic Practice*. Paris: Double Entendre. pp. 10–13.

Ingold, T. and Lee Vergunst, J. (eds) (2008) *Ways of Walking: Ethnography and Practice on Foot*. Aldershot: Ashgate.

Johnston, E. (2008) 'Beyond the clearing: Towards a dwelt animal geography', *Progress in Human Geography*, 32: 633–49.

Jones, O. and Cloke, P. (2002) *Tree Cultures: The Place of Trees and Trees in their Place*. Oxford: Berg.

Lorimer, H. (2006) 'Herding memories of humans and animals', *Environment and Planning D: Society and Space*, 24: 497–518.

Lorimer, H. (2008) 'Cultural geography: Nonrepresentational conditions and concerns', *Progress in Human Geography*, 32: 551–9.

Philo, C. and Wilbert, C. (2000) *Animal Spaces, Beastly Places: New Geographies of Human–Animal Relations*. London: Routledge.

Thrift, N. (1996) *Spatial Formations*. London: Sage.

Thrift, N. (2007) *Non-representational Theory: Space, Politics, Affect*. London: Routledge.

Thrift, N. and Dewsbury, J.D. (2000) 'Dead geographies – and how to make them live', *Environment and Planning D: Society and Space*, 18: 411–32.

Wylie, J. (2007) *Landscape*. London: Routledge.

27
Cindi Katz

Tracey Skelton, University of Otago

Biographical Details and Theoretical Context

Cindi Katz was born and spent her early childhood in the Bronx, New York, US. Later, her family relocated to Larchmont in Westchester County which was where Katz went to school. In 1975, aged 17 she left home to study for a BA in Geography, followed by a MA (1979) and PhD (1986), all at Clark University, Worcester, Massachusetts. Katz took up a visiting lectureship at Khartoum University in Sudan for one year (1979), and throughout 1980 and 1981 she conducted critical ethnography fieldwork in rural Sudan and specifically in a village with the fictitious name 'Howa' (Katz, 2004). This field research formed the basis of her PhD and has been an extremely important foundation for Katz's contribution to spatial and social theorisation, much of which has focused on social reproduction and the production of space, place and nature, as well as the politics of knowledge with specific attention to children, childhood and environments, and the consequences of global restructuring for everyday life.

Returning to New York, Katz began her long career with The Graduate School and University Center, City University of New York in 1987. She was promoted to Associate Professor in 1994 and in 1999 was conferred with a Professorship. In 2011–2012 Katz was the Diane Middlebrook and Carl Djerassi Visiting Professor of Gender Studies at the University of Cambridge. Katz became a Co-Director at the Dartmouth Institute on the Futures of American Studies in 2016. Throughout her career Katz has received many honours, fellowships and awards. In 2015 she was presented the James Blaut Memorial Award for Scholarship, Teaching, and Activism towards Social Justice, by the Socialist and Critical Geography Speciality Group of the American Association of Geographers (AAG). In 2021, Katz received the Distinguished Scholarship Honors from the AAG, with the citation noting Katz's immense impact on contemporary human geography:

> Her theoretical and empirical contributions to geography and beyond are of the highest quality. Her enthusiasm for her field and relentless efforts to identify and transform exploitative power relations have had a profoundly positive influence on geographic theory, practice, and education. Katz's scholarship is among the most significant in the discipline. (AAG, 2020)

Katz's scholarship has a remarkable global reach. Her books, articles and book chapters are highly cited, and many articles have been reprinted or translated into non-English languages. Her most cited publication – *Growing Up Global: Economic Restructuring and Children's Everyday Lives* (2004) – won the Meridian Book Award for Outstanding Scholarly Work in Geography from the AAG.

In 2022, Katz's contributions were reassessed in the *Interventions* section of the journal *Political Geography* by eight feminist political geographers: Sarah Hughes, Amber Murrey, Sneha Krishnan, Karolien van Teijlingen, Patricia Daley, Mel Nowicki, Maria Fanin, and Julie MacLeavy. They revisited *Growing Up Global* (2004) in order to consider Katz's concepts of resilience, resistance, and reworking, in the context of the relentless revanchism that had occurred since the book's publication (Hughes et al., 2022). The authors acknowledged the substantial conceptual and critical analytical writing that Katz has contributed spatially and globally, and recognised her commitment to feminist futurity in political geography. This is a very fitting acknowledgement of Katz's scholarly interpretations of her now 'famous' all-important 3Rs – resilience, resistance and reworking – which play out at different scales and among different actors. Katz (2022) offered her own reflections on these concepts, and on minor theory, in her response: 'Grateful reckonings with the political geographies of resistance: A response'.

Spatial Contributions

Katz has made a range of theoretical and empirical interventions in geographical debates. She has had an important influence in scholarship around ethnography, fieldwork and the production of knowledge (for example, Katz, 2013). Her work has also made important contributions to feminist geography, including around questions about the lifecourse (e.g., Katz and Monk, 1993), influencing work on the geographies of age and intergenerationality, in particular. However, what Katz may be best known for is her longitudinal engagement with the conceptualisation of social reproduction: its practices, spaces, places and materialities. Much of this work has centered around children and childhood and hence Katz is an exceptionally important influence in the field of children's and young people's geographies.

Katz's work in 'Howa', Sudan, and later in Harlem, New York, sparked a long-term intellectual and political commitment to making the processes and spatialities of social reproduction visible and recognised in critical geographical contributions to political economic theory, globalisation and development (Katz, 2004; 2012a). Katz's work on social reproduction worked to bring it, and in particular the social reproductive work of children, into the mainstream of geopolitical and social theoretical debate where it rightfully belongs. As Katz has famously pointed out, social reproduction is the messy, 'fleshy, open-ended stuff of everyday life' (2001: 711), and yet it is hidden, frequently taken for granted or ignored in theory, empirical work and by actors in the everyday social relations that would not survive without this 'life's work'. However, Katz does not only call for social reproduction to be taken seriously because of its profound impact on everyone's lives and its integration with every facet of globalised capitalist production, but also because it holds the possibility of profound structural, material and relational change (Katz, 1991).

Katz determinedly positions children as important social, political, spatial actors, and central to understanding the practices of neoliberal capitalism in crisis. Her work insists such practices are formed, and touch down, in spaces and places where there are always children. Indeed, their impact on children is profound whether through: war, state violence and displacement; resource extraction; the diversion of funds away from social welfare; the retreat of state support for health, education and housing; collapses in food security; or environmental pollution and destruction. However, so many of the consequences of neoliberal capitalist production are hidden, especially in urban space. Geographical analysis of such dispossessive processes have often served to hide the material social practices because of an excessive focus on finance, the economy and production (2011; 2021). Katz calls upon radical geographers, and particularly feminist geographers, to examine what it means for space to hide such consequences and to recognise the complexities of scale through which neoliberalised capitalist production acts. Examination of local, neighbourhood and domestic spaces is, Katz argues, central to opposing, and reworking, these acts. Such analytical and political interventions are crucial in rendering the importance of social reproduction visible.

The spatial disconnect between capitalist production and social reproduction, and the deliberate withdrawal and reduction of a 'social wage' impoverishes people's lives and intensifies insecurity and vulnerability for the vast majority. Katz argues that akin to, and integrated within globalisation and neoliberalism, social reproduction has political-economic, cultural and environmental aspects. Consequently, analysis of neoliberalism and capitalist production cannot be effective unless social reproduction is considered, and vice versa (see Katz's work on New Orleans and Hurricane Katrina in Skelton, 2011: 266–7). Through Katz's detailed attention to place, global capital, and social reproductive practices and materialities in sites including rural Sudan, and Harlem, New York, she developed spatial metaphors – *topographies* and *counter-topographies* – that offered a theoretical, organisational and methodological foundation for developing resistive political responses.

Here, topographies are conceptualised by Katz as detailed descriptions of place and location produced and utilised by those in power to maintain and advance uneven development. Consequently, they can be used as tools by the powerless. Topographies allow an examination of processes at play in certain places and specifically the effects of their encounters with social relations of reproduction and production; they illustrate the landscapes of social life. The common effects of capitalism can then be examined as they are experienced in different places in different ways. The connections can be made through the use of counter-topographies: a geographical and political response to contemporary processes of neoliberal capitalism in crisis, excess accumulation by dispossession, and decades of privatisation, commodification and financialisation create negative contexts and effects for children, childhood and youth.

Katz's project *Childhood as Spectacle* connects with counter-topographies of risk and waste, accumulation, aspiration management, childhood, and excess (2008; 2011; 2018). Katz consistently works to render visible the:

> [...] symptomatic silences around social reproduction [that] remain all too common in analyses of capitalism. Working through these issues and their occlusion, I offer what I hope is a useful and vibrant theoretical framework for examining geographies of children, youth and families. (Katz, 2011: 47)

Katz's project *Childhood as Spectacle* is multifaceted and centres her concern around social reproduction. She envisages childhood as spectacle in a way that faces a wide range of negative impacts on children, young people and childhood-youth. These include neoliberal capitalism, crises of accumulation, insecurities around social reproduction and futures, children as waste, and the school-to-prison pipeline (Katz, 2019a). Katz utilises counter-topographies to work towards associative politics and a more positive geographical imagination. Katz emphasises that it is essential to pay much more attention to children and young people and to recognise the deep and visceral inequalities they face and the consequences of precarity, neglect, exclusion and poverty.

During her visiting professorship as the Diane Middlebrook and Carl Djerassi Visiting Professor of Gender Studies at the University of Cambridge, Katz wrote a piece for the *CAM*, the University of Cambridge's alumni magazine, titled 'Lost Youth' (Katz, 2012b). On two cover pages Katz is photographed surrounded by, or overlooking, a huge pile of toys. Upon reading Amy Chua's (2011) book, *Battle Hymn of the Tiger Mother*, Katz challenges the author's form of brutal control and parenting:

> Looking at childhood [...] is a way to gauge some of the broader anxieties and futurity [...] As the social wage shrinks and three decades of neoliberal globalism reconfigure people's expectations of government and business, militarised violence takes on new forms and places, environmental change goes global, and whole lifeworlds seem to be disappearing, long-held assumptions about coming of age alter and ramp up ontological insecurity. (Katz, 2012b: 21)

Katz's engagements with Chua's book allowed her to muse upon clashes around social reproduction, a central theme in her article, 'The angel of geography' (2018). As in all of her writing, Katz lays bare the corrupt, appalling, deficient practices that provide so little for adults and children. These include reductions of the social wage, militarisation and violence, war, and lifeworlds (both human and non-human) that are intensely at risk. However, with Katz, there is always a sense of hope; a future that might be good, inclusive, and caring. Katz calls for reimagining ways that time and space can be positive for childhood: to make it possible that social reproduction could be enriching and supported; to provide and encourage children and young people to take time to play, do nothing, hang out, and create playfulness (Katz, 2019b). As Katz argues, these affordances could dramatically change children's and young people's lives for the better.

A salient element of Katz's writing is that she works within iterative spirals that move forward and return to themes that twist and rewind to look back, but always with an impetus to go forward. Katz has a strong determination to keep the politics of social reproduction visible, whether successful, failed, and possibly, importantly, 'in-between'. For example, Katz (2001: 709) argues that 'focusing on social reproduction allows us to address questions of the making, maintenance, and exploitation of a fluidly differentiated labour force, the productions (and destructions) of nature, and the means to create alternative geographies of opposition to globalised capitalism'.

Key Advances and Controversies

Katz established the centrality of social reproduction as a theoretical, conceptual, analytical and methodological construction through which to explore, examine and understand globalised, neoliberal capitalist production and its effect on everyday life, and she has gone on to tackle a range of arguments and debates animating urban studies. For example, in her article on 'Splanetary Urbanization' (2021) she states of the prevailing discourse on planetary urbanisation that 'the whole thing seems to be a masculinist theoretical move to define the field' and that comes with a lot of mansplaining (Katz, 2021: 598). Through her focus on agency, difference and everyday life she counterposes planetary urbanisation because it ignores agency, subjectivity, and all forms of difference, arguing that this is not a productive way forward and that it is essential to reinstate difference, everyday life and social reproduction (see also **Jennifer Robinson** on 'ordinary cities'). Urbanisation impacts social actors in very different ways. For Katz there should be conversations about and for the recognition and inclusion of material social practices of social reproduction in everyday life and positive possibilities of careful, respectful and achievable theory-making.

Katz has then questioned the ways in which geographers and geography construct theory and knowledge, notably through her elaboration of the concepts of *minor* and *major theory* (Katz, 1995; 1996). Katz here recognises the inseparable significance of the personal and the political by pursuing a minor feminist theory that explores socio-spatial spaces of betweenness, possibility and living things differently. For Katz, 'minor theory' is a project and practice which creates space for questions, categories and analyses; which enables understanding and engagement around inequitable relations of production and reproduction. She argues that minor theory is about displacement, creating the spaces that new subjectivities, spatialities, and temporalities afford as alternative ways of thinking, learning and taking time over. This kind of minor approach highlights the limitation that major theory faces. Minor theory has a certain finesse and openness which illuminates the ways in which gendered racial capitalism can be engaged with, encountered, known and felt – minor theory works more flexibly, inclusively and intimately. Katz (2015; 2017) has revisited minor theory in keynote lectures and papers, underlining the significance of efforts to re-see and keep open new spaces of politics of knowledge, and importantly, creative action and intention.

Katz's notion of minor theory and its interstitial engagement with major theory aims to reconfigure the production of geographical knowledge in order to develop a political academic project, hence '*Childhood as spectacle*' (Katz, 2008). Katz's political commitment allows the movement between abstract grand theory and global political processes through to the small-scale, interstitial, local everyday life experiences examined through ethnography. Katz's work is an ongoing project of challenge, critique and creativity used to tackle global political processes and the lived realities of social reproduction *in combination,* and boldly carries the word 'children'. This is testament to a theorist's political commitment to render visible, central and placed those who feel most keenly the negative consequences of neoliberal, global, capitalist vagabondage but who *force* us to have *hope for the future*.

Katz's Key Works

Katz, C. (1995) 'Major/minor theory: Theory, nature and politics', *Annals of the Association of American Geographers*, 85 (1): 164–8.

Katz, C. (1996) 'Towards minor theory', *Environment and Planning D: Society and Space*, 14: 487–99.

Katz, C. (2001) 'Vagabond capitalism and the necessity of social reproduction', *Antipode*, 33 (4): 709–27.

Katz, C. (2004) *Growing up Global: Economic Restructuring and Children's Everyday Lives*. Minneapolis, MN: University of Minnesota Press.

Katz, C. (2008) 'Childhood as spectacle: Relays of anxiety and the reconfiguration of the child', *Cultural Geographies*, 15 (1):15–29.

Katz, C. (2011) 'Accumulation, excess, childhood: Towards a countertopography of risk and waste', *Documents d'Anàlisi Geogràfica*, 57 (1): 47–60.

Katz, C. (2012a) 'Work and play: Economic restructuring and children's everyday learning in rural Sudan', in M. Bourdillon and G. Spittler (eds), *African Children at Work*. Berlin: LIT. pp. 227–48.

Katz, C. (2012b) 'Lost youth', *CAM*, University of Cambridge, 65 (Lent): 18–21.

Katz, C. (2013) 'Playing with fieldwork', *Social & Cultural Geography*, 14 (7): 67–72.

Katz, C. (2017) 'Revisiting minor theory', *Environment and Planning D: Society and Space*, 35 (4): 396–9.

Katz, C. (2018) 'The angel of geography: Superman, Tiger Mother, aspiration management, and the child as waste', *Progress in Human Geography*, 42 (5): 723–40.

Katz, C. (2019a) 'Accumulation, dispossession, and waste in childhood and children's everyday lives', in T. Skelton and S. Aitken (eds), *Establishing Geographies of Children and Young People. Geographies of Children and Young People* (Vol. 1), Singapore: Springer. pp. 129–46.

Katz, C. (2019b) 'Children and childhood', in Antipode Editorial Collective (eds), *Keywords in Radical Geography: Antipode at 50*. Oxford: John Wiley and Sons. pp. 40–4.

Secondary Sources and References

AAG (2020) '2021 Honors announced', *AAG*, December 14. https://www.aag.org/2021-aag-honors-announced/ (accessed November 6, 2023).

Chua, A. (2011) *Battle Hymn of the Tiger Mother*. New York, NY: Penguin.

Hughes, S.M., Murrey, A., Krishnan, S., van Teijlingen, K., Daley, P.O., Nowicki, M., Fannin, M. and MacLeavy, J. (2022) 'Interventions in the political geographies of resistance: The contributions of Cindi Katz, 15 years on', *Political Geography*, doi.org/10.1016/j.polgeo.2022.102666

Katz, C. (1991) 'Sow what you know: The struggle for social reproduction in rural Sudan', *Annals of the Association of American Geographers*, 81 (3): 488–514.

Katz, C. (2015) Keynote Lecture delivered at the Critical Geography Conference, University of Kentucky. https://vimeo.com/144656746

Katz, C. (2021) 'Splanetary urbanization', *International Journal of Urban and Regional Research*, 45 (4): 597–611.

Katz, C. (2022) 'Grateful reckonings with the political geographies of resistance: A response', *Political Geography*, 98. doi.org/10.1016/j.polgeo.2022.102713

Katz, C. and Monk, J. (eds) (1993) *Full Circles: Geographies of Women over the Life Course*. London: Routledge.

28
Audrey Kobayashi

Sarah de Leeuw, University of Northern British Columbia

Biographical Details and Theoretical Context

Audrey Kobayashi's geographic work exemplifies a lifelong commitment to intersectional social justice work informed by spatial theory and practice. Through activism, teaching, policy input, academic writing, public engagement, and community-oriented scholarship, Kobayashi has spent the better part of 45 years advancing discussions in Geography about racialisation, gender, class, dis/ability, citizenship, identity, and radical geographic history across multiple scales ranging from individual bodies and family homes to global cities and beyond.

Born in 1951 on the traditional unceded lands of the Sylix/Okanagan Peoples in the rural agricultural community of Okanagan Centre, British Columbia, to a Canadian-born mother and a Japanese-Canadian father (whose own father was a Haiku Master) who farmed in the region, Kobayashi had by 1978 completed both her Bachelor of Arts and Masters in Geography at the University of British Columbia in Vancouver, on the unceded traditional territories of the xʷməθkʷəy̓əm (Musqueam), Sḵwx̱wú7mesh (Squamish), and səlilwətaɬ (Tsleil-Waututh) Nations. After a two-year research fellowship between 1980 and 1982 at Kyoto University in Japan, during which time she published her first academic work (a short paper in the journal *Landscape* about poetic geographies and the haiku) (Kobayashi, 1980), Kobayashi completed her PhD in 1983 at the University of California Los Angeles (UCLA), on the traditional territory of the Chumash, Tongva, and Kizh peoples and nations, under the mentorship of ***Allan Pred.***

Kobayashi spent a year in 1982 as a sessional geography instructor at Carleton University in Ottawa (the traditional lands and territories of the Anishinabe Algonquin) before beginning her post-PhD academic career in 1983 in Geography at McGill University in Montreal (known as Tiohtià:ke by the Kanien'kéha Nation, the geography's original inhabitants). While geography's 'qualitative revolution' had, by the 1980s, ensured subjective, humanistic, behavioural, and creative approaches were widespread in the discipline, Kobayashi's early career years still unfolded amidst a highly white-male dominated discipline. Indeed, it would be more than a decade before the journal *Gender, Place & Culture* began, with Kobayashi authoring the journal's opening paper. Unique at that time as a Japanese-Canadian woman in geography, Kobayashi's unwavering commitment to equity issues unfolded in close dialogue with the methodological and institutional transformations driven by critical social theory and feminist activism unfolding in late twentieth century geography (see Bondi, 1990; Rose, 1993).

In 1994, ten years after starting with McGill University, after a decade of publishing almost exclusively in geographic-focused venues with a sustained focus on race, ethnicity, gender, place, and law, and having co-edited the text *Remaking Human Geography* (Kobayashi and Mackenzie, 2014 [1989]) Kobayashi joined Queen's University (on the traditional territories of the Anishinaabe and Haudenosaunee) as Director of the Department of Gender Studies, with a cross appointment in the Department of Geography. By 1999, Kobayashi was a Full Professor in the Department of Geography at Queen's University where, after having been appointed a Queen's Research Chair, she retired in 2022 as Professor Emeritus.

Over her 40 years as an academic geographer, Kobayashi authored or edited 17 books, including a second edition of the influential *Remaking Human Geography*, as well as more than 100 book chapters and academic journal articles. In 2001, Kobayashi became the first woman of colour to be elected President of the Canadian Association of Geographers (CAG) and in 2011, became the first Canadian woman of colour to become President of the American Association of Geographers (AAG). In both roles, she extended an ethos of relationship-building that characterised her research and pedagogy, actively engaging AAG members and representatives across regions less represented in the discipline: her unabashed goal was to extend the discipline's consideration of race and gender, both beyond high impact factor geographic research and large urban metropoles. Kobayashi also deployed a social-justice intersectional approach to editorial work, including as Associate Editor Cultural Geography for the *Encyclopedia of Contemporary Japanese Culture* (2002); the People, Place and Region editor of *Annals of the Association of American Geographers* (2002–10); and, in 1988, when Kobayashi was guest editor of the first special edition of its kind in *The Canadian Geographer*: 'Focus on Asian Migration to Canada'. She has been the editor and editor-in-chief of two major encyclopaedia projects, *The International Encyclopedia of Geography* (2017, 15 volumes) and the *International Encyclopedia of Human Geography* (2nd edition, 2020, 12 volumes). In 1995, the National Association of Japanese Canadians awarded Kobayashi their Biannual Award for Outstanding Contribution to the Association. In recognition of her pursuits in expanding anti-racism, concepts of spatiality, and social justice activism, Kobayashi was in 2011 named a fellow of the Royal Society of Canada and, in 2020, was awarded the American Association of American Geographer's Harold M. Rose Award for Anti-Racism Research and Practice.

In addition to developing geographic thinking around race and racism, Kobayashi's contributions to human geography included applied service, teaching, and grounded practice. In the late 1980s in Canada, and within the confines of a conservative and right-leaning federal government, Kobayashi worked toward redress for Japanese Canadians displaced and interned during the 1940s because of violent and racist national and provincial policies. She worked for over a decade with Japanese Canadian communities in Vancouver's Downtown East Side (DTES), including on archival and activist projects supporting the rights of low-income, under-housed tenants in single occupancy hotel rooms (Masuda et al., 2020). She similarly put her ideas into practice through her institutional positions within the discipline of Geography, as an editor of journals and encyclopaedias, a leader of disciplinary organisations, and as an award-winning teacher and supervisor (having overseen more than 45 graduate students).

Spatial Contributions

Kobayashi's scholarly contributions to geography (and beyond) are best summarised as broadening ways of thinking about how racism, sexism, gender, race, and anti-racism form, and are formed by, space and place. Her work has been particularly influential in differentiating space from spatiality, especially in reference to race, racism, and racialization (Kobayashi, 2004). While geographers have been critically scrutinising the social production of space and spatiality since at least the late 1970s, by the early 2000s many human geographers were actively working with **Michel Foucault's** insistence that space is an *active force* as opposed to a neutral or passive container of socio-cultural occurrences (see, for instance, Philo, 1992; Crampton and Elden, 2007). Kobayashi's scholarship instead turned to **Edward Said** and Jean Paul Sartre (including and especially Sartre's relationship to Frantz Fanon) to advance arguments about how racism is created by sociocultural hierarchies anchored in, and produced by, relationships defined through distance and proximity (Kobayashi, 2004; see also Kobayashi, 2014). Here, Kobayashi pushed the concept of spatiality beyond a conception of space and place as neutral containers and into theorising recursive *relationships* between people, space and place. Or, as Kobayashi (2016: 6) writes, 'Spatiality is the condition, the form, the becoming of human, concretized through and as human relationship. In other words, the shift is from relative space to relational spatiality'. For Kobayashi, spatiality is an active and always politicised conceptualisation that allows for a re/thinking of geographic imaginings of place and space themselves. Space itself, for Kobayashi, appeared a somewhat useless field of analysis:

> Geographers have bashed away at the concept of space [...] to the point where, as itself, it no longer carries much analytical utility, not simply because its meanings have become so diverse as to be almost meaningless [...] but because the discipline is more and more aware that it is precisely the spatial–social, the becoming of humanity, rather than the fictional quality that encloses or regulates humanity, that [should be the] major concern [of geographers]. (Kobayashi, 2016: 5–6)

While it is now almost axiomatic for critical social scientists, including geographers, to argue that human sociocultural organising and structuring occur spatially and in place, Kobayashi's work was at the vanguard of applying that logic to intersections of race and gender – and bringing spatiality to the fore of those theorisations and concepts. Kobayashi subsequently applied these same logics and theorisations to discussions about dis/ability (see, for instance, Kobayashi, 2003b).

While Kobayashi's research geographies of gender, race, racialisation, sexism, and racism have always been couched in geographic discourse, her work was also influential in gender and critical anti-racism studies (e.g., Essed et al., 2009). Moreover, her research consistently contested the dichotomy between episteme and ontology: in other words, through grounded examples located within specific places and spaces, she refuted that, either empirically or theoretically, material, embodied, or lived realities could be somehow decoupled from the discursive systems and structures within which those realities were embedded. Race, as Kobayashi (2014) shows, is always dialectic, in relationship with whiteness and the environments in which race is formed, a social process with clearly-embodied outcomes located in geography and history. By extension race is thus always inextricable from racialisation, sex and gender, and cannot be parsed apart from patriarchy or sexism and, in turn, that each and all of these are formed, and in turn (re)produce, the geographies in which each exists. For Kobayashi (2014: 1103) purely discussing the socially constructed nature of anything – race and gender included – is 'not a relevant task because social constructionism denies an ontological split between the essential and the constructed – much less the real and the ideational [and] such discussion masks the important philosophical and normative question of how social construction and social change occurs and what we can and should do about the results' (Kobayashi, 2014: 1103).

Kobayashi (2013: 62) noted that for much of geography's history, discussion about race occurred, methodologically, at a 'distance from the "sites of struggle" in which racialised discourse occurs'. Kobayashi herself was *living* the intersections of race and gender about which she was undertaking geographic scholarship. As a woman of colour in Geography, Kobayashi embodied a site of struggle in the discipline itself. In line with the momentum of second-wave feminism unfolding from the 1960s onward, Kobayashi, along with early (mostly white) feminist geographers, including Suzanne Mackenzie and Linda Peake, were active in inserting key concepts to advance feminist-informed antiracist methods and methodologies in human geography into geographic literatures.

Often using evidence from archives, primary texts, interviews, ethnographies, and in-situ observations, Kobayashi is, at heart, a social historical geographer. Drawing on methods of textual and discursive analysis to unearth overlooked or actively erased peoples and places, Kobayashi's scholarship consistently relied on *historicising* – whether by tracing biographies in the discipline, mapping the philosophical terrains upon which key geographic concepts rest, or contextualising power across varied geographies. Kobayashi's (2013: 61) foundational message was that 'the discipline of geography is dominated by Northern, Western, white scholars whose lives and careers have been constructed out of the very colonial systems that produced them'. Kobayashi's epistemological and ideological work pushed the discipline to critically assess and address geography's lack of Indigenous geographers and geographers of colour; the lack of writings and scholarship by Black, Indigenous, and geographers of colour in histories and teachings of geography; and the pressing need to reimagine geography as pivotal in futures of critical studies of race and racism.

Key Advances and Controversies

Kobayashi's scholarship played an early and enduring role in Geography's turn toward a critical feminist anti-racist ethos, ensuring disciplinary silos or hierarchies of marginality and otherness were no longer foundational. Twenty-first century geographies of gender do not, at least with the same level of normalcy that prevailed in the late years of the twentieth century, exclude considerations of race, sexuality, class, or colonialism (e.g., Valentine, 2007; de Leeuw et al., 2011). The white feminist geography of the mid-to-late 1990s has rightfully been disrupted by, for instance, Black, queer, Indigenous, non-Western, feminists of colour, including those from outside academia, the Global South and other places on the margins. Indeed, in one encyclopaedia entry on 'Feminist Geographies' (Ruwanpura and Gay-Antaki, 2023), the focus is consciously and adamantly upon non-academic voices, scholarship from the Global South, writings not in English or French, and the perspectives of women and women-identifying feminist geographers who will shape the discipline's future. It is notable, however, that despite the authors calling for a 'need to respect difference and diversity, moving away from homogenising second-wave, middle-class, white Western feminisms' (Ruwanpura and Gay-Antaki, 2023: 154) – a call Kobayashi has been making since the 1980s – none of Kobayashi's work is cited. Perhaps this signals just how successful Kobayashi's work has been: that feminist geographers of the early 2020s are able to write expressly with an 'aim of highlighting the work of black feminist geographers and feminists of colour' (Ruwanpura and Gay-Antaki, 2023: 154) without citing Kobayashi means, to an extent, that Kobayashi's work (along with other anti-racist feminist geographers of the late 1900s) may simply have been fully absorbed into geographic thought. It might also signal, however, that geographic knowledge and thinking remain somewhat geographically contained – as a woman geographer of colour working in Canada – or as geographer Cole Harris (2004) once put it, the 'edge of Empire' – it is possible Kobayashi's work does not circulate with the intensity of work by geographers in the heart of Empire (Britain, for instance) or in the larger metropoles of the US.

To be fair, pivotal works by Kobayashi (e.g., Kobayashi and Peake, 1994; 2000) continue to circulate in some of the most recent conversations about whiteness, white supremacy, and racism in geography: in a recent editorial (Hamilton, 2020) Kobayashi's early work on racism in the discipline and the spatiality of race and racism are referenced as early imaginings of a revolution in Geography. Other interrogations of whiteness and white privilege in geographic thought (e.g., Bonds and Inwood, 2016: 717) acknowledge Kobayashi's role in establishing the 'spatial contingency of the social construction of whiteness and privilege' and, by extension, conceptualise landscape not as reflective of racial patterns, but as active processes in racialisation itself.

In 2003, Kobayashi published 'GPC Ten Years On: Is Self-Reflectivity Enough?' in the journal *Gender, Place & Culture*. With respect to gender, she argued, geographers had much to celebrate in the opening years of the twenty-first century, including that 'gender as a fundamental aspect of social life has become accepted wisdom' (Kobayashi, 2003: 345). While clearly advancing the position that Geography had much room to grow in areas of critical anti-racist scholarship, Kobayashi also acknowledged, and celebrated, that a growing body of geographic work recognised intersections and simultaneities of sexism and racism, as well as classism, ableism, and homophobia. Kobayashi (2003b: 346) acknowledged that many of the advances made by (especially feminist) geographers could be traced to the 'polar shift away from the attempts by scholars a few decades ago to remain detached and neutral from their subjects'. Still, for Kobayashi, who had done so much to advance lived and embodied orientations to social justice and questions of power in geography, there existed a pressing need to 'address the demon of the personal in contemporary feminist/anti-racist geography' (Kobayashi, 2003b: 347). Kobayashi's paper 'Is Self-Reflectivity Enough' might be read as one of her boldest methodological advancements:

while for decades she had embraced the qualitative revolution in geography, by 2003 Kobayashi was concerned that geography had become too personal, especially in terms of self-reflexivity. This logic put (and continues to put) Kobayashi quite squarely at odds with much feminist geographic scholarship that suggests that positionality is a laudable means of navigating, or even eschewing, sociocultural and political power. As has always been the case for Kobayashi, a pressing question about any geographic method or methodology is its potential to improve the world. For Kobayashi (2003b: 345), positionality and self-reflexivity are 'perhaps not the best tool[s] at our disposal for taking us further towards social change' because both are rooted almost exclusively and purely in personal geographies, limiting their potential for broader transformative social and public change. Consistent with decades of unwavering commitment to work across scales in efforts to expand geographies of social justice, Kobayashi's critique of the focus on self-reflexivity and positionality – especially in the work of new generations of feminist geographers – remains true to the broadest narrative of all her geographic scholarship, activism, practice, and teaching: asking difficult questions of geography is precisely how the discipline might transform and, in turn, play a pivotal role in transforming the world.

Kobayashi's Key Works

Kobayashi, A. (2003a) 'The construction of geographical knowledge: Racialization, spatialization', in K. Anderson, M. Domosh, S. Pile and N. Thrift (eds), *Handbook of Cultural Geography*. London: Sage. pp. 544–56.

Kobayashi, A. (2005) 'Anti-racist feminism in geography: An agenda for social action', in L. Nelson and J. Seager (eds), *A Companion to Feminist Geography*. Oxford: Blackwell. pp. 32–40.

Kobayashi, A. (2006) 'Why women of colour in geography?', *Gender, Place & Culture*, 13 (1): 33–8.

Kobayashi, A. (2014) 'The dialectic of race and the discipline of geography', *Annals of the Association of American Geographers*, 104 (6): 1101–15.

Kobayashi, A. (2016) 'Spatiality', in D. Richardson, N. Castree, M.F. Goodchild, A. Kobayashi, W. Liu and R.A. Marston (eds), *International Encyclopedia of Geography: People, the Earth, Environment and Technology*. Chichester: Wiley. pp. 1–7.

Kobayashi, A. and Mackenzie, S. (eds) (2014 [1989]) *Remaking Human Geography* (2nd edition). Abingdon: Routledge.

Kobayashi, A. and Peake, L. (1994) 'Unnatural discourse. "Race" and gender in geography', *Gender, Place & Culture*, 1 (2): 225–43.

Kobayashi, A. and Peake, L. (2000) 'Racism out of place: Thoughts on whiteness and an antiracist Geography in the new millennium', *Annals of the American Association of Geographers*, 90 (2): 392–403.

Secondary Sources and References

Bondi, L. (1990) 'Progress in geography and gender: Feminism and difference', *Progress in Human Geography*, 14 (3): 438–45.

Bonds, A. and Inwood, J. (2016) 'Beyond white privilege: Geographies of white supremacy and settler colonialism', *Progress in Human Geography*, 40 (6): 715–33.

Crampton, J.W. and Elden, S. (eds) (2007) *Space, Knowledge and Power: Foucault and Geography*. Farnham: Ashgate.

de Leeuw, S., Kobayashi, A. and Cameron, E. (2011) 'Difference', in V. Del Casino, M. Thomas, P. Cloke and R. Panelli (eds),

A Companion to Social Geography. New York, NY: Blackwell. pp. 17–37.

Essed, P., Goldberg, D.T. and Kobayashi, A. (eds) (2009) *A Companion to Gender Studies* (Vol. 26). Chichester: John Wiley & Sons.

Hamilton, A.R. (2020) 'The white unseen: On white supremacy and dangerous entanglements in geography', *Dialogues in Human Geography*, 10 (3): 299–303.

Harris, C. (2004) 'How did colonialism dispossess? Comments from an edge of empire', *Annals of the Association of American Geographers*, 94 (1): 165–82.

Kobayashi, A. (1980) 'Landscape and the poetic act: The role of haiku clubs for the Issei', *Landscape*, 24 (l): 42–7.

Kobayashi, A. (2003b) 'GPC ten years on: Is self-reflexivity enough?', *Gender, Place & Culture*, 10 (4): 345–49.

Kobayashi, A. (2004) 'Geography, spatiality, and racialization: The contribution of Edward Said', *The Arab World Geographer*, 7 (1–2): 79–90.

Kobayashi, A. (2013) 'Critical "race" approaches', in N.C. Johnson, R. Schein and J. Winders (eds), *The Wiley-Blackwell Companion to Cultural Geography*. Oxford: Wiley-Blackwell. pp. 57–72.

Masuda, J.R., Franks, A., Kobayashi, A. and Wideman, T. (2020) 'After dispossession: An urban rights praxis of remaining in Vancouver's Downtown Eastside', *Environment and Planning D: Society and Space*, 38 (2): 229–47.

Philo, C. (1992) 'Foucault's geography', *Environment and Planning D: Society and Space*, 10 (2): 137–61.

Rose, G. (1993) *Feminism and Geography: The Limits of Geographical Knowledge*. Minneapolis, MN: University of Minnesota Press.

Ruwanpura, K.N. and Gay-Antaki, M. (2023) 'Feminist geographies', in F. Rowe, D. Demeritt and L. Lees (eds), *Concise Encyclopedia of Human Geography*. Chichester: Edward Elgar. pp. 150–55.

Valentine, G. (2007) 'Theorizing and researching intersectionality: A challenge for feminist geography', *The Professional Geographer*, 59 (1): 10–21.

29
Bruno Latour

Eric Laurier, University of Edinburgh

Biographical Details and Theoretical Context

Bruno Latour (1947–2022) was the youngest of eight children in a family of wine producers from Burgundy (*not* Bordeaux, home of 'Chateau Latour'). From the outset he took the less travelled path in French intellectual life. He was educated in the provinces of Dijon rather than Paris, and after starting training in the philosophy of religion, in a typical Latourian reversal, acquired his belief in social science and switched to anthropology. His initial anthropological fieldwork was in the Ivory Coast, which he visited while on French military service, followed up by what has become recognised as an iconoclastic study of a laboratory in California (Latour, 1987).

For the larger part of his career he lived anything but the life of the lone French intellectual, working instead in a collective laboratory at the *Ecole Nationale Supérieure des Mines* (1982–2006) and then creating *Médialab* at *Sciences Po* where he worked until his retirement in 2017. Throughout, he collaborated widely with researchers from social sciences, humanities, technological and natural sciences. Off campus, he worked with policymakers, organisational managers, theatre producers and artists. Drawing on his connections with arts communities, he curated three international social science and art exhibitions: 'Iconoclash', 'Makings Things Public' and 'Reset Modernity'. He is most widely recognised for the leading role he played, alongside Michel Callon and John Law, in the initiation and remarkable spread of actor-network theory (ANT). Though centred in Science and Technology Studies, he and his work travelled widely, passing through Sociology, Art History, Law, Ecology, Literature, Geography and Primatology amongst other disciplines. He received numerous honours, including the prestigious Holberg Prize in 2013, and the French Légion de honneur in recognition of his overall contributions to the social sciences.

Latour was a *constructivist*, not a *social* constructivist. This unexpected disavowal of the social is rooted in a reaction to the 'strong programme' in the sociology of scientific knowledge which sought to *symmetrically* explain successes *and* failures in scientific progress in terms of social factors (Bloor, 1976). The strong programme accordingly suggested that when certain scientific fields (Phrenology versus Neuroscience) or certain phenomena (such as X-rays versus N-rays) came to be taken as fact, or conversely, were discredited, this was not a purely scientific victory but was a victory explained by social and cultural forces. Conventionally, those forces were acknowledged as external biases that might delay or disrupt scientific progress. Truth though was internal to science, and thus guaranteed to come out and vanquish falsity. In contrast, the so-called symmetrical programme conceived of society as at one with the internal forces that give science the shape that it has, rather than a force that bent true science or technology out of shape from the outside. Latour took this radical programme a step further. Rather than allowing the social sciences a privileged vantage point he used scientific activity to *symmetrically* explain failures and successes in *society*. Latour showed that science is frequently a far better analyser of society than social science; in effect, he showed scientists to be practical sociologists.

In his late work, Latour became increasingly interested in re-working the Gaia hypothesis of the Earth as a system that produces and maintains itself. This interest came hand in hand with his urgency in addressing the earthly problems of climate change, habitat destruction and mass extinction. This later work built on both his ANT-based forays into imagining what a 'parliament of things' might look like as well as his latter inquiries into modes of existence (*An Inquiry Into Modes of Existence*, 2013), though his target was more political and public. In his early work, he had called attention to the 'missing masses' of things in order to understand society and it was those missing masses that Latour sought to bring into politics. They are the overlooked materials, technologies and, above all, the lowly and varied lifeforms that produce and maintain the planet. Environmental activism was a primary site for new class struggles where representation of these missing masses, he hoped, would emerge from.

Spatial Contributions

There are good reasons why Latour was enthusiastically received in Geography, not least that his work straddled a number of its divides: society/nature; human/physical; premodern/modern; language/action. At first taken up for his studies of science in action, Latour's work helped geographers cross into studying scientific problems (Hinchliffe, 1996), reconsidering the nature of the human (Whatmore, 1997) and extend speech beyond human actors (Ogborn, 2019). Latour's initial influence in Geography was via *We Have Never Been Modern* (1993), a book which argued against modernity itself as a separate age from the pre-modern (or non-modern). It turned geographers' attention to the proliferation of hybrid entities that combine nature and society in 'chemical reactions and political reactions' (Latour, 1993: 1).

In Latour's work geographers found an abiding attention to the connecting up, assembling, centring and distributing of a much wider assembly of things in space than had been traditionally recognised as agents (Bingham, 1996; Murdoch, 1998). In describing how actor-networks are gradually extended, stabilised and sometimes collapsed, Latour radically shifted from a Euclidean concept of space and time as universal abstract axes that contain and constrain events to what he latterly called modes of existence. Rather than axes or containers, for Latour space and time came about as consequences of the ways in which particular heterogenous elements are related to one another. Through this material and relational approach to modes of existence, Latour laid the groundwork for Assemblage Theory and its concern with multiple ontologies (Müller and Schurr, 2016).

Early Latour, in studying scientific discovery and technological invention, was concerned with how new things come to exist in the world. Later Latour turned more toward modes of existence of the many living things that make the world a place that we can live in. The existence of the new and what might seem old, like the geological, is in no way *inevitable*. Things may perish as quickly as they came to gain a foothold on the Earth (Latour, 1997; 2021). It is not only a question of bare living conditions, Latour argued in relation to technologies, that without them we would live in a socially *unstable* world. Our social world would be like that of baboons where trials of strength are resettled daily. Instead, the endless busy proliferation of things as mediators, delegates, boundaries, 'immutable mobiles' achieves the complicated conditions that we exist in. Geographers inspired by this attention to lowly devices and the emergence of new socio-technical-scientific agents investigated the BSE crisis (Hinchliffe, 2001), financial systems in the city of London (Thrift, 1996), GM crops (Bingham, 2006), climate change (Demeritt, 2006), political movements (Featherstone, 2007), urban ecologies (Hinchliffe et al., 2007), and transnational surrogacy (Müller and Schurr, 2016).

Latour has been a powerful influence on geographers pursuing the allied projects of the *more-than-representational* and *more-than-human* which are exploring the elements of places that are apparently outside of language. His work does not, unsurprisingly, draw a boundary between speech and/or texts and things; instead he examined how, for example, texts as planning documents and projects for a new train system are gradually transformed into a prototype and then may (or may not) eventually function as a transport system (Latour, 1997). Geographers tended to set aside Latour's examination of language and the influence of ethnomethodology and speech act theory in favour of his materialist semiotics. In *An Inquiry Into Modes of Existence* (2013) he returned to speech act theory with the ambition of widening what we might consider to be an utterance and so equally what kind of entity can be treated as a speaker (Ogborn, 2019). At the same time, Latour's work helped us see that if we consider utterances as action we come to understand the particular places where those performances occur because those settings provide the conditions for their success and failure.

Key Advances and Controversies

Latour delighted in controversial if also gnomic slogans, not least his attribution of *social agency* to 'actants' which are more often non-human than human. His democratisation of who can act, away from the anthropocentrism of the social sciences, raised an awareness of the 'agency of things' previously restricted to debates over animal rights and the nature of artificial intelligence. Latour's move is characteristic of the flattening out of subjectivity found elsewhere in post-structural thought which critiqued the Enlightenment's position of 'man' at a privileged level above all other life forms or modes of existence. Geographers were equally inspired and perplexed by Latour's extension of social agency, rights and obligations to automatic door closers, sleeping policemen, bacteria, public transport systems, sheep dogs, etc. Referring to diverse objects and life forms that make up the world, as 'the missing masses', Latour argues that they have been ignored socially, politically and philosophically, even as we clearly attend to, care for and depend on them in our everyday lives.

Relatedly, the longstanding problem of *structure–agency* is one to which Latour's work offers a novel reformulation. Where many social theorists, and political philosophers, from Hobbes onwards, set up a binary opposition between social structure and individual agency, Latour pursued impure entities that have characteristics of structure *and* agency. They are, in other words, actors *and* networks or actor-networks. Latour suggested that those who employ an empty gulf between agency and structure do so by ignoring the dark matter of material objects that articulate, embody, coordinate and, even, author actions.

Just as Latour used the 'actor-network' to fill the gap between agency and structure, so he used 'hybrids' to refer to the proliferating entities that are made and remade as mixes of what we usually treat as two distinct poles: *society* and *nature*. In doing so he responded to the endless controversies based in society versus nature that have been at their most symptomatic in the 'science wars'. Rather than accepting society or nature as explanations at face value, Latour, like many others in science and technology studies, turned them over from being explanations to being topics for his inquiries. Where the argument, at its starkest, uses, say 'bacteria' as a source of explanation, Latour made 'bacteria becoming an explanation for X happening' the topic of his inquiry (Latour, 1988). What we then find from his studies are the connections which associate specific explanations and ensuing courses of action (i.e., building the networks of pasteurisation, practice of sterilisation in hospitals, changes in food production etc.) with specific kinds of bacteria. His studies convincingly describe a world where there is no pure nature nor pure society. There are only fibrous webs gradually extending and contracting, erasing one another, copying one another and producing the shape of space and time in doing so.

As was noted at the outset, Latour's extension of the symmetry principle deprived society from being the *explanation* of successes and failures in science. *A priori* favourites of the social sciences like class, race, gender and politics cannot be assumed as relevant in scientific and technological events, nor can everyday un-explicated terms like 'hard facts', 'geniuses' nor 'bias', be brought into explain how the world moves or what moves the world. So what does Latour offer us, having denied the traditional explanatory terms for how the pure will of the subject or the blind force of the object gets bent out of shape by other effects? In typically elegant prose Latour delivers one of his credos on *irreductions*:

> Nothing can be reduced to anything else, nothing can be deduced from anything else, everything may be allied to everything else. (Latour, 1988: 163)

To make anything similar to or different from anything else requires translation, deformation, reformation or other forms of alteration. To make one thing identical to another, to make one place the same as another place requires building relations between them 'out of bits and pieces with much toil and sweat' (Latour, 1988: 162). With this leap away from the various reductions of various theories, Latour set the 'things' free to do what they do, to ally with what they ally. A move we can note that he later critiques for its political consequences in describing his approach to modes of existence (Latour, 2013). As analysts we can follow the movement of 'things': where, and when, they grow and shrink, associate, locate one another, become aligned, produce insides and outsides, subjectify and objectify. All of which sounds rather abstract but Latour was never far from perspicuous examples:

> We neither think nor reason. Rather, we work on fragile materials – texts, inscriptions, traces, or paints – with other people [...] The butcher's trade extends as far as the practice of butchers, their stalls, their cold storage, their pastures, and their slaughterhouses. Next door to the butcher – at the grocer's, for example – there is not butchery. It is the same with psychoanalysis, theoretical physics, philosophy, social security, in short all trades. However, certain trades claim that they are able to extend themselves potentially or 'in theory' beyond the networks in which they practice. The butcher would never entertain the idea of reducing theoretical physics to the art of butchery, but the psychoanalyst claims to be able to reduce butchery to the murder of the father and epistemologists happily talk of the 'foundations of physics.' Though all networks are the same size, arrogance is not equally distributed. (Latour, 1988: 187)

The actor-network of butchering reminds us that all actors only gain agency by being part of networks made of more or less durable materials. If we take the butcher out of the assembly of farmers, delivery companies, freezers, trucks, sharp knives and saws, cash registers and banking then we have a weak actor able to do very little for his trade. Latour, in his studies of scientists, brought them down from their privileged position to place them on a level with butchers and grocers. Whilst retaining the greatest respect for the toil of science, he dispels its idealised claims in favour of taking seriously its rootedness and routed-ness in practices and things.

Latour dismantles the traditional division of geographical explanation into *micro and macro levels*. Through an experimental 'virtual book' (Latour, 2003) on Paris, Latour grounds his critique in examining a site where there are practices that produce and then inspect the 'macro'. In Paris, these are urban control rooms in which 'very little can be seen at any time, but everything appears with great precision' (Latour, 2003: 35). Equally he looks at how a couple having a coffee in a café are transformed and in effect disappear when the desire is to trace the money moving through the café economy. In pursuing various chains of transformation between 'big' and 'small' he provides a remarkable series of sketches of how Paris holds together as a city, how numerous entities within it circulate and how as soon you try to zoom out for a macro-view or zoom-in for a micro-view quite what the connections are that constitute the city's very fabric begin to disappear. The macro-micro division is both shown to be a view produced and utterly reliant on zooming in and out and one that we are unwise to surrender authority to the macro over.

Latour's approach has an oblique relationship to the *language of theory* and its desire to build and use languages that 'zoom' out or upward, departing from ordinary language to become a metalanguage:

> There is no metalanguage, only infralanguages. In other words there are only languages. We can no more reduce one language to another than build the tower of Babel. (Latour, 1988: 179)

An *infralanguage* for Latour is both the languages used by other groups while also for an investigator using ANT and/or *An Inquiry*

Into Modes of Existence (AIME) to write and reveal things about science, engineering and society without claiming that they are laying foundations. Akin to Latour, ethnomethodology (Lynch, 1993) has described scientific and other practical activities while also returning to the reasoning inherent in the language of the places in their studies. Latour parts way with ethnomethodology because his 'infralanguage' of paths, connections, displacements, associations, topologies and networks displaces those of the groups and places that he studies.

Latour's Key Works

Latour, B. (1987) *Science in Action: How to Follow Scientists and Engineers Through Society*. Milton Keynes: Open University Press.

Latour, B. (1988) *The Pasteurization of France*. London: Harvard University Press.

Latour, B. (1993) *We Have Never Been Modern*. London: Harvester Wheatsheaf.

Latour, B. (1997) *Aramis, or the Love of Technology*. London: Routledge.

Latour, B. (2003) *Paris, Invisible City*, photographs by Natalie Hermant and web design by Patricia Reed. Explore on the site http://www.bruno-latour.fr/virtual/ (may not function in recent web browsers), PDF available at http://www.bruno-latour.fr/sites/default/files/downloads/PARIS-INVISIBLE-GB.pdf (accessed November 7, 2023).

Latour, B. (2005) *Re-Assembling the Social: An Introduction to Actor-Network Theory*. Oxford: Oxford University Press.

Latour, B. (2013) *An Inquiry Into Modes of Existence*. London: Harvard University Press. Explorable at the site http://modesofexistence.org (accessed November 7, 2023).

Latour, B. (2017) *Facing Gaia*. Cambridge: Polity.

Secondary Sources and References

Bingham, N. (1996) 'Object-ions: From technological determinism towards geographies of relations', *Environment and Planning D: Society and Space*, 14: 635–57.

Bingham, N. (2006) 'Bees, butterflies, and bacteria: Biotechnology and the politics of nonhuman friendship', *Environment & Planning A*, 38: 483–98.

Bloor, D. (1976) *Knowledge and Social Imagery*. London: Routledge and Kegan Paul.

Demeritt, D. (2006) 'Science studies, climate change and the prospects for constructivist critique', *Economy & Society*, 35 (3): 453–79.

Featherstone, D. (2007) 'Skills for heterogeneous associations: The Whiteboys, collective experimentation, and subaltern political ecologies', *Environment and Planning D: Society and Space*, 25: 284–306.

Hinchliffe, S. (1996) 'Technology, power and space – the means and ends of geographies of technology', *Environment and Planning D: Society and Space*, 14: 659–82.

Hinchliffe, S. (2001) 'Indeterminacy in decisions: Science, policy and politics in the BSE crisis', *Transactions of the Institute of British Geographers NS*, 26 (2): 182–204.

Hinchliffe, S., Kearnes, M., Degen, M. and Whatmore, S. (2007) 'Ecologies and economies of action – sustainability, calculations, and other things', *Environment and Planning A*, 39: 260–82.

Latour, B. (2021) 'The anthill and the beam: A response to Elden', *Dialogues in Human Geography*, 11 (2): 200–2.

Lynch, M. (1993) *Scientific Practice and Ordinary Action: Ethnomethodology and Social Studies of Science*. Cambridge: Cambridge University Press.

Müller, M. and Schurr, C. (2016) 'Assemblage thinking and actor-network theory:

Conjunctions, disjunctions, cross-fertilisations', *Transactions of the Institute of British Geographers*, 41 (3): 217–29.

Murdoch, J. (1998) 'The spaces of actor-network theory', *Geoforum*, 29: 357–74.

Ogborn, M. (2019) 'Uttering geographies: Speech acts, felicity conditions and modes of existence', *Progress in Human Geography*, 44 (6): 1124–40.

Thrift, N. (1996) *Spatial Formations*. London: Sage.

Whatmore, S. (1997) 'Dissecting the autonomous self', *Environment and Planning D: Society and Space*, 15: 37–45.

30
Henri Lefebvre

Rob Shields, University of Alberta

Biographical Details and Theoretical Context

Born in Hagetmau, southwest France, Henri Lefebvre (1901–1991) was a Neo-Marxist and existentialist philosopher, a sociologist of urban and rural life, and a theorist of the State. Over his lifetime, he witnessed the modernisation of everyday life, the industrialisation of the economy, and suburbanisation of France, noting how they combined to destroy family and rural life, not least in his native Pyrenees (see Lefebvre, 2022). Although not exclusively 'his', Lefebvre contributed so much to certain lines of inquiry that it is difficult to discuss concepts including everyday life, modernity, the 'right to the city', mystification, the social production of space, Humanistic Marxism, or alienation without retracing some of his arguments.

Lefebvre was profoundly affected by the post-First World War malaise of the French populace alienated by new, industrialised forms of work and bureaucratic institutions of civil society. This spurred him to focus on alienation and led him to the social criticism of Marx and Hegel. With Norbert Guterman, he published ground-breaking translations of the young Marx (1934), Hegel (1936) and Nietzsche (1939). These informed his introduction to *Dialectical Materialism* (Lefebvre 1968 [1939]). Before discovering the work of Hegel and Marx, Lefebvre was influenced by Schopenhauer, developing a romantic humanism which glorified 'adventure', spontaneity and self-expression. Lefebvre was part of the group 'Jeunes Philosophes' (including Politzer, Nisan, Friedman, and Mandelbrot) loosely connected with author André Gide, Surrealist artists (including Andre Breton) and Dada-ists (including Tristan Tzara). In turn, the Philosophes group's rejection of metaphysical solutions in favour of action influenced Sartre and existentialism (Short, 1966; Poster, 1975).

Through the 1930s, Lefebvre juggled commuting between Paris and Montargis where he taught at a college while serving as a councillor. With Guterman he not only disseminated Hegel and Marx, but also critiqued authoritarian discourses of Fascism (1938b). Their 'four-handed' theorising and writing (Trebitsch, 1992) provided a model of transatlantic collaboration between Paris and New York, and developed key concepts of alienation (see Marcus, 1989), land-rent (Lefebvre, 2022), everyday life and mystification (*la conscience mystifiée* – see Guterman and Lefebvre, 1999). During the Second World War he joined the resistance, before moving to a role as broadcaster in Toulouse in the late 1940s. Completing his PhD on the rural sociologies of Pyrenees communities in 1954, Lefebvre did not gain a permanent position until he was in his 50s (in Sociology at Strasbourg). At the end of the 1950s, Lefebvre's collaboration with the *Situationniste International* group, led by Guy Debord, was crucial in directing his attention to emergent, modern, urban environments as spectacular contexts for everyday life. In 1962 he founded the Institut de Sociologie Urbaine in Paris which expanded his research to urbanisation and the social conditions and contradictions of suburban new towns (Raymond et al., 2001 [1965]; Lefebvre, 1970).

Lefebvre moved to the new University of Nanterre in suburban Paris in 1965 and became an influential figure in the May 1968 student occupation of the Nanterre campus, and subsequently the Sorbonne, along with his then-assistant ***Jean Baudrillard***. While at Nanterre he developed his critique of alienation (Gamsby, 2022), which he argued was being obscured by the 'mystifications' of consumerism, suburban home-ownership and nationalistic heritage and tourism industries. These critiques of modernism became the basis for Lefebvre's investigation of cities, nature and regions, with his essay on the 'right to the city' originally published only weeks before the 1968 rising (see Lefebvre, 1996). During international travels from the early 1970s he developed one of the first theories of what came to be referred to as 'globalisation', later spawning the related idea of 'planetary urbanisation' (Wachsmuth and Brenner, 2014).

Although retired, he continued to write and to lecture internationally until his death in 1991, participating in the lively debates of the 'Groupe de Hagetmau' published in the magazine *M-Marxismes, Mensuels, Mensonges*. Publishing upwards of 50 books in his lifetime, many of which remain untranslated, Lefebvre's work was increasingly translated into English after his death. Translations of his work also became an important base for critical social science and geography in Latin America (see **Santos**, 1974; 2021; Friendly, 2013).

Lefebvre's works on space and on difference (the 'differential') have been rediscovered by a new generation of feminist, queer, ethnographic and Indigenous researchers and activists in the 2010s and 2020s (e.g., Razack, 2002; Buckley and Strauss, 2016; Lee, 2017; Alfaro, 2021). His notion of the 'Right to the City' was implemented in some national statutes (e.g., Brazil's 'Estatuto da Cidade', 2001) and in city proclamations (e.g., Mexico City) but has not been widely tested in the courts (Friendly, 2013).

Spatial Contributions

Lefebvre's initial interest in rural life (Lefebvre, 2022) switched in the 1950s and 1960s to cities. Here, he extended his critique of the domestic life of the household to neighbourhoods and urban life, posing the question: what constitutes 'the urban' (Lefebvre, 2003; 2016)? His answer was that the urban is not a certain population, a geographic size, a collection of buildings; nor is it a node, a transhipment point, a centre of production. Rather, it is all of these together, meaning any definition must explore how the urban holds different elements together in tension. Lefebvre understands 'the urban' from this phenomenological basis as one way such a dialectical tension can exist. He describes urban form as 'social centrality' where many elements of capitalism intersect, despite often merely being part of the place for only a short time (as is the case with goods or people in transit). This position can be contrasted with ***Manuel Castells***' dichotomy of place and the space of flows. 'City-ness' was for Lefebvre the simultaneous gathering and dispersing of goods, information and people: he argued some cities achieve this more fully than others – and hence our own perceptions of some as 'great cities'.

Famously insisting everyone has a 'right to the city' (1968, English translation 1996), Lefebvre presented the city as the pre-eminent site of social interaction and exchange. Conceiving the city-dweller's or community-member's 'right to the city' as a universalising principle, Lefebvre developed Marx's analysis of the citizen which offered a Hegelian synthesis holding the promise of resolving the thesis–antithesis opposition of Christian and Jew (Marx, 1844) while criticising Rousseau for reducing the citizen to a legal abstraction separate from the passions and beliefs of live persons. Lefebvre 'spatialised' this, linking the city as a technology of diversity to questions of inclusion, transcending the differences of communities and individuals (see also Ostrowetsky, 2008). This wonderfully illustrates Lefebvre's method of spatialising Marxist theory and his championing of the humanistic tradition of prioritising the study of authentically 'real' persons with agency as opposed to legally 'correct' or abstractly-theorised individuals.

The Production of Space (1991, first published 1974) forms the keystone of the all-important 'second phase' of Lefebvre's analysis of space. This dealt with social space itself as a national and 'planetary' expression of modes of production. Lefebvre used the term *l'espace* to express active creation: 'spatialisation' is a strong translation of this into English which expresses it as a mode of spacing and placing entities and social activities (Shields, 1999). In *De l'Etat* (Vol. 4, 1978, excerpted in Lefebvre, 2009) he moved his analysis of 'space' from the old synchronic order of discourses 'on' space (e.g., 'social space' as found in sociological and biological texts on 'human territoriality', 1950s planning texts and Chicago School social ecology) to the manner in which understandings of geographical space, landscape, and property change over time. Rather than discussing a particular theory of social space, he examined struggles over the meaning of space and considered how relations across territories were given cultural meaning. In the process, Lefebvre attempted to establish the importance of 'lived' experiences and argued that geographical space is fundamentally social. This was not just a critique of academic conceptions of space but also the everyday attitude that took spatiality for granted.

Lefebvre's open and experimental writing in *The Production of Space* resisted what he detected as capitalism's method of control, which is to fix identities in a static manner and to flatten differences (Lefebvre, 1991 [1974]: 23; 423; Soja, 1996: 8–9; Snart, 2001). Lefebvre's multidimensional thesis was in direct contrast to the more customary conception of space solely as part of production, exchange or accumulation (as in Castells, 1977). In addition, Lefebvre argued that space constituted a fourth and determining realm of social relations – one in which the production, exchange and accumulation of wealth and surplus value take place. Space is not a given, nor is the city an object – hence the difficulties with defining them. Rather space and the urban are both intangible but constitutive aspects of society: its virtual image, so to speak (Lefebvre, 2003: 45).

Historical spatialisations were related by Lefebvre to Marxist modes of production such as feudalism and industrial capitalism. Marxist historical periods and temporal thinking structure this second phase of Lefebvre's work. He examined these on three axes which are dialectically synthesised in a shifting balance or 'trialectic'. These three aspects were explained in different ways by Lefebvre at different times. Simplifying, we might say that the practices and perceptions of space (*le perçu*, 'practices of space') blend popular action and outlook but this axis of everyday life is often ignored in the professional codes and abstractions of 'conceived space' (*le conçu*, 'discourses on space') by cartographers, planners, or property speculators. Nonetheless, any person who is fully human (*l'homme totale*) also dwells in a 'lived space' (*le vécu*). Although suppressed in the 'abstract space' of contemporary capitalist societies, lived space remains visible in art, literature and fantasy. This 'lived space' is a frame for the other two axes, establishing the limits within which we imagine reality and potential changes. It has the power to refigure the balance between popular 'perceived space' and official 'conceived space' (Shields, 1990). Lefebvre cites Dada, the work of the Surrealists, and particularly the works of René Magritte as examples challenging taken-for-granted frameworks of space. Also included in this axis are clandestine and minority spatial practices hiding in plain sight, for example archaic or alternative non-capitalist modes of the production of space (spatialisations). This reservoir of everyday alternatives prompts revolutionary restructurings of institutionalised space and new modes of spatial praxis. Examples include squatting, illegal migration, and Third World informal housing, which all fashion a spatial presence outside of the norms of the prevailing (hegemonic) social spatialisation.

From the 1980s, English-language geographers and urban thinkers such as ***Edward Soja*** (1996: 60-61) tentatively envisioned Lefebvre's trialectic (*dialectique de triplicité*) as not 'an additive combination of its binary antecedents but rather [...] a disordering, deconstruction, and tentative reconstitution of their presumed totalisation producing an open alternative that is both similar and strikingly different.' The 'triplicity' that he derives from Lefebvre's position as 'Thirding,' 'decomposes the dialectic through an intrusive disruption that explicitly spatializes dialectical reasoning [...] [and] produces what might best be called a cumulative *trialectics* that is radically open to additional othernesses, to a continued expansion of spatial knowledge' (Soja, 1996: 61). In a related vein, **Neil Smith** and others drew on the *Production of Space* (1991 [1974]) to launch a critique of the denial of individuals' and communities' 'rights to space.' In capitalist societies, they noted, geographical space is 'spatialised' as private property, with property anchoring all capitalist societies. **David Harvey** similarly took up Lefebvre's thinking about urban social life in both its economic and symbolic dimensions, adding his own analysis of the temporality of urban life both in the sense of long-term accumulation in the cycles of finance, industry and infrastructure and, in the shorter term, individual memory and meaning.

The third phase of Lefebvre's thinking was a consideration of tempo and space-time in *Rhythmanalysis* (in Lefebvre and Régulier, 2003). Here, the history of 'modes of production of space' or spatialisations gained more nuance, completing Marx's vision in urban and environmental terms. A true Communist revolution, Lefebvre suggested, must not only change the relationship of labourers to the means of production, but also create a new spatialisation, shifting the balance away from abstract, 'conceived space' of which private property, city lots and the surveyor's grid are prime examples. Embracing the 'lived space' of everyday life and of art *avant gardes* was a step toward harnessing its potential and redirecting the 'perceived space' of everyday practice. This theory provides a bridge between the European Marxist thought of the mid-twentieth century to the environmental activism and countercultures of the 1970s–90s and anti-neoliberal globalisation protests of the 2000s.

Key Advances and Controversies

Lefebvre originally developed his stance on alienation and false consciousness, or 'mystification' of popular and consumer culture, together with Norbert Guterman in the 1930s. In 1947 he published the first of what were to be three volumes of *Critique of Everyday Life* (2014). In this series, Lefebvre presented a Marxist materialist remedy to 'everydayness' (*quotidienneté, Altäglichkeit* or banality) as a soul-destroying feature of modern culture, social interaction and the material environment. Opposing Heidegger's dismissal of everyday life as banal boredom (Gamsby, 2022), Lefebvre argued that the everyday is 'what we are first and most often' (Blanchot cited in Davis, 2009), the reality we live when we embark on each new day as both full of potential and disciplined by needs and obligations.

Arguing against 'mystification' and the banality of the '*métro-boulot-dodo*' ('subway-work-sleep') life of the suburban commuter, Lefebvre proposed that we seize and act on all 'Moments' of revelation, emotional clarity and self-presence as the basis for becoming more self-fulfilled (*'l'homme totale'* – see Lefebvre, 1959). This concept of 'Moments' reappears throughout his work as a theory of presence and the foundation of a practice of emancipation. By definition, 'Moments' have no duration, but can be relived. Lefebvre argues that these cannot easily be reappropriated by capitalist processes of commodification or codification.

After the failure of the student occupations of May 1968 to secure political-economic reform and work–life balance, Lefebvre's oeuvre was eclipsed by Louis Althusser's 'Scientific Marxism' in which the base-superstructure division was a privileged element of a structural analysis of the repressive forces and institutions of capitalist states. Ironically, Lefebvre first became well known to English-speaking theorists through the critiques of his work by Althusserians, including ***Manuel Castells***, who, in *The Urban Question* (1977), criticised Lefebvre's work for its 'anti-structuralist' bias.

Lefebvre's gender blindness and celebration of heterosexuality limit his theories of the body and sexuality that are part of his analysis of spacing. He had bitter personal quarrels with other radical thinkers such as Debord and the founder of ecofeminism, Françoise d'Eaubonne (1974, English translation 2022). Lefebvre predates or was untouched by Anglophone writers' queer theories as a 'third' alternative outside a heterosexual dualism (A/not-A/neither) (Blum and Nast, 1996). The importance of everyday life and his rural sociology for ecological and queer thought in France are only now being explored (e.g. Pruvost, 2021). Although he was an active critic of French colonial oppression in Algeria, he did not engage with critiques of slavery, racism, and colonialism as a 'spatial fix' improving the profits of European capitalists offered, for example, by postcolonial French writers such as Frantz Fanon and Aimée Césaire. Relatedly, Maria Lugones (2003) proposed the feminist figure of the marginalised 'street-walker theorist' as a critical development of Lefebvre's own intellectual abstraction.

However, Lefebvre wrote perhaps the first critique of ethnonationalism (Lefebvre, 1988, first published 1937) and warned against the risks of ethnic identification and xenophobia. But he did not use the contemporary language of discrimination, of 'insiders and outsiders' and the ethics of their relationships. He tended to conceive of the state as a once-authentic instrument of a single people which has been seized by the capitalist class for itself against citizens. He showed their dreams and sense of identity were manipulated for the purposes of capitalist accumulation (Lefebvre, 2023).

Lefebvre went beyond twentieth-century philosophical debates on the nature of space which considered people and things merely as being 'in' space to present a coherent theory of relational systems of spatiality in different historical periods. These are not just physical arrangements of things, but spatial patterns of embodied routine, as well as historical conceptions of the world (such as the idea of a flat Earth). These regimes operate at all scales. At the most personal, we think of ourselves in spatialised terms as an ego contained within an objectified body. People extend themselves – mentally and physically – into space much as a spider extends its limbs in the form of a web. We become as much a part of these extensions, as they are of us. Arrangements of objects, labour, landscapes and architecture are the concrete instances of this spatialisation. Equally, ideas about regions, media images of cities and perceptions of 'good neighbourhoods' are other aspects of the space necessarily produced by each society as it makes its mark.

Lefebvre used the changing spatial order or spatialisation to explain why capitalistic accumulation did not occur earlier, even in those ancient economies which were commodity and money-based, which were committed to reason and science, and which were based in cities (see Merrifield, 1993). One well-known explanation is that slavery stunted the development of wage-labour. Lefebvre found this unconvincing. Instead, he postulated that the development of secular space – commodified as private lots, quantified by surveyors and stripped of the old local gods and spirits of place – was a necessary precondition for the separation of people from the means to their own subsistence other than by work in return for wages.

Lefebvre was an experimental writer, poet and playwright. The core of Lefebvre's humanism was his critique of the alienating conditions of everyday life. The greatness of *The Production of Space* (1991 [1974]) is that it transposes Marx into a method inviting readers to make their own critical synthesis of the book and of their world. As well as being a *product*, Lefebvre reminds us that space is a relation: a *medium*. Changes in the way we understand and live spatially provide clues to how our capitalist world may eventually give way to unanticipated geopolitics at all scales, producing a new sense of our relation to our own bodies and our world as a changing space of distance and difference (see Leary-Ohwin and McCarthy, 2019).

Lefebvre's Key Works

Lefebvre, H. (1939) *Nietzsche*. Paris: Editions Sociales Internationales.

Lefebvre, H. (1947) *Critique de la vie Quotidienne*, I: Introduction. Paris: Grasset.

Lefebvre, H. (1959) *La Somme et le Reste*. Paris: La Nef de Paris.

Lefebvre, H. (1968a) *Sociology of Marx*. Trans. N. Guterman. New York: Pantheon.

Lefebvre, H. (1968b [1939]) *Dialectical Materialism*. Trans. J. Sturrock, London: Cape.

Lefebvre, H. (1978) *Les Contradictions de L'Etat, De L'Etat*, Vol. 4. Paris: UGE.

Lefebvre, H. (1988 [1937]). *Le Nationalisme contre les nations*. Paris: Méridiens Klincksieck.

Lefebvre, H. (1991 [1974]) *The Production of Space* trans. N. Donaldson-Smith), Oxford: Basil Blackwell.

Lefebvre, H. (1996) *Writings on Cities*. Trans. and ed. E. Kofman and E. Lebas. Oxford: Blackwell.

Lefebvre, H. (2003) *The Urban Revolution* (trans. R. Bononno). Minneapolis, MN: University of Minnesota Press.

Lefebvre, H. (2014) *The Critique of Everyday Life (One Volume Edition)*. London: Verso.

Lefebvre, H. (2016) *Marxist Thought and the City*. Minneapolis, MN: University of Minnesota Press.

Lefebvre, H. (2022) *On the Rural. Economy, Sociology, Geography*. Minneapolis, MN: University of Minnesota Press.

Lefebvre, H. (2023) *Everyday Life in the Modern World*. London: Routledge.

Lefebvre, H. and Guterman, N. (1934) *Morceaux choisis de Karl Marx*. Paris: NRF.

Lefebvre, H. and Guterman, N. (1936) *La Conscience Mystifiée*. Paris: Gallimard.

Lefebvre, H. and Guterman, N. (1938a) *Morceaux choisis de Hegel*. Paris: Gallimard.

Lefebvre, H. and Régulier, C. (2003) *Henri Lefebvre: Key Writings* (trans. and ed. S. Elden, E. Lebas and E. Kofman). London: Continuum.

Lefebvre, H., Brenner, N. and Elden, S. (2009) *State, Space, World: Selected Essays*. Minneapolis: University of Minnesota Press.

Secondary Sources and References

Alfaro, C.F. (2021) 'Feminist Lefebvre? Understanding urbanization through the global intimate', *ACME: An International Journal for Critical Geographies*, 20 (4): 366–86. Available from https://acme-journal.org/index.php/acme/article/view/1891 (accessed November 7, 2023).

Blum, V. and Nast, H. (1996) 'Where's the difference? The heterosexualization of alterity in Henri Lefebvre and Jacques Lacan', *Environment and Planning D: Society and Space*, 14: 559–80.

Buckley, M. and Strauss, K. (2016) 'With, against and beyond Lefebvre: Planetary urbanization and epistemic plurality', *Environment and Planning D: Society and Space*, 34 (4): 617–36.

Castells, M. (1977) *The Urban Question: A Marxist approach*. London: Edward Arnold.

Estatuto da Cidade, Pub. L. No. L10257, Code Civil (2001). Congresso Nacional, Brazil. https://www.planalto.gov.br/ccivil_03/leis/leis_2001/l10257.htm (accessed November 7, 2023).

Davis, T. (2009) '"What true project has been lost?" Modern art and Henri Lefebvre's Critique of Everyday Life', in S. Ross (ed), *Modernism and Theory: A Critical Debate*. New York, NY: Routledge. pp. 65–79.

D'Eaubonne, F. (2022 [1974]) *Feminism or Death: How the Women's Movement Can Save the Planet*. London: Verso.

Friendly, A. (2013) 'The right to the city: Theory and practice in Brazil', *Planning Theory and Practice*, 14 (2): 158–79.

Gamsby, P. (2022) *Henri Lefebvre, Boredom, and Everyday Life*. Lanham, MA: Lexington.

Guterman, N. and Lefebvre, H. (1999) *La Conscience Mystifiée*. Paris: Éditions Syllepse.

Leary-Owhin, M. and McCarthy, J.P. (2019) *The Routledge Handbook of Henri Lefebvre, the City and Urban Society*. London: Routledge.

Lee, S.P. (2017) 'Ethnography in absentia: Applying Lefebvre's rhythmanalysis in impossible-to-research spaces', *Ethnography*, 18 (2): 257–26.

Lefebvre, H. (1970) 'Introduction à l'etude de l'habitat pavillonnaire', in M. Gavira (ed) *Du rural à l'urbain*. Paris: Anthropos. pp. 179–80.

Lefebvre, H. (2009) *State, Space, World: Selected Essays*. Minneapolis, MN: University of Minnesota Press.

Lefebvre, H. and Guterman, N. (1938b) *Hitler au pouvoir, bilan de cinq années de fascisme en Allemagne*. Paris: Bureau d'Editions.

Lugones, M. (2003) *Pilgrimages/Peregrinajes: Theorizing Coalition against Multiple Oppressions*. Lanham, MD: Rowman & Littlefield.

Marcus, G. (1989) *Lipstick Traces*. Cambridge, MA: Harvard.

Marx, K. (1844) 'The Jewish question', H.J. Stenning (trans.), *Selected Essays*. Originally published 1844. Revised translation online: www.marxists.org/archive/marx/works/1844/jewish-question (accessed November 7, 2023).

Merrifield, A. (1993) 'Space and place: Lefebvrian reconciliation', *Transactions of the Institute of British Geographers*, 18: 516–31.

Ostrowetsky, S. (2008) 'L'Identité des Droits'. Paper presented to Colloque de la pensée d'Henri Lefebvre, Vienna 2000, *La Somme et le Reste: Etudes Lefebvriennes* 13, pp. 2–14. cep.cl/Cenda/Cen_Documentos/Varios/S&R-13.pdf

Poster, M. (1975) *Existential Marxism in Postwar France: From Sartre to Althusser*. Princeton, NJ: Princeton University Press.

Pruvost, G. (2021) *Quotidien Politique*. Paris: La Decouverte.

Raymond, H., Haumont, N., Dezes, M.G. and Haumont, A. (2001 [1965]) *L'habitat pavillonaire*. Paris: L'Harmattan.

Razack, S. (2002) 'Introduction: When place becomes race', in S. Razack (ed), *Race, Space and the Law: Unmapping a White Settler Society*. Toronto: Between the Lines. pp. 1–20.

Santos, M. (1974) 'Geography, Marxism and underdevelopment', *Antipode*, 6 (3): 1–9.

Santos, M. (2021) *The Nature of Space* (trans. B. Baletti). Durham, NC: Duke University Press.

Shields, R. (1990) *Places on the Margin: Alternate Geographies of Modernity*. London: Routledge.

Shields, R. (1999) *Lefebvre: Love and Struggle: Spatial Dialectics*. London: Routledge.

Short, R.S. (1966) 'The politics of surrealism 1920–1936', *Journal of Contemporary History*, 1: 3–26.

Smith, N. (1984) *Uneven Development; Nature, Capital and the Production of Space*. Oxford: Basil Blackwell.

Snart, J. (2001) 'Entropy in Pynchon's "Entropy" and Lefebvre's The Production of Space'. *CLCWeb: Comparative Literature and Culture*, 3 (4). doi.org/10.7771/1481-4374.1136

Soja, E. (1996) *Third Space*. Oxford: Blackwell.

Trebitsch, M. (1992) 'Correspondances d'intellectuels. Le Cas Des Lettres d'Henri Lefebvre à Norbert Guterman (1935–1947).' *Les Cahiers de l'IHTP. Sociabilités Intellectuelles. Lieux, Imlliux, Reseaux*, 20 (March): 70–84.

Wachsmuth, D. and Brenner, N. (2014) 'Introduction to Henri Lefebvre's "Dissolving City, Planetary Metamorphosis"', *Environment and Planning D: Society and Space*, 32 (2): 199–202.

31
Akin Mabogunje

Ruth Craggs, Kings College London
Taibat Lawanson, University of Lagos, Nigeria

Biographical Details and Theoretical Context

Akinlawon Ladipo Mabogunje (1931–2022), known as 'the father of African Geography' (Abler, 2023), became the first Nigerian Professor of Geography at the age of 34 in 1965 (Mabogunje, 2011). Mabogunje was born in Kano, but was educated in the south of Nigeria, becoming one of the first cohort of students studying Geography at the newly opened University College, Ibadan in 1949.

After studying for an ordinary Geography degree at Ibadan, Mabogunje enrolled at University College, University of London (UCL) for his honours degree (winning the Parry Prize for his performance). He followed this with a Masters at UCL, during which he impressed so much he was offered an Assistant Lectureship at Liverpool University, which he turned down to take up a position at University College Ibadan, becoming the first Nigerian lecturer in the Geography Department in 1958 (Mabogunje, 2011). He completed his PhD in Geography, also at UCL (at a distance) in 1961.

Mabogunje's education was heavily influenced by its British colonial context, as he later reflected: given the staffing and aims of the university colleges like Ibadan which were set up as late colonial development projects, it was 'not surprising that the British tradition in geography formed the foundation of much of the training of geographers' in Nigeria (Mabogunje, 2004: 63). Moreover, study in the UK cemented this – particularly through Mabogunje's engagement with historical methods. He recalled that 'the historical geography angle', imparted by his MA and PhD supervisor H.C. Darby, was 'one of the lasting impacts of studying at UCL in my career' (Akin Mabogunje, personal communication, July 7th, 2016).

Mabogunje's early research, including that for his MA and PhD, focused on urbanisation in Nigeria, combining historical and contemporary data and drawing on the regional tradition in geography. However, he soon engaged with new methods as the so-called Quantitative Revolution swept through the discipline from the late 1950s. His first major book, *Urbanization in Nigeria* (1968) drew on his doctoral work, but utilised new quantitative methods to introduce what he described as more 'sophisticated analytical methods' (Mabogunje, 1968: 6). Mabogunje learned these primarily in North America, courtesy of a Rockefeller grant, which also facilitated access to the computing technology necessary to perform the analysis. This was not a passive process: Mabogunje negotiated the terms of his travel, requesting access to specific theoretical specialisms. As he recalled:

> [...] with independence, Ford Foundation, Rockefeller Foundation, then started sending teams of their respected academics to come out to countries in Africa to find out in what way they can help. And so, in 1961, a team came from Rockefeller, it comprised [...] Professor [*Gilbert*] White of Chicago [...] And so they came to Ibadan. [...] they were asking what else can we do to help. And of course I was saying I wanted to go to a university in the U.S. where I can supplement my quantitative and theoretical capabilities. And of course Gilbert White comes from the University of Chicago, but he thought the best teacher at that point in time was at Northwestern, that was Bill Garrison [...] I had nine months [...] This travel was transformative: 'I would say what Northwestern did was throw me into orbit, you know. That anytime you then had something quantitative, and you were looking for a spread round the world my name always came up. (Akin Mabogunje, personal communication, July 7th, 2016)

The funding that allowed Mabogunje and many other African academics to travel to the US – and to learn new analytical methods – was tied to the wider priorities of the Cold War, the geopolitics of the era shaping the opportunities and funding streams available to geographers globally (Craggs and Neate, 2020; see Barnes and Farish, 2006 for impacts in US geography).

Mabogunje's approach in his early work blended historically-careful reconstructions of urban change with an interest in policy for the present and future. As he explained, his historical training provided an understanding of 'the substantive development process of a people over time in space', but in order to move beyond the description of landscape change and towards an understanding of process and explanation of change, quantitative approaches were required (Akin Mabogunje, personal communication, July 7th, 2016). Together, historical and quantitative methods could provide insights into the development process, something which would be a focus – in different ways – of Mabogunje's whole career. In 1980, he published his second major monograph, *The Development Process: A Spatial Perspective* (1980a).

As his career developed, Mabogunje became more involved in applied geographical work, focusing primarily on getting spatial thinking on the national development agenda, and putting these ideas into practice through regional and urban planning. He was critical of Nigeria's 1970 National Development Plan because of its lack of regional orientation, and was well-connected enough to be able to argue successfully for a greater emphasis on this in future government plans (Mabogunje, 2011; see, for example, Mabogunje, 1971a). This was part of a wider trend – what Hougham (1981: 527) called a 'dramatic surge in spatial planning' – in national development plans across Africa in the 1960s and 1970s. This replaced the solely sectoral planning which located projects only on the basis of what was best for that specific industry. Instead, spatial planning focused on projects within a wider spatial system, and in Africa, often aimed for 'the furthering of social justice (meaning, usually, a more egalitarian distribution of the benefits of modernization) and the fostering of national unity' (Hougham, 1981: 528). In his career, Mabogunje increasingly moved outside of the university, becoming involved with large-scale development projects, including the evaluation of the Kainji Dam project and the creation of the new capital at Abuja (replacing Lagos – and an example of spatial planning in action) (Mabogunje, 1973; Abumere, 1998).

Whilst he was committed to development, even in its most large-scale and top-down forms, Mabogunje's writings about the dam – both his academic evaluations published at the time and his autobiographical reflections published half a century later – highlight a sensitivity to processes of social change and the devastation that could take place at a local level as a result of development projects (Mabogunje, 1973; 2011). Much of his later work focused on community-level and community-led approaches to development.

Mabogunje's commitment to social justice and inclusive development earned him a number of awards including the Certificate of Merit of the Ijebu-Ode (his home town) Development Association in 1993, the Commander of the Order of the Federal Republic of Nigeria in 2011, and the 1998 UN Centre for Human Settlements Scroll of Honour for outstanding contributions to human settlement development. He was also recognised from within the national and international geographical community. For example, he was elected President of the Nigerian Geographical Society in 1968, and President of the International Geographical Union in 1980, and received the Vautrin-Lud Prize, the discipline's highest award, in 2017.

Maobogunje was a prolific scholar and published over a dozen books, mainly on urbanisation, planning and development, for a range of audiences over his lifetime. Prompted by his own interests, as well as the structural and political problems besetting Nigerian Universities at the time, Mabogunje resigned from his post at the University of Ibadan in 1981, setting up a consultancy with a colleague from the Economics Department (Mabogunje, 2011), and became increasingly focused on policy work afterwards.

Spatial Contributions

Akin Mabogunje's PhD was adapted to become his first major monograph *Urbanization in Nigeria* (Mabogunje, 1968), which was one of the first accounts to address *African,* rather than solely *colonial,* urbanisation in Nigeria. Beginning in the pre-colonial period, the book is rich in detail about the historical development of Nigerian cities and evocative of Southern Nigerian urban life before, during and after colonisation. As Oyalowo (2022: n.p.) notes, the book presents 'colourful, endearing and poignant descriptions of the streets of Ibadan and Lagos' with readers 'invited to walk alongside him through history', but also to understand 'the tenacity of those cities' modern problems'. The book made two significant contributions.

First, it provided the first documented hierarchy of human settlements in Nigeria, and this work remains important today. In describing urban growth through densification, Mabogunje (1964) established Lagos Island as the core Central Business District for Lagos, with additional activity centres. Despite intense growth of the city over time, the typologies outlined remain the same today (Wolak, 1984; Odunbaku, 2017).

Second, at a time when, as the book notes, urban theory tended 'to regard urbanization as a peculiarly European and American phenomena', it took urban growth unrelated to the West seriously, and analysed the patterns of life and urban forms of Nigerian cities (Mabogunje, 1968: 33). Crucially, Mabogunje's study provided a counterpoint to the large number of studies, carried out primarily by white anthropologists and sociologists, focusing on the (mal)adjustment of 'the African' to modern (colonial settler) cities, and to the broader notion than urban space was alien to 'traditional' African life (Mabogunje, 1968). The book was well reviewed and remains important today for scholars researching Nigeria's towns and cities. *Urbanization in Nigeria* provided new insights by overturning colonial historical geographies of Africa which started with colonisation, and in doing so, paralleled innovations in the study of history led by the first generation of historians in independent Nigeria (Falola, 2004). Uniquely, *Urbanization in Nigeria* 'connects the history of urbanisation, the heritage of coloniality and the missed opportunities for changes in Nigeria's urban areas' (Oyalowo, 2022: n.p.).

Mabogunje's second major monograph, *The Development Process: A Spatial Perspective* (1980a) brought together economic geography perspectives and literature from Development Planning – with an eye always on policymakers and on the role of space and geography within urban, regional and national development. According to one reviewer, the book was a 'welcome relief from the parade of high quality, concept-laden, heavy-weight economic geography texts that have marched, clone-like, from Universities in the developed world in recent years [...]' (Barker, 1983: 173).

Beyond his published work, Akin Mabogunje contributed significantly to development policy and social development across Africa. This work transcended scale from community to global, from public to private and from academia to practice, but always with a focus on improving people's lives, especially in his home country. When structural adjustment hit Nigeria in the mid-1980s, Mabogunje worked with the military government of Babangida, becoming Deputy Chair of the new Directorate of Food, Roads and Rural Infrastructure established in 1986, which aimed to bring about community-led rural development to mitigate the worst impacts of currency devaluation on food prices (Mabogunje, 1988; 1990a; 2011).

Mabogunje also served as member of the Nigerian Presidential committee on Housing and Urban Development, and as Chair of the redevelopment of the Lagos Megacity Region (2005) which recommended reforms including the Lagos-Ogun Joint Development Commission which was finally established in 2021. He was also the initiator and Chairman of the Ijebu Development Initiative on Poverty Reduction (IDIPR), which entailed remittances, microfinance, vocational training and catalysing local economic development in his hometown of Ijebu Ode Nigeria (Mabogunje, 2007). This model has become the blueprint for various community development initiatives across Nigeria, including policy action at the national level.

Akin Mabogunje's influence therefore transcended geography and reshaped policy across Nigeria and beyond. His methodological approach in the early days of his career shaped geography into the technical discipline it is today (Craggs and Neate, 2020), but his influence was far wider, and he was also considered the father of urban planning and housing in Nigeria (Mabogunje, 1990b). Furthermore, his robust engagement with history (1971b), policy analysis (1992; 1998), participatory planning (Mabogunje and Kates, 2004) and quantitative analysis (1994) also set the stage for mixed methods research design in areas such as social development, urban studies and public administration.

Key Advances and Controversies

In 2001, Alan Mabin eloquently described Mabogunje as 'the single best-known African voice on urbanism and cities', given his profound influence on firmly placing the African context within the global discourse on urban planning, policy and development (Mabin, 2001: 180). His work pushed back against the uncritical embrace of westernised theories to analyse African urban trajectories. According to Briggs (2022), he was arguably the first geographer to 'Africanise' thinking and to break out of the colonial mindset and mode of thought – setting the stage for the work of Vanessa Watson, Shuaib Lwasa, Franklin Obeng-Odoom and other African proponents of 'Southern' urban theory (see also the entry for **Jennifer Robinson**).

Despite this innovation, Mabogunje was critiqued, particularly by those in the growing tradition of Marxist and critical development geography in the late 1970s and '80s. Despite the innovation of his first book, overall Akin Mabogunje sought a geography combining western and African knowledge, and focused more on practical and applied geographical research for development on a liberal rather than radical model. In *The Development Process,* the focus was, according to one critical reviewer, on handing 'politicians the "tools" with which to further their own ends within the overall development domain', 'leaning on a technical approach' cast – problematically – as 'value free': 'It pretends to be value neutral, when it should serve an emancipatory interest' (Friedmann, 1982: 901, 903).

Mabogunje collaborated with Federal and State governments in Nigeria on large development projects that could be controversial, both politically and intellectually. Mabogunje and his colleagues were working on the plan for the new capital Abuja in the same year that the radical Brazilian geographer **Milton Santos** (1977: 86), drawing on his experiences in Tanzania, was publishing his critique of theories of central places and growth poles as worsening under-development and inequality by allowing the 'rapid and brutal intrusion of capital in underdeveloped countries' in the journal *Antipode*. Indeed, increasingly from the early 1970s, some of the key concepts in new geographical studies of development: spatial differentiation, spatial diffusion, and spatial integration, and the closely connected ideology of modernisation, were coming under criticism from radical geographers, who viewed them as ahistorical, ethnocentric and neglectful of the role of colonialism and capitalism in producing and perpetuating inequalities (Slater, 1973). Fundamentally, this work critiqued the inherent assumption that modernisation 'leads to improvement and progress' for poorer areas, with regional planning strategies taking for granted a capitalist economic structure, rather than evaluating the possibilities of alternatives (Slater, 1973: 26). From this perspective, development geography – including that produced by Mabogunje in the 1970s – could be cast as irrelevant and inadequate (Slater, 1977: footnote 14).

The intersections between the independence of great swathes of Africa between 1957–1965 and the rise of modern urban and regional planning provided a potent mix, and additional impetus – ethical, political and financial – for involvement in development projects. Mabogunje's career was shaped by this context, and he played a leading role in positioning geography as a key discipline for postcolonial development in Africa. In this his work can be seen as part of a project of African decolonisation and independence, even if one aligned to what Al-Bulushi (2022: 259) describes as 'statist conceptions of liberation', rather than more radical non-sovereignty based alternatives.

Fundamentally, throughout his career, he remained optimistic about what an engaged, geographical understanding of development might offer to Nigeria and more broadly; his writing always hopeful as well as practical (Oyalowo, 2022: n.p.). And through his role as the first African President of the International Geographical Union, Mabogunje made a name for geography on the continent, and for African geographers in the discipline worldwide

Mabogunje's Key Works

Mabogunje, A.L. (1962) *Yoruba Towns*. Ibadan: Ibadan University Press.

Mabogunje, A.L. (1964) 'The evolution and analysis of the retail structure of Lagos, Nigeria', *Economic Geography*, 40 (4): 304–23.

Mabogunje, A.L. (1968) *Urbanization in Nigeria*. London: University of London Press.

Mabogunje, A.L. (1971a) *Growth Poles and Growth Centres in the Regional Development of Nigeria*. United Nations Research Institute for Social Development (Report No. 17/3).

Mabogunje, A.L. (1980a) *The Development Process: A Spatial Perspective*. London: Hutchinson.

Mabogunje, A.L. (1980b) 'The dynamics of centre-periphery relations: The need for a new geography of resource development', *Transactions of the Institute of British Geographers*, 5 (3): 277–96.

Mabogunje, A.L. (1981) 'Geography and the dilemma of rural development in Africa', *Geografiska Annaler: Series B, Human Geography*, 63 (2): 73–86.

Mabogunje, A.L. (1983) 'The case for big cities', *Habitat International*, 7 (5–6): 21–31.

Mabogunje, A.L. (1988) 'Coping with structural adjustment: The Nigerian experience', in R. Cohen (ed), *Satisfying Africa's Food Needs: Food Production and Commercialization in African Agriculture*. Boulder, CO: Carter Studies on Africa, Lynne Rienner. pp. 191–208.

Mabogunje, A.L. (1992) 'A new paradigm for urban development', *Proceedings of the World Bank Annual Conference on Development Economics 1991*: 191–208.

Mabogunje, A.L. (2007) 'Tackling the African "poverty trap": The Ijebu-Ode experiment', *Proceedings of the National Academy of Sciences*, 104 (43): 16781–6.

Mabogunje, A.L. (2011) *A Measure of Grace: The Autobiography of Akinlawon Ladipo Mabogunje*. Ibadan: Book Builders.

Secondary Sources and References

Abler, R.F. (2023) 'Akinlawon "Akin" Mabogunje (1931 to 2022): The father of African geography', *Proceedings of the National Academy of Sciences*, 120 (23).

Abumere, S. I. (1998) 'Abuja: Planning and development of the new federal capital territory', in O. Areola and S.I. Okafor (eds), *50 Years of Geography in Nigeria: The Ibadan Story*. Ibadan: Ibadan University Press. pp. 87–410.

Al-Bulushi, Y. (2022) 'Thinking racial capitalism and Black radicalism from Africa: An intellectual geography of Cedric Robinson's world-system', *Geoforum*, 132: 252–62.

Barker, D. (1983) 'Review of *The Development Process: A Spatial Perspective*', *Social and Economic Studies*, 32 (3): 171–3.

Barnes, T. J. and Farish, M. (2006) 'Between regions: Science, militarism, and American geography from World War to Cold War', *Annals of the Association of American Geographers*, 96 (4): 807–26.

Briggs, J. (2022) 'Professor Akin Mabogunje (1931–2022)', *Scottish Geographical Journal*, 138 (3–4): 389–90.

Craggs, R. and Neate, H. (2020) 'What happens if we start from Nigeria? Diversifying histories of Geography', *Annals of the American Association of Geographers,* 110 (3): 899–916.

Falola, T. (2004) *Nationalism and African Intellectuals*. Rochester, NY: University of Rochester Press.

Friedmann, J. (1982) 'Review of *The Development Process: A Spatial Perspective*', *Economic Development and Cultural Change,* 30 (4): 897–903.

Hougham, D. (1981) 'The spatial components of Anglo-African development plans, 1943–1956', *Canadian Journal of African Studies/La Revue canadienne des études africaines*, 15 (3): 527–38.

Mabin, A. (2001) 'Contested urban futures: Report on a global gathering in Johannesburg, 2000', *International Journal of Urban and Regional Research*, 25: 180–5.

Mabogunje. A.L. (1971b) 'Lands and people in West Africa', in M. Crowder, and J.F. Ajayi (eds), *The History of West Africa*. London: Longman. pp. 1–32.

Mabogunje, A.L. (ed) (1973) *Kainji a Nigerian Man-made Lake, Vol. 2*. Ibadan: NISER.

Mabogunje, A.L. (1990a) 'Mobilizing Nigeria's grassroots for increased food production: Reaching out from the centre', *Food Policy*, 15 (4): 306–12.

Mabogunje, A.L. (1990b) 'Urban planning and the post-colonial state in Africa: A research overview', *African Studies Review*, 33 (2): 121–203.

Mabogunje, A.L. (1994) 'Overview of research priorities in Africa', in R. Stren (ed), *Urban Research in the Developing World, vol.2 – Africa*. Toronto: University of Toronto Press. pp. 19–46.

Mabogunje, A.L. (1998) 'The struggle for a second independence: Sociopolitical construction of space in Africa', 46 (184): 1–74.

Mabogunje, A.L. (2004) 'Geography in the Nigerian public policy domain: The impact and influence of American geographers', *GeoJournal*, 59 (1): 63–7.

Mabogunje, A.L. and Kates, R.W. (2004) *Sustainable Development in Ijebu-Ode, Nigeria: The Role of Social Capital, Participation, and Science and Technology*. Center for International Development Working Paper Series 2004.102. Cambridge, MA: Harvard University.

Odunbaku, O. (2017) *Travel Behaviour of Patrons and Location of Business Districts in Lagos, Nigeria*. Unpublished PhD thesis, Urban and Regional Planning, University of Lagos.

Oyalowo, B. (2022) 'Akin Mabogunje: Nigerian urban geographer who mapped the origin and trends of African cities', *The Conversation*. https://theconversation.com/akin-mabogunje-nigerian-urban-geographer-who-mapped-the-origin-and-trends-of-african-cities-190378 (accessed November 8, 2023).

Santos, M. (1977) 'Planning underdevelopment', *Antipode*, 9: 86–98.

Slater, D. (1973) 'Geography and underdevelopment – Part 1', *Antipode*, 5: 21–32.

Slater, D. (1977) 'Geography and underdevelopment – Part 2', *Antipode*, 9: 1–31.

Wolak, Z. (1984) *Master Plan for Metropolitan Lagos, Implementation: Nigeria. Final Report*. Nigeria: UN Centre for Human Settlements.

32
Doreen Massey

Marion Werner, University at Buffalo, State University of New York

Biographical Details and Theoretical Context

Doreen Massey (1944–2016) grew up on a local authority housing estate in Wythenshawe, south of Manchester in North-West England. From an early age, Massey suffered from a debilitating osteogenic disease, and received treatment through the newly-established National Health Service, which she would vociferously support throughout her lifetime. From a working-class family, Massey passed a highly-selective entrance exam to enter Manchester High School for Girls in 1955. In 1962, she won a scholarship to attend Oxford University where she read Geography. Ninety percent of students were male, along with most of the lecturers except those who taught in the handful of women's colleges (McDowell, 2018). The move from Wythenshawe to Oxford consolidated her class politics and spurred her feminism, while catalysing her lifelong dedication to questions of uneven development (Peck et al., 2018). As Massey once explained, '[I] have lived with, through and kind of in combat with regional inequality [since] my childhood' (Massey and HGRG, 2009: 405).

Massey began her research career at the Centre for Environmental Studies (CES), founded in 1967 to address urban and regional development under the Labour government of Harold Wilson. The CES sought progressive reforms through the application of mathematical economics and analytical modelling. She would go on to complete her Masters at the University of Pennsylvania in regional science and would return to a CES that was also in transition towards more progressive research and policy (Peck, 2024). In 1973, Massey published her first critical academic paper 'Towards a critique of industrial location theory', in the radical geography journal *Antipode*, where she laid out the limits of the dominant firm-centred neoclassical approach.

Over the course of the next decade, Massey would reconceptualise space, place, and methods of inquiry, marking a watershed for economic geography, urban and regional studies and beyond. The main thrust of her intervention was to shift the concept of 'space' from a neutral, static backdrop for the distribution of economic activity to a dynamic set of relations that actively (re)made economic organisation and broader social relations. This was the essence of why *geography matters*, a phrase closely associated with Massey's oeuvre, in a material, political and social sense. Key to this move was to rethink 'the region'. A region's economy was not the property of a given, bounded area but rather 'a complex result of the combination of its succession of roles within the series of wider, national and international, spatial divisions of labour' (Massey, 1979: 235). The relationship between the dynamics of accumulation and the shifting geographies of work and production was one of mutual interaction and adaptation. To grasp the constitutive role of spatial variegation, geography and allied fields would have to dispense with 'reading' industrial development off of superficial phenomena like changes in employment or investment numbers. With CES colleague Richard Meegan, Massey (1982) showed that seemingly similar empirical regularities like job loss could in fact be attributable to distinct processes and their combination. This approach required a patently different set of conceptual tools and methods.

Massey would elaborate these in her transformative monograph, *Spatial Divisions of Labour* (1995 [1984]). Remarkably prefiguring later global production debates, Massey identified 'spatial structures of production' that organised the functional division of labour in an industry (a term she used to refer to all paid employment). These structures did not simply reflect the desires of owners or shareholders; rather they were formed through changes in social relations, both reflecting, conditioning and remaking patterns of regional difference. Articulation would become a key concept for Massey, betraying the early influence of French philosopher Louis Althusser (Massey, 1995 [1984]). Economic, political and ideological forms and practices articulated regional conjunctures. Causality, therefore, had to be interrogated given the combination of elements, and the shifting social relations that reproduced regions differently over time. Massey argued, for example, that histories of trade unionism and changing gendered participation in labour markets were central to how regions were being restructured. She wrote, '[s]uch shifts in spatial structures are a response to changes in class relations, economic and political, national and international [...] [T]he geography of industry is an object of struggle. The world is not simply the product of capital's requirements' (Massey, 1995 [1984]: 7).

Massey's interventions clearly hinged on her tireless commitment to Left politics and her insistence on the usefulness of spatial concepts for advancing social justice. Critical debates on space and place were not a scholastic exercise, but instead key for thinking through political possibilities. After joining the faculty of the Open University in Milton Keynes in 1982 (where she would remain until her retirement in 2009), Massey advised the Greater London Council (GLC) on practical efforts to 'restructure for labour' (rather than capital; Peck, 2024). She also participated in solidarity work for the historic 1984–5 Miners' Strike, a labour struggle *cum* regional restructuring process that figured prominently in *Spatial Divisions of Labour*. Around this time, Massey began her lifelong engagement with Latin American Left movements and governments in the spirit of learning from and thinking with the Global South (Zusman, 2018). As Thatcherism consolidated, the government shut down spaces for progressive policymaking, including the GLC. Massey maintained her steadfast commitment to socialist politics and radical spatial theory in a fast-changing political conjuncture, doing so in a variety of venues and for numerous audiences. She contributed regularly to *Marxism Today*, a high-circulation Left magazine; she formed part of the Ariel Road Group with lifelong political comrades, including Ken Livingstone and Hilary Wainwright; and she participated in the Hegemony Group with Stuart Hall, Chantal

Mouffe and Ernesto Laclau to read and discuss radical democracy. Massey would go on to establish the progressive journal *Soundings* with **Stuart Hall** and Michael Rustin in the mid-1990s. In these and other endeavours, Massey's influence extended to municipal socialism, the New Left, feminism, and British cultural studies (Featherstone and Kelliher, 2022).

In the ensuing decades, Massey would marshal the critical spatial thinking and feminist epistemology she had pioneered in the 1980s on UK restructuring to broader debates on the meaning and status of the watchwords of the day: globalisation and postmodernism. While it is perilous to collapse her many contributions on these debates, three broad, inter-related themes emerge: gender and flexibility, place and anti-essentialist identity, and space as a relation of power (see Callard, 2011). First, the nexus of capitalism and patriarchy, and the consequences of their articulation for regional trajectories of investment and government policy, were longstanding themes of Massey's work. From the vantage point of regional cultures, capitalist relations of production could cement or disrupt gender hierarchies; the outcome was not foretold and it very much shaped successive developments (Massey, 1995 [1984]). In the 1990s, Massey turned to the high-tech sector, widely celebrated as a regional solution in an era of globalisation. Her study of Cambridge's (UK) high-tech cluster showed how the industry engendered a masculinised workforce whose ability to fill these 24/7 flexible occupations depended upon their 'service wives' (1995a; 1995b). Foreshadowing arguments in her book *World City* (2007) and reflecting her critique of the so-called regional problem, Massey emphasised south-east England's relational construction. The concentration of creative and/or managerial jobs there reproduced class identities that reinforced place-based race and class segregation while consigning other regions to devalued positions (1995a; 1995b).

Second, with rising nationalist violence across the former Soviet Union and ethnic cleansing in Yugoslavia in the 1990s, Massey sought to reclaim a place for progressive ends in debates on globalisation (Massey and HGRG, 2009). Regressive class, gender, and racial biases were built in and through local/global dualisms, not just in liberal, but also in Left discourses. Drawing on a range of anti-essentialist approaches, from feminist, post-structuralist, postcolonial, and later queer theory, Massey would re-envision place as 'a particular constellation of relations, articulated together at a particular locus' (Massey, 1993: 66), relations constructed on a far larger scale than the locus itself. This progressive or global sense of place was neither de facto 'good' – i.e., a romanticised redoubt against global capitalism – nor 'bad' – i.e., a homogeneous, exclusionary enclave. Rather, in deconstructing the global/local dualism, Massey foregrounded place-as-process, as the site of politics, full of differences and conflicts, of potential to build solidarities and to navigate and negotiate historic and emerging tensions (1994). She insisted that identity premised upon 'a deep connection to place does not have to be insular. It can be "place-based" but not "place-bound"' (Featherstone and Kelliher, 2022: 11).

Third, Massey (1992b) expanded her interventions on place through a sustained critique of the concept of space in Western philosophy, including postmodern and post-structuralist theories. She frequently wrote against a common tendency to portray space as inert, a mere surface, or a projection of historical sequence (as in 'develop*ing* countries') in Western thought. Instead, she proposed a deeply relational view:

> The spatial [...] can be seen as constructed out of the multiplicity of social relations across all spatial scales, from the global reach of finance and telecommunications, through the geography of the tentacles of national political power, to the social relations within the town, the settlement, the household and the workplace. It is a way of thinking in terms of the ever-shifting geometry of social/power relations, and it forces into view the real multiplicities of space-time [...] The spatial is both open to, and a necessary element in, politics in the broadest sense of the word. (Massey, 1994: 4)

Key to this conception was Massey's analysis of power. Space was shot through with the symbolic and material conditions of exclusion and oppression, what she called 'power-geometries', on the one hand, and the potential of forging progressive relations, on the other (Massey, 1991a; 1993). Massey would propose a three-fold conceptualisation of space in her germinal Hettner Lectures (1999a), which she would later elaborate in her widely-cited monograph, *For Space* (2005). First, she argued space was the production of interrelations, and, second, that these constituted a contemporaneous plurality, or what she called a multiplicity. Third, these relations, as 'embedded material practices which have to be carried out', were always becoming; or in sum, space could be imagined 'as a simultaneity of stories-so-far' (Massey, 2005: 8). In her numerous contributions on space, Massey offered not a model, but rather a set of conceptual provocations against dominant mobilisations of space as a static, passive surface. This is what it meant to be 'for space'.

Reflecting the long-time influence of Marxist philosopher Antonio Gramsci on British New Left politics, Massey would go on to develop an analysis of the fissures and cracks in neoliberal hegemony that had emerged in the wake of the global financial crisis of 2008. She undertook this work with her *Soundings* collaborators, **Stuart Hall** and Michael Rustin. Together they published *The Kilburn Manifesto* (Hall et al., 2015), named after their north London neighbourhood (also a key locus for Massey's earlier framing of a global sense of place, 1991a). More than a text, *Kilburn* was a political project that involved large public meetings, contributions to popular media, and other activities. The project's aim was to offer a conjunctural analysis of neoliberalism's persistence and the possibilities for rupture of this 'common sense' in order to change it. In this and a host of other undertakings across multiple disciplines, and especially in public engagement involving all manner of interlocutors, Massey would continue to explore generative connections between relational thinking, place-based progressive identities, and power relations across sites and scales until her death in 2016.

Spatial Contributions

Massey's contributions are both far-ranging and remarkably consistent in their overall orientation. In the throes of industrial restructuring and emerging financialisation in 1980s Britain, Massey not only pushed back against anaemic neoclassical location theory, but also rejected mechanistic formulations of uneven development that circulated in orthodox Marxism, both quick to dismiss or devalue phenomena that did not easily fit their frameworks. What her approach to space and place insisted upon was the significance of unique outcomes, as neither outliers nor distractions, but rather as the very stuff of capitalist restructuring and our theories about it. The point was not the unmoored pursuit of idiographic description, but rather the construction of a spatial theory that embraced the openness and potential of social and spatial categories and the need to determine, through the analytical process, their form and significance. 'The "localities" with which the analysis starts', she wrote, 'do not have the same validity or boundaries by the end' (Massey, 1995 [1984]: 189). Her approach opened the field to much less deterministic accounts of how socio-economic structures, multiple axes of social difference and regional trajectories contingently came together to produce uneven geographical development along existing and new contours.

In questioning the very boundaries of regions and places, and in developing a notion of space as process rather than static background, Massey drew upon and extended poststructuralist and feminist epistemologies. Influenced by socialist feminism as well as thinkers like **Donna Haraway**, and in dialogue with feminist geographers like Jane Jenson, Susan Hanson, and **Linda McDowell**, Massey adopted a critical anti-essentialist stance:

> [W]hat an anti-essentialist position [...] represented is an optimism of ontological assumptions with particular reference to feminism. It has subsequently been invested with other advantages, most particularly the way in which, through the constant recognition of articulation, it enables an approach to the contingencies and uniquenesses, as well as the broad structures of space-time. (Massey, 1995[1984]: 353)

Articulation was a way to analyse the coming together of different social forces and dynamics that did not presume their form or relative importance *a priori*. In spatial terms, Massey operationalised articulation as a relational approach. Relational thinking would become a key tool for poststructuralist geographies, including Gill Hart's work on neoliberal restructuring in South Africa (2002) and **J.K. Gibson-Graham**'s contributions on the political potential of diverse economies (2008).

Massey's relational approach bridged the normative, often abstract realm of political theory (e.g., Rawls' *Theory of Justice*, 1971) with economic geography, which tended to position normative questions beyond its remit. A relational understanding of space and place necessarily called for a progressive ethics of responsibility of 'place beyond place' (Massey with HGRG, 2009). As Massey would return to in *World City* (2007), not all places were 'subject to' globalisation. The responsibility of places to other locales rested upon their relative positions of power, exploitation and dispossession. 'The nature of the resubjectivation required, and of the responsibility implied in consequence', Massey wrote, 'also varies between places' (Massey, 2004: 10). For Massey then, relationality indexed dimensions of power and positionality, and of the possibilities to form progressive, political solidarities across constituencies and among different social movements. Massey would later disagree with 'post' post-structural uses of relationality in the early 2000s that effectively denied material realities and power relations. She attributed this turn to a disjuncture between theory and politics. 'The real two-way movement', Massey insisted, 'between the conceptual and the political has always, for me, been very important' (Featherstone et al., 2013: 260). Her work on geographies of responsibility spoke directly to her lifelong commitment to internationalism. In 2007, she was invited to Venezuela to revisit her notion of power-geometries in the context of the government's efforts to address long-standing uneven development in the country. 'My appreciation for the forms and nuances of power was vastly enriched', she reflected, '[and] the concept of power-geometries itself was dynamised – precisely because it was being used in a political *process*' (Featherstone et al., 2013: 263).

Key Advances and Controversies

Massey's interventions on capitalist restructuring were part and parcel of a nascent radical geography. Thus, controversy surrounding her work was largely about the terrain and stakes of this emerging field, including who and what was, as Susan Christopherson (1989) put it, 'outside the Project'. **David Harvey** argued that Massey's emphasis on place-based historical specificity and differentiated outcomes of capitalist restructuring effectively evacuated Marxist categories and reduced theory to empiricism (Harvey, 1987). Massey's response to Harvey's charge that *Spatial Divisions of Labour* was weighed down with contingency and complexity was characteristically forthright: 'it certainly was and [...] I meant it to be!' (Martin et al., 1993: 71). Early on, Massey embraced Marxism as method not doctrine, one that proceeded through a mix of critique of existing categories (the region, gender, class) and their subsequent reconceptualisation through rigorous empirical engagements that embraced articulation and social difference.

Controversies over the study of regional restructuring in the UK continued in the form of the 'locality debates' of the late 1980s and early 1990s. The debates ensued from a large research program, inspired by Massey's framework, to examine socio-spatial change in seven UK regions. Participants in these debates wrangled over the balance between theory and empirical research, the utility of comparison among diverse case studies, the relationship between Marxism and postmodernism, and more (Duncan and Savage, 1991). With characteristic acuity, Massey's intervention pushed back against the underlying epistemological assumption of the debate: 'Those who conflate the local with the concrete [...] are confusing geographical scale with processes of abstraction in thought' she argued (Massey, 1991b: 270). This key insight, simple but powerful, would shape subsequent debates on globalisation, where Massey insisted on the global as both a particular coming together of relations in place and a perception that reflected people's socio-spatial positionality. In her widely read critique of two major monographs on postmodernism, **David Harvey**'s *Condition of Postmodernity* (1989) and *Edward Soja*'s *Postmodern Geographies* (1989), Massey took both authors to task for their monolithic view of modernity, which universalised the particular perspective of the white, male, heterosexual, Western subject (1991c). In her grounded, engaged style, Massey extended her critique of this privileged position in her subsequent essay, *A place called home?* (1992a). Drawing on **bell hooks'** reading of Toni Morrison's *Beloved* (1991), Massey wrote: '[t]hose who today worry about a sense of disorientation and a loss of control, must once have felt they knew exactly where they were, and that they had control [...] [For the colonised] the security of boundaries of the place one called home must have dissolved long ago' (Massey, 1992a: 9–10).

After her death, Massey's scholarship has continued to resonate in multiple spheres. In the discipline of Geography, her excavations of common ground between critical human geography and geomorphology anticipated and partly inspired efforts to bridge the divide, most recently with critical physical geography (Massey, 1999b; Lave, 2014). The political conjuncture that produced the UK vote to leave the European Union (so-called 'Brexit') saw those disinvested peripheries that had long been at the centre of Massey's research agenda reject an establishment politics that had discounted them (Beynon and Hudson, 2018). Overall, Massey's generative conceptions of space and place for a progressive class politics of alliances across differences, locally and internationally, are more salient than ever.

Massey's Key Works

Massey, D. (1973) 'Towards a critique of industrial location theory', *Antipode*, 5 (3): 33–9.

Massey, D. (1979) 'In what sense a regional problem?', *Regional Studies*, 13 (2): 233–43.

Massey, D. (1991a) 'A global sense of place', *Marxism Today*, June: 24–29.

Massey, D. (1992a) 'A place called home?', *New Formations*: 3–15.

Massey, D. (1992b) 'Politics and space/time', *New Left Review*, 196: 65–84.

Massey, D. (1993) 'Power-geometry and a progressive sense of place', in J. Bird, B. Curtis, T. Putnam and L. Tickner (eds), *Mapping the Futures: Local Cultures, Global Change*. London: Routledge. pp. 59–69.

Massey, D. (1994) *Space, Place and Gender*. Cambridge: Polity.

Massey, D. (1995 [1984]) *Spatial Divisions of Labour: Social Structures and the Geography of Production*. New York, NY: Routledge.

Massey, D. (1999b) 'Space-time, "science" and the relationship between physical geography and human geography', *Transactions of the Institute of British Geographers*, 24 (3): 261–76.

Massey, D. (2004) 'Geographies of responsibility', *Geografiska Annaler*, 86B (1): 5–18.

Massey, D. (2005) *For Space*. London: Sage.

Massey, D. (2007) *World City*. Cambridge, UK: Polity.

Secondary Sources and References

Beynon, H. and Hudson, R. (2018) 'The road to Brexit on the British coalfields', in M. Werner, J. Peck, R. Lave and B. Christophers (eds), *Doreen Massey: Critical Dialogues*. Newcastle: Agenda. pp. 161–72.

Callard, F. (2011) 'Doreen Massey', in P. Hubbard and R. Kitchin (eds), *Key Thinkers on Space and Place* (2nd edition). Thousand Oaks, CA: Sage. pp. 299–306.

Christopherson, S. (1989) 'On being outside "the project"', *Antipode*, 21 (2): 83–9.

Duncan, S. and Savage, M. (1991) 'New perspectives on the locality debate', *Environment and Planning A*, 23 (2): 155–64.

Featherstone, D. and Kelliher, D. (eds) (2022) *Doreen Massey: Selected Political Writings*. London: Lawrence & Wishart.

Featherstone, D., Bond, S. and Painter, J. (2013) '"Stories so far": A conversation with Doreen Massey', in J. Painter and D. Featherstone (eds), *Spatial Politics: Essays for Doreen Massey*. Chichester: John Wiley and Sons. pp. 253–66.

Gibson-Graham, J.K. (2008) 'Diverse economies: Performative practices for "other worlds"', *Progress in Human Geography*, 32 (5): 613–32.

Hall, S., Massey, D. and Rustin, M. (2015) *After Neoliberalism? The Kilburn Manifesto*. London: Lawrence & Wishart.

Hart, G. (2002) *Disabling Globalization: Places of Power in Post-Apartheid South Africa*. Berkeley, CA: University of California Press.

Harvey, D. (1987) 'Three myths in search of a reality in urban studies', *Environment and Planning D: Society and Space*, 5 (4): 367–76.

Harvey, D. (1989) *The Condition of Postmodernity*. Oxford: Blackwell.

hooks, bell (1991) *Yearning: Race, Gender, and Cultural Politics*. London: Turnaround.

Lave, R. (2014) 'Engaging within the academy: A call for critical physical geography', *Acme: An International E-journal for Critical Geographies*, 13 (4): 508–15.

Martin, R., Markusen, A. and Massey, D. (1993) 'Classics in human geography revisited', *Progress in Human Geography*, 17 (1): 69–72.

Massey, D. (1991b) 'The political place of locality studies', *Environment and Planning A: Economy and Space*, 23 (2): 267–81.

Massey, D. (1991c) 'Flexible sexism', *Environment and Planning D: Society and Space*, 9: 31–57.

Massey, D. (1995a) 'Masculinity, dualisms and high technology', *Transactions of the Institute of British Geographers*, 20 (4): 487–9.

Massey, D. (1995b) 'Reflections on gender and geography', in T. Butler and M. Savage (eds), *Social Change and the Middle Classes*. London: Routledge. pp. 330–44.

Massey, D. (1999a) 'Philosophy and politics of spatiality: some considerations. The Hettner-Lecture in Human Geography', *Geographische Zeitschrift*, 87 (1): 1–12.

Massey, D. and Human Geography Research Group (2009) 'The possibilities of a politics of place beyond place? A conversation with Doreen Massey', *Scottish Geographical Journal*, 125 (3–4): 401–20.

Massey, D. and Meegan, R. (1982) *The Anatomy of Job Loss: The How, Why and Where of Employment Decline*. London: Methuen.

McDowell, L. (2018) 'North and South: Spatial divisions in a life lived geographically', in M. Werner, J. Peck, R. Lave and B. Christophers (eds), *Doreen Massey: Critical Dialogues*. Newcastle: Agenda. pp. 41–52.

Peck, J. et al. (2018) 'Out of place: Doreen Massey, critical geographer', in B. Christophers, R. Lave, J. Peck and M. Werner (eds), *The Doreen Massey Reader*. Newcastle: Agenda. pp. 1–38.

Peck, J. (2024) 'Open horizons: Doreen Massey', in N. Castree, T.J. Barnes and J. Salmond (eds), *Making Geography Matter: The Past and Present of a Changing Discipline*. London: Routledge.

Rawls, J. (1971) *A Theory of Justice*. Cambridge, MA: Belknap Press.

Soja, E.W. (1989) *Postmodern Geographies: The Reassertion of Space in Critical Social Theory*. London: Verso.

Zusman, P. (2018) 'Doreen Massey and Latin America', in M. Werner, J. Peck, R. Lave and B. Christophers (eds), *Doreen Massey: Critical Dialogues*. Newcastle: Agenda. pp. 303–12.

33
Achille Mbembe

Amber Murrey and Patricia Daley, University of Oxford

Biographical Details and Theoretical Context

Achille Mbembe is a widely-cited Cameroonian historian, political philosopher, and critical theorist. He was born in the town of Otélé in Central Cameroon in 1957, in the twilight of the French colonial period. He described his childhood as one 'in a community where each home was by definition a home for others' (Oboe, 2010: 1). On 13 September 1958, Mbembe's uncle, Pierre Yem Mback, was assassinated by the French army alongside the anti-colonial leader and teacher, Ruben Um Nyobè, and six other militants. His grandmother, Suzanna Ngo Yem, was also an independence fighter with the Cameroonian maquisards, one of the most powerful resistance movements against French colonial occupation of the mid-twentieth century. He credits her for 'forging his intellectual sensibility' from a young age and recalls singing independence songs as a child (Juompan-Yakam, 2021).

Mbembe attended a Catholic boarding school, Jeunesse Étudiante Chrétienne (JEC), where he was the editor-in-chief of the student journal, *Au Large*. As a young student, he published several articles critical of the complicity of the Catholic church in the oppressive practices of the Cameroonian postcolonial state. His master's thesis on the topic of anti-imperial resistance in the Sanaga-Maritime region, supervised by the influential historian Martin Zachary Njeuma, was prohibited by government administrators and was never publicly defended. While studying at the JEC and later at the Université de Yaoundé, Mbembe travelled around Cameroon and visited Zimbabwe and Tanzania, where he learned of the Pan-African socialism of Julius Nyerere. Mbembe's first published book was an edited volume of Ruben Um Nyobé's speeches and writing, published by L'Harmattan in 1984. The book was banned in Cameroon and Mbembe was blacklisted, unable to return to the country until 1992 (Juompan-Yakam, 2021).

Mbembe completed his PhD at the Université de la Sorbonne in Paris in 1989, under the supervision of the French historian and Africanist Catherine Coquery-Vidrovitch. He then held academic posts at several US universities, including Columbia, Brookings Institution, University of Pennsylvania, University of California Berkeley, and Yale. From 1996 to 2000, Mbembe was the Executive Secretary of the Council for the Development of Social Science Research in Africa (CODESRIA) in Dakar, Senegal, the leading institution supporting social science research on the African continent. In 2000, he moved to South

Africa, where he has held various posts at Witwatersrand University and Stellenbosch University. While he describes South Africa as his current 'center of gravity, [his] watchtower into the global world, the primary location of [his] political and intellectual reflection', he understands himself as 'belonging to nowhere in particular' (Oboe, 2010: 1). In a 2021 interview, Mbembe spoke of his enjoyment of the 'radical inner freedom' (radicale liberté intérieure) of being an eternal voyager (éternel passant) (Juompan-Yakam, 2021).

Mbembe's scholarship is broad and transdisciplinary. His thinking is grounded in a critique of Eurocentrism and it is oriented to topics of sovereignty, knowledge, violence, and Self/Other; the conditions of living together, humanness, inhumanity, and thinghood; racism, racialisation, and planetary political economies of racial capitalism, with a particular focus on the Black Atlantic and the place of Africa. His writing tends towards the figurative, capacious, nested, and poetic. He describes his task as mobilising language 'to return to life what had been abandoned to the powers of death' (Mbembe, 2019: 8). While his work speaks to planetary phenomena, 'the African world [...] occup[ies] the forefront of [his] reflection[s]' (Mbembe, 2019: 7). He understands writing from Africa as a process of repair against histories of 'suffering, trauma and victimization' (Mbembe, 2016: 30). He also seeks to mobilise a critical and creative reimagination of Africa. In *Critique of Black Reason,* he explains that he seeks to 'inhabit several worlds at the same time, not in an easy gesture of fragmentation, but in one of coming and going, able to authorize the articulation, from Africa, of a *thinking of circulation and crossings'* (Mbembe, 2017: 8, original emphasis). Attuned to the geopolitics of knowledge and struggles to decolonise knowledge, Mbembe describes the contemporary university as being 'reframed and rescaled' with the 'purpose of better turning it into a springboard for global markets in an economy that is knowledge based, innovation based' (Mbembe, 2016: 38).

Spatial Contributions

The withdrawal of European powers from Africa transformed, but did not end, domination and repression. Mbembe (2001) asserts that the resulting postcolony is subjected to a kind of colonialism, albeit in revised form. The postcolony is characterised by the persistence of colonial economic relations, the supremacy of authoritarian politics, and novel forms of violence and inequality, with postcolonial domination networked between various actors beyond the nation-state. For Mbembe, the postcolony is delineated by struggles for power, resources, and recognition, both within the confines of the nation-state and on the global stage.

When *On the Postcolony* was published, the study of Africa tended towards antagonistic readings between the state and society (Adebanwi and Orock, 2022). In contrast, Mbembe focused his analysis on spectacle, symbol, fetish, and carnivalesque in post-colonial African political expressions, primarily in Cameroon and, to a lesser extent, Togo. Political domination in the postcolony, according to Mbembe, is pervasive, both highly symbolic and yet materially concrete. Influenced by the work of Hannah Arendt, Mbembe theorised the state as a key site of the 'banality of power' and violence. For Mbembe, the body is also a site of resistance, violence, excess, vulgarity, and spectacle. The postcolony is not restricted to a specific geographical or temporal context, instead serving as a broader framework for apprehending the conditions of power, subjectivity, and resistance in other societies across the globe. Mbembe's article introducing some of these theorisations, 'Provisional notes on the postcolony' (Mbembe, 1992), remains (as of mid-2023) the most cited article in the history of the journal *Africa*.

Mbembe's subsequent reconceptualisation of the politics of power and racial violence set the agenda for a generation of scholars working in global critical theory. Expanding upon Frantz Fanon's theorisation of colonial violence, he conceptualised the colony and the plantation as spaces of exceptional violence where racial capitalism was refined and a toxic political milieu created 'death-worlds' (Mbembe, 2019: 92). Mbembe coined the term 'necropolitics' to offer a genealogy of contemporary exclusions emerging from the planetary continuation of colonial relations of racist violence, predation, and extraction during the early years of the twenty-first century. In *Politiques de l'Inimitié* (first published in 2016, and translated into English in 2019 as *Necropolitics*), he analysed the converging and perilous situations of the age: fascism, racism, nationalism, terror, militarisation, and the inversion of democracy. 'War', Mbembe (2019: 3) explained, had 'become both remedy and poison—[the] pharmakon of our time'. For Mbembe, states are produced through determinations and evaluations of who dies and who is targeted for death. Necropolitics encompass physical violence as well as the production of cultural subjectivities and the political and economic conditions that negate life and render it unliveable – thus it draws from Fanon's notion of the 'zone of non-being' and **Foucault's** biopolitics. Foucault described the significance of biopolitics within contemporary statecraft as the organising mechanisms and institutional forms of knowledge and power that erect authorities and influence and enforce power over life (often referred to as the 'will to life') (Foucault, 2008 [1979]). Within his examination of modern violence and statecraft, Mbembe critiqued the destructive sibling of biopolitics, necropolitics, as a form of power that administers, optimises, and multiplies death (or the 'will to death').

For Mbembe, racism and marginalisation operate through necropolitical means, fostering environments in which certain groups do not have access to resources (including, sometimes, the imaginative resources) necessary for living dignified lives. People are therefore suspended between life and death in 'death-worlds' (Mbembe, 2003: 40): they experience 'expulsion from humanity altogether' (Mbembe, 2003: 21). Mbembe (2019: 5) writes, 'the colonial machinery sought not merely to block their desire to live. It aimed to affect and diminish their capacities to consider themselves moral agents'. Mbembe's theorisations of spaces of the postcolony has influenced geographical research on border regimes, border enforcement, and refugee and immigration detention camps. Mbembe's necropolitics have informed research on the neglect of racialised bodies in public health emergencies and on the prison complex.

Mbembe's theorisation of the relationship between space and violence has helped shape historical and contemporary work by postcolonial and decolonial geographers and has informed studies of Black geographies in the Americas. Critical geographers have expanded his observations regarding the racialisation of disposability within necro-economies of global racial capitalism (Chari, 2015; Neely and Lopez, 2022). For example, Wright builds upon Mbembe's necropolitics to examine femicide in Ciudad Juárez, arguing that the politics of gender are central (Wright, 2011).

McKittrick's (2011: 960) work on the geographies of transatlantic slavery is influenced by Mbembe's analysis of the plantation as a 'materializ[ation of] modern terror'. These ideas are enriched by **Gilmore**'s (2007) examinations of the normalisation of premature death in US racial capitalism. Scholars have also drawn greater attention to the capacity of people to resist and offer alternative imaginaries to those necropolitical relations which would seek to render entire populations disposable or 'surplus'. Mbembe himself speaks of the need to avoid the trap of nihilism. In an interview with David Theo Goldberg, he comments that '[t]o recall past horrors does not need to necessarily end with, as the only alternative, a politics of impossibility', articulating the need for a holistic notion of justice that is reparative, regenerative and transformational (cited in Goldberg, 2018: 218).

Mbembe has also made an important intellectual contribution through *Critique of Black Reason* (2017), a philosophical study of the meaning of Blackness within its historical development. He writes, 'Since the beginning of the eighteenth century, Blackness and race have constituted the (unacknowledged and often denied) foundation, what we might call the nuclear power plant, from which the modern project of knowledge – and of governance – has been deployed. Blackness and race [...] represent twin figures of the delirium produced by modernity' (Mbembe, 2017: 2). For Mbembe, Black reason includes those forms of thought that emerge from the experience of being excluded, subjugated, and oppressed within and by dominant forms of knowledge and (white) power. It is characterised by the ways in which it resists and subverts dominant discourses and ways of knowing. Black reason thus fosters alternative knowledges that emerge from lived experiences of Blackness and through which repair might be sought. Black reason is not merely reactionary or oppositional; it is dynamic, fluid and creative. In this, we see the connections between his notion of Black reason and Afropolitanism (Mbembe, 2021).

Throughout his career, Mbembe has made pivotal contributions to the fundamental criticisms of Eurocentric Africanist scholarship, criticisms which have reverberated within various strands of African geographies. For example, he has been at the forefront of critiques against ethnographic presentism (that freeze Africa in time and deny its co-evalness) as well as studies that perpetually reiterate the existence of a gap or absence in African being-in-the-world. Mbembe reflected that he wrote *On the Postcolony* 'at a time when it appeared to many as if the study of Africa was caught in a dramatic analytical gridlock' (Mbembe, 2005: 5). He continues, 'Africa is not only perpetually caught and imagined within a web of difference and absolute otherness [...] Africa epitomizes the intractable, the mute, the abject' (Mbembe, 2005: 6–7). His critique of Africanist scholarship demands recognition of how two dominant frameworks – Marxist-nationalist and nativist – have long framed (and limited) scholarship, so that even critical texts remain embedded in the conceptual structures of the West, what he calls 'arche-writing'.

Mbembe coined the philosophical concept of 'Afropolitanism' to refer to the ways in which Africa and Africans always-already belong in the world, explaining that 'there is no world without Africa and there is no Africa that is not part of it' (Balakrishnan, 2016: 29). He also distinguishes the concept from Pan-Africanism, referring to the latter as a 'racial ideology' and cautions that Afropolitanism is a 'planetary' rather than an 'Afrocentric reading'. Mbembe's move away from Pan-African thought may reflect his disillusionment with the African nationalist movements of the 1950s and 1960s, including the oppressive regime of Cameroon's first president, Ahmadou Ahidjo. In his elaboration of the concept of Afropolitanism, Mbembe asserts that the world's future will be 'played out' on the continent. (These sentiments are echoed in his engagement with the related intellectual projects of 'Southern Theory' à la Raewyn Connell and 'Theory from the South' à la Comaroffs). These siblings of critical thought shy away from anti-racist praxis and action, tending towards philosophical rumination and ethical mediation in response to racial capitalism.

Key Advances and Controversies

Several of Mbembe's speaking engagements have been met with controversy. He was disinvited from the Ruhrtriennale Cultural Festival in Germany in March 2020 due to criticism that he had 'relativised the Holocaust' by drawing comparisons between it and the apartheid regime in South Africa (Sznaider, 2021). In October 2021, Mbembe co-organised the 'New Africa–France Summit' with French President Emmanuel Macron. The summit was criticised for its lack of engagement with the neo-imperial relations that continue to influence the relationship between France and its former colonies, such as CFA monetary colonialism and troop presence in Mali and Burkina Faso. Some critiqued Mbembe for participating in a 'clean-up' of French empire and its 'toxic colonial relations' (Bisoka et al., 2021). Hundreds of activists, movement leaders, and Pan-Africanists organised a 'counter-summit', denouncing la Françafrique.

Mbembe's collected works have advanced understandings of the intersection of race, space, and power (see Hafiz, 2020). His exploration of the genealogy of Blackness in Africa and the Atlantic World advances the debates of postcolonial scholars of the Black Atlantic (Mbembe, 2017). At the same time, his work has not thoroughly considered worlds prior to the historical construction of Blackness. In his attempt to create a 'world beyond race', he dismisses studies of precolonial Africa as 'racial mythology' or 'Africanity' (Mbembe, 2019: 177, 81). For him, it is problematic to look for ontological existence prior to Blackness. Some passages of *Critique of Black Reason* have been interpreted as objectifying and essentialising Black people, inadvertently repeating the weaknesses of the writings that it is critiquing. Apata explains that for Mbembe:

> Blackness becomes a unique intrinsic property that all black people share, like a blood type that runs through the veins of every black person and acts like cordage that binds them all together. (Apata, 2017: 288)

Yet, for example, colour gradations play a major role in the racial hierarchies of the Caribbean and Latin American states. Colourism matters precisely because Black people have diverse experiences in the different places in which they live (Daley, 2020). The subdiscipline of Black Geographies has critically reiterated the heterogeneities of Blackness and Black experiences and centred the significance of care and shared joy in being human.

Scholars have drawn attention to Mbembe's methodological and theoretical orientations and groundings in several key Western theoretical resources, including Bayat, Freud, Arendt, **Foucault**, and more (Adebanwi and

Orock, 2022: 45). African feminists have critiqued Mbembe's work for following the Eurocentric tradition of focusing on the 'Black Man' as the object of analysis. In *Critique of Black Reason*, it is argued that every pronoun he uses is 'he' (Apata, 2017). This authorship reiterates the enlightenment position of masculine reason; for example, Mbembe (2019: 178) writes that 'to be African is to be first and foremost a free man'. The absence of a role for African women amongst his 'politics of humanity' or his 'conception of justice and responsibility' is an omission, a tendency that many Black feminists would term 'misogynoir'. This lacuna is sometimes reflected in the literature he centres and references. Kharnita Mohamed writes:

> none of the scholarship, fiction or writing by Black women is accorded the careful, respectful and loving attention given to canonical Black male writers' reasoning. What is an archive, if Black reason is delimited to male writing? [...] How are we to understand the stakes of racialised logic if we only bring men (White or Black) into conversation with each other? Are there no worthwhile texts by Black women to include in the making of Black Reason? (Mohamed, 2018: 243)

Mbembe's work has influenced how critical scholars understand contemporary power, politics, and violence in the postcolony and the African Diaspora. Yet, a sustained critique of his work has been that his articulation of modern racialised power effects a form of Afro-pessimism, from which there seems to be no escape. This is partially because his work engages less with concrete struggles against racial capitalism, apartheid, and other forms of oppression that have shaped Black people's lives on the African continent. Mbembe tends to engage superficially with Pan-African intellectual movements, their various tendencies, and range of geographical imaginaries, some of which embraced a unified Africa with its diverse peoples, as Mbembe seeks to do. Mbembe (2019: 178) yearns for Black people 'to escape the status of victimhood', yet his imagination of liberatory futures has tended to overlook the concrete practices of existing struggles to produce this self and communal actualisation.

Achille Mbembe has made significant contributions to ideas of space and place, including embodiment. His scholarship offers a reinterpretation of the border, the city, the plantation, and the postcolony as governed by specific racialised and embodied relations. Against the violence and destruction of necropolitics, Mbembe sketches a nascent project of transfiguration: emboldened with the 'ethics of the passerby' (le passant), the voyager can move away from the false promises of birth, nationality, and citizenship and embrace the possibilities of 'becoming-human-in-the-world' through simultaneous 'solidarity and detachment, but never indifference' (Mbembe, 2019: 187). Working from Fanon and feminist thought, Mbembe sees potential within the pursuit of relations of care for other beings *'in-common'* (Goldberg, 2018: 207). His pursuit of the humanity and 'radical openness' of the eternal voyager explicitly turns away from other Black intellectual projects like fugitivity and marronage. He argues that 'freedom consists in full habitation of the world', including its contradictions, ambiguities, and uncertainties, and, therefore, freedom is 'not a matter of running away to form some separate space of seclusion' (Goldberg, 2018: 226). For those interested in concrete practices of care and freedom, the specific interplays between action and thought within this political struggle need yet to be refined.

Mbembe's Key Works

Mbembe, A. (1992) 'Provisional notes on the postcolony', *Africa*, 62 (1): 3–37.

Mbembe, A. (2001) *On the Postcolony*. Berkeley, CA: University of California Press. Originally published in 2000 as *De la Postcolonie. Essai sur L'imagination Politique dans l'Afrique Contemporaine*. Paris: Karthala.

Mbembe, A. (2003) 'Necropolitics' (trans. L. Meintjes), *Public Culture*, 15 (1): 11–40.

Mbembe, A. (2016) 'Decolonizing the university: New directions', *Arts and Humanities in Higher Education*, 15 (1): 29–45.

Mbembe, A. (2017) *Critique of Black Reason*. Durham, NC: Duke University Press. Originally published in 2013 as *Critique de la Raison Nègre*. Paris: Éditions La Découverte.

Mbembe, A. (2019) *Necropolitics* (trans. S. Corcoran). Durham, NC: Duke University Press.

Mbembe, A. (2021) *Out of the Dark Night: Essays on Decolonization*. New York, NY: Columbia University Press.

Secondary Sources and References

Adebanwi, W. and Orock, R. (2022) 'Rethinking Achille Mbembe's "Provisional Notes on the Postcolony"', *Africa*, 92 (1): 42–48.

Apata, G.O. (2017) 'Review: Achille Mbembe, trans. Lauren Dubois, Critique of Black Reason', *Theory, Culture & Society*, 34 (7–8): 283–91.

Balakrishnan, S. (2016) 'Pan-African legacies, Afropolitan futures: A conversation with Achille Mbembe', *Transition (Kampala, Uganda)*, 120: 28–37.

Bisoka, A.N., Mwambari, D. and Ndlovu-Gatsheni, S.J. (2021) 'From summit to counter-summit: Imperialism, Françafrique and decolonisation', *ROAPE*. https://roape.net/2021/12/13/from-summit-to-counter-summit-imperialism-francafrique-and-decolonisation/ (accessed November 8, 2023).

Chari, S. (2015) 'African extraction, Indian Ocean critique', *The South Atlantic Quarterly*, 114 (1): 83–100.

Daley, P. (2020) 'Lives lived differently: Geography and the study of black women', *Area*, 52 (4): 794–800.

Foucault, M. (2008 [1979]) *The Birth of Biopolitics: Lectures at the Collège de France, 1978–1979*. Basingstoke: Palgrave MacMillan.

Gilmore, R.W. (2007) *Golden Gulag: Prisons, Surplus, Crisis and Opposition in Globalizing California*. Berkeley, CA: University of California Press.

Goldberg, D.T. (2018) '"The reason of unreason": Achille Mbembe and David Theo Goldberg in conversation about *Critique of Black Reason*', *Theory, Culture & Society*, 35 (7–8): 205–27.

Hafiz, M. (2020) 'Smashing the imperial frame: Race, culture, (de)coloniality', *Theory, Culture & Society*, 37 (1): 113–45.

Juompan-Yakam, C. (2021) 'Achille Mbembe: Macron, Thuram et la pensée décoloniale', *Jeune Afrique*, August 8. https://www.jeuneafrique.com/1202857/culture/achille-mbembe-macron-thuram-et-la-pensee-decoloniale/ (accessed November 8, 2023).

Mbembe, A. (2005) 'On the postcolony: A brief response to critics', *Qui Parle*, 15 (2): 1–49.

McKittrick, K. (2011) 'On plantations, prisons, and a black sense of place', *Social & Cultural Geography*, 12 (8): 947–63.

Mohamed, K. (2018) 'Critique of black reason', *Anthropology Southern Africa*, 41 (3): 242–4.

Neely, A.H. and Lopez, P.J. (2022) 'Toward healthier futures in post-pandemic times: Political ecology, racial capitalism, and black feminist approaches to care', *Geography Compass*, 16 (2): 1–15.

Oboe, A. (2010) 'Africa and the night of language: An interview with Achille Mbembe', *The Salon*, 2: 1–6.

Sznaider, N. (2021) 'The Summer of discontent: Achille Mbembe in Germany', *Journal of Genocide Research*, 23 (3): 412–19.

Um Nyobé, R. (A. Mbembe, ed.) (1984) *Le problème national kamerunais*. Paris: L'Harmattan.

Wright, M.W. (2011) 'Necropolitics, narcopolitics, and femicide: Gendered violence on the Mexico–US border', *Signs*, 36 (3): 707–31.

34
Linda McDowell

Sarah Hall, University of Nottingham

Biographical Details and Theoretical Context

Over her career, Linda McDowell has played a leading role in developing feminist theoretical and methodological approaches within human geography, producing multiple research monographs, significant journal papers and edited collections. Throughout, her work has been based on rigorous empirical research and analysis that focuses on processes of social and economic restructuring in Britain. This has contributed significantly to conceptual understandings of labour market dynamics, the changing nature of work and the role of different forms of migration associated with work in advanced capitalist economies. In addition to her own research, McDowell has mentored a considerable number of graduate students, many of whom have gone on to develop successful academic careers of their own. More broadly, she has played an important role in documenting the gendered nature of the academy and the social practice of geography as a discipline. These research interests are heavily influenced by her own biography of growing up and being educated in the UK in the second half of the twentieth century, something that she herself has reflected on (McDowell, 2003; 2007).

McDowell was born in Stockport, just south of Manchester. After attending a girl's grammar school, she studied at a women's college at the University of Cambridge and was awarded a BA (Hons) degree in Geography in 1971. Here, her early interest in the ways individuals experience economic restructuring was already evident in her undergraduate dissertation in which she conducted interviews with male workers from a number of Eastern European countries who were employed in the steel industry in the English Midlands. These men had come to Britain following the end of the Second World War as EVWs (European volunteer workers) and were employed as labourers in heavy industries such as steelmaking in order to contribute to the rebuilding of Britain's industrial base at the time. Reflecting her commitment to studying the gendered nature of the economy, she subsequently returned to the subject of post Second World War European migration through a study of the lives of Latvian women who were also part of the same EVW scheme established by the British Government at that time (McDowell, 2005).

McDowell was awarded an MPhil degree in Town Planning from University College London in 1973. She then held research positions at the Centre for Research in the Social Sciences, University of Kent, and the

Institute of Communities Studies in London. In 1978, McDowell moved to a Lectureship in Urban Studies and Geography at the Open University. In 1986 she was promoted to Senior Lecturer at The Open University and was also awarded her PhD from University College London. The disjuncture between these early experiences of working in the academy and her formative familial and educational experiences in which 'if you were prepared to work hard, then, it seemed nothing was impossible' (McDowell, 2007: 60) proved to be an important stimulus for her work both as a feminist scholar and her concern about the social practice of Geography as a discipline. As she wrote:

> When I became a university teacher in the 1970s, I found an academy dominated by men. My earlier life, before entering the labour market, however, had not totally resonated with my practical knowledge of men's superiority. (McDowell, 2003: ix)

These experiences gave rise to several papers exploring gendered power relations within the academy (McDowell, 1990a; 1990b). McDowell also extended her interests in this area beyond academic writing by founding, with several other feminists, the Women in Geography Working Party (subsequently the Women in Geography Study Group) of the Institute of British Geographers in 1979, a group that she went on to chair.

In 1992, she moved to the University of Cambridge. In addition to her duties in the Department of Geography, she played a significant role in the management of Newnham College, becoming Vice-Principal between 1997 and 1999. It was during her time at Cambridge that she completed her hugely influential study of gendered identities in the rapidly-changing merchant banking industry in the City of London. In addition to several papers (McDowell, 1994a; 1994b), this study was published as the monograph *Capital Culture: Gender at Work in the City* (McDowell, 1997a). This book was significant across a number of geographical sub-disciplines in addition to the social sciences more widely. At one level, her analysis revealed the persistence of gendered occupational segregation within financial services while, on another, the research made significant contribution to conceptual understandings of the embodied nature and performative qualities of gender identities in the workplace, developing the work of **Judith Butler** (1990). In so doing, this research was one of the earliest and most detailed demonstrations of the need to take issues of culture within economic geography seriously. This followed the 'cultural turn', a research agenda that subsequently expanded substantially (McDowell, 1997b; see also Amin and Thrift, 2004; Pryke and Du Gay, 2007).

After leaving Cambridge in 1999, McDowell held Chairs at the London School of Economics, University College London and University of Oxford. Yet the importance of McDowell's biography in shaping her research interests remained constant. This was particularly evident in her research into the labour market experiences of young white males in Britain who had been identified as 'underachieving' at school. This extended McDowell's interest in the embodied nature of gendered identities to include the construction of multiple masculinities as young men negotiated low-skilled labour markets. Part of the inspiration for this research stemmed from McDowell's observations of her (then) teenage son's classmates as they left education to enter the world of paid work. As she wrote, 'while many of the young women in his peer group achieved good results in school-leaving exams, too many of his male friends lost their commitment to academic success and left school to look for work in a local labour market increasingly dominated by low-paid service-sector jobs' (McDowell, 2003: x). This research, and the methodological issues associated with it, resulted in a number of journal papers (McDowell, 2001a; 2001b; 2001c) and in the monograph *Redundant Masculinities: Employment Change and White Working Class Youth* (McDowell, 2003).

In her later work, McDowell developed these interests further. First, through her book *Working Bodies: Interactive Services Employment and Workplace Identities* (McDowell, 2009) she examined the ways work in advanced economies increasingly draws on the embodied identities of workers. Second, she developed a new research strand that provides what she terms an 'autobiography' of the changing nature of the UK labour market post-1945 as millions of women, including migrants and mothers, joined paid work (McDowell, 2013; 2016). Whilst this research developed her longstanding interest in the working lives and identities of women with different socio-economic backgrounds, it extended the temporal reach of her work, situating UK labour market changes in their longer post World War II histories.

Following her retirement in 2016, McDowell remained an active researcher, becoming Emeritus Professor at the University of Oxford. In the Queen's New Year Honours list of 2016 she was appointed Commander of the Order of the British Empire (CBE) for services to Geography and higher education. The significance of her contributions has also been recognised by the wider higher education community: she was awarded the Victoria Medal by the Royal Geographical Society – Institute of British Geographers in 2008 for her work in feminist and socio-economic geography, was made a Fellow of the British Academy and, from 2017, became a Leverhulme Fellow Emerita.

Spatial Contributions

McDowell's research is characterised by an interdisciplinary theoretical approach grounded in rigorous, largely qualitative, research. The contribution her work has made to advancing feminist scholarship within human geography (McDowell, 1993; 1999) and developing conceptual and empirical understandings of the gendered nature and geographical specificity of both labour markets and workplace identities, is considerable. These contributions can be summarised focusing on three main themes.

First, McDowell has been at the forefront of work in Geography that theorises the relationships between the embodied nature of gendered identities and the changing nature of service work. In so doing, she has moved beyond her earlier work that adopted a materialist approach to gendered occupational segregation (see, for example, McDowell and Massey, 1984). This shift reflects broader changes in the nature of feminist scholarship that:

> Turned from the measurement and explanation of sex segregation in the labour market towards approaches emphasising 'doing' gender, the construction of embodied performances in the labour market, and the fluidity and mutability of gendered identities in different sectors, places and occupations where everyday interactions as well as institutional rules and regulations are part of the construction of workplace-based identifies. (McDowell, 2004: 46)

By adopting such a perspective McDowell's work challenged powerful and enduring discourses concerning the gendered nature of the workplace. For example, her research into the gendered identities performed by merchant bankers in London's financial district was conducted in the wake of significant deregulatory changes, collectively known as 'Big Bang'. Through this research, she countered straightforward public narratives that the arrival of US investment banks into London from the mid-1980s onwards had led to a more meritocratic City labour market for women. In so doing, McDowell's work provided a powerful reminder of the unequal opportunities that still existed within finance, in common with other business services. Similarly in her work on the experiences of male school leavers with few or no qualifications, she sought to counter popular understandings of such young men in Britain as irresponsible 'yobs'. Rather, through interview-based research conducted in Sheffield and Cambridge, she has explored the commitment often demonstrated by such men to 'acquire and hold down a job and to construct lives imbued with the value of domestic respectability, while negotiating the complex and often contradictory expectations associated with working class masculinity' (McDowell, 2003: 2).

This emphasis on the intersection between embodied identities and the transformation of the structures that co-constitute contemporary labour markets was also pursued in *Working Bodies* (McDowell, 2009). Setting out how labour markets are operating through the commodification of previous forms of unpaid labour, she argued that 'what we are witnessing is the commodification of many of the types of work that were previously undertaken mainly in private homes and for "love" – in the sense of not for wages' (McDowell, 2009: 6). She has continued to examine the embodied and gendered nature of various forms of work in her more recent research. For example, *Working Lives* (2013) provides a detailed study across time and space of the changing nature of women's employment in post-war Britain. *Migrant Women's Voices* (2016) likewise developed her interest in women's employment by giving voice to migrant women's labour market experiences in the UK over sixty years in a range of workplaces including care homes, banks, supermarkets and universities.

As this latter example shows, McDowell's work does not simply demonstrate the purchase of a performative and embodied analysis of gendered identities, it also reveals the co-constitutive relationship between the economic structures that reproduce occupational segregation for men and women and the dynamic performance of gendered identities.

Consequently, the second theme characterising her geographical contribution is a commitment to combining both structural and post-structuralist perspectives to understand identity formation and its reproduction in the workplace. She has written several predominantly theoretical pieces that explore this issue through an examination of the relationship between identity, inequality and labour market change (see, for example, McDowell, 2001d; 2004). In particular, she has developed these theoretical interests through a focus on the intersections between gendered identities and class relations (McDowell, 2006; 2008).

In this respect, McDowell played a leading role in challenging 'conventional assumptions made by geographers and other theorists [...] about the nature and location of class struggles' (McDowell, 2006: 826). In particular, she problematised the binary separation in which class relations are understood to be a function of the public sphere whilst gender relations are limited to the preserve of the private sphere, particularly the home. Here she critiqued post-structuralist approaches focused on (gendered) difference in and of itself, drawing attention to the continued importance of the intersection between structural inequalities and gender through a focus on class divisions *between* women. This work takes as its starting point the relatively familiar statistics on the increasing participation of women in the paid labour market in the UK throughout the second half of the twentieth century, in common with other advanced capitalist economies. However, McDowell focused on the uneven class differences that are emerging amongst women working in different geographical and sectoral labour markets, most notably the consumption by middle-class women, particularly mothers, of domestic services provided by working-class women. For McDowell, such developments raised important theoretical questions. She developed these themes through work with migrant women working in a range of workplaces in the UK in her 2016 book *Migrant Women's Voices*. As she had earlier summarised:

> Just as women's entry into the labour market raises questions about the overall structure of the class hierarchy and the degree of polarisation between classes, the growing importance of commodified domestic labour – especially childcare in which middle class women buy services from typically working class women to facilitate their own labour market participation – also raises new questions about class divisions and class contacts between women [...] both within and outside the space of the home. (McDowell, 2006: 840)

In making these arguments, McDowell contributed to the renewed interest in class relations in geography by forwarding a relational understanding of class that took gender issues seriously in its analysis. This was central to her 2014 Roepke Lecture in *Economic Geography* entitled 'The lives of others: body work and the production of difference and labour geographies'.

The third theme running through McDowell's work is a commitment to advancing feminist methodological approaches within geographical research and reflecting on the gendered way this research is regarded within the academy (McDowell, 1992a). Methodologically, her work predominantly uses detailed qualitative research typified by interviews and oral histories. These methodological choices reflect the fact that her work is firmly grounded in feminist critiques of 'scientific' method and an appreciation of the need for researchers to be reflexive about the situated nature of knowledge produced through academic research (McDowell, 1992b). This commitment is reflected in the detailed methodological appendices that are included in all her monographs and her openness in terms of acknowledging possible weaknesses in her own research approach (McDowell, 2003). In so doing, her work has been particularly important in highlighting the methodological possibilities within economic geography in the wake of the 'cultural turn' as interviews, participatory and ethnographic methods have become increasingly widely used. However, McDowell not only played an important role in developing the research practices of contemporary human geography, but became an influential commentator on the practice of geography as an academic discipline and the nature of work in the academy more generally (McDowell, 1990b). This continues to be important given the disproportionately small number of women working in UK university geography departments (particularly at higher levels) when compared to the popularity of geography amongst undergraduate women in the UK.

Key Advances and Controversies

McDowell's commitment to ensuring that feminist perspectives and methodologies are placed more centrally within geographical scholarship was sometimes tested. As she writes:

> It has been tough, especially initially, when feminist work was derided as 'political' and 'biased' [...]. I became used to criticism from referees who saw feminist theories and methods as an anathema, to sexism and hostility from audiences, to colleagues who thought married women had no need of promotion as their husband should be the main breadwinner, to being the only woman on all-male committees. (McDowell, 2007: 61)

However, by working through such challenges, McDowell's work developed and extended several contemporary research agendas and debates in both Geography and the wider social sciences. First, her work on the embodied and performative qualities of identities provided an early example of work at the interface of economic, social and cultural geography. This area of research became quite vibrant as economic geographers sought to theorise the entangled nature of economy and culture. Whilst McDowell's empirical focus in this regard was primarily on labour market change and the workplace, work by others extended this agenda to include issues ranging from individual experiences of transnationalism to a growing attention to consumption practices within the economy (see Jackson, 2002; and Ley, 2004 respectively). Moreover, McDowell's work on performativity has been extended through interdisciplinary theoretical debates in cultural economy. This latter work privileges understandings of performativity that follow philosophical work on performative speech acts and is significantly influenced by insights from Science and Technology Studies (see, for example, MacKenzie, 2006), rather than McDowell's approach that drew heavily on the work of **Judith Butler**. However, cultural economy research in Geography and beyond is beginning to examine the possibilities for a rapprochement between these different approaches to performativity (Berndt and Boeckler, 2009).

Second, McDowell's work on the complex and place-specific nature of identity construction and (re)production from a feminist perspective has been debated and developed by geographers and social scientists more generally. McDowell herself noted the possibilities for developing this work further by focusing on issues of intersectionality. For instance, she suggested that her research on working-class masculinities underlined the importance of interrogating questions of race in studies of working-class identities (McDowell, 2003). Meanwhile, whilst McDowell remained committed to combining structural and post-structural approaches to identity formation, work in geography has been more heavily influenced by post-structural thought. This has given rise to feminist scholarship within and beyond geography that privileges detailed, ethnographic research methods to focus on issues of meaning and representation (see, for example, Pratt, 2004). In response, McDowell (2006) argued strongly that structural inequalities must not be erased from social scientific analysis, particularly the intersections between class and gender, echoing debates in feminist scholarship beyond geography (Adkins, 2002).

Finally, McDowell's research agenda and the way in which she 'practises' geography have played an important role in the academy beyond geography. Intellectually, she is one of a relatively small number of geographers whose research not only draws on debates beyond geography but contributes to the broader social sciences, most notably by emphasising the critical role played by space and place in shaping the gendered nature of workplace identities and labour market change. Similarly, in addition to drawing attention to the gendered nature of academia's own labour markets, her own research practices have been influential and inspirational to many. In particular, throughout her career, research monographs have been central to the ways in which McDowell has written up and disseminated her research. Choosing to publish her research findings in book form has been significant given often-voiced concerns regarding this form of publication in a metric-driven academy, especially in the UK (see *Progress in Human Geography,* 2009).

McDowell's Key Works

McDowell, L. (1991) 'Life without father and Ford: The new gender order of post-Fordism', *Transactions of the Institute of British Geographers,* 16: 400–19.

McDowell, L. (1992a) 'Doing gender: Feminism, feminists and research methods in human geography', *Transactions, Institute of British Geographers,* 17: 399–416.

McDowell, L. (1993) 'Space, place and gender relations: Identity, difference, feminist geometries and geographies', *Progress in Human Geography,* 17: 305–18.

McDowell, L. (1997a) *Capital Culture: Gender at Work in the City.* Oxford: Blackwell.

McDowell, L. (1999) *Gender, Identity and Place.* Cambridge: Polity.

McDowell, L. (2003) *Redundant Masculinities? Employment Change and White Working Class Youth.* Oxford: Blackwell.

McDowell, L. (2005) *Hard Labour: The Forgotten Voices of Latvian Migrant 'Volunteer' Workers.* London: UCL Press.

McDowell, L. (2009) *Working Bodies: Interactive Service Employment and Workplace Identities.* Oxford: Wiley Blackwell.

McDowell, L. (2013) *Working Lives: Gender, Migration and Employment in Britain 1945–2007.* London: Sage.

McDowell, L. (2014) 'Roepke lecture in economic geography – the lives of others: body work, the production of difference, and labor geography', *Economic Geography,* 91(9): 1–23.

McDowell, L. (2016) *Migrant Women's Voices: Talking About Life and Work in the UK since 1945.* Oxford: Wiley Blackwell.

Secondary Sources and References

Adkins, L. (2002) *Revisions: Gender and Sexuality in Late Modernity.* Buckingham: Open University Press.

Amin, A. and Thrift, N. (2004) *The Blackwell Cultural Economy Reader.* Oxford: Blackwell.

Berndt, C. and Boeckler, M. (2009) 'Geographies of circulation and exchange: Constructions of markets', *Progress in Human Geography,* 33: 535–51.

Butler, J. (1990) *Gender Trouble: Feminism and the Subversion of Identity.* London: Routledge.

Jackson, P. (2002) 'Commercial cultures: Transcending the cultural and the economic', *Progress in Human Geography,* 26: 3–18.

Ley, D. (2004) 'Transnational spaces and everyday lives', *Transactions, Institute of British Geographers,* 29: 151–64.

MacKenzie, D. (2006) *An Engine not a Camera: How Financial Models Shape Markets.* Cambridge, MA: MIT Press.

McDowell, L. (1990a) 'Women in British Geography revisited', *Journal of Geography in Higher Education,* 14: 19–32.

McDowell, L. (1990b) 'Sex and power in academia', *Area,* 22: 323–32.

McDowell, L. (1992b) 'Multiple voices: Speaking from inside and outside the project', *Antipode,* 24: 6–72.

McDowell, L. (1994a) 'Gender divisions of labour in the post-Fordist economy: The maintenance of occupational sex segregation in the financial services sector', *Environment and Planning A,* 26: 1397–418.

McDowell, L. (1994b) 'Performing work: Bodily representations in merchant banks', *Environment and Planning D: Society and Space,* 12: 727–50.

McDowell, L. (1997b) 'A tale of two cities? Embedded organisations and embodied workers in the City of London', in R. Lee and J. Wills (eds), *Geographies of Economies.* London: Arnold. pp. 118–29.

McDowell, L. (2001a) 'It's that Linda again: Ethical and practical issues in research with young men', *Ethics, Place and Environment,* 4: 88–100.

McDowell, L. (2001b) 'Father and Ford revisited: Gender, glass and employment change in the new millennium', *Transactions of the Institute of British Geographers,* 24: 448–64.

McDowell, L. (2001c) 'Masculine discourses and dissonances: Strutting "lads", protest masculinity and domestic respectability', *Environment and Planning D: Society and Space,* 20: 97–119.

McDowell, L. (2001d) 'Rethinking scale: Some reflections on the significance of research on gender and organisations for economic geography', *Journal of Economic Geography,* 1: 227–50.

McDowell, L. (2004) 'Masculinity, identity and labour market change: Some reflections on the implications of thinking relationally about difference and the politics of inclusion', *Geografiska Annaler B,* 86: 45–56.

McDowell, L. (2006) 'Reconfigurations of gender and class relations: Class differences, class condescension and the changing place of class relations', *Antipode,* 38 (4): 825–50.

McDowell, L. (2007) 'Sexing the economy, theorizing bodies', in A. Tickell, E. Sheppard, J. Peck and T. Barnes (eds), *Politics and Practice in Economic Geography.* London: Sage. pp. 60–70.

McDowell, L. (2008) 'Thinking through class and gender in the context of working class studies', *Antipode,* 40: 20–4.

McDowell, L. and Massey, D. (1984) 'A women's place?' in D. Massey and J. Allen (eds), *Geography Matters.* Cambridge: Cambridge University Press. pp. 128–47.

Pratt, G. (2004) *Working Feminism.* Philadelphia, PA: Temple University Press.

Prkye, M. and Du Gay, P. (2007) 'Take an issue: Cultural economy and finance', *Economy and Society,* 36: 339–54.

Progress in Human Geography (2009) 'Forum: The future of research monographs: an international set of perspectives', 33: 102–26.

35
Katherine McKittrick

LaToya E. Eaves, University of Tennessee, Knoxville

Biographical Details and Theoretical Context

Katherine McKittrick is professor of Gender Studies and Canada Research Chair in Black Studies at Queen's University in Kingston, Canada. The radical possibility of interdisciplinary spatial thinking that characterises her work is reflective of her upbringing and career. She was 'born outside of Toronto, Ontario, and grew up in small Ontario towns – on Georgian Bay and near the Grand River and in and around the borders of the Niagara Escarpment' (Hudson and McKittrick, 2014: 233). McKittrick studied History and English at the University of Ottawa (BA), followed by a BA (Hons) in Women's Studies at York University. She continued at York, receiving an MA in Women's Studies, followed by a PhD in Women's Studies from York University in 2004, supervised by feminist geographer Linda Peake. Her dissertation was the basis for her first book – *Demonic Grounds: Black Women and the Cartographies of Struggle* (2006).

McKittrick writes that, 'Description is not liberation' (2021: 44), a refrain that appears throughout the text of her paradigm-shifting *Dear Science and Other Stories*. Its repeated appearance, calling attention to the relational, multidisciplinary approach that she insists on, serves as an introduction to her intellectual influences. While reading and working with the writings of scholars such as **bell hooks**, Patricia Hill Collins, Sylvia Wynter, Bessie Head, Paul Gilroy, **Stuart Hall**, Toni Morrison, NourbeSe Philip, Robbie McCauley, and Octavia Butler, she began to develop an interest in how black theorists and creatives attended to the production of space. As a PhD student, McKittrick thought about the connections between creative works, Black Studies, and Geography, noticing how Black women, especially, offered innovative and anti-colonial projects that were deeply tied to placemaking practices that subverted or overturned colonial maps.

Her research continues to draw on many of the thinkers noted above, alongside the writings of geographers Clyde Woods, **Ruth Wilson Gilmore**, **Gillian Rose**, and **Doreen Massey**. Her work centres and develops key concepts such as a Black sense of place, the ungeographic, plantation futures, demonic grounds, more humanly workable geographies, Black livingness, and more. Methodologically, she has always been an interdisciplinary Black Studies scholar and employs creative-critical methodologies, writing:

> Black scholars work with multimodal and interdisciplinary texts and ideas. If we read across black studies, from slave narratives to the present, we can observe that black creatives and intellectuals and organizers thread together the arts, the hard sciences, history, sociology, music, math, ecology, architecture, dance, economics, and more. In bringing together multiple sets of ideas, these thinkers and artists challenge the hierarchical organizing of knowledge. (McKittrick, 2022a: 37)

In doing so, she emphasises the importance of collaboration, which she also sees as a form of liberation. Connections with and between people in different worlds – such as scientists and writers and practitioners and theorists – offer one way to 'center Black livingness and Black life rather than preventable death', and to address the harm that has been done to vulnerable communities (Prescod-Weinstein, 2021: n.d.).

Spatial Contributions

Over the course of her career, now spanning two decades, McKittrick has substantially shaped a new generation of geographers and spatial thinkers. Based on current knowledge, it is McKittrick who introduced the term 'black geographies' in her transformative text *Demonic Grounds: Black Women and the Cartographies of Struggle*. In the book, McKittrick places Black women's geographies at the centre and asserts her intellectual interests in what becomes possible when Blackness is examined in relationship with and to human geography. She argues that such a reflection on the practice of Human Geography is 'not only needed because existing cartographic rules unjustly organize human hierarchies in place and reify uneven geographies in familiar, seemingly natural ways' (2006: x). She continues, writing that the:

> 'where' of black geographies and black subjectivity, then, is often aligned with spatial processes that *apparently* fall back on seemingly predetermined stabilities, such as boundaries, color-lines, 'proper' places, fixed and settled infrastructures and streets, oceanic containers. If space and place *appear* to be safely secure and unwavering, then what space and place make possible, outside and beyond tangible stabilities, and from the perspective of struggle, can potentially fade away. (McKittrick, 2006: xi)

In framing Black geographies, she troubles the white androcentric mappings of Blackness, arguing that European cartographic processes have furthered colonial thought by seeking to fix the 'where' of Black geographies. In this way, McKittrick critiques the structures that underpin uneven geographies. As she states, the term 'Black geographies':

> is an intervention into the assumption that human geography is, first and foremost, an academic and colonial way of knowing; it offers a somewhat clunky way (the term itself is cumbersome and imprecise to me) to centre and really contemplate these brilliant black geographic knowledges that circulate as acts of liberation and resistance. I am not abandoning geography, but instead illuminating different kinds of geographic practices that cannot be seen, or are rendered inadequate-problematic, within the context of white supremacy. (cited in Makonnen, 2021: 255)

It is in the critique of the white supremacist, anti-Black structuring of the world, a structuring that maintains and normalises dominant geographic patterns, that McKittrick asserts the possibilities – and indeed the availability – of more humanly workable geographies (McKittrick, 2006). At the centre of more humanly workable geographies are Black women, as McKittrick argues that the contours of a normalised spatial order are challenged through Black women's spatial knowledges and stories. She insists,

> Black women's geographies disclose that black femininity or body-histories are connected to a long and interhuman geographic story; they illustrate the ways in which the act of expression reveals why these geographies are important and how they express new forms of life that contest our historically present, and uneven, genres of human geography. (McKittrick, 2006: 146)

Consistent with her entire body of scholarship, McKittrick theorises more humanly workable geographies through a transdisciplinary approach, examining a range of texts from material geographies and cultural landscapes to Octavia Butler's fiction and Sylvia Wynter's cultural theory. She argues that dominant geographic patterns normalise spatial hierarchies, flattening complex interhuman relationships into seemingly static geographic sites. She asserts that Black women's knowledges and expressions disrupt these normalised spatial orders, saying that Black women's geographies 'express new forms of life that contest our historically present, and uneven, genres of human geography' (McKittrick, 2006: 146). McKittrick is deeply influenced by Wynter, as reflected in how she theorises more humanly workable geographies. She says that 'Wynter's anticolonial vision is not, then, teleological – moving from colonial oppression outward and upward toward emancipation – but rather consists of knots of ideas and histories and narratives that can only be legible in relation to one another' (McKittrick, 2015: 2). Further, she suggests that Wynter opens up 'the possibility of a new sciences of human discourse' (McKittrick, 2015: 3) or what it means to be human. The two concluding chapters of *Demonic Grounds* take up these possibilities, using what Wynter calls a 'third perspective' (McKittrick, 2006: 143) to assert that black women's geographies are expressions of new forms of life that reshape and respatialise dominant spatial order.

A year after the release of *Demonic Grounds*, McKittrick published the volume *Black Geographies and the Politics of Place*, co-edited with Clyde Woods, which convened a conversation that was 'a reflection of [their] continuing interest in developing and sustaining questions about the intersections between race, blackness, and spatial politics in the diaspora' (McKittrick and Woods, 2007: vii). For McKittrick and Woods, Black geographies were and are necessary for more just geographies. Centring Black geographies functions as a counternarrative to an understanding of Blackness in monolithic terms and to the persistent rhetorical and physical displacement of Black people, communities, and geographic stories. They use Black geographies to reconfigure the mapping of Blackness, showing instead how relationalities, identifications, networks, and power of and about Black subjects critique the contours of conquest and relegation to the 'ungeographic'. Resisting the idea of 'measureable facts' and the 'where' of Blackness as engaged through quantitative and traditional cartographic methods (McKittrick and Woods, 2007), Black geographies open up crucial conceptual, methodological, and material possibilities.

The interventions from these two works make possible a capacious form of liberationist praxis, commanding attention to the intersection of Blackness and geography in ways that are both rigorous and illuminating. McKittrick goes further with her imperative, extending her analyses to 'black sense of place'. The concept of a Black sense of place is perhaps one of McKittrick's most widely circulated interventions in spatial thinking. The term continues her longstanding intellectual project linking Blackness and geography and is introduced by a focus on two very particular, often racialised, geographies in the Americas – the plantation and the prison. In framing the concept, McKittrick offers that:

> a black sense of place can be understood as the process of materially and imaginatively situating historical and contemporary struggles against practices of domination and the difficult entanglements of racial encounter. Racism and resistance to racism are therefore not the sole defining features of a black sense of place, but rather indicate how the relational violences of modernity produce a condition of being black in the Americas that is predicated on struggle. (McKittrick, 2011: 949)

Building on a Black sense of place and continuing her work on troubling the normalised production of a spatial order rendering Black life as invisible, McKittrick (2013) brings us to 'plantation futures', arguing (further) that the plantation offers a particular time–space context from which to think both through the interconnectedness of modernity and Blackness, a relationship that produces uneven racial geographies and sites of violence and terror and that also roots ontological expressions of survival (creolisation, the blues, etc). Put differently, 'plantation futures' guides us through the afterlives of slavery. This coupling of contexts and expressions yields a theorisation of what McKittrick calls 'black livingness' (McKittrick, 2016; 2021).

In *Dear Science and Other Stories,* McKittrick addresses the ways that Black people and communities have and continue to upend present logics and structures through inherently interdisciplinary ways of knowing, feeling, and resisting. McKittrick argues that while the realities of scientific racism and biological determinism dominate how we understand science, they do not define Black life or ways of living. Through this line of thinking, McKittrick argues for Black methodologies that are liberated from descriptive-data-induced answers, arguing again that 'description is not liberation'. Instead, she argues that Black methodologies are necessarily rebellious and disobedient, undoing disciplinary boundaries and making room for liberatory praxis.

Key Advances and Controversies

In the years that followed their respective publication dates, *Demonic Grounds* and *Black Geographies and the Politics of Place* were taken up in limited ways by geographers, with more attention being paid to both works since the mid-2010s. This timing corresponds with broader structural changes in the discipline that generated interest in Black geographies, starting with a group of graduate students and early career faculty organising to create the Black Geographies Specialty Group for the American Association of Geographers (AAG). Beginning with the 2015 Annual Meeting of the AAG, in a series of sessions entitled 'Black Matters are Spatial Matters', both texts were placed in regular discussion in papers and on panels. At the 2016 Annual Meeting, a session titled 'Legacies of Black Feminisms' honoured the tenth anniversary of the publication of *Demonic Grounds*. In the same time period, an interdisciplinary/transdisciplinary audience, particularly in the social sciences, began looking towards geographic concepts through Black geographies (see, for example, Cox, 2015; Perry, 2016; Reese, 2018). Feminist scholarship, scholarship on women and gender, and work on and by Black women are often relegated to margins, in spite of **bell hooks'** (2000) urging that 'feminism is for everybody' and McKittrick's (2006) own thinking through the ungeographic. Consequently, geographers appear to be reticent to consider the gendered ideas apparent in *Demonic Grounds*, instead opting for 'Black sense of place' and 'plantation futures' as McKittrickian frameworks (King, 2019). As it were, McKittrick's emphasis on the erasure and omission of Black feminism and Black women seems fulfilled by the persistent oversight of her theorisation of and engagement with Black feminist geographies and more humanly workable geographies.

There is one exception, however, which is the term 'black geographies' itself. It should be noted that most scholarship refers back to *Black Geographies and the Politics of Place* as the origin story of the term. While the empirical and theoretical foundations of Black geographies extend deeper (decades, centuries) into intellectual and disciplinary histories (Hawthorne, 2019; Bledsoe et al., 2020), the term solidified a particular set of methods and approaches as exemplified in McKittrick's body of scholarship. Black geographies attend to the ways in which geographic traditions can be rendered differently through a Black sense of place. The tradition resists epistemological stances that resort to quantitative, positivist measures through the discussion of Blackness and, more broadly, race and racialisation. Rather, by using a Black geographies framework, scholars and activists draw attention to the practices of Black life, agency, resistance, and resilience that are hegemonically generated as un-thought spatialities. In *Demonic Grounds* McKittrick (2006: xii) writes 'black matters are spatial matters'. Since then, the phrase has been used as a foundation for the study of Blackness and place and what is now referred to as Black geographies. Camilla Hawthorne's (2019) paper, aptly titled 'Black Matters are Spatial Matters', has compounded McKittrick's impact, emphasising McKittrick's recognition of the relationship between Black spatial knowledge and geographies of domination: 'this relationship reveals the ways in which Blackness has been central to both the production of space and to the formation of Western, Eurocentric forms of geographic knowledge. At the same time, the modes of Black geographic knowledge that have emerged in resistance to domination provide alternative pathways toward new understandings of space and to the undoing of violent practices of geographic organization' (Hawthorne, 2019: 3).

Regarding 'plantation futures', Priscilla McCutcheon suggests that McKittrick 'challenges us to move away from the simplistic binaries that we use to view the systems that Black people created in response to the oppressive landscape of slavery, and instead view them as a dialectic where each is needed to fully understand the other' (McCutcheon, 2019: 209). Such positioning of 'plantation futures' mirrors reception to McKittrick's 'livingness'. Tiffany Lethabo King asserts that McKittrick's recent work, particularly 'Mathematics Black Life' (2014) and 'Diachronic Loops/Deadweight Tonnage/Bad Made Measure' (2016), has expanded spatial thinking 'to create a methodology, ethics of care, and way of "noticing" the "other possibilities" in the midst of Black death and Black degradation'

(King, 2019: 30). Such a methodology and methodological approach, furthered in *Dear Science and Other Stories* (2021), reflects McKittrick's continuous, career-long commitment to interdisciplinary/transdisciplinary spatial thinking consistent with work in Black studies more broadly (McKittrick, 2022b). Her approach to spatial thinking has influenced scholars throughout the Americas, such as on the racialised geographies of major cities (Shabazz, 2015; Summers, 2019), the pluralities of place (Bledsoe and Wright, 2019; Gardner, 2021), and the urgent ecological crises transforming Black places (Purifoy, 2021; Bruno, 2022).

McKittrick's work has also been taken up widely outside of geography, with presence in a range of social science and humanities disciplines, including Black Studies, Sociology, Feminist and Gender studies, Religious Studies, Environmental Studies, Education, Anthropology, History, and Philosophy of Science, with burgeoning interest from geosciences and other physical sciences (see, for example, Ducre, 2018; McTighe and Haywood, 2018; Nxumalo, 2020; Okellow and Morgan, 2022; Clemons and Grieser, 2023). Non-geographers have looked to McKittrick in a spatial turn, emphasising the role of place in the production of knowledge. For example, Religious Studies scholar Laura McTighe and activist Deon Haywood (2018) use McKittrick to query how Black feminist leaders can respatialise post-Katrina New Orleans. Education scholar Fikile Nxumalo (2020) looks to McKittrick in placing Black and Indigenous children in the Anthropocene, as part of a practice of environmental education, saying 'Black feminist practices of what I am calling testifying-witnessing make visible the complexities of Black geographies, beyond stories of damaged place relations, surveillance, and absence' and instead offer 'modes of revealing the inequitable differentiations of Black children's situatedness within anthropogenic inheritances, while simultaneously affirming Black children's humanity and generative nature relations' (Nxumalo, 2020: 548–9). Importantly, these scholars have kept to McKittrick's emphasis on alternative mappings of place by engaging with indigenous ways of knowing, and outside of white European methods of being. Such a transdisciplinary presence and impact are an explicit nod to her commitment to a Black studies approach to liberation.

Mckittrick's Key Works

McKittrick, K. (2006) *Demonic Grounds: Black Women and the Cartographies of Struggle.* Minneapolis, MN: University of Minnesota Press.

McKittrick, K. (2011) 'On plantations, prisons, and a black sense of place', *Social & Cultural Geography*, 12 (8): 947–63.

McKittrick, K. (2013) 'Plantation futures', *Small Axe: A Journal of Criticism*, 17 (3): 1–15.

McKittrick, K. (2014) 'Mathematics Black Life', *The Black Scholar*, 44 (2): 16–28.

McKittrick, K. (2015) *Sylvia Wynter: On Being Human As Praxis*. Durham, NC: Duke University Press.

McKittrick, K. (2016) 'Diachronic loops/deadweight tonnage/bad made measure', *cultural geographies*, 23 (1): 3–18.

McKittrick, K. (2021) *Dear Science and Other Stories*. Durham, NC: Duke University Press.

McKittrick, K. (2022a) 'Respite. Quiet. A House of Dreams', *CR: The New Centennial Review*, 22 (3): 37–55.

McKittrick, K. (2022b) 'Dear April: The aesthetics of Black miscellanea', *Antipode*, 54 (1): 3–18.

McKittrick, K. and Woods, C.A. (eds) (2007) *Black Geographies and the Politics of Place*. Toronto: Between the Lines.

Secondary Sources and References

Bledsoe, A. and Wright, W.J. (2019) 'The pluralities of Black geographies', *Antipode*, 51 (2): 419–37.

Bledsoe, A., Wright, W.J., and Eaves, L. (2020) 'Black geographies', in *Encyclopedia of Human Geography* (2nd edition). Kidlington: Elsevier. pp. 347–50.

Bruno, T. (2022) 'Ecological memory in the biophysical afterlife of slavery', *Annals of the American Association of Geographers*, 113 (7): 1543–53.

Clemons, A.M. and Grieser, J. (2023) 'Black womanhood: Raciolinguistic intersections of gender, sexuality and social status in the aftermaths of colonization', *Dædalus, the Journal of the American Academy of Arts & Sciences*, 152 (3): 115–29.

Cox, A.M. (2015) *Shapeshifters: Black Girls and the Choreography of Citizenship*. Durham, NC: Duke University Press.

Ducre, K.A. (2018) 'The Black feminist spatial imagination and an intersectional environmental justice', *Environmental Sociology*, 4 (1): 22–35.

Gardner, T. (2021) '(Tiny) houses and Black feminist geographic praxis: Building more humanly workable geographies', in B. Gökarıksel, M. Hawkins, C. Neubert and S.H. Smith (eds), *Feminist Geographies Unbound: Discomfort, Bodies, and Prefigured Futures*. Morgantown, WV: West Virginia University Press. pp. 116–38.

Hawthorne, C. (2019) 'Black matters are spatial matters: Black geographies for the twenty-first century', *Geography Compass*, 13: article #e12468. doi.org/10.1111/gec3.12468

hooks, b. (2000) *Feminism is for Everybody: Passionate Politics*. Cambridge, MA: South End.

Hudson, P.J. and McKittrick, K. (2014) 'The geographies of Blackness and anti-Blackness: An interview with Katherine McKittrick', *The CLR James Journal*, 20 (1/2): 233–40.

King, T.L. (2019) *The Black Shoals: Offshore Formations of Black and Native Studies*. Durham, NC: Duke University Press.

Makonnen, B. (2021) 'Interview with Katherine McKittrick: Black Human Geographies', in G. Aloi and S. McHugh (eds), *Posthumanism in Art and Science: A Reader*. New York, NY: Columbia University Press. pp. 249–55.

McCutcheon, P. (2019) 'Fannie Lou Hamer's freedom farms and Black agrarian geographies', *Antipode*, 51 (1): 207–24.

McTighe, L. and Haywood, D. (2018) 'Front porch revolution: Resilience space, demonic grounds, and the horizons of a Black feminist otherwise', *Signs: Journal of Women in Culture and Society*, 44 (1): 25–52.

Nxumalo, F. (2020) 'Situating Indigenous and Black childhoods in the anthropocene', in A. Cutter-Mackenzie, K. Malone and E. Barratt Hacking (eds), *Research Handbook on Childhoodnature*. Cham: Springer. pp. 535–56.

Okello, W.K. and Morton, C.S. (2022) 'Speaking our imaginings into existence: Poetry as a contestation of Black erasure in academia', *Journal of Diversity in Higher Education*, doi.org/10.1037/dhe0000411

Perry, K.-K.Y. (2016) 'Geographies of power: Black women mobilizing intersectionality in Brazil', *Meridians*, 14 (1): 94–120.

Prescod-Weinstein, C. (2021) 'Public thinker: Katherine McKittrick on Black methodologies and other ways of being'.

https://www.publicbooks.org/public-thinker-katherine-mckittrick-on-black-methodologies-and-other-ways-of-being/ (accessed November 9, 2023).

Purifoy, D. (2021) 'The parable of Black places', *Transactions of the Institute of British Geographers*, 46: 829–33.

Reese, A.M. (2018) '"We will not perish; we're going to keep flourishing": Race, food access, and geographies of self-reliance', *Antipode*, 50 (2): 407–24.

Shabazz, R. (2015) *Spatializing Blackness: Architectures of Confinement and Black Masculinity in Chicago*. Urbana, IL: University of Illinois Press.

Summers, B.T. (2019) *Black in Place: The Spatial Aesthetics of Race in a Post-Chocolate City*. Chapel Hill, NC: University of North Carolina Press.

36
Richa Nagar

Caroline Faria, Mahboobe Safaei Mehr, and Dominica Whitesell, University of Texas at Austin

Biographical Details and Theoretical Context

Richa Nagar was born in 1968 in Lucknow, the capital of the Indian state of Uttar Pradesh. Raised in a multi-generational, middle-class household, her family often hosted scholars, artists, singers, filmmakers, journalists, labourers, and activists. In her home, and via the accounts of these varied guests, family members, and friends, Nagar heard stories of caste, class, and gender violence, of colonial pasts and presents, of partition, of everyday survival, joy, and resistance. Their stories were told in a mix of languages and accents – Hindi, Urdu, Gujarati, Awadhi, and Brah and interwoven with the literary influences of Marathi, Bangla, and Russian thinkers. Nagar's (2016: 74) childhood was, she retells, like 'growing up simultaneously in "the university" and "the field"' (2016: 74). This embodied approach to understanding the world infuses Nagar's scholarship and is a key hallmark of her work in, and beyond, Geography.

Nagar received a Bachelor of Arts in English Literature, Geography, and Anthropology from Lucknow University in 1986 and a Masters in Geography from the University of Pune a few years later. She moved to the US and completed a doctoral degree in Geography from the University of Minnesota in 1995. Following a brief period at the University of Colorado, Boulder, Nagar took up a position in the Department of Women's Studies (later the Department of Gender, Women, and Sexuality Studies) at the University of Minnesota. She became a full professor in 2007 and has also taken honorary and adjunct positions at: Kadir Has University, Istanbul; Savitribai Phule Pune University, India; York University, Toronto; and Rhodes University, South Africa. Her work spans the genres of poetry, performance, ethnography, and oral history; linking and disrupting the disciplinary boundaries of geography, women's studies, and performance studies. She publishes in several languages, notably Hindi/Hindustani, English, and Awadhi.

Nagar's early research explores the postcolonial diasporic politics of South Asians in Dar es Salaam, Tanzania. Between the mid-1990s and early 2000s, she examined their social and cultural lives, place-making in the city, philanthropic work, and their engagement with the state. This was a richly-empirical body of scholarship, mapping particular Hindu halls, temples, schools, and cemeteries, detailing their layered cultural politics, and tracing the intense debates around bodies, power, and place (Nagar, 1995; 1996; 1997a; 1997b;

1998; 2000a). Nagar's feminist attention to women's lives and gender politics is formative, exemplified in '"I'd rather be rude than ruled": Gender, place, and communal politics among South Asian communities in Dar es Salaam' (2000a). But while the struggles faced by women are centred, she refuses a singular analysis of gender power or a focus on women alone. Instead, Nagar's work provided an early analysis in geography of how poor, lower caste women and men are disciplined by, but also disrupt, upper caste, elitist patriarchal norms.

There is a notable shift in Nagar's work from the late 1990s. She began to form relationships with women's organisations in Lucknow and Hyderabad and centred her research in northern India. Although a marked geographical move from Tanzania, her intellectual concern with the ties between casteism, classism, and Hindu Brahmanical patriarchy endured and deepened at this time. A key piece here was 'Mujhe Jawab Do! (Answer me!): Women's grass-roots activism and social spaces in Chitrakoot (India)' (2000b; see also 2000c), which examined violence in rural northern Indian women's lives. Nagar here analysed the problems of domestic abuse, entrenched poverty, dowry murders, suicide and other forms of femicide, and the persistent experience of hunger. But beyond a purely victimised story of women's lives, Nagar highlighted community-based educational responses led by grassroots women's groups.

Layered into this work was a growing engagement with critical development geographies, transnational feminisms, and the position of the 'Third World Woman' in both international and state-based interventions. Nagar formed important connections with women organisers from the Sitapur District of Uttar Pradesh who worked for the Mahila Samakhya program. This is an organisation launched by the Indian government in the late 1980s to educate and empower primarily rural, lower caste, and poor women. They were troubled by the 'NGOization' of women's empowerment in response to such violence, and the co-optation of once radical and transformative organising around women's empowerment by the development industry. Attentive to their intellectual concerns, Nagar collaborated with them to form the 'Sangtin Collective'. They co-authored *Sangtin Yatra: Saat Zindgiyon Mein Lipta Nari Vimarsh* in Hindi and Awadhi (Anupamlata et al., 2004) and its translation *Playing with Fire: Feminist Thought and Activism across Seven Lives in India* (Sangtin Writers and Nagar, 2006). In these books, they wove together the life histories of seven rural members of the Sangtin Collective. Their stories provided complex accounts of the oppressions Dalit and other low caste women, like those in Chitrakoot, faced, but also highlighted their capacity to respond. Interwoven throughout was a geopolitical critique of development. The Collective traced how that marginality was produced through their relationship to the Indian state, and to international, governmental, and non-governmental agencies: those very development institutions aimed at their improvement and modernisation, their integration into capitalist circuits, or their salvation from violent men. This seemingly contradictory relationship between development and oppression is, at its heart, a formative critique of capitalism and its reliance on intersectional power asymmetries. Building on transnational feminist critique of development (e.g., Kabeer, 1994; Mohanty, 2003), the work of Nagar and the Sangtin Collective offers a deeply embodied insight into Third World women's political subjectivity and the gendered, classed, and caste-based politics of development.

The Sangtin Collective is significant not only for these intellectual strides but also for their solidarity-based approach to knowledge production – the difficult work of thinking, connecting, and presenting as a 'fractured but blended "we"' (Sangtin Writers and Nagar, 2006: xxxiv). Nagar builds here on *Sangtin Yatra/Playing with Fire* in two other influential books *Muddying the Waters: Co-authoring Feminisms Across Scholarship and Activism* (2014) and *Hungry Translations: Relearning the World Through Radical Vulnerability* (Nagar in journeys with Sangtin Kisan Mazdoor Sangathan and Parakh Theatre, 2019), along with a host of sister articles. This body of work uses innovative collaborative story-telling, reflections, performance, and writing techniques to create a trailblazing example of transnational feminist praxis centred on hungry translation through radical vulnerability. For Nagar:

> A hungry translation […] is distinguished by its insistence on a collective ethic of radical vulnerability that refuses to assume that it can arrive at a truth; it commits itself to grappling with an always fluid and unresolvable set of incommensurabilities in ongoing relationships where the meaning of 'justice', 'ethics', or 'politics' can only emerge in the shifting specifics of a given moment in an ongoing struggle – a particular convergence of subjectivities and articulations that is itself located at a unique confluence of time and place. The political potential of hungry translations lies in this yearning to keep the translations – as well as the relationships that energise and authorise those translations – open and flowing. (Nagar, 2018: 20)

Or, as the Sangtins make clear: 'It is only these collective energies that can give us the strength to turn our desire for justice into a hunger for justice' (Sangtin Kisan Mazdoor Sangathan, 2018: 19–20 cited in Nagar, 2018). Via these hungry translations, Nagar forges truly disruptive, alternative intellectual pathways for academic research across difference.

Spatial Contributions

Three spatial threads weave through the tapestry of Nagar's work: an intersectional understanding of power – casteism, patriarchy, elitism, and colonial-capitalism; a concern with journeying, specifically intellectual and methodological border-crossing and translation across difference; and a commitment to social justice, change, resistance, and other world-making – both via and beyond the limits of academic knowledge-production.

Attention to the interlocking work of gender, racial, heteronormative, and class-based power emerged via the Black feminist thought and the writings of, for example, Anna Julia Cooper (1988 [1892]), Patricia Hill Collins (2000), Kimberley Crenshaw (1993), **bell hooks** (1984), and the Combahee River Collective (1995 [1977]). Nagar extends this work by examining how intersectional racial power is grounded in South Asian diasporic and sub-continental histories, contexts, spaces, and subjects, and tracing how caste politics, religion, homophobia, and Brahmanical patriarchy, for example, are also racial-gendered formations of power. Even in her earliest work in Dar es Salaam, oppressive patriarchal structures were always understood in relation to those of class, caste, and coloniality. There is a fine-grained engagement with difference, and in particular the overlapping and intersecting politics of power through spaces like Hindu temples, halls, streetscapes, and homes. In her work in India, this intersectional analysis of power is sustained. The dowry suicides and murders she examines in places like Chitrakoot are linked to the immobilities of poverty, patriarchy, and anti-caste prejudices. Here, a keen attention to space and scale clarifies the contours of this intersectional power. Nagar's granular and visceral ethnography of spaces of violence and of the spatial tactics of women's resistance (for example via street theatre spaces) shows how intersectional power operates through the bodies of women, the place of the home, the politics of the village, and the state. Elsewhere, the erasures of binary spatial imaginaries like the 'Global North' and 'Global South', or the rural and urban, are disrupted by Nagar's stories of relational politics across place and her scalar analysis of the ways that colonial geopolitical and localised Indian power relations manifest in NGO operations via their 'fieldsite' locations. Perhaps most notably, and aligned with transnational feminist critiques, Nagar's engagement with power destabilised the monolith of the 'Third World Woman', complicating this powerful geopolitical trope with rich accounts of difference within and between. Nagar's contributions to intersectional theory delineates the differences, yet connections, to other intersectional, colonial racial formations of anti-Blackness and/or anti-Indigeneity, for example, providing us with other layers through which we can understand power.

Intersectional work is, at its heart, sensitive to relationships and co-production, to the ways that patriarchy, capitalism, and casteism are connected, and how they rely on and feed into one another. This approach demands a spatial movement – both a methodological and intellectual border crossing and translation across power. Such journeying is a second key spatial contribution of Nagar's work. As she argues:

> The challenge for postcolonial and feminist geographers, then, is to conceptualise border-crossings that are committed to forming collaborative partnerships with academic and non-academic actors in "other" worlds, in every sense of the term–partnerships in which the questions around how power and authority would be shared cannot be answered beforehand, but are imagined, struggled over and resolved through the collaborative process itself. (Nagar in consultation with Ali, F. and the Sangatin women's collective, 2003: 368)

In the research process, Nagar's work journeys across the asymmetrical social locations of herself and her collaborators, building knowledge that travels, 'between "The 'I's" and "We's", of here and there, of past and present, of human and non-human–through modes of connecting and trusting, living and being, troubling and creating' (Nagar, 2018: 4; see also Chowdhury et al., 2016). In doing so, she offers a radical spatial epistemology via collaboration that entails struggle, discomfort, unease, and a loss of personal power as 'expert', resulting in more complex and transformative understandings of the world. This mode of knowledge is translatable to local, perhaps illiterate or less highly educated, communities, and creates more openings for engagement, dispute, co-thinking, and change.

This methodological insistence on research-in-solidarity is deeply tied to the third key contribution of her work: forging antithetical scholarship oriented to transformation 'on the ground'. The work of the Sangtin Collective combines an intellectual critique of patriarchy, casteism, the state, capitalism, and the development industry with the nurturing of a social movement. The Sangtin Kisaan Mazdoor Sangathan (SKMS) or the Sangtin Peasants and Workers Organization emerged from the collective reflections of the Sangtin

Collective. It now includes more than 8000 small kisans (farmers) and mazdoors (labourers) who work across 150 villages on labour rights, access to pensions, environmental management, health issues, and a host of issues fundamental to well-being. This movement building is also evident in Nagar's knowledge production process. She is committed to fostering space for exchange, even across steep power differentials. A particularly important example is her work with community theatre. Detailed in *Hungry Translations* and elsewhere, Nagar worked with twenty actors, the SKMS community, and writer-director Tarun Kumar to develop the play 'Hansa, Karo Puratan Baat' (see Nagar and Kumar, 2011). Based on Munshi Premchand's 1935 short story 'Kafan' (the Shroud), the piece examines gendered, classed, and caste-based oppression and resistance amongst those so often 'dismissed as rural, Dalit, uneducated and poor' (Nagar, 2018: 4). The performance fostered many instructive moments of exchange, including the six-month period of workshop-based preparation, the performance itself, and the many lengthy post-performance conversations it prompted.

Key Advances and Controversies

Nagar's sustained attention to the lives of low caste, poor Indian women exemplifies her long-standing commitment to typically under-researched and marginalised subjects and spaces. This is deeply instructive for geographers. She has made clear that we cannot understand colonialism, capitalism, development – and the terrors of hunger or violence, for example – without listening to the stories of those most marginalised, silenced, impacted, and exploited. To do so, she has deployed a transnational feminist scalar analysis that attends to the embodied intimacies of formal, abstracted and typically masculinised processes. This particular sensitivity to the subjects, spaces, and scales of power is developed in a manifesto of sorts in the piece 'Locating globalization: Feminist (re)readings of the subjects and spaces of globalization' (Nagar et al., 2002). It has become a foundational text for developing transnational and otherwise antiracist feminist approaches to colonial and contemporary capitalist processes, calling attention to elisions in analyses of capitalism, and offering instructive ways to understand capitalism otherwise. While it has paved the way for significant shifts in the sub-discipline of economic geography, it remains an ongoing struggle, with persistent resistance to an understanding of capitalism that is *co-produced* through, not just constructive of, patriarchies, heteronormativities, and racisms.

Nagar's use of performance, co-writing, workshopping, and other modes of participatory and collaborative research has been deeply influential in fostering ethical and radical feminist methodologies that transgress, stretch, and undo the academy. Moreover, feminist collectives and solidarity-based efforts at co-authorship have become more common in the wake of her example. But one of Nagar's most notable methodological interventions centres on the thorny concept of 'positionality'. Feminist scholars have led the discipline in critiquing scientific, distanced, objective and expert knowledge production. Positionality statements – an acknowledgment of one's social location, and recognition of the modesty and partiality of the research, or a paralysis in work across power-geometries at all – emerged as common responses. Nagar notes:

> My approach to narrating the intersections of my own journeys on the one hand, and the lives and journeys of those I have become entangled with on the other, emerges from my own disappointments with dominant ways of 'doing positionality' in the U.S. academy – ways that often fail to peel themselves away from the conceptual and stylistic monolingualism of academic knowledge making and writing. (Nagar, 2016: 73)

Nagar, amongst other feminist scholars of colour such as **Audrey Kobayashi** (2005), refuses these moves as inadequate, highlighting the ways that these reinforce the privileges of western thinkers (and elite global south scholars). She models instead efforts to centre injustices of power where they occur, to understand the fraught intersectionalities of power that infuse them – even where the researcher may themselves be implicated – and to commit to lifelong relationships of openness, critique, action and shared, collaborative knowledge production in situated solidarity. Social location remains key, as Nagar notes:

> The stories of our own lives and trajectories: the languages, yearnings, and hauntings that we come from and that breathe inside us, mold in critical ways the 'projects' that we come to and how we shape those projects as scholars, thinkers, writers, learners, educators, and creative or intellectual community builders. (Nagar, 2016: 73)

She does not shy away from revealing the difficulties of this work: the moments of praxis that are embarrassing, painful, humbling, reproachful. But these have provided a guide for feminist geographers and other critical scholars. Her path is a radically different approach from that typically undertaken in geography. It relies upon lifelong relationships, sustained work to build trust, co-developing projects, and resistance to more typically extractive forms of academic knowledge production. This is not an easy path, and it perhaps requires its own kinds of privilege (for example of tenure and class mobility). But as Nagar has often noted, this is an ongoing struggle. Even in critical fields like feminist geography, where ethical research relationships have now long been emphasised and worried over, Nagar provides an inspiring example.

Last, *Playing with Fire, Hungry Translations,* and Nagar's wealth of articles, commentaries, and reflections valuably complement and extend Black feminist intersectional work on power. Via Nagar's work, we see the ways in which caste is also implicated in intersectional racial power geometries, the linked imperatives of capitalism and development, and the interwoven historical genealogies of patriarchy, religious oppression, racism, and colonialism. Nagar is a leading feminist geographer tracing the ways that grassroots and radical women's empowerment have been co-opted for colonial-capitalist development and, more complexly still, how gender-caste-racialised expectations and norms of women and men can be put to work to further these agendas. This is richly informative for scholarship linking oppressions in and beyond South Asia, such as those developed by Malini Ranganathan (2022) of Hindu casteism as a mechanism of global racial capitalism. A meaningful engagement with intersectionality is an ongoing and fraught challenge for the discipline of geography. But Nagar's body of work is now foundational for a host of Black feminist, Indigenous, Latinx, and otherwise post- and/or decolonial scholarship committed to this project. Indeed, we see Nagar's interventions in a wide body of embodied and intersectional analyses of development and geopolitics. But we also see this influence far beyond Nagar's own empirical focus on the development industry, to work on gentrification, extraction, environmental management, and degradation, and a host of related imperatives of capitalism and modernisation that invoke, devalue and rely on poor people of colour. And in keeping with the imperatives of Black, transnational, Third World, and Indigenous foundational thinkers, Nagar's work refuses a simple victimisation of poor rural women in this context. Instead, her work exemplifies an inspiring life's work narrating and practising resistance, refusal, and other-world making.

Nagar's Key Works

Nagar, R. (1997a) 'The making of Hindu communal organizations, places and identities in postcolonial Dar es Salaam', *Environment and Planning D: Society and Space*, 15 (6): 707–30.

Nagar, R. (2000a) '"I'd rather be rude than ruled": Gender, place, and communal politics among South Asian communities in Dar es Salaam', *Women's Studies International Forum*, 23 (5): 571–85.

Nagar, R. (2000b) 'Mujhe Jawab Do! (Answer me!): Women's grass-roots activism and social spaces in Chitrakoot (India)', *Gender, Place & Culture*, 7 (4): 341–62.

Nagar, R. (2002) 'Footloose researchers, "traveling" theories, and the politics of transnational feminist praxis', *Gender, Place & Culture*, 9 (2): 179–86.

Nagar, R. (2013) 'Storytelling and co-authorship in feminist alliance work: Reflections from a journey', *Gender, Place & Culture*, 20 (1): 1–18.

Nagar, R. (2014) *Muddying the Waters: Co-authoring Feminisms Across Scholarship and Activism*. Champaign, IL: University of Illinois Press.

Nagar, R. (2015) 'Women's theater and the redefinitions of public, private, and politics in North India', *ACME: An International Journal for Critical Geographies*, 1 (1): 55–72.

Nagar, R. in journeys with Sangtin Kisan Mazdoor Sangathan and Parakh Theatre (2019) *Hungry Translations: Relearning the World Through Radical Vulnerability*. Champaign, IL: University of Illinois Press.

Nagar, R., Lawson, V., McDowell, L. and Hanson, S. (2002) 'Locating globalization: Feminist (re) readings of the subjects and spaces of globalization', *Economic Geography*, 78 (3): 257–84.

Sangtin Writers and Nagar, R. (2006) *Playing with Fire: Feminist Thought and Activism Through Seven Lives in India*. Minneapolis, MN: University of Minnesota Press.

Secondary Sources and References

Anupamlata, Ramsheela, Ansari, R., Bajpayee, V., Shashibala, Vaish, S., Surbala, Singh, R. and Nagar, R. (2004) *Sangtin Yatra: Saat Zindgiyon Mein Lipta Nari Vimarsh*. Sitapur: Sangtin.

Class War University (2013) 'Desiring alliance and complex translations in activist research: An interview with Richa Nagar'. https://classwaru.org/2013/07/25/desiring-alliance/ (accessed November 9, 2023).

Chowdhury, E., Pulido, L., Heynen, N., Rini, L., Wainwright, J., Inayatullah, N. and Nagar, R. (2016) 'Muddying the waters: Coauthoring feminisms across scholarship and activism', *Gender, Place & Culture*, 23 (12): 1800–12.

Collins, P. (2000) *Black Feminist Thought: Knowledge, Consciousness, and the Politics of Empowerment*. Revised, 10th Anniversary, 2nd Edition. London: Routledge.

Combahee River Collective (1995 [1977]) 'The Combahee River Collective statement', in B. Guy-Shefthall (ed), *Words of Fire: An Anthology of African-American Feminist Thought*. New York, NY: The New Press. pp. 231–40.

Cooper, A.J. (1988 [1892]) *A Voice from the South by a Black Woman from the South*. New York, NY: Oxford University Press.

Crenshaw, K. (1993) 'Mapping the margins: Intersectionality, identity politics, and violence against women of color', *Stanford Law Review*, 43 (6): 1241–99.

hooks, b. (1984) *Feminist Theory from Margin to Center*. Boston, MA: South End Press.

Kabeer, N. (1994) *Reversed Realities: Gender Hierarchies in Development Thought*. London: Verso.

Kobayashi, A. (2005) 'Anti-racist feminism in geography: An agenda for social action', in L. Nelson and J. Seager (eds), *A Companion to Feminist Geography*. Malden, MA: Blackwell. pp. 32–40.

Mohanty, C. (2003) *Feminism without Borders: Decolonizing Theory, Practicing Solidarity*. Durham, NC: Duke University Press.

Nagar, R. (1995) *Making and Breaking Boundaries: Identity Politics among South Asians in Post Colonial Dar es Salaam*. PhD dissertation, University of Minnesota.

Nagar, R. (1996) 'The South Asian diaspora in Tanzania: A history retold', *Comparative Studies of South Asia, Africa and the Middle East*, 16 (2): 62–80.

Nagar, R. (1997a) 'Communal places and the politics of multiple identities: The case of Tanzanian Asians', *Ecumene*, 4 (1): 3–26.

Nagar, R. (1997b) 'The politics of gendered boundaries: South Asian communities in Tanzania', *Ghadar: A Bimonthy Publication of the Forum of Indian Leftists*, 1: 6–7.

Nagar, R. (1998) 'Communal discourses, marriage, and the politics of gendered social boundaries among South Asian immigrants in Tanzania', *Gender, Place & Culture*, 5 (2): 117–39.

Nagar, R. (2000c) 'Religion, race, and the debate over Mut'a in Dar es Salaam', *Feminist Studies*, 26 (3): 661–90.

Nagar, R. (2016) 'Editor's Interview with Richa Nagar', *Journal of Narrative Politics*, 2 (2): 73–80.

Nagar, R. (2018) 'Hungry translations: The world through radical vulnerability: The 2017 Antipode RGS-IBG Lecture', *Antipode*, 51 (1): 3–24.

Nagar, R. in consultation with Ali, F. and the Sangatin women's collective (2003) 'Collaboration across borders: Moving beyond positionality', *Singapore Journal of Tropical Geography*, 24 (3): 356–72.

Nagar, R. and Kumar, T. (2011) 'Theater of hopes', *Environment and Planning D: Society and Space*, 29 (3): 54–5.

Ranganathan, M. (2022) 'Caste, racialization, and the making of environmental unfreedoms in urban India', *Ethnic and Racial Studies*, 45 (2): 257–77.

37
Gunnar Olsson

Martin Gren, Linnaeus University

Biographical Details and Theoretical Context

Gunnar Olsson was born in Ekshärad, Sweden, on September 11, 1935. By all accounts, his parents gave him a safe and happy childhood, which provided him with a good dose of self-confidence, a large portion of curiosity, and perhaps also with his inimitable sense of humour. His mother was a midwife who kept the door open for people from all walks of life, particularly those worse off: the homes of Olsson and his wife Birgitta, in Ann Arbor (US) and Uppsala (Sweden), have similarly served as alternative academic asylums, places for human interaction at a relatively safe distance from the collective institutional power that Olsson has always tried to keep at bay. Individual human beings have always been more important to Olsson, a position consistently reflected in his academic works.

In 1957, after completing military service, Olsson moved to Uppsala, where he studied Geography, Economic History and Political Science. He earned his bachelor's degree in 1960, and in the same year was accepted as a licentiate student, on stipend, at the Department of Geography at Uppsala. During his first year (1960–61), Olsson's supervisor Gerd Enequist (the first female professor of Human Geography in Sweden) introduced him to a young man from England, with whom he then shared an office: **David Harvey**. They have remained very close friends ever since.

While working on his licentiate thesis he spent a year (1963–64) as a visiting fellow at the University of Pennsylvania (US) and its Regional Science Department (at the time a major hub of quantitative spatial analysis). Olsson often shared lunch with geographer Michael Dacey, and participated in the year-long course offered by Walter Isard, a founding father of Regional Science. Olsson was awarded a licentiate degree in Human Geography at the University of Uppsala in 1965. Soon after, he accepted a job offer from the University of Michigan, Ann Arbor, as Assistant Professor of Geography (1966). After only three weeks he became Associate Professor with tenure, and was promoted to full Professor in 1971. During his stay in Ann Arbor, Olsson also completed his doctoral thesis – *Distance, Human Interaction and Stochastic Processes: Essays in Geographic Model Building* – which he successfully defended at the University of Uppsala in May 1968.

Olsson continued to investigate human interaction during his eleven years at the Department of Geography at Ann Arbor (1966–1977), but his former quantitative approach shifted toward qualitative studies centred on human thought-and-action, soaked in philosophical problems, firstly *Birds in Egg* (Olsson, 1975) and then *Birds in Egg / Eggs in Bird* (Olsson, 1980). In 1976 Olsson was appointed to the chair of Economic Geography and Planning at the Nordic Institute in Urban and Regional Planning (NORDPLAN, Sweden). During his 20 years at NORDPLAN Olsson was responsible for an ambitious PhD program before the NORDPLAN institute was reorganised and his appointment was transferred to the Department of Geography at Uppsala University in 1997, closing the geographical circle of his academic affiliations. In 2000, Olsson became Professor Emeritus. Having felt an urge to stage some kind of *retro*spective, he instead opted for a 'forward-looking *pro*spective' (Olsson, 2007: ix): *Abysmal: A Critique of Cartographic Reason*, a magnum opus of more than 500 pages (Olsson, 2007).

Spatial Contributions

Olsson knew early on that he wanted to use quantitative approaches, a formative moment being his 1958 reading of a licentiate thesis by Esse Lövgren, 'a brilliant man obsessed with the idea of translating the vagaries of human behavior into the precise language of mathematics' (Abrahamson and Gren, 2012: 8). He also wanted to learn more about the gravity model, and, more precisely, about the question of whether migration and central place theory are special cases of a general theory of human interaction.

In his licentiate thesis of 1964, *Distans och mänsklig interaktion* (in English: 'Distance and human interaction'), Olsson investigated two sets of regression models, one with distance as the independent variable, the other as the dependent. The issue was whether the slopes of the regression lines exhibited systematic variations or whether they are stable. In analogy with Newton's law of universal gravitation, a geographical locality will 'attract' human interaction, such as migration, with a 'gravitational force' that is proportional to its 'mass' (population size), and which decreases with increasing geographical distance.

However, what Olsson found was that the regression lines are not stable but vary with the hierarchical level of the places of in- and out-migration. The conclusion, bluntly put, was that the model 'says less about migration than about the spatial prison in which our interactions are taking place' (Olsson 2022: 9). Lurking was a suspicion that the gravity model was too deterministic, and more appropriate for studies of aggregate human behaviour than for the 'thought-and-actions' of individuals. As a possible remedy, Olsson began translating various human interaction models into the language of probability, where individual variance is allowed to play a more prominent role than the mean value.

Olsson's doctoral thesis *Distance, Human Interaction and Stochastic Processes: Essays on Geographic Model Building* (Olsson, 1968) reflects the fact that by then he was at the forefront of the theoretical, quantitative, and model-building approach in Geography, searching for the laws of human spatial interaction underlying the social. In one of Olsson's (2002: 246) retrospective formulations he asked 'for what is the glue of the social if it is not human interaction, and what is the fundamental variable of geography if it is not distance?' Olsson's thesis hence addressed the classic geographic inference problem between form and process: is spatial form an *effect* of other processes, or is it to be regarded as a *cause* for other processes? Is it reasonable to draw unambiguous inferences from a spatial form (e.g., a mapped geometrical pattern) about the (social) processes generating that form? Part of the problem was that correlation models measure statistical co-variances, paving the way to the trap of spatial autocorrelation, whereas regression models measure causality. In turn, this leads to two quite different ways of understanding the inference problem between (spatial) form and (social) process.

In his doctoral thesis Olsson experimented with deterministic and stochastic geographic model-building. He used distance sometimes as an explanatory variable (*explanans*), and at other times as that which was to be explained (*explanandum*), and he also allowed the aggregation levels to vary in the models. Olsson found that the same spatial form can be generated through drastically different processes, suggesting that one cannot draw conclusions with any certainty about how and why a given spatial form has been generated on the basis of a perfect description of it. Furthermore, although it is sometimes possible and justified to reason from process to form, Olsson claimed moving from form to process is never appropriate. In other words, most spatial science was flawed because of the dubious internal translation 'code' in which it was written.

What for Olsson initially had begun as an attempt to project human interaction onto the reference plane of space, translated through the languages of mathematics and statistics, thereafter transformed into an in-depth engagement with the philosophy and theory of science, evident in the volume *Birds in Egg* (Olsson, 1975). However, this did not mean leaving the old behind, as Olsson's excursions into alternative logics, epistemology, and ontology were further investigations into the possibilities and limits of translating, and thereby understanding, *human interaction*.

One particular, and critically important, vehicle of translation that Olsson began investigating was the equal sign (=), a sign that holds together and apart whatever is placed on either of its sides. Central to all quantitative geographic model-building, Olsson revealed it as central to the language of positivism and science more generally, where the equal sign serves as a spatial marker that often privileges a translation of what 'is', tied to a preservation of the status quo. Even though it is possible to alter the translation function of the equal sign – as one can switch the positions of the X and Y axes in a mathematical formula – one may also replace the equal sign with another sign. In Olsson´s case, the slanted line of dialectics (/) became particularly important. It is premised on internal relations in processes of continuous becoming, and thereby allowing for change to take place. Whereas two plus two always equals four, capital could signify a house, money, and many other things and internal self-referential relations. Dialectics thus opened a door to a human 'thought-and-action' that transgresses seemingly objective scientific descriptions of human interaction embedded in spatial science models.

Olsson hence presented a fundamental conflict, and incommensurability, between the two languages of (positivist) science and human-thought-and-action, involving deep existential power-struggles between identity and difference, ambiguity and certainty, the individual and the collective. The dangers here, Olsson argued, increase as science mutates into models that become part of the taken-for-granted, and are applied in practice (e.g., models of human spatial interaction in planning). For Olsson, this application was a form of social engineering that inevitably involved a reduction of individual variety and integrity, hence his persistent critical stance toward planning and power (as shown in his later Swedish-language book *Antipasti*, with the Swedish welfare state presented as a case) (Olsson, 1990).

After only a few months *Birds in Egg* (1975) had sold out and Olsson was approached by a publisher (Pion) with respect to the book. He eventually added a part that explicitly addressed the issue of the self-referential nature of language, reflected in the new title; *Birds in Egg/Eggs in Bird* (Olsson, 1980). Inspired by modern artists and authors, among them Samuel Beckett, Marcel Duchamp, James Joyce and Stéphane Mallarmé, Olsson sought to transgress the limits of language by writing in such a way that the text was not only about something, but that it was that something itself. It asked if it was possible to minimise, or even erase, the distance between the language one speaks *in* and the phenomena one speaks *about*? It is the added part, *Eggs in Bird,* that contributed most to Olsson's reputation as someone involved in 'linguistic experiments', someone who tried to break out from the 'prison-house of language' by writing in such a unique way as to become imitable. In fact, it was the dialectical slash (/) of *Birds in Egg / Eggs in Bird* that was most important.

Hence, *Lines of Power/Limits of Language* (Olsson, 1991a) was a collection of essays that reconnected with *Birds in Egg/Eggs in Bird* and foreshadowed his conceptualisation of 'cartographic reason'. Power and language were here creatively condensed through the abstractions of points, lines, and planes. For those asking if this was Geography, Olsson would reply: 'Of course it is! For what is geography, if it is not the drawing and interpretation of lines. The only quality that makes my geography unusual is that it does not limit itself to the study of visible things. Instead, it tries to foreshadow a cartography of thought' (Olsson, 1991a: 113). At the end of *Lines of Power/Limits of Language* the ontological transformations that constitute human action were graphically depicted as a 'torpedo', where the fluctuation and inevitable distance between signifier and signified gave rise to its continuous spiraling movement (Olsson, 1991b: 199). In this case, the torpedo had a previous spatial existence as a self-made prototype of steel-wire and wooden matches: eventually, Olsson's attempts to give spatial form to the processes of human-thought-and-action materialised in the glass sculpture *Mappa Mundi Universalis* (Jensson, 2000a).

This attempt to depict the lines of power and the limits of language were retraced and expanded in *Abysmal: A Critique of Cartographic Reason* (Olsson, 2007), and brought to a conclusion in *Arkography: A Grand Tour through the Taken-for-Granted* (Olsson, 2020). The title of the latter alludes to the biblical ark, the wooden container that 'held the commandments and social relations that came with them' (Olsson, 2020: viii). Consequently, the task for Olsson, now appearing as a self-styled 'arkographer', was to conduct 'a minimalist attempt to codify the taken-for-granted', searching for some 'present-day counterparts to the biblical ark' (Olsson, 2020: viii). The excavation begins with his mappings of the creation epics of Enuma elish, Gilgamesh, Genesis and Exodus, and then proceeds to Saussure's Bar, Plato's Republic and Kant's Island of Truth. All not only offer stories about what it means to be human (another take on human interaction), but also reveal invaluable insights about the underlying structures in which their stories have been written. Simplified, these 'arks' are steeped in a cartographic reason that simultaneously *maps* (and therefore delimits) the Human Territory *and* the taken-for-granted ramifications of Human Interaction.

After a long journey, with Pallas Athene as a caring guide, and after having been transformed into both arkologist and arkonaut,

Olsson finally turns into an 'arkotect' ready for a house-tour through the Crystal Palace. This is one of the names of the sculpture of three glass walls on a granite base, designed by Olsson and his former student Ole Michael Jensen, and formally known as *Mappa Mundi Universalis* (Jensson, 2000a). The spatial sculpture is a conceptual depiction of the Human Territory and the fundamental geometric elements of the signifying structure of human understanding, distilled into points, lines and planes. For Olsson, human beings are semiotic animals where human interaction is held together and apart by their usage of signs, understood as both making distinctions (geo*metry*) *and* naming distinctions (geo*graphy*). To the extent that human understanding is patterned on Euclidean geometry, it follows that reasoning which conforms to this mode carries a high degree of credibility. Here, the inference problem between (spatial) form and (social) process is revisited and reimagined.

The Crystal Palace rests on a granite base where a centrally-placed ruby *points* out the original distinction of the human, a difference that once made enough of a difference for setting the violent human world of signification in motion. Over time, three walls of the Crystal Palace grew out from its base, and these are the three primary invisible projection *planes* of Science, Religion, Arts (where the cornerstones are one with Charles Saunders Pierce's semiotic triad of index/symbol/icon turned into the signs of =, /, and –). All planes thus provide central and different translations of and for the beamed *lines* of human significations. Such is the bare spatial and geometrical skeleton of the Crystal Palace, that which, according to Olsson, makes semiotic and cartographic(al) humans human.

Inside the Crystal Palace, true to Olsson's own abstract spatial minimalism (his cartographic reason), there are four floors and an attic to visit: 1) the Basement with the primordial red ruby mark; 2) the Sharing Cross Station and the Prophets' Hall, where the relations between Religion, Arts, and Science are displayed; 3) the Ballroom, the only floor full of people, busy conversing about the prepositions *of*, *at*, and *in*; 4) the scary Attic where the three signs of =, /, and – have gone berserk, the horrors of the present erased from, but at the same time reproduced by, the taken-for-granted of past constitutional orders; 5) finally, a power-filled Penthouse for secular communion, and where, lo and behold, the *Mappa Mundi* sculpture and its catechesis are co-incidentally found on the marble top of an altar.

It is possible to imagine that the Crystal Palace (i.e., *Mappa Mundi Universalis*) is Olsson´s own reimagined version of the gravity model of human interaction, now at last freed from its original Newtonian physical ontological confinement. A cartographic(al) imagination where the codification of the taken-for-granted of human interaction is still significantly informed by geometry, and filled with the inherently violent translations that always come with lines of power: 'Damned be chaotic difference, blessed the identities of cosmic order. ERGO', and 'as imagination bodies forth, the poet´s pen gives to airy nothing a local habitation and a name. FILIOQUE, SIMSALABIM' (Olsson, 2022: 11–12).

On life's way, inside the Crystal Palace, a never-ending power-filled game of ontological transformations and violent translations by which 'we become so obedient and so predictable' (Olsson, 2007: 437). However, in Olsson's understanding, human thought-and action is never to blame, hence his insistence on the importance of Greek tragedy: in the beginning all is well, yet it ends in a terrible mess. From beginning to end, the characters have merely been travelling on a road paved with their own good intentions. As the inhabitants of the Crystal Palace live forward and understand backward (to paraphrase Kierkegaard´s saying, which Olsson often does), they may eventually be able to better understand the limits of their own taken-for-granted world of, and for, human interaction.

Key Advances and Controversies

Olsson has made several key advances into thinking about space and place. Firstly, he made important contributions to spatial science, mapping its limits and dangers through works that identify inherent problems in translating human (spatial) interaction onto the reference plane of space through quantitative models. More generally, this includes advancing the understanding of the geographic inference problem between (spatial) form and (social) process. As a runaway 'space cadet', Olsson opened the door to what it means to be human, and thus his later works contributed to the development of a humanistic geography critically reacting to the reduction of the human in positivistic spatial science.

Secondly, Olsson has, more than any other geographer, significantly advanced understanding of language as a power-filled medium for human interaction. He has consistently made original contributions to the critique of 'power' (for short), by showing how it penetrates all places and spaces of, and for, human interaction, including questions about power in relation to language, meaning, politics, science, thought-and-action, planning, ideology, religion, arts, culture, reason, and the discipline of geography itself. Whatever the knowledge, it is never exempt from language and power considerations.

Thirdly, Olsson has set an example in how knowledge and disciplines can be creatively transgressed, by foregrounding boundaries and limits in a body of work that does not subscribe to a rigid distinction between science and the arts, and that has experimental and artistic ingredients.

Finally, Olsson has substantially, and in his own unique way, advanced the knowledge and understanding of humans as geographical and cartographical beings. More precisely, he has developed a conceptual and artistic depiction of humans as semiotic and cartographic creatures whose world (of human interaction) is held together and apart by their usage of spatial signs, and where they *also* in their invisible realm of culture (the Human Territory) are finding their way(s) by being dependent upon the mapping translation function of *as-if*, as if navigating in culture were like wayfinding on the Earth´s surface.

As for controversies, there has been little direct criticism of Olsson´s works. He has been distinguished as a sort of 'maverick' in geography, a spatial scientist turned poet, someone who writes obscure essays classified as difficult to comprehend. However, anyone vaguely familiar with the social, cultural, and political construction of geographical knowledge, is likely to be rather provoked by an attempt to construct a *Mappa Mundi Universalis*. Any universal knowledge claim emanating from a white, western man who has been sitting behind his desk thinking-and-writing with his masculine colonial gaze about the universal structure of the human world, is quite incommensurable with today's taken-for-granted discourse in human geography.

As true as that might be, Olsson has never hidden from the fact that it is his own understanding that he puts on display in writing. In that sense, *Abysmal* is his minimalist guide to the territory of Western culture, and his critique of cartographic(al) reason. That Olsson has always wanted it his way is likely to be more provoking for those who favour the politics of the collective than for those who cherish the individual. This, however, is not a deflecting defence, because the critical and justified question can still be asked: Does the language Olsson speaks *in* say more of his own language than the phenomena he speaks *about*? In addition, there may well be aspects that are missing in the Olssonian depiction of the taken-for-granted (the *Mappa Mundi Universalis*). Notably, the granite base of the Crystal Palace is no longer a stable Earth surface, a passive foundational background for the human semiotic theatre. The new *Earth of the Anthropocene* is mutating in real time, and inside the Crystal Palace the level of atmospheric carbon dioxide and the temperature is now dangerously high, even moving toward the limit of the human niche. As for the future life of the inhabitants of the Crystal Palace, in the time of planetary climate and ecological emergency, their cartographic reason may be on the verge of becoming obsolete. More generally, this conceptually challenges all thinkers of place and space who have taken-for-granted the Earth as Earth's *surface*.

Olsson's Key Works

Jensson, G. (2000a) *Mappa Mundi Universalis*. Uppsala: Museum Gustavianum. [Glass tetrahedron on granite base, 64 x 64 x 49 cm. Mixed material (Kalmar granite, Weissglass, gold, ruby), designed by Gunnar Olsson and Ole Michael Jensen]

Olsson, G. (1964) *Distans och mänsklig interaktion* ['Distance and human interaction']. Unpublished licentiate thesis, Department of Human Geography, University of Uppsala.

Olsson, G. (1965) *Distance and Human Interaction: A Review and Bibliography*. Philadelphia, PA: Regional Science Research Institute.

Olsson, G. (1968) *Distance, Human Interaction and Stochastic Processes: Essays on Geographic Model Building*. Doctoral thesis, Faculty of Social Sciences, University of Uppsala.

Olsson, G. (1975) *Birds in Egg*. Ann Arbor, MI: Michigan Geographical Publications, No. 15.

Olsson, G. (1980) *Birds in Egg/Eggs in Bird*. London: Pion.

Olsson, G. (1990). *Antipasti*. Göteborg: Bokförlaget Korpen.

Olsson, G. (1991a) *Lines of Power/Limits of Language*. Minneapolis, MN: University of Minnesota Press.

Olsson, G. (2007) *Abysmal: A Critique of Cartographic Reason*. Chicago, IL: University of Chicago Press.

Olsson, G. (2020) *Arkography: A Grand Tour through the Taken-for-Granted*. Lincoln, NE: University of Nebraska Press.

Secondary Sources and References

Abrahamsson, C. and Gren, M. (2012) *GO: On the Geographies of Gunnar Olsson*. Farnham: Ashgate.

Olsson, G. (1991b) 'Invisible maps: A prospectus', *Geografiska Annaler*, Series B, 73: 85–91.

Olsson, G. (2002) 'Glimpses', in P. Gould and Pitts (eds), *Geographical Voices: Fourteen Autobiographical Essays*. New York: Syracuse University Press. pp. 237–268.

Olsson, G. (2022) 'Notes from the dome', *Human Geography*, 15 (1): 7–13.

38
Aihwa Ong

Tim Bunnell, National University of Singapore

Biographical Details and Theoretical Context

Aihwa Ong is Professor Emerita of Anthropology at the University of California, Berkeley, US. Born in Penang, Malaysia, she completed undergraduate studies at Barnard College in New York City and then a PhD in Anthropology at Columbia University. Her doctoral research on women industrial workers in Selangor, Malaysia was completed in 1982. She first joined UC Berkeley two years later.

Over four decades based in the Department of Anthropology at UC Berkeley – though spending time at many other universities overseas and travelling regularly to Asia – Ong has built a body of work encompassing issues of citizenship, globalisation, multiculturalism, sovereignty, transnationalism and the anthropology of bioscience. The influence of her oeuvre has been felt well beyond Anthropology, including in Sociology, Political Science, Science and Technology Studies (STS), Urban Studies, and Asian Studies, as well as in Human Geography.

Ong's early publications established her reputation as an anthropologist and ethnographer. Her first book is widely regarded as a classic ethnography of industrial transformation and shifting cultural practices in the context of globalisation. *Spirits of Resistance and Capitalist Discipline: Factory Women in Malaysia* was developed from doctoral research that involved Ong returning to a demographically Malay-dominated part of her national homeland she had not previously known as an ethnic Chinese Malaysian. Drawing on 14 months of fieldwork in and around a Free Trade Zone, the book examines shifting relations of power associated with the freedoms and unfreedoms experienced by 'neophyte' Malay industrial workers (Ong, 1986).

The theme of people negotiating norms in a new environment continued in Ong's second detailed ethnographic study, *Buddha is Hiding: Refugees, Citizenship, The New America* (published in 2003). Set mainly in the Bay Area around Ong's academic home in California, this work focused on the institutional and self-definitional manoeuvrings of Cambodians who were resettled to Oakland and San Francisco as refugees in the 1980s. In contrast, Ong's third ethnographic book, *Fungible Life: Experiment in the Asian City of Life*, concerns bioscience. Centred on analysis of a biomedical research ecosystem in Singapore, *Fungible Life* (Ong, 2016) explores issues of politics, power and profit associated with specifically Asian bioscientific development.

In between the publication of those three ethnographic monographs, Ong wrote and co-edited a series of articles and essays that were oriented more to observation and conceptualisation of emergent milieux than to documenting the memories, experiences, aspirations or anxieties of individual research subjects. Most famous of all is *Flexible Citizenship: The Cultural Logics of Transnationality* (Ong, 1999). This collection of essays centres on the proliferation of transnational practices through which people seek to avail themselves (and their family members) of economic opportunities in one political context while retaining or actively seeking citizenship rights elsewhere. Some of the arguments and ideas first discussed in *Flexible Citizenship* were extended in *Neoliberalism as Exception: Mutations in Citizenship and Sovereignty*. Here, the focus is on increasingly variegated modes of government and political (de)regulation, within as well as across national territories (Ong, 2006). A co-edited volume the previous year saw Ong (together with Stephen J. Collier) prescribe an analytics of assemblage for examination of socio-spatial phenomena as at once situated and global (Ong and Collier, 2005). Among Ong's other influential co-edited works are volumes on the 'worlding' of Asian cities (Roy and Ong, 2011), Chinese transnationalism (Ong and Nonini, 1997) and the gendered politics of labour in Southeast Asia (Ong and Peletz, 1995).

Across the two different genres of Ong's published contributions and the wide-ranging thematic scope of her research are important threads of scholarly continuity. Above all is Ong's deep suspicion of structuralist theorisation and its purported explanatory power. Ong repeatedly distances herself from Marxist political economy approaches out of a preference for examining situated and open-ended or continually unfolding outcomes. Reference is made to the poststructuralist later work of **Michel Foucault** on 'governmentality' even in Ong's earliest writing from her doctoral research (e.g., Foucault, 1979 in Ong, 1986). The attention that she gave in that writing to the cultural, gendered, religious, moral and political dynamics of labour in a 'global' industrial zone in Malaysia was partly about recuperating what she considered to have been lost in Marxist macro-scale preoccupation with the capitalist mode of production. Subsequent phases of Ong's research and writing drew upon and contributed to the proliferation of poststructuralist transdisciplinary scholarship on governmentality (Rose, 1999).

In an interview in 2015 (Andersen, 2015), Ong identified two foundational underpinnings to her scholarly approach, both of which arose from a reaction to her late-twentieth century encounters with hegemonic Western cultural framings and socio-political categorisation of Asia(ns). The first concerned an effort to do more than merely challenge EuroAmerican stereotyping of other places and people: 'it's about troubling, about demanding a more complex, contingent understanding of people, living conditions, and pressing issues in Asia' (in Andersen, 2015: 14). Second, and related, was conscious resistance to historical notions of the West as a universal modern telos. In Ong's own words, this drove her attempts to 'disrupt overarching theories that view other places as particular steps in or instantiations of Western trajectories of liberal democratic modernization' (in Andersen, 2015: 14: see also Ong, 1986).

While being (seen as) Asian in the American academy clearly had a foundational bearing on Ong's scholarly approach and politics, her own relationship with Asia changed over the course of her career. In methodological terms, not only has her post-PhD work understandably been based mostly on shorter stints of fieldwork, it has also often been more peripatetic (or less site-specific) within Asia. Ong's (2016) monograph is a notable exception to that trend in that it was focused largely on Singapore's Biopolis complex, where she conducted interviews and observational fieldwork during a decade of trans-Pacific trips. In an even longer-term biographical frame, Ong's trans-Pacific shuttle changed from one involving the return to Asia of a Malaysian citizen, into visits by a long-time resident and citizen of the US.

Spatial Contributions

The degree to which Ong has engaged issues of place, space and spatiality has varied. Geography features most explicitly in *Flexible Citizenship* (1999) and *Neoliberalism as Exception* (2006) – in the form of 'zones' of graduated sovereignty and variegated territories of neoliberal governmentality – as well as in the conceptualisation of 'global assemblages' (Ong and Collier, 2005) and the relational 'worlding' of cities through 'inter-referencing' practices (Ong, 2011). Unsurprisingly, these are among Ong's published contributions that have had widest appeal and greatest impact in Geography. In addition, selected geographers have long engaged Ong's work through a shared position within area studies (as in Malaysian Studies, Southeast Asian Studies, East Asian Studies or Asian Studies).

'Asia' features in Ong's scholarship in ways that extend far beyond area studies and associated preoccupation with more or less bounded regional analysis – indeed, this may be considered as one of the key constitutive geographies of her research. Ong has herself described Asia as 'a shape-shifter, constantly invoked as malleable space, imaginary, referent, intervention, *difference*' (Ong, 2017, original emphasis). In Ong's work, Asia invariably confounds both Western-derived imagined pathways of modernisation and normative expectations of the future (Ong, 1995). Asia clearly also exceeds Western stereotypes of oriental otherness, although people and states in Asia are, as Ong (1999) has shown, known to embrace Western stereotypes in strategic forms of self-orientalisation. She has analysed shifting practices in Asia, Asians' border-crossing relations with elsewhere (especially across the Pacific) and imaginaries of Asia as means of resisting EuroAmerican-derived theoretical over-reach. It is also through work on Asia – as a series of sites and as referent – that Ong has undertaken counter-theorisation and (re)conceptualisation. Significantly, her self-described 'concept work' is, in turn, not limited to Asia in any fixed areal, regional or continental geographical sense, but has travelled (and been taken up in the thinking of many geographers working elsewhere).

A second set of geographies concerns the identification and conceptualisation of distinct spaces and associated technologies of government, both within Asia and extending elsewhere. Here the key term is 'zone', especially in the way that this has been deployed in Ong's work from the late 1990s. Practices in and around a Malaysian Free Trade Zone were central to her much earlier work on export-oriented industrial development (Ong, 1986). However, in later writing, manufacturing workforce disciplining and regulation became merely one of a wider range of zones governing different segments of population in different ways (Ong, 1999). Ong's consideration of zoning 'technologies' in China has been particularly impactful, not only because of burgeoning interest in that national political economic context since the early 2000s but also because of Ong's associated engagement with wider debates on neoliberalism (Ong, 2006). She argues that a key part of China's response to globalisation has been to create zones of neoliberal economic as well as political exception to normalised party rule, enabling highly selective integration into the global economy, often mediated through ethnic Chinese populations and associated territories outside the PRC. Ong's (2004) essay on 'The Chinese Axis' involves recognition of a variety of zones of exception, mostly related to economic investment or special administrative regions.

A third key geography – perhaps the most pervasive across Ong's oeuvre – concerns her attentiveness to the entanglement of global forms and technologies on the one hand, and locally-grounded institutions, practices and regimes on the other. This is evident from her earliest work in terms of how the globalisation of manufacturing investment manifested in contextually distinctive ways, including through workers' active remaking of themselves and their social conditions. Ong's interest in the cultural situatedness of global forms and technologies has long been couched in a spatial vocabulary of 'arenas', 'ecologies', 'environments', and 'milieus'. Her use of (global) 'assemblages', meanwhile, also indexes Ong's interest in relational geographies, and how that has been bound up with her shifting methodology. In a piece published in a leading geography journal, Ong (2007: 7) highlighted what she termed 'the limits of stand-alone micro analysis' and corresponding efforts to 'track the diverse vectors of global forms [...] and their unstable alignments with situated constellations'. This, in turn, becomes a spatially expansive means of engaging difference and resisting structuralist universalisation: 'a mode of comparative analysis across situations without reducing them to mass produced equivalents of some master signifier' (Ong, 2007: 7).

The fact that Ong has published in geography journals shows a degree of reciprocal engagement around issues of space and spatiality, as well as overlapping empirical, thematic or theoretical interests. Geographers have recognised the strong underlying spatial sensibility of Ong's best-known works but have also been disciplinarily disposed to lament the overly metaphorical geographies of her analysis and writing. It was in this regard, for example, that one otherwise very positive review of *Neoliberalism as Exception* compared that book unfavourably with **Doreen Massey**'s *For Space* (Castree, 2007, citing Massey, 2005). Ong is first and foremost an anthropologist, albeit one who has said (acknowledging the influence of **Michel Foucault** via **Gilles Deleuze** and Félix Guattari) that she considers 'anthropologists as cartographers [...] looking at how mobile contemporary practices configure new spaces of human action' (cited in Andersen, 2015; see also Ong, 2006, Chapter 9 on 'bio-cartography', situating (female) migrant workers in transnational labour and cultural networks, rather than just in the territorial regulatory context of their work).

Invitations to speak at Geography conferences (including a plenary lecture at the Annual Meeting of the Association of American Geographers in 2007), as well as publications in Geography journals, also gave Ong the opportunity to explicitly address her 'geography' and its relation to her perception of geographers' methods and epistemologies. This has largely meant Ong contrasting the approaches of structuralist economic and political geographers – whose work on neoliberalisation she characterises as following 'administrative units' – with her own 'anthropology of mutating political cartography' that 'puts human agency at the centre of analysis' (Ong, 2016: 118–120). In Ong's terms, her approach is distinguished from that of (a small but highly influential subset of) geographers working on issues of political economy in that she pays attention to 'an endless game of spatializing strategies and mappings of power' (Ong, 2016: 120).

Key Advances and Controversies

The influence and high impact of Ong's work is attributable to her ability to encapsulate ideas in memorable concepts with potential for extensive application and reach. In keeping with her oft-declared scepticism towards grand, epochal claims and structural explanation, Ong has differentiated her approach from theory building: 'I don't come up with theories, I do concept-work' (cited in Andersen, 2015: 17–18). On other occasions, she has likened her conceptual work to 'mid-level theorising' that 'stays close to the ground'. That has not always been met with approval within her own discipline. The essays and books through which Ong made many of her most well-known conceptual contributions have been subjected to critique by fellow anthropologists for not being sufficiently ethnographically grounded or empirically substantiated. Yet in a wider biographical frame, it seems likely that the mid-level formulations in the more conceptually-driven genre of Ong's writing were made possible through her having adopted a range of approaches and methods, including (but not limited to) the kind of ethnographic work that anthropologists have conventionally expected of each other. Ong's self-description as a 'bricoleur' – 'fashioning concepts from things lying around me' (cited in Andersen, 2015: 18) – is therefore an apt distillation of how she has both made and spurred conceptual advances through observation of practices in the public realm and critical engagement with a variety of documentary sources (from policy reports to media articles to inflight magazines), as well as through long-term field-based ethnography.

Pre-eminent among the concepts arising from Ong's bricolage is 'flexible citizenship'. Ong coined this term to refer to the cultural logics and effects of ethnic Chinese life strategies to benefit from investing, working or relocating family members across multiple nation-states. In terms of cultural analysis, flexible citizenship has advanced understanding of ways in which the experience and negotiation of multiple identities has been bound up with efforts to take advantage of globalising political and economic conditions. In terms of the political terrain of globalisation, flexible citizenship catalysed debates on how the form, meaning and spatiality of citizenship and nationalism need to be rethought in light of increasing cross-border mobility and multi-local family formations. Set in relation to pre-existing landmark scholarly contributions, Ong explicitly cast the concept as building from **David Harvey's** 'flexible accumulation' (Harvey, 1989) in the realm of capital – foregrounding 'human agency' (Ong, 1999: 3) and giving greater attention to 'the effects of political economy on culture' (Ong, 1999: 265, note 5). Importantly, although the archetypal agent in Ong's writing on flexible citizenship is the Chinese professional or executive, the concept is not limited to cosmopolitan and hypermobile Chinese subjects, nor indeed to Chinese people at all. The material and symbolic resources (as well as the personal and familial costs) associated with efforts to exercise flexible citizenship are unevenly distributed among the huge diversity of 'Chinese' people, among other groups in the Asia Pacific region and, in late capitalism, elsewhere.

A second set of key advances made by Ong revolves around her conceptualisation of state strategies to calibrate the management of sovereignty in an era of global capital. To the extent that this 'graduated sovereignty' encompasses ways in which governments respond to and seek to capitalise upon transnational practices – including efforts to attract or retain valuable talent from overseas – it is effectively the other side of the coin of flexible citizenship. But Ong conceptualises graduated sovereignty much more widely to denote how places and populations – including citizens as well as non-citizens – are differentially valued and governed. Citizenship itself thus comes to be understood as variegated within and across borders in that 'populations subjected to different regimes of value enjoy different kinds of rights, discipline, caring, and security' (Ong, 1999: 217). Global capital looms large in the calculus of differential value, and control of populations in some domains or enclaves are given over to corporations (or supranational organisations in the case of some refugees). Yet Ong cast graduated sovereignty in terms of the reconfigured, not necessarily diminishing, role of the nation-state. What she termed 'a checkered geography of governing' results 'not from an anaemic state apparatus but from a deliberative neoliberal calculation as to which areas and which populations are advantageous or not advantageous in appealing to global markets' (Ong, 2006: 77). These arguments spurred work on socio-spatially selective forms of (neo) liberalisation in 'postdevelopmental' Asia, and on shifting relations between sovereignty, space and governmentality much more widely.

Through having formulated concepts from Asia that have been taken up and gained traction elsewhere – including in research in Western Europe and North America – Ong is among a small number of social scientists whose research inverts established Western-centred geographies of knowledge production. While this might be considered a profound contribution to postcolonialising Anglophone social theory, however, Ong has repeatedly distanced herself from postcolonial approaches. She has critiqued postcolonial theory along four main lines. First, it is too sweeping to make legacies from the European colonial era apply to heterogeneous non-Western societies and spaces, especially given that some of them – including much of China – never directly experienced this form of colonisation. Second, even in those territories that were formally colonised (such as the part of present-day Malaysia where Ong herself was born a British subject), experiences of colonial rule often ended a long time ago, and residual effects have been reworked through entanglement with other forms of oppression that cannot simply be attributed to Western colonialism(s). Third, by seeking to incorporate contemporary inequities as well as diverse histories into a single analytic, postcolonial theory is itself a form of 'theoretical imperialism' that distant academics and students can mobilise 'without seriously engaging the scholarship of faraway places' (Ong, 1999: 33). Fourth, even in work that consciously strives to take local conditions and traditions of scholarship seriously, a focus on 'subaltern' agency privileges a particular form of subjectivity to the neglect of a wider array of actors and emergent forces involved in global assemblages and 'worlding' practices (Ong, 2011).

Ong's views on postcolonial theory run against the grain of recent scholarly partitions and nomenclature, especially in relation to postcolonial urban studies. It is ironic that a scholar who has done so much to post- or decolonialise spatially-attuned social science – whose mostly Asia-sited global bricolage may be said to have generated 'new geographies of theory' (Roy, 2009) – rejects the label 'postcolonial' and has said that she does not do theory. However, perhaps the defining feature of Ong's French poststructuralist-inspired approach to geographies of power over four decades has been to question the explanatory reach of encompassing theoretical framings, whether centred on patriarchy, capitalism, or colonialism.

Ong's Key Works

Ong, A. (1986) *Spirits of Resistance and Capitalist Discipline: Factory Women in Malaysia*. Albany, NY: SUNY Press.

Ong, A. (1999) *Flexible Citizenship: The Cultural Logics of Transnationality*. Durham, NC: Duke University Press.

Ong, A. and Collier, S.J. (eds) (2005) *Global Assemblages: Technology, Politics and Ethics as Anthropological Problems*. Malden, MA: Wiley-Blackwell.

Ong, A. (2006) *Neoliberalism as Exception: Mutations in Citizenship and Sovereignty*. Durham, NC: Duke University Press.

Roy, A. and Ong, A. (eds) (2011) *Worlding Cities: Asian Experiments and the Art of Being Global*. Malden, MA: Wiley-Blackwell.

Secondary Sources and References

Andersen, N.T. (2015) '"I don't do theory, I do concept-work": An interview with Aihwa Ong', *Women, Gender and Research (Kvinder, Køn & Forskning)*, 24 (1): 12–19.

Castree, N. (2007) 'Review of "Neoliberalism as Exception: Mutations in Citizenship and Sovereignty"', *Progress in Human Geography*, 31 (4): 579–80.

Foucault, M. (1979) 'On governmentality', *Ideology and Consciousness*, 6 (1): 5–21.

Harvey, D. (1989) *The Condition of Postmodernity*. Oxford: Blackwell.

Massey, D. (2005) *For Space*. London: Sage.

Ong, A. (1995) 'Anthropology, China, and modernities: The geopolitics of cultural knowledge', in H. Moore (ed), *The Future of Anthropological Knowledge*. New York, NY: Routledge. pp. 60–92.

Ong, A. (2003) *Buddha is Hiding: Refugees, Citizenship, the New America*. Berkeley, CA: University of California Press.

Ong, A. (2004) 'The Chinese axis: Zoning technologies and variegated sovereignty', *Journal of East Asian Studies*, 4: 69–96.

Ong, A. (2007) 'Neoliberalism as a mobile technology', *Transactions of the Institute of British Geographers*, 32 (3): 3–8.

Ong, A. (2011) 'Introduction: Worlding cities, or the art of being global', in A. Roy and A. Ong (eds), *Worlding Cities*. Oxford: Wiley-Blackwell. pp. 1–26.

Ong, A. (2016) *Fungible Life: Experiment in the Asian City of Life*. Durham, NC: Duke University Press.

Ong, A. (2017) 'Ahiwa Ong on her book "Fungible Life: Experiment in the Asian City of Life"', *Rorotoko*, January 18. https://rorotoko.com/interview/20170118_ong_aihwa_on_fungible_life_experiment_asian_city_life/ (accessed November 10, 2023).

Ong, A. and Nonini, D. (1997) (eds) *Ungrounded Empires: The Cultural Politics of Modern Chinese Transnationalism*. London: Routledge.

Ong, A. and Peletz, M.G. (1995) (eds) *Bewitching Women, Pious Men: Gender and Body Politics in Southeast Asia*. Berkeley, CA: University of California Press.

Rose, N. (1999) *Powers of Freedom: Reframing Political Thought*. Cambridge: Cambridge University Press.

Roy, A. (2009) 'The 21st-century metropolis: New geographies of theory', *Regional Studies*, 43 (6): 819–30.

39
Anssi Paasi

Rhys Jones, Aberystwyth University

Biographical Details and Theoretical Context

Anssi Paasi is one of the world's foremost regional and political geographers. He was born on 14 September 1955 and was subsequently educated at the University of Joensuu in Eastern Finland (at Master's, Licentiate and Doctoral levels) between 1979 and 1986. Following his doctorate, he was appointed to the post of Docent in the University of Joensuu and was subsequently promoted to a Chair in Geography in the University of Oulu in 1990. He is currently Emeritus Professor of Geography at the same institution. He has, additionally, been a visiting scholar in the University of Nijmegen (Netherlands), Loughborough University (UK) and UCLA (US), and holds an Honorary Chair at Aberystwyth University (Wales). He has published over 150 papers, reviews and chapters, as well nineteen books and monographs in both Finnish and English. His work has centred on understanding key concepts within Human and Political Geography, most notably the region, nationalism and territories/boundaries. The empirical focus for Paasi's research throughout his career has been on Finland and its relationship with the broader Baltic region.

Such a brief account of Paasi's career fails to acknowledge the significance of the geographies of his life history. The fact that Paasi is a Finnish geographer has had a profound effect on his career. Perhaps most prosaically, Paasi's nationality has meant that he has conducted the vast majority of his empirical research on Finland. More significantly, Paasi (2022a) himself maintains that much of his career has been spent working from a series of 'peripheries'. First, the fact that Paasi is a Finnish geographer has meant that he has always worked on the relative 'margins' of a Human and Political Geography that is dominated by geographers living and working within an Anglo-American sphere. Such peripherality has arguably made it more challenging for Paasi to influence international debates within Human Geography. The fact that he has succeeded in doing so is testament to the originality of his conceptual contributions on varied themes.

Second, Paasi has published a large proportion of his research in Finnish-language publications, especially during the early stages of his career. He has noted how his appointment to a position in the University of Oulu, along with the fact that the Geography Department of that university was located in the Faculty of Science, encouraged him to shift his language practices (Passi, 2022a: 285) and publish more through the medium of English. However, it is clear that the dominance of the English language within international debates in Human Geography – and the way in which such a dominance is supported by citation practices and metrics – has been a source of some frustration (Paasi, 2022a).

A third kind of peripherality derives from the type of Human Geography taught and researched in Finland during Paasi's formative years. Paasi (2022a) has noted that the character of Human Geography within Finland during the 1970s was affected by its academic affiliation to Faculties of Science within the vast majority of universities and by the dominance of an older generation of human geographers who were proponents of quantitative and positivist approaches. These twin influences made it difficult for a new generation of critical human geographers, such as Paasi, to gain a foothold within academic geography in Finland, especially during the 1970s and 1980s. Paasi's early work on regions, for instance, was deemed by certain individuals within an older generation of Finnish human geographers to be marginal to Finnish Human Geography and led him to downplay the critical and theoretical aspects of his work when applying for research funding (Paasi, 2022a: 278). Paasi, in this regard, considers himself to be a geographer who was in the 'wrong place' (pers. comm.).

A final theme of peripherality arises in the context of Paasi's specific lifepath within Finland. Paasi's formal geographic education in the University of Joensuu has had two important implications for his career. First, Paasi has maintained that the Department of Geography in the University of Joensuu acted as a key location for the 'social scientification' of the discipline of Geography within Finland. 'Close connections with fields such as history, economics, sociology and social policy', according to Paasi (pers. comm.), enabled human geographers within the University of Joensuu to engage more fully with some of the important philosophical currents that were affecting the humanities and social sciences internationally from the 1970s onwards. The 'social scientification' of Human Geography taking place in the University of Joensuu during this period also contributed to the relative dis-enfranchisement of Paasi and other researchers at a broader Finnish scale. Second, one needs to note the position of the University of Joensuu within Finland. Much of Paasi's research has centred around developing conceptual and empirical understandings of the Finnish territory and border, most notably in the context of the disputed borderlands of Karelia. The University of Joensuu lies in Finnish North Karelia and, in this respect, Paasi's time within the University can be seen to have contributed to some of his most notable and innovative work.

Taken together, therefore, Paasi's career has been influenced by his own personal lifepath in key ways: on the margins of a Human Geography dominated by Anglo-American academic imperatives; on the academic margins of an academic project within Finnish Geography (at least during the late 1970s and early 1980s); and on the geographical margins of Finland (during the formative stages of his career). Despite and/or because of this positioning, Paasi has made significant international contributions to three key and interrelated debates: the institutionalisation of regions; the socio-spatial processes that help to reproduce nationalism; and the broader significance of borders and territories as a way of articulating a socio-spatial consciousness.

Spatial Contributions

Paasi's (2022a: 274) work is a response to and elaboration of research on different aspects of socio-spatial theory that emerged during the late 1970s and 1980s. His own contributions have examined the way in which understandings of key geographical concepts, such as the region, borders and territories, can be enhanced through insights from social theory, while at the same time demonstrating how a spatial sensitivity can augment conceptualisations of various social processes and forms. In essence, therefore, all of Paasi's contributions to geographical knowledge have been spatial or, at least, spatially-informed.

Paasi's work on the institutionalisation of regions has cemented his reputation as one of the world's foremost contemporary geographers. He has admitted that the inspiration for this work derived from academic debates concerning structuration and social constructionism (Giddens, 1984), which were emerging during the late 1970s, as well as more geographical contributions by ***Allan Pred*** (1984), **Doreen Massey** (1984) and **Nigel Thrift** (1983) which were beginning to interrogate the meaning of place and region in more theoretically sophisticated ways. Paasi (pers. comm.) believed, nonetheless, that these various contributions failed to explain fully how 'regions became what they were'. It was with this context in mind that Paasi (1986a; 1991) sought to provide a new theoretical and methodological way of comprehending how regions – understood as a socio-spatial unit and a version of a 'higher-scale' history into which individuals are socialised – are institutionalised. Four mutually constituting stages are key to the process of institutionalising regions. The first is the 'territorial shape' of a region or, in other words, the degree to which the territorial extent of a region becomes embedded in individual or collective consciousness and in various social practices. The second is the 'symbolic shape' of a region, which refers to the process through which particular symbols come to represent the region in question. Third, a region may derive 'institutional shape' from a broad range of institutions, which give some formal structure to social and spatial life within the region. Finally, Paasi describes the way in which regions may become 'established' within the socio-spatial consciousness or, in other words, the way in which a region takes on a meaningful identity.

Viewed as a whole, Paasi's work examines the way in which regions are in a constant process of 'becoming' in a variety of different contexts. Regions may emerge and become consolidated or 'established' within society. By the same token, they may become marginal to people's lives and, indeed, they may even disappear. Equally as significant as Paasi's efforts to devise a conceptual and methodological framework for understanding regions, though far less evident in an international and specifically English-medium context, has been his empirical study of the institutionalisation of four Finnish regions: Lapland, Southern Ostro-Bothnia, Northern Karelia and Uusimaa (Paasi, 1986b; 1986c). Paasi has subsequently refined and expanded his engagement with regions. He has co-edited volumes that have sought to bring additional conceptual and analytical clarity to understandings of regions, particularly in the context of globalisation (Jones and Paasi, 2017; Paasi et al., 2018a). He has also examined new and emerging aspects of regionalism, not least in relation to regional planning and identity (Paasi, 2013) and the relationship between regions and climate change (Paasi, 2023).

Paasi's other two key areas of academic intervention have followed on from his earlier work on the institutionalisation of regions. These two areas are linked and focus on the reproduction of national territories and borders. Paasi's research on the reproduction of nations and national territories – specifically in the context of Finland – can be viewed as an attempt to understand the institutionalisation of the nation (Paasi, 1996; see also MacLeod, 1998). This work shows how nations and national territories are reproduced through a combination of a socio-spatial consciousness, centred on the activities of the state's agents at the national scale, and a social representation, which lies within the domain of more ordinary people within localities. Equally as interesting within this work has been Paasi's empirical focus on Finnish national identity, particularly the role of the border region of Karelia in reflecting and shaping the socio-spatial consciousness and social representation associated with Finnish nationalism. Paasi has been able to show how the reproduction of a Finnish nation and national territory has come about as a result of a combination of formal state-building projects and the more mundane and day-to-day practices of ordinary people. This work has also enabled him to make important contributions in relation to the interplay between hot and more banal versions of nationalism (e.g., Paasi, 2016).

Paasi's concern with the role of the Finnish–Russian border in defining the Finnish nation has also encouraged him to engage in a second, related area of research on the broader significance of boundaries in a contemporary globalising world. His first major foray to develop a more conceptually-attuned understanding of borders and boundaries occurred in collaboration with political geographer David Newman (Newman and Paasi, 1998) and this important paper contributed to a flourishing of research on borders in Geography and beyond. Part of this project has revolved around the need to view borders and boundaries as geographical concepts that are produced. Paasi has shown how boundaries should be viewed as contingent processes in motion. They are, moreover, geographical entities that are produced as a result of a mixture of processes; some are undoubtedly state-driven but other actors within civil society are also implicated in the process of 'narrating boundaries' (Jones and Fowler, 2007).

More recently, Paasi has examined the impact of globalisation on borders and boundaries. His work has shown that globalisation has not undermined the significance of boundaries but has merely made the various processes associated with the reproduction of contemporary borders more complicated and contingent. Empirically, while some borders may be weakened as a result of global processes, others become more significant (Paasi, 2022b). And such statements take on additional ethical saliency in the context of the unequal ability of individuals and groups to be mobile and borderless in their own lives (Paasi et al., 2018b). In more conceptual contexts, these developments open up a series of issues around the connection between mobility and relationality, and bordering and territoriality, within the contemporary world (Paasi and Zimmerbauer, 2016).

Key Advances and Controversies

Paasi's work has formed a remarkably coherent academic project over a number of years. He has drawn inspiration from new developments in social and cultural theory to re-think the significance of key concepts in

Human and Political Geography, most notably those relating to regions, nations/territories and boundaries. The impact of his work has been widespread, as can be seen by the extremely high citations for his key publications on these themes (such as Paasi, 1986a, 1996; Newman and Paasi, 1998). While such quantitative measures are impressive, Paasi (2022a: 276) maintains that his career in the 'Nordic periphery' has put him at a relative disadvantage, particularly given the tendency more broadly within academia to believe that academic theories are 'produced in the academic cores and consumed in the margins' (see also Müller, 2021; Korf, 2021), though he identifies a series of factors that have enabled him to buck this trend or perception (Paasi, 2022a: 287–8).

First, he refers to a series of 'brokers and boundary spanners' who have enabled his work to move 'against the tide' from the 'Nordic periphery' to an Anglo-American Geography 'core'. For instance, ***Torsten Hägerstrand*** helped Paasi's (1986a) article on regions to 'travel' extensively within Nordic countries (Paasi 2022a: 288), something that 'opened the first steps to internationalisation, visits to various departments and the path to interdisciplinary Nordic projects and conferences'. The support of other prominent academics enabled Paasi's ideas to travel more broadly still. Ron Johnston's (1991) discussion of the institutionalisation of regions and places in *A Question of Place* expanded the reach of this early work on the institutionalisation of regions (Paasi, 1986a) to Anglo-American audiences. Similar support was provided by Anne Gilbert (1988), whom Paasi (2022a: 184) credits with writing the 'first major international response to the idea of the institutionalisation of regions'. A new generation of regional geographers has taken much inspiration from Paasi's work and continues to bring his ideas to new international audiences, with scholars such as Martin Jones, Gordon MacLeod and John Harrison being notable examples (Jones and MacLeod, 2004; Harrison and Growe, 2014).

Second, it may be that Paasi has evolved from being a geographer 'in the wrong place' during the early stages of his career to being a geographer who was writing 'in the right time' or at least on the 'right topics'. For instance, he admits that his work on regions resonated particularly with international audiences because 'political scientists, historians and many others were just beginning to find regions interesting, or perhaps, because there was an acute need for socio-spatial theory in the critical studies of bounded spaces' (Paasi, 2022a: 285). Similarly, Paasi's growing interest in understanding territories and borders during the 1980s and 1990s coincided with an increased significance of borders in the context of the collapse of the Soviet Union (Paasi, 2022a: 286). Timing aside, it is clear that Paasi's work has helped to inspire a new generation of political geographers to examine notions of territoriality and borders in critical and processual ways, thus putting paid to any simple notions of the emergence of a borderless, globalised world (e.g., Amoore, 2006; Mountz, 2010; Jones, 2016).

Third, we should not underestimate the agency of Paasi himself. In a revealing section of a recent, autobiographical, chapter, he discusses (drawing on Plato) the notion of *thymos* (the hunger for recognition) (Paasi, 2022a: 289–90). Despite the challenges of finding himself at the periphery of Anglo-American geographical debates, it is evident that Paasi, throughout his career, has attempted to make connections with high-profile academics, to reach out to new audiences, and to ensure that his work has been as impactful as possible. Key examples include travelling as a visiting scholar to UCLA and discussing his research on borders with John Agnew before publishing the co-written article with David Newman (Newman and Paasi, 1998). A similar proactive engagement on Paasi's part with Nordic and British academics helped to raise the profile of his earlier work on the institutionalisation of regions.

In all of this, we witness an academic who has striven to produce a series of conceptual and empirical contributions on themes that are central to Human Geography. Given the significance of regions and particularly borders for contemporary research and public policy, his work is likely to remain crucial for scholars and practitioners alike for many years to come.

Paasi's Key Works

Newman, D. and Paasi, A. (1998) 'Fences and neighbours in a postmodern world: Boundary narratives in political geography', *Progress in Human Geography*, 22: 186–207.

Paasi, A. (1986a) 'The institutionalization of regions: A theoretical framework for understanding the emergence of regions and the constitution of regional identity', *Fennia*, 164: 105–46.

Paasi, A. (1991) 'Deconstructing regions: Notes on the scales of human life', *Environment and Planning A*, 23: 239–56.

Paasi, A. (1996) *Territories, Boundaries and Consciousness*. Chichester: Wiley.

Paasi, A., Harrison, J. and Jones, M. (eds) (2018a) *Handbook on the Geographies of Regions and Territories*. Cheltenham: Edward Elgar.

Paasi, A., Prokkola, E.K., Saarinen, J. and Zimmerbauer, K. (eds) (2018b) *Borderless Worlds for Whom? Ethics, Moralities and Mobilities*. London: Routledge.

Secondary Sources and References

Amoore, L. (2006) 'Biometric borders: Governing mobilities in the war on terror', *Political Geography*, 25: 336–51.

Giddens, A. (1984) *The Constitution of Society*. Cambridge: Polity.

Gilbert, A. (1988) 'The new regional geography in English and French-speaking countries', *Progress in Human Geography*, 12: 208–28.

Harrison, J. and Growe, A. (2014) 'When regions collide: In what sense a new "regional problem"?', *Environment and Planning A*, 46: 2332–52.

Johnston, R.J. (1991) *A Question of Place? Exploring the Practice of Human Geography*. Oxford: Blackwell.

Jones, M. and MacLeod, G. (2004) 'Regional spaces, spaces of regionalism: Territory, insurgent politics and the English question', *Transactions of the Institute of British Geographers*, 29: 433–52.

Jones, M. and Paasi, A. (eds) (2017) *Regional Worlds: Advancing the Geography of Regions*. London: Routledge.

Jones, R. (Reece) (2016) *Violent Borders: Refugees and the Right to Move*. London: Verso.

Jones, R. (Rhys) and Fowler, C. (2007) 'Where is Wales? Narrating the territories and borders of the Welsh linguistic nation', *Regional Studies*, 41: 89–101.

Korf, B. (2021) '"German theory": On cosmopolitan geographies, counterfactual intellectual histories and the (non)travel of "German Foucault"', *Environment and Planning D: Society and Space*, 39: 925–44.

MacLeod, G. (1998) 'In what sense a region? Place hybridity, symbolic shape, and institutional formation in (post-)modern Scotland', *Political Geography*, 17: 833–63.

Massey, D. (1984) *Spatial Divisions of Labour*. London: Macmillan.

Mountz, A. (2010) *Seeking Asylum: Human Smuggling and Bureaucracy at the Border*. Minneapolis, MN: University of Minnesota Press.

Müller, M. (2021) 'Worlding geography: From linguistic privilege to decolonial anywhere', *Progress in Human Geography*, 45: 1440–66.

Paasi, A. (1986b) 'The institutionalization of regions: Theory and comparative case studies', *University of Joensuu Publications in Social Sciences No.* 9. Joensuu.

Paasi, A. (1986c) 'Four provinces in Finland: Geographical study of the development of regional consciousness (in Finnish with a summary in English)', *University of Joensuu, Publications in Social Sciences No.* 8. Joensuu.

Paasi, A. (2013) 'Regional planning and the mobilization of "regional identity": From bounded spaces to relational complexity', *Regional Studies*, 47: 1206–19.

Paasi, A. (2016) 'Dancing on the graves: Independence, hot/banal nationalism and the mobilization of memory', *Political Geography*, 54: 21–31.

Paasi, A. (2022a) 'The institutionalization of regions: An autobiographic view on the making of socio-spatial theory in the Nordic periphery', in P. Jakobsen, E. Jönsson and H.G. Larsen (eds), *Socio-Spatial Theory in Nordic Geography: Intellectual Histories and Critical Interventions*. Cham: Springer. pp. 273–93.

Paasi, A. (2022b) 'Examining the persistence of bounded spaces: Remarks on regions, territories, and the practices of bordering', *Geografiska Annaler Series B: Human Geography*, 104: 9–26.

Paasi, A. (2023) 'Regional geographies of climate change', *Tijdschrift voor Economische en Social Geografie*, 114: 71–8.

Paasi, A. and Zimmerbauer, K. (2016) 'Penumbral borders and planning paradoxes: Relational thinking and the question of borders in spatial planning', *Environment and Planning A*, 48 (1): 75–93.

Pred, A. (1984) 'Place as historically contingent process: Structuration and the time-geography of becoming places', *Annals of the Association of American Geographers*, 74: 279–97.

Thrift, N. (1983) 'On the determination of social action in space and time', *Environment and Planning D: Society and Space*, 1: 23–57.

40
Jamie Peck

Christian Berndt, University of Zürich

Biographical Details and Theoretical Context

Jamie Peck began his academic career as a geographer and political economic institutionalist at the University of Manchester, UK. After completing his undergraduate degree, Peck secured a doctoral studentship closely aligned with the Tameside Metropolitan Borough Council, a local authority in Greater Manchester. During this time (1983–1988) the conservative government led by Margaret Thatcher had just introduced a key labour market policy, the Youth Training Scheme, and it was this policy instrument and its implementation in Eastern Manchester that Peck's doctoral project focused on (Peck, 1988).

Peck was interested in the Youth Training Scheme as an example of a new generation of social policy interventions. The scheme symbolised a profound policy shift, away from the traditional Keynesian welfare state and municipal socialism towards an emerging formation driven by anti-unionism, extensive privatisation, and deregulation. Geographically, this transformative shift manifested as regional restructuring, leading to glaring inequalities between London, Southern England, and the rest of the country, referred to as the 'North–South divide'. Within this context, Peck developed his empirical project within the conceptual triad of labour market, space and state. However, dissatisfied with the way these key elements had been theorised, he sought inspiration from a new generation of scholars emerging in the field of (economic) geography and neighbouring disciplines.

Within British economic geography, a key debate at that time evolved around what became known as 'locality studies' (see also entry on **Doreen Massey**). These originated from research programs funded by the UK's Economic and Social Research Council that looked at the impact of restructuring on particular places and regions, as well as uneven development in the UK (Painter, 2009). Arguing that it provided him with 'bridges into an emergent kind of human geography', Peck was influenced by the debate around regional restructuring and critical realism, above all by Massey's (1984) *Spatial Divisions of Labour* and ***Derek Gregory*** and **John Urry**'s (1985) edited volume *Social Relations and Spatial Structures* (Peck et al., 2022; Peck, 2020a: 294). However, there was also a certain distance between his work and these debates, both due to his affiliation with a local authority, where he was conducting policy evaluations alongside doctoral research, and because there was no economic geography research on the type of issues at the heart of his project. Peck also missed a deeper engagement with the role of the state at different scales. Help was provided by two literatures that would prove extremely influential for his subsequent career. One concerned institutionalism and what would later come to be known as socioeconomics, mainly inspired by the work of Karl Polanyi and labour market segmentation theory. The other was regulation theory and its interest in questions of longer-term institutional change theorised around the wage–labour nexus. These both afforded a more meso-level perspective vis-à-vis the finely grained one informing 'locality studies' and studies of 'regional restructuring' and resulted in early influential papers, some co-authored with Adam Tickell (Peck and Tickell, 1992; Tickell and Peck, 1992), that grounded regulation theory locally (Peck, 2020a; 2022; 2023a).

These early years laid the foundation for an intellectual career that made Peck a highly influential scholar in Geography and related fields. After a one-year postdoctoral fellowship at the University of Melbourne, he spent the 1990s at the University of Manchester before leaving the UK to take up a professorship at the University of Wisconsin-Madison (2000–2008). From 2008, he has held a Professorship at the University of British Columbia, Canada. Continuing to work on the broad interdisciplinary terrain offered by heterodox economic geography, urban-regional restructuring, labour regulation and statecraft, there are three broader fields in which Peck has established himself as a leading scholar, setting the tone for generations of younger researchers within geography and beyond.

The first concerns his continuing work on labour relations and labour markets. By focusing on labour markets as socially instituted and emphasising the role of the state, he introduced a meso-level approach to a debate that had turned to the agency of workers (see Herod, 1997). This work culminated in his 1996 book *Work-Place: The Social Regulation of Labor Markets*. Peck wrote his book against the dominant neoclassical economic mainstream and the idea that labour markets are governed by universal economic processes. Against this he argued for a more empirically grounded approach foregrounding variation, noting labour markets develop in tandem with social and political institutions. All this was set in the context of the unfettered dismantling of welfare state protections during the 1980s and then the so-called 'third way' policies associated with centre-left governments in the UK and the US. *Work-Place* was followed by a second book, *Workfare States* (2001), which read neoliberal 'welfare-to-work' policies as deeply racialised and gendered and as forcing recipients into a spiral of precarious, low-paying and unstable jobs. Crucial empirical material grounding these analyses was gathered with Nik Theodore on the late 1990s Chicago's temporary labour market (Peck and Theodore, 2001).

The second major area of influence is Peck's long-standing engagement with neoliberalism. Witnessing the brutal assault on the welfare state in 1980s Britain under Thatcher and the dramatic consequences for regions on the wrong side of the 'North–South divide', Peck was sceptical of the more optimistic accounts of the so-called California school and the literature on the 'new industrial spaces' (e.g., Scott, 1988). Peck took inspiration from Bob Jessop's theorisation of the emerging post-Fordist constellation as 'Schumpeterian workfare state' and in particular Jessop's delineation of different trajectories (neoliberal, neocorporatist, neostatist) (Peck and Jones, 1995). For Peck, the UK clearly went down the market-radical neoliberal route. In his subsequent writings the term 'post-Fordist' was gradually replaced by 'neoliberal', with increasingly pessimistic undertones regarding the situation in the UK (Peck, 1994: 170).

In the 2000s, Peck established himself as a leading figure in debates about neoliberalism by foregrounding neoliberalisation as a historically contingent and geographically uneven process. An early example is the influential analytical distinction between 'roll-back' and 'roll-out' neoliberalism in response to the emerging new political constellations outlined above (Peck and Tickell, 2002). When outlining his understanding of the neoliberal project, Peck – in collaboration with Theodore and Brenner (Brenner et al., 2010) – adopted a characteristic middle ground between those who work with a small n, pluralised version of the concept (neoliberalisms) and those who insist on an expansive capital N-version (Neoliberalism). Peck's style of reasoning is accommodative, adopting an approach that has been characterised as a compromise between 'a fluid and variegated appreciation of contextual difference while maintaining a structural approach that recognises the ways in which local differences and contextually embedded forms are shaped by the *context of context*' (Peck, 2020a: 290; original emphasis).

This intellectual work found a preliminary endpoint in his 2010 book *Constructions of Neoliberal Reason*. In this book Peck excavates neoliberalism as a socially constructed phenomenon shaped by a range of actors and institutions, including think tanks, business groups, political parties, and international organisations. For Peck the focus is on neoliberalisation as a contested, evolving and uneven process of restructuring rather than neoliberalism as an order, era, 'system' or epoch (Peck, 2020a: 295). In this context, the highly selective rolling out of workfare policies that characterised the 1990s and 2000s gave way to the 'progressive neoliberalism' of the Obama years (Fraser, 2019), only to be replaced by a return of austerity and authoritarianism (Peck, 2012a). But again, for Peck the ugly authoritarian face of right-wing populism is far too contradictory and complex to 'be reduced to a question of whether this is, or is not, (still) neoliberalism' (Peck, 2020a: 305; Peck and Theodore, 2019).

A third area of influence has to do with Peck's distinctive intellectual style. Inspired by Polanyian substantivism and the relational-geographical thinking advanced by Massey and Gillian Hart (e.g., Massey, 1991; Hart, 2018), he developed what he refers to as conjunctural analysis. Conceptualising capitalism as always socially unequal and spatially uneven, this approach combines rigorous empirical work with reflexive theorisation. It eschews abstract Weberian ideal-type thinking for the kind of 'dirty hands' analysis 'on the ground' that starts methodologically from socioeconomic diversity and is characteristic for the discipline of geography (Peck, 2020a: 294–96; see also Peck, 2020b). In so doing, he adopted a methodological middle ground between the 'sanitised' and restrictive comparative analysis (characterising the Varieties of Capitalism approach, for example), and a similarly one-sided overemphasis on differentiation and fragmentation in parts of economic geography (Peck, 2023a). Peck (2020a: 301) did this in a spirit that is integrative, acknowledging that 'no one theoretical or methodological approach is ever going to be entirely sufficient', arguing for a collaborative heterodox project that combines political economy in its different stripes with socioeconomic thinking and also cultural economy (Peck, 2023b). This integrative approach informed his PhD dissertation that he characterised as a dialogue between 'regulation theory, Polanyian socioeconomics, Gramscian state theory,

Stuart Hall-style cultural studies and neo-Marxist economic geography' (Peck, 2020a: 293); it also influenced Peck's later theorisation of actually-existing capitalist markets and marketisation as always internally diverse, unequal and contested (see Peck, 2020b; Berndt et al., 2020).

Spatial Contributions

The way Peck engages with space and place has much to do with the debates in Anglo-American economic geography in the 1980s. While sympathetic to the 'local turn', he criticised what he saw as an overemphasis on the uniqueness of place and a lack of attention to the internal differences of the regions and localities studied. In contrast, he pointed to the spatial unevenness of economic restructuring. 'If we are to really understand the extent to which "geography matters"', he argued in his PhD dissertation, 'the argument that places are unique is surely not the best starting point' (Peck, 1988: 4). Peck subsequently refined his position, influenced by the work of Massey and her relational theorisation of space and place. Peck's understanding of institutional diversity thus cannot be separated from geographical unevenness. He rejects an understanding of 'the local' as a delimited and bounded entity set against a seemingly unbounded global space. The various places that he studied are understood relationally as sites of 'articulation, intersection, configuration and (re)combination' (Peck, 2020a: 294), always 'asymmetrically situated and scaled' (Peck, 2020b: 66).

There are two main ways this relational conceptualisation of space resulted in distinct contributions to the wider discipline. The first concerns the mobile geographies of economic models and policies in our global age. At the heart of Peck's work on neoliberalisation is an understanding that seemingly global processes such as marketisation are made and remade in a tangle of local sites, distanciated networks and grounded practices. It is the context-specific translation of the models, concepts and policies associated with neoliberalism writ large that creates the mundane worlds of variegated, actually-existing neoliberalism (Peck, 2020a). This was the starting point for a more general examination of how neoliberal policies travel and how the relationship between state, market and civil society changes with increasing policy mobility. This is most succinctly illustrated in the book *Fast Policy: Experimental Statecraft at the Thresholds of Neoliberalism*, co-authored with Theodore (Peck and Theodore, 2015). In this and other publications, Peck and collaborators pointed to a new kind of experimental government thriving on flexibility and agility, bypassing the democratic processes and institutional checks and balances that characterised the era of the Fordist welfare state. Flexibly adapted to the respective context, experimental governance enables political and economic decision-makers to impose a class project that goes hand in hand with growing inequality and the marginalisation of large parts of the population. This is evident in a variety of policy areas, ranging from economic development and the labour market (workfare) to areas such as education and offshore outsourcing, which has completely reshaped the international division of labour (Peck, 2017). The key geographical insight is that 'fast policies' take form by linking often distant locations and through a process of translation and adaptation ensuring they are more than mere replications (Peck, 2020a: 300).

The second contribution follows from Peck's conceptualisation of capitalism as always multifaceted not only historically but also spatially. Peck insists on integrating questions of geography and space into critical political economy from the outset rather than as a mere after-thought (Peck, 2023a: 9). His conceptual lens is 'variegation', a critical response to the so-called 'Varieties of Capitalism' approach that became very popular during the 2000s as part of a wider institutional economics literature that grappled with the sweeping end-of-history and end-of-geography claims during the early 1990s. For Peck variegation carries a richer and more complex meaning, the kind of dynamic 'unity in difference' (**Hall**, 1985) that the more static connotation of 'variety' is unable to capture (Peck, 2023a: 237; see also Jessop, 2014).

In an influential paper co-authored with Theodore, Peck took issue with the overtly stylised distinction between Anglo-American liberal market and Continental European coordinated capitalism by 'Varieties of Capitalism' scholars (Peck and Theodore, 2007). Influenced by his long-standing fascination with the 'older', socially-richer institutionalism of Karl Polanyi, Peck's emphasis is on institutional diversity and actually-existing capitalist formations rather than a theorisation 'from above' that classifies economies according to their perceived distance from ideal-types (Peck, 2013b; 2016). But 'variegation' is much more than a simple reference to empirical variation across time and space. In its application to neoliberalism, in particular, it reflects Peck's processual and conjunctural understanding of capitalist development:

> As a tendential process, neoliberalization is not trending toward a state of completion or uniformity, but produces, reproduces, and responds to uneven geographical development as it carves its zigzagging way. There is no original, pure, or template form against which 'others' can be judged; there is only and can only be a world of conjuncturally situated and coevolving processes of neoliberalization, which over time have evidently become increasingly interdependent and mutually referential. (Peck, 2023a: 15)

Such an understanding of capitalism also underlies Peck's work on China. Avoiding unproductive debates about whether China is capitalist or not, he instead proposes a more nuanced understanding of China's transformation in recent decades. In two papers co-authored with Jun Zhang, Chinese 'state capitalism' is presented as a challenge to established theories of capitalist development that defies easy categorisation (Peck and Zhang, 2013: 385–86). The focus is on internal heterogeneity and fragmentation, on interconnected regional formations that are 'embedded within different configurations of extra-local relations' (Zhang and Peck, 2016: 75). Peck's work on capitalist diversity also made an important contribution to the revival of work on state capitalism (Alami et al., 2023) and the Trotskyist notion of 'uneven and combined development'. Connecting it with his conjunctural method, Peck has invigorated the latter concept as a 'constitutive facet of "spatialized" social theory-building' (Peck, 2023a: 17). Peck makes explicit connections to the work of **David Harvey** and **Neil Smith**, and in particular to **Doreen Massey**'s relational geography, bringing us full circle to the early formative years of his PhD research in 1980s Manchester.

Key Advances and Controversies

Throughout his career, Peck has refused to take a narrow theoretical and methodological position. His commitment has always been to Economic Geography as a whole, as the long-time editor-in-chief of the journal *Environment and Planning A* and as the founder and organiser of the Summer Institute in Economic Geography, which has been important for generations of young scholars in the discipline since its inception in 2003 (Barnes and Christophers, 2018: 166–7).

Peck established several research agendas that have strongly influenced Economic Geography and related disciplines. His role in this context is typically a supportive and inclusive one, arguing from a position of dialogue and pluralistic engagement. Arguably it is his work on neoliberalism that is most influential across the social sciences in general. A good example is a debate in the journal *Social Anthropology* in 2012, which included initial interventions by Löic Wacquant and Mathieu Hilgers as well as commentaries by Stephen Collier and by Peck and Theodore. The original interventions roughly represent two opposing views on neoliberalism – Wacquant's (2012) macrosocial emphasis on neoliberalism as Leviathan combined with a Janus-faced carceral state, and Hilgers' (2012) approach from a global South perspective that emphasises variety and difference. Having long been dissatisfied with an 'insufficiently reflexive political economy' and an 'excessively fluid, diffuse governmentality' approach to neoliberalism, Peck and Theodore (2012: 180) offer an additional analytical dimension: attention to 'the distinctive spatialities implied in these readings of neoliberal historicity' (Peck and Theodore, 2012: 177). This allows them to move beyond simply presenting both interventions as competing accounts and instead conceptualise them as 'alternate analytical poles', both necessary to better understand 'the unevenly developed yet globalising process of neoliberalisation' (Peck and Theodore, 2012: 177).

Collier (2012: 186) endorses Peck and Theodore's 'via media' as an attempt at reconciliation, but he does so with a healthy scepticism about whether such an accommodation can be achieved. While somewhat valid, this critique of Peck's approach from a post-structural, Foucauldian perspective overlooks that Peck's position remains a political-economic one, his openness to particularistic readings of small-n neoliberalisms always checked by his insistence on neoliberalism as a global system with a common set of rules and norms (Peck, 2013a).

Debates such as these reveal frictions that are perhaps inevitable when taking the kind of mediating position that comes with Peck's intellectual and methodological style. As he himself acknowledges 'relational comparison and conjunctural analysis can only be truly realised through critical application and ongoing, iterative development' (Peck, 2023a: 17). With his rejection of an intellectual culture that is austere, monist and 'intolerant of alternative theories and methods' (Peck, 2020b: 57), this insight is consistent with calls for an 'engaged pluralism' in the discipline. Initiated by ***Trevor Barnes*** and Eric Sheppard (2010), Peck strongly supports this inclusive project (e.g., Peck, 2012b: 219; 2023a).

There has been a broader discussion of the implications of this project in the discipline, with feminist economic geographers objecting to a hidden logic of *laissez faire* and a liberal politics that seems to assume that it is enough for 'individual scholars to become better listeners' (Rosenman et al., 2020: 517). As Emily Rosenman, Jessa Loomis and Kelly Kay point out, this neglects structural inequalities in knowledge production. They argue that more senior scholars have much greater potential to be heard and recognised, calling for a more radical, collective and collaborative agenda that includes recognition of different positionalities in academic knowledge production. In doing so, they take Peck's pluralist heterodox economic alternative to orthodox neoclassical economics as a positive example and call for extending this position to the politics of knowledge production itself.

Peck's Key Works

Peck, J. (1996) *Work-Place: The Social Regulation of Labor Markets*. New York, NY: Guilford.

Peck, J. (2010) *Constructions of Neoliberal Reason*. Oxford: Oxford University Press.

Peck, J. (2012b) 'Economic geography: Island life', *Dialogues in Human Geography,* 2 (2): 113–33.

Peck, J. (2013) 'For Polanyian economic geographies', *Environment and Planning A,* 45 (7): 1545–68.

Peck, J. (2023a) *Variegated Economies*. New York, NY: Oxford University Press.

Peck, J. and Theodore, N. (2007) 'Variegated capitalism', *Progress in Human Geography,* 31 (6): 731–72.

Peck, J. and Theodore, N. (2015) *Fast Policy: Experimental Statecraft at the Thresholds of Neoliberalism.* Minneapolis, MN: University of Minnesota Press.

Peck, J. and Tickell, A. (2002) 'Neoliberalizing space', *Antipode,* 34 (3): 380–404.

Tickell, A. and Peck, J. (1992) 'Accumulation, regulation and geographies of post-Fordism – missing links in regulationist research', *Progress in Human Geography,* 16: 190–218.

Secondary Sources and References

Alami, I., Whiteside, H., Dixon, A.D., and Peck, J. (2023) 'Making space for the new state capitalism, part II: Relationality, spatiotemporality and uneven development', *Environment and Planning A: Economy and Space,* 55 (3): 621–35.

Barnes, T.J. and Christophers, B. (2018) *Economic Geography: A Critical Introduction.* Hoboken, NJ: Wiley-Blackwell.

Barnes, T.J. and Sheppard, E. (2010) '"Nothing includes everything": Towards engaged pluralism in Anglophone economic geography', *Progress in Human Geography,* 34 (2): 193–214.

Berndt, C., Peck, J. and Rantisi, N.M. (eds) (2020) *Market/Place: Exploring Spaces of Exchange*. Newcastle upon Tyne: Agenda.

Brenner, N., Peck, J. and Theodore, N. (2010) 'Variegated neoliberalization: Geographies, modalities, pathways', *Global Networks,* 10 (2): 182–222.

Collier, S.J. (2012) 'Neoliberalism as big Leviathan, or…? A response to Wacquant and Hilgers', *Social Anthropology/ Anthropologie Sociale,* 20: 186–95.

Fraser, N. (2019) *The Old Is Dying and the New Cannot Be Born.* New York, NY: Verso.

Gregory, D. and Urry, J. (eds) (1985) *Social Relations and Spatial Structures*. New York, NY: St. Martin's Press.

Hall, S. (1985) 'Signification, representation, ideology: Althusser and the post-structuralist debates', *Critical Studies in Media Communication,* 2 (2): 91–114.

Hart, G. (2018) 'Relational comparison revisited: Marxist postcolonial geographies in practice', *Progress in Human Geography,* 42 (3): 371–94.

Herod, A. (1997) 'Labor as an agent of globalization and as a global agent', in K. Cox (ed), *The Global and the Local: Making the Connections*. New York, NY: Guilford/ Longman. pp. 167–200.

Hilgers, M. (2012) 'The historicity of the neoliberal state', *Social Anthropology,* 20 (1): 80–94.

Jessop, B. (2014) 'Capitalist diversity and variety: Variegation, the world market, compossibility and ecological dominance', *Capital & Class,* 38 (1): 45–58.

Massey, D. (1984) *Spatial Divisions of Labour: Social Structure and the Geography of Production*. London: Macmillan.

Massey, D. (1991) 'A global sense of place', *Marxism Today* (June): 24–29.

Painter, J. (2009) 'Locality', in D. Gregory, R. Johnston, G. Pratt, M. Watts and S. Whatmore (eds), *The Dictionary of Human Geography*. Oxford: Blackwell. p. 425.

Peck, J. (1988) *The Structure and Segmentation of Local Labour Markets: Aspects of the Geographical Anatomy of Youth Employment in Great Britain*. PhD dissertation, University of Manchester.

Peck, J. (1994) 'Regulating labour: The social regulation and reproduction of local labour-markets', in A. Amin and N. Thrift (eds), *Globalization, Institutions and Regional Development in Europe*. Oxford: Oxford University Press. pp. 147–76.

Peck, J. (2001) *Workfare States*. New York, NY: Guilford Press.

Peck, J. (2012a) 'Austerity urbanism', *City*, 16 (6): 626–55.

Peck, J. (2013a) 'Explaining (with) neoliberalism', *Territory, Politics, Governance*, 1 (2): 132–57.

Peck, J. (2016) 'Polanyian pathways', *Economic Geography*, 92 (2): 226–33.

Peck, J. (2017) *Offshore: Exploring the Worlds of Global Outsourcing*. Oxford: Oxford University Press.

Peck, J. (2020a) 'Contextualizing neoliberalism: An interview with Jamie Peck', in S. Dawes and M. Lenormand (eds), *Neoliberalism in Context: Governance, Subjectivity and Knowledge*. Cham: Palgrave Macmillan. pp. 289–309.

Peck, J. (2020b) 'Where are markets?' in C. Berndt, J. Peck and N. Rantisi (eds), *Market/Place: Exploring Spaces of Exchange*. Newcastle upon Tyne: Agenda. pp. 49–68.

Peck, J. (2022) 'Confessions of a recovering régulation theorist', in B. Hillier, R. Phillips and J. Peck (eds), *Regulation Theory, Space, and Uneven Development: Conversations and Challenges*. Vancouver: 1984 Press. pp. 169–88.

Peck, J. (2023b) 'Practicing conjunctural methodologies: Engaging Chinese capitalism', *Dialogues in Human Geography*. doi.org/10.1177/20438206231154346

Peck, J. and Jones, M. (1995) 'Training and Enterprise Councils: Schumpeterian workfare state, or what?', *Environment and Planning A: Economy and* Space, 27 (9): 1361–96.

Peck, J. and Theodore, N. (2001) 'Contingent Chicago: Restructuring the spaces of temporary labor', *International Journal of Urban and Regional Research*, 25 (3): 471–96.

Peck, J. and Theodore, N. (2012) 'Reanimating neoliberalism: Process geographies of neoliberalisation', *Social Anthropology*, 20 (2): 177–85.

Peck, J. and Theodore, N. (2019) 'Still neoliberalism?', *South Atlantic Quarterly*, 118 (2): 245–65.

Peck, J. and Tickell, A. (1992) 'Local modes of social regulation? Regulation theory, Thatcherism and uneven development', *Geoforum*, 23 (3): 347–63.

Peck, J. and Zhang, J. (2013) 'A variety of capitalism ... with Chinese characteristics?', *Journal of Economic Geography*, 13 (3): 357–96.

Peck, J., Werner, M. and Jones, M. (2022) 'A dialogue on uneven development: A distinctly regional problem', *Regional Studies*, 57 (7): 1392–403.

Rosenman, E., Loomis, J. and Kay, K. (2020) 'Diversity, representation, and the limits of engaged pluralism in (economic) geography', *Progress in Human Geography*, 44 (3): 510–33.

Scott, A.J. (1988) *Flexible Production Systems and Regional Development: The Rise of New*

Industrial Spaces in North America and Western Europe. Toronto: Centre for Urban and Community Studies, University of Toronto.

Wacquant, L. (2012) 'Three steps to a historical anthropology of actually existing neoliberalism', *Social Anthropology*, 20 (1): 66–79.

Zhang, J. and Peck, J. (2016) 'Variegated capitalism, Chinese style: Regional models, multi-scalar constructions', *Regional Studies*, 50 (1): 52–78.

41
Jasbir Puar

Sneha Krishnan, University of Oxford

Biographical Details and Theoretical Context

Growing up in the 1970s and '80s in New Jersey, Jasbir Puar credits her Sikh upbringing with her instinctive disidentification with the nation-state. With the cry for Khalistan – a separate state for Sikhs – sounding loud in the wake of anti-Sikh pogroms in India in the early 1980s, Puar came of age in a moment of disillusionment with the liberal project of the postcolonial state: India was not a nostalgic home to return to, and postcolonial theory was inadequate to make sense of the violence her community endured. The US, where she had grown up, was also violently racist, and in the aftermath of 9/11 would turn particularly dangerous for Sikhs. Exploration of the connections of race, violence and terror has hence been central to her work, which is best known for addressing the biopolitics of life and death in heteronormative global times.

Puar is far from the first to put questions of sex, intimacy, and liberal feminism at the heart of imperialism expansionism: indeed, her writing in some ways bridges two bodies of scholarship that have rarely been put in dialogue. On the one hand, feminist scholarship on the imperial world has long drawn attention to the ways in which ideas about sex and sexuality, as well as about normative gender and family, have been deployed as justifications for imperial projects past and present (Sinha, 1994; Najmabadi, 2005). The other body of scholarship to which Puar's work speaks is the queer studies corpus which, by the early 2000s, increasingly worried about the assimilation by liberal gay rights agendas (Bacchetta, 2002; Edelman et al., 2004; **Halberstam**, 2011). Addressing this, Lisa Duggan (2002) coined the concept of 'homonormativity' to describe the normative practices of social reproduction that lies at the heart of gay culture in its mainstream form.

Terrorist Assemblages (2007) is perhaps Puar's most important work, exploring the links between sexuality, race and imperialism. Published in the same year as Joseph Massad's 'Desiring Arabs' it drew attention to the irony that nineteenth and early twentieth century colonial imaginative geographies positioned the Middle East as too sexually licentious and thus backward, whilst now, as the West has embraced its sexual revolution, this logic is reversed and the Middle East is seen as too socially conservative to be in step with the time of modernity. As Puar (2007: 13) puts it in her book, within the secular logic of US homonormative formations, 'Religion, in particular Islam, has now supplanted race as one side of the irreconcilable binary between queer and something else'. The US Exceptionalism that drives this rendering of Islam into the illiberal *bête noire* of the West, Puar (2007) shows, is a fantasy that fuels domestic homonormativity. For Puar, this is materialised in the notion that appropriately-consuming queers, who demonstrate patriotism – in the demand to repeal 'Don't Ask, Don't Tell' for instance – will be assimilated into full citizenship in the Global War on Terror. This fantasy of 'homonationalism', of patriotic queers seeking to defend the liberal nation against its illiberal aggressors, Puar (2007: 78) further argues, is a fantasy too-easily belied: 'homonationalism is also a temporal and spatial illusion, a facile construction that is easily revoked, dooming the exceptional queers to insistent replays and restagings of their exceptionalisms'.

Using **Gilles Deleuze's** framework of 'assemblages' in this argument allows Puar to account for the ways in which political speech, military practice, media discourse, technologies of policing, affects, and performative inhabitations of gender and sexuality come together in formations of 'homonationalism'. For instance, reading the turbaned body as an assemblage allows Puar (2007: 174) to unpack 'the foreclosure of a queer diasporic turbaned Sikh, male or female – a subject that is distinct from the queernesses that have often been attributed to Sikh masculinities – and the pathological queernesses endowed upon terrorist populations that Sikh gurdwara communities seek to evade'. Queer bodies, in Puar's writing of them as assemblages, matter less in the individualist terms of liberal gay rights, and more in their juxtaposition against affects of national belonging, especially in relation to projects of secular citizenship and policing.

In addition, Puar draws on the work of Michael Warner and Lisa Duggan on heteronormativity and homonormativity respectively, in conjunction with Deleuze's 'control societies, when conceptualising 'homonationalism'. Here, Puar's reading of assemblage was furthered by her work on intersectionality. In a 2005 essay on 'Queer times, queer assemblages', Puar had written that:

> the Deleuzian assemblage, as a series of dispersed but mutually implicated networks, draws together enunciation and dissolution, causality and effect. As opposed to an intersectional model of identity, which presumes components – race, class, gender, sexuality, nation, age, religion – are separable analytics, and can be thus disassembled, an assemblage is more attuned to interwoven forces that merge and dissipate time, space, and body against linearity, coherency, and permanency. (Puar, 2005: 127)

Puar's next major project focused on how experiences of attrition and debilitation figure in Empire's calculus. In 'The Right to Maim', Puar (2017) focuses on the slow attrition of bodies, and the making of disabled subjects under liberal imperialism. If 'homonationalism' builds a bridge from **Foucault**'s biopolitics to Deleuze's control societies, in 'The Right to Maim', Puar pauses on the dyad bargain of biopolitics – making live and letting die – asking about the temporalities of abandonment inherent in allowing life under the brutally violent conditions of settler colonialism. Focusing on Israel's maiming of Palestinians in the 'shoot to cripple' attacks on Gaza in the early 2000s that left many severely disabled but not dead, Puar (2007: 16) asks how debilitation seeks to foreclose futurity – 'to preemptively debilitate the resistant capacities of another intifada' – and control the space-time of occupation through its slow violence.

This work builds from Puar's anti-imperialist critique in *Terrorist Assemblages* and connects to the scholarship of Black feminists like **Katherine McKittrick** and Christina Sharpe, both of whom draw on the foundational writing of cultural theorist, Sylvia Wynter. It asks how the human subject of liberal citizenship is constructed under the securitising logic of contemporary imperialisms, settler colonial and otherwise, through the body-politics of disablement. Crucially, Puar's writing on debility thinks both about the slow violence of imperialism – the normalised everyday wounding and debilitation that enacts attritions that fail to register as spectacular events – and about 'slow life'. As she puts it in a conversation with Ezekiel Dixon-Romàn, 'the expansion of labour time, the refusal or withholding of temporal simultaneity so coveted in our connective technologies that signal modernity, but most significantly, the cordoning off (and thus the creation) of space through time' (Dixon-Romàn and Puar, 2021). Space, for Puar, is cordoned off through time via architectures of the carceral in settler colonial spaces: checkpoints, surveillance infrastructures that make small spaces feel larger, more difficult to traverse, but also constrict space, interrupting flows and rhythms of movement. She writes that the Israeli occupation in Palestine 'works in part by titrating control over temporality: by foreclosing or suspending access to speed, the immediacy of forms of contact, and the space-time compression so coveted in modernity and crucial to the circulation of goods, ideas, and bodies' (Puar, 2021: 404) The control society that results is, after Omar Jabary Salamanca, 'asphyxiatory' (Salamanca, 2011; Dixon-Romàn and Puar, 2021): debilitating, rather than killing.

In this work on debility, Puar also returns to the question of queer tourism taken up in her early work: the portrayal of contested geographical sites as queer-friendly, to attract gay tourism, typically from the West. Specifically, she focuses on the reinscription of sites of settler colonial violence as making a home for queers, in a way that colonised spaces like Palestine are imagined not to. In an interview with Kathryn Medien, Puar (Puar and Medien, 2018: 96) writes: 'Palestinian queer activists [...] understand pinkwashing as a form of internal colonialization, an introjection and reification of the discourse of "Palestinian homophobia" into the psyches and souls of Palestinians, thus suturing the occupation to yet another vector of inferiority'. This argument echoes a criticism articulated by feminist geographers like Walaa Alqaisiya (2018; 2020), writing on queer decoloniality in Palestine. Alqaisiya (2018) argues that critiques of settler colonial pinkwashing still tend to centre refusals by settler subjects to allow the instrumentalisation of their queerness for violent occupation, rather than the decolonial potentials of queering as a political practice for the colonised. Drawing on Palestinian registers of queer, Alqaisiya's (2018; 2020) work takes off in some ways where Puar's leaves off, doing the careful work of showing how being distinctly Palestinian and queer enacts decolonial praxis in its refusal of the settler narrative of indigenous cultural inferiority.

Spatial Contributions

Puar's work is both self-consciously interdisciplinary and distinctly geographical – much of her writing is concerned with the dynamics of territorialisation and deterritorialisation, and with how bodies and affects are apprehended within regimes of geopolitical security. Whilst Puar has never held a position in a Geography department, she has published widely within the field, and is extensively read in fields ranging from feminist geopolitics to digital geographies and studies of hybridity, to decolonial reckonings with the human subject in Geography.

Much of Puar's work has concerned the difficulty of being 'at home': a metaphor she has also applied to the interdisciplinarity of her scholarship. Writing in the preface to *Terrorist Assemblages,* for example, she argues that she works with an 'unhomed interdisciplinarity' (Puar, 2007: xvi). The metaphor of being 'unhomed' of course resonates uncomfortably with the terms of the postcolonial theory that Puar's project sets out to critique: it is not clear how conscious this is, and she has never addressed this particular genealogy. However, for scholars like ***Homi Bhabha*** (1992), the uncanny experience of 'unhoming' is central to the pervasive, embodied violence of the postcolonial condition: the sense of the familiar rendered strange. Arguably, complicating easy liberal platitudes is what Puar does: her work effectively *unhomes* the place where postcolonial theory has come to settle in its late twentieth-century normative form.

In an early essay Puar wrote about the impossibility of writing 'home', whilst critiquing the class and gender politics of postcolonial hybridity:

> The attraction of hybridity is that it initially appears to offer home. Though the flow of Bhangra music from Britain to South Asia and North America has broken the country-of-origin to country-of-displacement equation, it has done so through its (re)claiming of home(s) via its consolidation as authentic hybrid, as another call of authenticity for the second generation. (Puar, 1994: 103)

This critique of postcolonial cosmopolitanism as a centrepiece of globalisation, emerging from her own experiences as a queer Sikh woman in the US, runs through much of Puar's writing. Crucially, Puar's critique has turned to the ways in which 'home' is figured within practices of travel and queer mobility, through studies in which she herself 'returns' to India through an organisation that arranges such tours for diasporic queers (Puar, 2002a, 2002b). Nation and intimate practices of family-making circulate together for Puar as the terms by which 'homemaking' becomes possible for queer subjects under regimes of globalisation: for instance, in geographies of transnational adoption, where the liberatory project of queer family-making clashes against the inequalities of late twentieth-century globalisation (Puar, 2002b).

Overall, then, Puar's main contributions are to a geographical scholarship that grapples with the ways in which bodies, affects, and their relation to territory are continually refigured within relations of colonial, gendered, and sexual power. Her rendering of race, sex, and gender into an 'assemblage' with imperial discourse and geopolitical practice resonates with other geographical scholarship that emerged during the time period, notably Arun Saldanha's (2007) *Psychedelic White: Goa Trance and the Viscosity of Race,* which seeks to make sense of the hippie Western tourist's body and its affects through the lens of assemblage theory, asking how race registers alongside religion, psychopharmacology, coloniality, and the subcultural aesthetics of trance. Others have also built on and extended Puar's framework of homonationalism, particularly developing intersections of queerness with **Achille Mbembe**'s 'necropolitics' (Mbembe, 2003) – a reversal of the 'making live-letting die' dynamic of **Foucault**'s biopolitics in the colony, in which the state's power over life centres on killing (Haritaworn et al., 2014; Snorton and Haritaworn, 2022). Scholars writing in this vein trace the 'murderous inclusions' that result as 'current modes of power and control give rise to gender and sexual normativities that interpellate white and racialized populations unevenly' (Bacchetta et al., 2015: 770). Crucially, this scholarship has put forward 'queer people of colour critique' (Bacchetta et al., 2015; 2018) as a translocal formation that materialises correctives to liberal framings of gay rights, drawing attention instead to 'gay imperialism' (Ahmed, 2011; Haritaworn et al., 2008) and the affective politics of queerness in debates on race and the Global War on Terror.

Key Controversies and Advances

Puar's work has made important contributions to thinking on the ways in which bodies and affects are implicated in national and geopolitical processes of territory-making, during a time when liberal agendas of feminist and queer rights are widely used as justifications for imperial expansion. Puar's work has spanned an eventful period in the history of 'queer' as a formation in the public discourse globally. Since *Terrorist Assemblages* was written, regimes of security, surveillance, and anti-immigrant policing have continued to proliferate, if anything with newfound vigour in the context of a right-wing resurgence globally. In an essay on trans history, Susan Stryker (2008) writes on the circulation of the term 'homonormativity' in queer circles in the 1990s, where it was used as a critical term by trans communities to signal the cis privilege of non-transgender queers. This reading has in some sense come full circle in the backlash against trans communities globally, a lot of which maps security anxieties onto bodies whose being otherwise reads simultaneously as threatening, and as oriented away from objects of happiness. As trans geographies has emerged as a field (see **Halberstam** entry), Puar's critique of the subject in her writing at the intersection of trans experience and disablement positions trans less as fixed identity and more as movement – undoing categorical certainties in its reorganisation of bodies and affects. Becoming trans, for Puar, shifts not only how the body is read in terms of gender but also race. If, as Puar's reading of intersectionality suggests, race and gender are not fully-formed and distinct axes of identity that intersect, but are co-constituted such that whiteness is made material in differently gendered and classed bodies in distinct ways, then butch white womanhood represent an experience of racialisation that is necessarily different from that of being a white man.

Yet Puar's critique of intersectionality has met with resistance, particularly because of anxieties about the separation of Black feminism from questions about the ways in which race and gender shape each other. As Jennifer Nash (2014) argues, intersectionality theory has travelled across disciplines because it can, as Puar has argued in her critique, stand for questions of diversity more broadly, and has also been reclaimed for Black feminist politics from scholars who criticise the depoliticising of intersectionality in liberal projects of inclusion. In this context, Nash (2014) reads Puar's critique not as a call to abandon intersectionality, so much as to rearticulate it within a vocabulary that accounts for workings of power that are erased in the circulation of intersectionality theory outside the disciplinary bounds of socio-legal feminist scholarship.

Intersectionality, for Puar (2005), relies on a conception of identity as sovereign, fully-known to the identity-bearer, and non-shifting. It is, in her critique, easily appropriated by the temporality of progress that Puar's reading of 'homonationalism' seeks to critique. In a later essay on 'becoming intersectional in assemblage theory', Puar (2012b) clarifies that she sees her critique as intersectional in that it is interested in the co-constitution of race, sex, gender, and geopolitics. However, she argues that the 'poststructuralist fatigue' with a liberal politics of diversity demands the complication of intersectionality theory, which Puar (2012b: 49) argues assemblage provides. In so doing, she contests the idea that the non-representational and non-subject-oriented framework of the assemblage lacks the political teeth of intersectionality theory. Rather, for her, 'the convivial crossings of these two differentiated but not oppositional genealogies', allows for both a productive complication of intersectionality theory, and the explicit politicisation of the assemblage as a framework (Puar, 2012b: 50).

Crucially, at the heart of Puar's critique is the subject itself, which she argues is a normative construct. For Puar (2012b: 63), to fail to interrogate the subject – 'to dismiss how societies of control tweak and modulate bodies as matter, not predominantly through signification or identity interpellation but rather through affective capacities and tendencies' – is to fail feminism's demand to unsettle its own normativity. Further, for Puar, what is at stake is the re-centring of whiteness in the uses of intersectionality theory. Puar (2012b) writes that while intersectionality theory emerged from Black feminist critiques of the normatively white woman subject at the heart of feminist projects, contemporary deployments of intersectionality have re-centred the white woman subject at its heart.

Puar's work is perhaps most compelling in its relentless refusal of 'feel-good' liberalisms. Writing in the aftermath of Rutgers University gay undergraduate Tyler Clementi's suicide, prompted by cyberbullying, Puar (2012a) commented on the notion of 'It Gets Better', popularised by columnist Dan Savage's campaign to address suicide by queer teenagers with the message that things improve with age. 'It gets better' has been criticised for its implication that progressive well-being comes as queers move socially and physically into urban and more queer-friendly communities. In this writing, which bridges across her work on home, homonationalism, and debility, Puar (2012a) asks what is lost when we pay attention only to the spectacular instances of suicides, leaving by the wayside less eventful debilitations that materialise in failures to be enfolded, as the Savage campaign suggests, into middle-class, white, and cis spaces of homonormativity. In some ways, presaging the work of disability studies scholars like Eli Clare (2017), Puar's writing on queer suicide extends the question of how the 'making live' in Michel Foucault's dyad of 'making live and letting die' is often not so much a scene of flourishing but of debilitating labour to get back on the conveyer belt of progressive recovery. 'Getting better', Puar (2012a) argues, is often a zone of discipline: its cost is the abandonment of the affects, capacities, and selves that are incommensurate with the logic of neoliberal progress.

Puar's Key Works

Puar, J.K. (1994) 'Writing my way "home"', *Socialist Review,* 24 (4): 75–108.

Puar, J.K. (2002a) 'Circuits of queer mobility: Tourism, travel, and globalization', *GLQ: A Journal of Lesbian and Gay Studies,* 8 (1): 101–37.

Puar, J.K. (2002b) 'A transnational feminist critique of queer tourism', *Antipode,* 34 (5): 935–46.

Puar, J.K. (2005) 'Queer times, queer assemblages', *Social Text,* 23 (84–5): 121–39.

Puar, J.K. (2006) 'Mapping US homonormativities', *Gender, Place & Culture,* 13 (1): 67–88.

Puar, J.K. (2007) *Terrorist Assemblages: Homonationalism in Queer Times*. Durham, NC: Duke University Press.

Puar, J.K. (2012a) 'Coda: The cost of getting better suicide, sensation, switchpoints', *GLQ: A Journal of Lesbian and Gay Studies,* 18 (1): 149–58.

Puar, J.K. (2012b) '"I would rather be a cyborg than a goddess": Becoming-intersectional in assemblage theory', *PhiloSOPHIA,* 2 (1): 49–66.

Puar, J.K. (2017) *The Right to Maim: Debility, Capacity, Disability*. Durham, NC: Duke University Press.

Puar, J.K. (2021) 'Spatial debilities: Slow life and carceral capitalism in Palestine', *South Atlantic Quarterly,* 120 (2): 393–414.

Puar, J.K. and Medien, K. (2018) 'Thinking life, death, and solidarity through colonized Palestine', *Journal of Middle East Women's Studies,* 14 (1): 94–103.

Secondary Sources and References

Ahmed, S. (2011) 'Problematic proximities: Or why critiques of gay imperialism matter', *Feminist Legal Studies,* 19: 119–32.

Alqaisiya, W. (2018) 'Decolonial queering: The politics of being Queer in Palestine', *Journal of Palestine Studies,* 47 (3): 29–44.

Alqaisiya, W. (2020) 'Palestine and the will to theorise decolonial queering', *Middle East Critique,* 29 (1): 87–113.

Bacchetta, P. (2002) 'Rescaling transnational "queerdom": Lesbian and "lesbian" identitary – positionalities in Delhi in the 1980s', *Antipode,* 34 (5): 947–73.

Bacchetta, P., El-Tayeb, F. and Haritaworn, J. (2015) 'Queer of colour formations and translocal spaces in Europe', *Environment and Planning D: Society and Space,* 33 (5): 769–78.

Bacchetta, P., El-Tayeb, F., Haritaworn, J., Hernandez, J., Smythe, S., Thompson, V. E. and Willoughby-Herard, T. (2018) 'Queer of color space-making in and beyond the academic industrial complex', *Critical Ethnic Studies,* 4 (1): 44–63.

Bhabha, H. (1992) 'The world and the home', *Social Text,* (31/32): 141–53.

Clare, E. (2017) *Brilliant Imperfection: Grappling with Cure*: Durham, NC: Duke University Press.

Dixon-Român, E. and Puar, J. (2021) 'Mass debilitation and algorithmic governance', *E-Flux,* 123. https://www.e-flux.com/journal/123/436945/mass-debilitation-and-algorithmic-governance/ (accessed November 10, 2023).

Duggan, L. (2002) 'The new homonormativity: The sexual politics of neoliberalism', *Materializing Democracy: Toward a Revitalized Cultural Politics,* 10: 175–94.

Edelman, L., Barale, M.A. and Goldberg, J. (2004) *No Future: Queer Theory and the Death Drive*: Durham, NC: Duke University Press.

Halberstam, J. (2011) *The Queer Art of Failure*. Durham, NC: Duke University Press.

Haritaworn, J., Kuntsman, A. and Posocco, S. (2014) *Queer Necropolitics*. London: Routledge.

Haritaworn, J., Tauqir, T. and Erdem, E. (2008) 'Gay imperialism: Gender and sexuality discourse in the "war on terror"' in A. Kunstman and V. Miyake (eds), *Out of Place: Interrogating Silences in Queerness/Raciality*. York: Raw Nerve. pp. 71–95.

Massad, J.A. (2007) *Desiring Arabs*. Chicago, IL: Chicago University Press.

Mbembe, A. (L. Meintjes, trans.) (2003) 'Necropolitics', *Public Culture,* 15 (1): 11–40.

Najmabadi, A. (2005) *Women with Mustaches and Men without Beards: Gender and Sexual Anxieties of Iranian Modernity*. Berkeley, CA: University of California Press.

Nash, J. (2014) *The Black Body in Ecstasy: Reading Race, Reading Pornography*. Durham, NC: Duke University Press.

Salamanca, O. J. (2011) 'Unplug and play: Manufacturing collapse in Gaza', *Human Geography*, 4 (1): 22–37.

Saldanha, A. (2007) *Psychedelic White: Goa Trance and the Viscosity of Race*. Minneapolis, MN: University of Minnesota Press.

Sinha, M. (1994) 'Reading Mother India: Empire, nation, and the female voice', *Journal of Women's History*, 6 (2): 6–44.

Snorton, C.R. and Haritaworn, J. (2022) 'Trans necropolitics', in S. Stryker and D.M. Blackston (eds), *The Transgender Studies Reader Remix*. New York, NY: Routledge. pp. 66–76.

Stryker, S. (2008) 'Transgender history, homonormativity, and disciplinarity', *Radical History Review*, 2008 (100): article #14557.

42
Laura Pulido

Cristina Faiver-Serna, University of New Hampshire

Biographical Details and Theoretical Context

Laura Pulido was born in 1962 into a Mexican American household in Los Angeles, California. From an early age Pulido was curious about the why of where. As a child she was interested in nature and her physical environment, noticing the lack of trees, the polluted air, and the arid environment of Los Angeles. She also experienced anti-Mexican racism and observed anti-Black racism; and was aware of stark social and political divisions along racial lines. She dropped out of high school, but eventually pursued her education at community college, propelled by questions about power, place, environment, and race. She credits her early retail work experience for giving her pause about sexism in the workplace and gendered power dynamics, leading her to actively challenge patriarchal norms as a Mexican American woman – a Chicana – in historically white and male spaces throughout her academic career.

It was a community college class on the geography of California that sparked her interdisciplinary academic pursuits. Pulido was a first-generation college graduate. She earned her BA from California State University, Fresno (1984), and her MSc from the University of Wisconsin, Madison (1987). With one foot firmly planted in Geography and the other in Chicano Studies, she earned her doctorate in Urban Planning from the University of California, Los Angeles (1991). She wrote her dissertation on environmental racism faced by Chicano and Mexicano communities in the Southwestern US in the early 1990s on the heels of the 1987 groundbreaking study 'Toxic Wastes and Race in the United States' by the United Church of Christ Commission for Racial Justice (1987). This study helped define the concept of 'environmental racism' and drew scholars, like Pulido, to the burgeoning field of environmental justice studies. Pulido subsequently taught and held appointments at California State Fullerton, University of Southern California, and the University of Oregon, with visiting or adjunct positions at the University of California, Santa Barbara and Simon Fraser University.

Pulido's experiences as an undergraduate and graduate student advanced her perspective on critical inquiry about race and racism, and gave her the language to explore, understand, and describe the complex relationships between power and place. It is important to understand that Pulido trained distinctly in Chicano Studies, and in Geography and Planning. Her training and experiences in these separate spheres of academia led her to make two important observations that have been foundational to the trajectory of her research and scholarly contributions. First, early on in her career Pulido observed that race, a central organising theme of modern society, was not foregrounded in her Geography training, and did not centrally figure into geographical inquiry about space and place. As such, Pulido (2002) has argued that this has put the discipline behind many other social sciences in advancing cutting-edge research on questions about the spatiality of social justice struggles, particularly in the field of environmental justice. Second, Pulido observed that she was often the only Chicana and/or person of colour in her geography classes and at academic conferences. She has argued that the pervasive whiteness of North American geography thus lacked the 'critical mass' of people of colour necessary to critically inquire and innovate on questions of race, racism, and racialisation squarely from inside the discipline.

Laura Pulido's cutting-edge research has expanded and intertwined the fields of critical human geography, critical race studies, comparative ethnic studies, Chicanx and Latinx studies, Black studies, and environmental justice studies over the course of her career. Her commitment to antiracist praxis and wide-ranging interdisciplinary influences are driving forces behind her broad influence as a geographer, an environmental justice and ethnic studies 'scholar/activist', as well as a public intellectual. Pulido's unique and grounded perspective on the spatiality of power relations has contributed greatly to understanding processes of racialisation and the structural forces of racism in US society. She has drawn wide acclaim as a critical and influential scholar whose works have signalled important shifts and fostered vibrant growth within her fields.

Spatial Contributions

Laura Pulido has applied her antiracist lens to question how hegemonic power operates across scale and becomes embedded in our everyday landscapes. Her scholarship engages many themes, subfields, and regions, demonstrating both breadth and depth. As a scholar from and of Los Angeles, California, she has reinvigorated the kinds of questions asked, methodologies engaged, and stories told of the region through a critical geographic lens. Through her environmental justice research, she has examined how different modalities of structural racism operate across various scales to produce and sustain inequalities over time. Her geographic scholarship is antiracist and critical, grounded in a radical ethnic studies tradition that comes out of the US-based movement in the 1960s to build and grow academic spaces and interdisciplinary lines of inquiry by and for scholars of colour that were, and continue to be, institutionally excluded by Western academia. In turn, Pulido's critical race scholarship has also expanded the reach and utility of traditional geographic methods and socio-spatial perspectives into ethnic studies. As a self-identified scholar/activist (Pulido, 2008) her activism has inspired her scholarship, and her scholarship has critically inquired about the intersection of race, class, and power in space and place, with a goal of demonstrating that people on the ground have the power to alter their conditions for the better.

Her first book innovates at the forefront of environmental justice studies, drawing heavily on Third World environmentalism and social movement theory. In *Environmentalism and Economic Justice: Two Chicano Struggles in the Southwest,* Pulido (1996) engages with the concept of environmental racism through an exploration of how economic systems and structural racism co-produce subaltern environmental subjects, experiences, and places. She draws on Ranajit Guha's (1988) framework of subalternity to contend with the ontological totality of unequal power relations that produces the material conditions of environmental racism and economic injustice. Pulido demonstrates how subalternity is a useful framework for analysing the spatial scales of dominant and subordinate relationships, particularly two of the most serious problems at the end of the twentieth century: racial inequality and environmental degradation. While early work in environmental justice studies debated whether race or class was more influential in affecting people's exposure to environmental degradation, Pulido observed that the race versus class debate missed the point about how power relations produce material conditions. Instead, she asserts that it is not race *or* class, but race *and* class, and that the relationship between power, space, and place requires critical analysis in order to strive toward justice.

In her article 'Rethinking environmental racism: White privilege and urban development in Southern California', Pulido (2000) asserts that the spatial study of environmental racism is useful for understanding *how* racism operates. Drawing on critical race theory (Omi and Winant, 1994) she observes that environmental racism occurs beyond discrete acts of racist hostility and the proximity of environmentally hazardous sites to communities of colour. Pulido argues that scholars must treat scale critically in order to interrogate the spatiality – i.e., the relationship between social space and society (Soja, 1989) – of racism. She comparatively examines white health, safety, and distance from environmental hazards and non-white experiences of dangerous and unsafe environmental conditions in Southern California through a framework of white privilege in order to analyse it as a structuring form of racism across socio-spatial scales.

Pulido observed that in the 2000s, during the height of 'post-racialism' in the US, white privilege was a useful framework for analysing structural racism because it did not require racial animus or overt discrimination. In other words, examining one's privileges as a historical process, regardless of racial bias or racist intent, was a useful framework for studying structural racism. However, Pulido argued white privilege is also limited by passivity. White supremacy is a complementary and more precise framework because it includes the racist behaviour of people of colour, critically engages spatial and temporal scale, engages with racial domination as an action, and refers to a global system. In turn, Pulido (2015) names three elements of white supremacy that operate on the ground to create conditions of environmental vulnerability for people of colour: awareness of harm, taking of environmental resources, and a belief in the inferiority of people of colour – explicitly, that the lives of people of colour are expendable for profit.

Twenty years after the publication of her first book on the struggle for environmental justice, Pulido et al. (2016a: 524) observed that 'it is questionable if the environments of vulnerable communities have actually improved through environmental justice'. Pulido et al. (2016) also declared that the environmental justice movement in the US had failed to improve the lives and environments of vulnerable groups due to its proximity, collaboration, and complicity with a US regulatory state interested in upholding the socioeconomic system of capitalism ahead of repairing and ensuring healthy and just living conditions for all. Pulido (2016a) contends that environmental racism *is* state-sanctioned violence, and heeds scholars to push this line of inquiry further, in line with critical environmental justice scholar David Naguib Pellow (Pellow, 2018).

In her case study of Flint, Michigan, Pulido examines city-wide lead poisoning through a framework of racial capitalism (Pulido, 2016b). State of Michigan officials enacted austerity measures by choosing a cheaper source of water from a river well-known for being contaminated by the automotive industry. Pulido asserts that the state's violent act of poisoning the predominantly Black city of Flint was a result of the state's history of investment in and reliance on capitalist projects, whereby capitalism itself is a highly racialised, and violent, socioeconomic system. Pulido argues that racial capitalism has positioned Flint residents outside of, and as a threat to the local, regional, and global processes required for capitalism to reproduce itself, turn a profit, and benefit the State of Michigan and its interests. Pulido looks to the Black Radical Tradition (Robinson 2021 [1983]), that is, the long history of Black struggle for justice, and particularly grassroots organising at the intersection of race and class, as a potential solution beyond the neoliberal logics of the state. Pulido and De Lara (2018) argue that multiracial environmental justice organisations extend the legacy and commitment to racial equality of the Black Radical Tradition. Using both conceptual and physical mapping, they look at relations between Latinx and Black organising to advocate for abolitionist interracial collaboration that challenges the conditions needed for racial capitalism to persist.

The reader can draw a line from Pulido's social movement and Third World environmentalism framework for understanding environmental racism (1996) to her investigation of political consciousness and interethnic relations of the Third World Left in Los Angeles during the 1960s and 1970s. In *Black, Brown, Yellow, and Left: Radical Activism in Los Angeles* (2006), Pulido contextualises regional processes of racism, racialisation, and racialised political identity in the US within a global context. Pulido states she has always been interested in the group of activists that generationally preceded her, and during this period US leftist people of colour conceptualised themselves as oppressed nations. In a time of escalating global climate injustice, where peoples of the Global South are experiencing disproportionate climate impacts that are not of their making, Pulido's (1996) framework of subalternity and Third World environmentalism finds renewed relevance, especially in light of the call for multiracial organising toward justice (Pulido and De Lara, 2018). Pulido is a strong advocate in the public sphere for understanding that processes of climate change are deeply intertwined with regional and local processes of racial injustice, and that global climate injustice has long been in the making (see Malik, 2023).

Pulido's first two books laid a tremendous amount of groundwork for examining how social processes of racism and racialisation occur specific to place, with particular attention to various kinds of social activism, in both environmental and political contexts. *Black, Brown, Yellow, and Left* propelled her involvement as an editor for *Racial Formation in the Twenty-First Century* (Hosang Martinez et al., 2012) and *Black and Brown in Los Angeles: Beyond Conflict and Coalition* (Kun and Pulido, 2013) and inspired *A People's Guide to Los Angeles* (Pulido et al., 2012). In *A People's Guide* Pulido, Barraclough, and Cheng offer an alternative and radical geographical history of Los Angeles, 'flipping the script' on hegemonic narratives of Los Angeles perpetuated by traditional guidebooks. Pulido found herself collecting stories about places long forgotten in popular histories of L.A. while researching and writing *Black, Brown, Yellow, and Left*, and *A People's Guide* was born through collaboration at every stage:

data collection, naming the project, sharing research findings locally through posters, and then writing the book. Pulido's methods and processes demonstrate myriad ways that the line between scholar/activist can be blurred. *A People's Guide* takes its reader to 115 sites around Los Angeles, critically viewing local and everyday places through fresh eyes, literally putting history and geography of historically marginalised people on the map. This radical tour guide challenges and reimagines the local and everyday landscape and contributes to a powerful subgenre of critical scholarship on the region. The book also inspired similar projects elsewhere, igniting *A People's Guide* book series with the University of California Press that Pulido, Barraclough, and Cheng co-edit.

A People's Guide demonstrates that how and what we know about a place matters a great deal in shaping our relationship to space and place, and that the erasure of the histories and contributions of people of colour in the landscape speaks to the spatiality of unjust power relations. Pulido's scholarship also demonstrates how white supremacy imbues the landscape through cultural memory, that is, popular understandings of the US racial past. For example, Pulido (2022) examines the commemoration of the Mexican-American War in Los Angeles, and critiques US imperialism as part of racial capitalism's long history and its attempt to erase the violence of settler colonialism in cultural memory. Pulido (2023) argues that national historic landmarks are a way to study how cultural memory shapes our landscape and perpetuates white supremacy through the myth of white innocence.

Key Advances and Controversies

Pulido's 'Rethinking environmental racism' (2000) was a critical intervention in geography and environmental justice studies. Her survey of geography literature on environmental racism reveals a preoccupation with *proving* the existence of racism. As she lays out in 'Reflections on a white discipline' (Pulido, 2002), such preoccupation with proving racism exists, rather than seeking to understand how it works (in order to undo it) has had to do with the discipline of Geography's tenuous (at best) connections with ethnic studies and critical race theory, as well as the overwhelming racial homogeneity of the discipline, particularly in North America. Pulido lamented that she was tokenised for being a racialised minority who studied environmental racism, and moved away from environmental justice research for a number of years. She observed the need for a critical mass of geographers of colour in order to shift research agendas across the discipline, and for geographers to take seriously the task of studying processes of racialisation, and how racism structures society and our relationship to the environment. Pulido's research has laid the groundwork for the necessary shift in how geographers and scholars across the social sciences approach environmental racism through critical frameworks. Her sustained attention on the sociospatial and intersectional issues of racial and environmental injustice has only become increasingly urgent as climate change accelerates, and more scholars, activists, and the general public come to study, understand, and grapple with what it means for the reality of climate change to take place within a deeply racially-unjust world.

Pulido's scholarship has also effectively strengthened ties between Chicanx Studies and Human Geography. Since its inception and increasingly in recent years, the field of Chicanx Studies has been concerned with relationship to land and place, the physical and cultural landscape of the US–Mexico borderlands, geographies of migration and settlement, and Aztlán, the folkloric homeland of Chicano/as in what is now the southwestern United States (see Guidotti-Hernández, 2011; Valerio-Jiménez, 2013; Sagarena, 2014; Saldaña-Portillo, 2016; also entry on **Gloria Anzaldúa**). Pulido has innovated at the intersection of these fields, though this has not stopped her from critiquing how each area of study has treated settler colonialism. She observes that geography continues to be limited by its pervasive homogeneity in mostly considering whites' relationship to settler colonialism; while Chicanx Studies is ambivalent towards it, and critical engagement with settler colonialism has the potential to 'disrupt core elements of Chicana/o political subjectivity' by highlighting the role of coloniser. She calls for deeper critical engagement of how white supremacy operates in the ideologies and research agendas of both fields (Pulido, 2017).

Pulido's critique of the whiteness of the discipline has proven to be a powerful call. She has produced and co-produced scholarship that deeply engages questions of race, racism, and racialisation through geographic perspectives. Her work has inspired and influenced younger generations of scholars who have followed her lead in theorising and studying race from both within and beyond geography (see Bruno and Faiver-Serna, 2022). Her wide-ranging body of scholarship at the intersection of Chicanx and Latinx studies and Human Geography has paved the way for the birth and development of Latinx geographies (see Cahuas et al., 2023); and her comparative and collaborative work (see Woods et al., 2017) has proven foundational to Black geographies as well. The rise of these subfields, coupled with the resurgence of Indigenous geographies in recent years, has yet to reach a 'critical mass', but it has indeed helped to grow a vibrant scholarly community of critical race geographers, thanks in large part to Laura Pulido.

Pulido's Key Works

Pulido, L. (1996) *Environmentalism and Economic Justice: Two Chicano Struggles in the Southwest*. Tucson, AZ: University of Arizona Press.

Pulido, L. (2000) 'Rethinking environmental racism: White privilege and urban development in southern California', *Annals of the Association of American Geographers*, 90 (1): 12–40.

Pulido, L. (2002) 'Reflections on a white discipline', *Professional Geographer*, 54 (1): 42–9.

Pulido, L. (2006) *Black, Brown, Yellow, and Left: Radical Activism in Los Angeles*. Berkeley, CA: University of California Press.

Pulido, L. (2008) 'FAQs: Frequently asked questions on being a scholar/activist', in C. Hale (ed), *Engaging Contradictions: Theory, Politics and Methods of Activist Scholarship*. Berkeley, CA: University of California Press. pp. 341–66.

Pulido, L. (2015) 'Geographies of race and ethnicity I: White supremacy vs white privilege in environmental racism research', *Progress in Human Geography*, 39 (6): 809–17.

Pulido, L. (2016a) 'Geographies of race and ethnicity II: Environmental racism, racial capitalism and state-sanctioned violence', *Progress in Human Geography*, 41 (4): 524–33.

Pulido, L. (2016b) 'Flint, environmental racism, and racial capitalism', *Capitalism Nature Socialism*, 27 (3): 1–16.

Pulido, L. (2017) 'Geographies of race and ethnicity III: Settler colonialism and nonnative peoples of color', *Progress in Human Geography*, 42 (2): 309–18.

Pulido, L., Barraclough, L., and Cheng, W. (2012) *A People's Guide to Los Angeles*. Berkeley, CA: University of California Press.

Pulido, L., Kohl, E., and Cotton, N. (2016) 'State regulation and environmental justice: The need for strategy reassessment', *Capitalism Nature Socialism*, 27 (2): 12–31.

Secondary Sources and References

Bruno, T. and Faiver-Serna, C. (2022) 'More reflections on a white discipline', *The Professional Geographer*, 74 (1): 156–61.

Cahuas, M.C., Douglass-Jaimes, G., Faiver-Serna, C., González Mendoza, Y., Martinez-Lugo, D. and Ramírez, M.M. (2023) 'Latinx geographies: Opening conversations', *ACME: An International Journal for Critical Geographies*, 22 (6): 1462–89.

Guha, R. (1988) 'Preface', in R. Guha and G. Chakravorty Spivak (eds), *Selected Subaltern Studies*. Oxford: Oxford University Press. pp. 35–6.

Guidotti-Hernández, N.M. (2011) *Unspeakable Violence: Remapping U.S. and Mexican National Imaginaries*. Durham, NC: Duke University Press.

Hosang Martinez, D., LaBennett, O., and Pulido, L. (2012) *Racial Formation in the Twenty-First Century*. Berkeley, CA: University of California Press.

Kun, J. and Pulido, L. (2013) *Black and Brown in Los Angeles: Beyond Conflict and Coalition*. Berkeley, CA: University of California Press.

Malik, M. (Host) (2023) Climate Change and Race (No. 4) [Audio podcast episode]. *Climate Change: America and the World Podcast*, June 30. London School of Economics and Political Science. Available from https://www.lse.ac.uk/united-states/climate-change/climate-change-america-and-the-world-podcast (accessed November 10, 2023).

Omi, M. and Winant, H. (1994) *Racial Formation in the United States*. New York, NY: Routledge.

Pellow, D.N. (2018) *What is Critical Environmental Justice?* Cambridge: Polity.

Pulido, L. (2022) 'Erasing empire: Commemorating the Mexican-American war in Los Angeles', in S. Koshy, J. Byrd, L. Cacho, and B. Jefferson (eds), *Colonial Racial Capitalism*. Durham, NC: Duke University Press. pp. 284–308.

Pulido, L. (2023) 'Cultural memory, white innocence, and United States territory: The 2022 *Urban Geography* plenary', *Urban Geography*, 44(6): 1059–83.

Pulido, L. and De Lara, J. (2018) 'Reimagining "justice" in environmental justice: Radical ecologies, decolonial thought, and the Black radical tradition', *Environment and Planning E: Nature and Space*, 1 (1–2): 76–98.

Robinson, C. (2021 [1983]) *Black Marxism: The Making of the Black Radical Tradition*. Chapel Hill, NC: University of North Carolina Press.

Sagarena, R.R.L. (2014) *Aztlán and Arcadia: Religion, Ethnicity, and the Creation of Place*. New York, NY: New York University Press.

Saldaña-Portillo, M.J. (2016) *Indian Given: Racial Geographies across Mexico and the United States*. Durham, NC: Duke University Press.

Soja, E. (1989) *Postmodern Geographies*. London: Verso.

United Church of Christ, Commission for Racial Justice (1987) *Toxic Wastes and Race in the United States*. New York, NY: United Church of Christ.

Valerio-Jiménez, O.S. (2013) *River of Hope: Forging Identity and Nation in the Rio Grande Borderlands*. Durham, NC: Duke University Press.

Woods, C., Pulido, L. and Camp, J.T. (eds) (2017) *Development Drowned and Reborn: The Blues and Bourbon Restorations in Post-Katrina New Orleans*. Athens, GA: University of Georgia Press.

43
Paul Robbins

Phil McManus, University of Sydney

Biographical Details and Theoretical Context

Paul Robbins was born in 1967 and raised in Denver, Colorado. He obtained a bachelor's degree in Anthropology from the University of Wisconsin-Madison (1989), where his initial strong interests lay in questions of archaeology. During his degree, however, he conducted fieldwork that was to shape his subsequent research career:

> I had the opportunity on a professor's grant to go to the National Science Foundation-funded project in Gujarat, India, and what I learned, like many scholars, is that I did not want to be an archaeologist, but I did fall in love with India, and with the environment, and the landscape, and the farmers, and the herders, and everything that was going on in air quality, and water. And that's when I knew that the career was gonna be about the environment. (cited in Lawrence, 2021: n.p.)

He followed his undergraduate studies with a master's degree (1994) and a doctorate in Geography from Clark University (1996), with his PhD dissertation entitled 'Negotiating Ecology: Institutional and Environmental Change in Rajasthan, India'.

After graduating, Robbins taught at Ohio State University from 1998–2004 before becoming the Director of the School of Geography and Development at the University of Arizona from 2005 to 2012. He returned to the University of Wisconsin-Madison in 2012 as Director and then Dean of the Nelson Institute for Environmental Studies. He describes the latter as 'a huge, sprawling interdisciplinary experiment that's been going on since 1970', arguing 'you cannot solve the world's problems with ecology alone [...] we have everyone from atmospheric scientists to historians, from engineers to authorities on Native American land rights – all looking at ways to meet people's needs while still protecting the environment' (cited in Hilder, 2022: n.p.).

Robbins has been a major intellectual force in advancing the multi-disciplinary project of political ecology, which examines the power relations which infuse and shape the nature–society dialectic, not least in relation to environmental conflicts and the management of resources. Indeed, Robbins was identified by Desvallées et al. (2022) as the most prolific author in the field based on his number of publications within their sample of over 3,000 publications during the period 1950–2019 that had used the term 'political ecology'. Especially significant here has been his text *Political Ecology: A Critical*

Introduction (originally published in 2004, with revised editions in 2012 and 2020) as well as co-edited texts with other key figures who have helped develop political ecology as a healthy and vibrant field of study, including Richard Peet and ***Michael Watts*** (e.g., Peet et al., 2011).

An anthropologist by training, and sometimes self-described as a 'recovering archaeologist', Robbins is widely known for his interdisciplinarity, bringing Geography into dialogue with other disciplines. In various interviews and his own publications, he integrates Geography with Environmental Studies, Demography and Economics, particularly in relation to labour and energy, to address important questions relating to the future of the planet and the various humans and other species who inhabit the world. His research includes everyday subjects as varied as the moral economy of lawns, the management of mosquitos, and coffee cultivation, and he has published research on locales ranging from the American (US) Southwest and Yellowstone National Park to rural India.

Spatial Contributions

The theoretical context of Robbins' work is provided by the broad field of political ecology, a critical take on nature and science that brings critical theory and Marxist thinking into dialogue with Environmental Studies. He has both emerged in this field and has been a leading contributor to its development, notably in publications such as *Political Ecology* in which he contrasts traditional, localised approaches to resource management with contextual accounts that consider the political economy in which decisions to cultivate crops or conserve natural resources are taken:

> The difference between this contextual approach and the more traditional way of viewing problems like this is the difference between a political and an apolitical ecology. This is the difference between identifying broader systems rather than blaming proximate and local forces; between viewing ecological systems as power-laden rather than politically inert; and between taking an explicitly normative approach rather than one that claims the objectivity of disinterest. (Robbins, 2004: 13)

Likewise, *Global Political Ecology*, co-edited with Peet and Watts (2011), begins with a picture of Richard Branson and Al Gore, using the problematic climate change negotiations in Copenhagen, Denmark, at COP 15 (2009) as a window into political ecology. In their introduction, Peet et al. (2011a: 5) claim that 'global climate change – as science, policy and politics – reveals starkly the sorts of problems that a global political ecology – the subject matter of this book – must confront'. They subsequently unpack the production of knowledge and the conditions under which it is produced, the intentional creation of ignorance and uncertainty, the corporate rebranding that is often at odds with company actions, and the situated nature of knowledge that is presented as universal. These themes in relation to global climate change resonate with earlier strands in Robbins' geography, in particular his understanding of, and contributions to, political ecology which were informed by both close attention to ecology and a critical awareness of changing global politics.

While studying for his PhD at Clark University in Massachusetts, Robbins 'hung dried samples of every variety of Massachusetts grass above his desk' (Woodward King, 2017: 20).

As Steinberg (2018) noted, this may not seem unusual now given the 'material turn' in geography, but in the early 1990s for a human geography graduate student to engage with the materiality of native grasses and to take classes on biogeography was unusual and showed a commitment to the ecology in political ecology. Robbins' interest in native grasses developed into an important study of lawns in America (Robbins, 2007a), in part based on his personal experiences of owning a house with a lawn in Columbus, Ohio, while working at Ohio State University. *Lawn People: How Grasses, Weeds, and Chemicals Make Us Who We Are* highlights Robbins' conceptions of space and urban ecology, as he explores the ecological, social and economic context in which heavily-fertilised front yard lawns are promoted as a defining and desirable feature of North American suburban living. As Robbins (2007a: xv) noted, it is 'a book about the North American lawn [...] its ecological characteristics and its political economy' which, in keeping with the political ecology approach outlined by Robbins (2004), extended far beyond the actual lawn and what could be seen as proximate causes to include wider chemical production economies and the moral values of specific middle-class communities. In this book he hence 'posits a larger argument, that the lawn is a system that produces a certain kind of person – a turfgrass subject' (Robbins, 2007a: xvi). This is crucial for his own positionality as an author, rejecting the idea that the American middle-class are merely pawns for capital, noting that he writes 'as a lawn person myself [...] with no more scorn for my neighbor than I hold for myself' (Robbins, 2007a: xx).

While he is widely known for this work and is sought after for comment by media outlets on issues relating to lawn management (and the sometime replacement of lawns by artificial grass) within academia and specifically in the field of political ecology, it is his early work in rural India that brings forth some of the key themes relevant to space and place. In particular, Robbins found that the daily local struggle over resources generates conflicting accounts of nature and environmental change, and that it is 'alliances between state and local actors that render certain claims powerful and so determine natural resource management policy and direct landscape change' (Robbins, 2000: 126).

This theme can be seen in his research on battles over science, uncertainty, and wild animals such as elk and wolves in Yellowstone, the US's first National Park. Robbins (2006: 185) argues that while 'often dismissed as "barstool biology", the ecological knowledges of local hunters in the Northern Yellowstone ecosystem are rooted in environmental experience and situated politics'. What interests Robbins here, and more widely, is how this local knowledge intersects with that of other actors such as state officials, environmentalists and ranchers to form discourse coalitions that may influence wildlife management policy in this case, and environmental policy more generally in other scenarios. His deep simultaneous engagement with theory and the pragmatics of everyday life renders Robbins' research particularly noteworthy, and informs his editorial work, including his role as General Editor of the five-volume *Encyclopedia of Environment and Society* (Robbins, 2007b).

For Robbins, nature outside of society does not exist. In fact, an apparently apolitical view of nature as separate, for Robbins, has in part led to the development of an exceedingly narrow 'constituency for "the environment" [...] [that] comes from a wilderness tradition that is heavily white, older, middle-class' in the US and elsewhere, and he notes that 'if we don't fix that, all of the other problems are not going to get fixed' (cited in Allen, 2020: 24). The entwinement of environment and society, with its political, ecological, discursive causes and impacts, is a consistent feature of Robbins' understanding of space. This can be seen in early work on the neoliberalisation of nature (Heynen and Robbins, 2005) and in later work about climate change. Climate is the major concern for Robbins, because with 'Catastrophic climate change and the collapse of global biodiversity [...] there's no going back is what we say around here. I mean, minimally, you're talking about a degree and a half. That's a fundamentally altered planet' (cited in Allen, 2020: 28).

Robbins does not subscribe to a separation of humans and nature, nor does he blame poor people, who form most of the world's population, for detrimental impacts on the environment or climate change. Yet his work shows that the societal causes and implications of environmental crisis are manifest most prominently at the local scale. Whether it be the native grasses of Massachusetts or the invasive land cover in India, Robbins understands that people care about their local environment, but importantly that the local environment is impacted by distant events, processes and power relations. This was evidenced by the Wisconsin Initiative on Climate Change Impacts (WICCI) that was developed prior to his return to Madison,

Robbins' spatial thinking hence includes a recognition that so-called Third World political ecology should not be separate from so-called First World approaches and that concepts and methods should thus be similar (Robbins, 2000). However, acknowledging the relational construction of 'Third' and 'First' worlds, Robbins advocates for an approach that looks to the centre of power, but emphasises the need to 'not dispense with "local" political ecology in favour of the study solely of powerful organizations' and instead to imagine those organisations as connected to local struggles (Robbins, 2002: 1511). This approach appears to summarise the spatial contributions of Robbins' research and to reflect the general approach that Robbins has taken in fostering research through his editorial roles, notably as editor of *Geoforum* (2004–2007) and his service as an editorial board member on numerous journals that bring together human and physical environmental concerns.

Robbins has also shaped scholarship around the meaning of planetary population growth, and continuing debates about the (neo)Malthusian argument that population is growing inexorably and that this requires limits on human growth. Robbins and Smith (2016: 200) have critiqued the relative absence of demographic considerations in most critical social science, including Geography and political ecology, and they issue a 'call for a more vigorous engagement of critical human geography with issues of demographic change, but one that is explicitly and self-reflexively political'. They also note that demographic trends are essentially about labour, and they are key to understanding the future because 'social, ecological, political and economic futures will always be entangled with birth, death, and the presence and absence of laboring bodies' (Robbins and Smith, 2016: 213). Here, they build on demographic studies to argue that the end of population growth is happening in some countries and regions already and is inevitable at a global scale. In a further contribution, Robbins (2020) identifies notions of scarcity as coming from a position of privilege and notes that calls for a reduction in population and consumption can hardly be considered emancipatory, an observation that puts Robbins at odds with some aligned with the degrowth movement.

Key Advances and Controversies

Robbins' political ecology work puts him at odds with people and entities who exploit the environment and people for profit, with environmentalists who separate humans from nature, and to an extent with ideas for which he has some sympathy, such as degrowth. That said, his work has been well received, and in 2009 he received The James M. Blaut Innovative Publication Award from the Cultural and Political Ecology Specialty Group of the Association of American Geographers (AAG) for *Lawn People*. Bassett and Peimer (2015) have likewise identified Robbins (2007) as an exemplar of the 'co-production of nature' approach (people make lawns, and lawns make people) and contrast this approach with others that place humans and other parts of nature on a more equal footing. Co-production does not overlook power relations, instead focusing on power at different levels and how they interact. Further, as Bassett and Peimer (2015: 163) noted, political ecologists employ a co-production framework to 'combine social science methods [...] with positivist biophysical data collection methods and analytical techniques'. This is consistent with Robbins' long history of studying biogeography and becoming familiar with native grasses and so on, such that the 'ecological' in political ecology is based on sound science and on an intimate knowledge of the environment.

Robbins is also known for his role in recent debates around the notion of degrowth, although the antecedents of these debates are present in his earlier publications. Sultana (2021: 1725) noted that the idea of degrowth 'remains an arena of considerable debate within political ecology', citing Robbins (2020). Desvallées et al. argue that this debate emanates from 'two impulses running through political ecology: one leaning more towards critical "deconstruction", the other towards radical "advocacy"' (Desvallées et al., 2022: 309). While they note that both impulses stem from political ecology's origins in a Marxist framework, they are diverging as part of political ecology's 'turn towards post-structuralist approaches including questions of culture and representation' (Desvallées et al., 2022: 309). According to Desvallées et al. (2022: 311), Robbins' position is that 'political ecology must be critical of any approach, including any consistent with its values – taking a more deconstructivist line'.

Robbins (2020: 3) has also argued that there are elements of degrowth that do 'not sit as well in political ecology, at least [in] some forms of this diverse field'. While Robbins (2020: 4) has expressed guarded support for locally-based experiments and deemed it 'entirely unnecessary to denounce local electricity coops and other actually-existing experiments in decentralised collectivist environmental action as all equivalent to romantic green consumerism', he has been critical of degrowth scholarship in two key areas.

First, he has argued that 'political ecology continuously and convincingly has made strong arguments against natural limits' noting examples of deconstructing the notion of carrying capacity, undermining Malthusianism 'or challenging the political assumptions of planetary boundaries' (Robbins, 2020: 3). In essence, he argued that 'scarcity is a construct that is allied with elite power, not emancipatory process' and 'degrowth's insistence on the urgency of less, as a discourse, puts it in friction with political ecology's rejection of elitist arguments and policies that advocate forms of self-denial' (Robbins, 2020: 3).

Second, and less forcefully, Robbins (2020: 3) has argued that within degrowth thinking there is a 'tacit distrust in modern technology', while modern technology 'is viewed with more ambivalence in political ecology'. Rather than leading to the inevitable collapse of capitalism or the environment, Robbins (2020: 1) analyses automation on a dairy farm in Wisconsin, seeing technological innovation on this farm as a 'surreal cyborg affair' with the farmer being 'repeasantized' and the robotic technology allowing him to both 'maintain and raise wages for his very small remaining workforce' and free up 'time for him to see his grandchildren' while not having to get up early in winter to milk cows.

Later work by Robbins (2020) hence offers a somewhat sympathetic critique of the degrowth movement, while rejecting its Malthusian-inspired elements, adding to his contributions in defining a relentlessly critical political ecology. Robbins has provocatively characterised political ecology as 'a kind of troublemaker, effectively adopting the role of Trickster' (Robbins, 2015: 89–90). In this sense his work is characterised by its troublemaking approach, in contrast to what he identifies as 'apolitical ecology' (Robbins, 2004: 13).

Robbins' Key Works

Peet, R., Robbins, P. and Watts, M. (eds) (2011b) *Global Political Ecology*. Abingdon: Routledge.

Robbins, P. (2004) *Political Ecology: A Critical Introduction*. Malden, MA: Blackwell.

Robbins, P. (2007a) *Lawn People: How Grasses, Weeds and Chemicals Make Us Who We Are*. Philadelphia, PA: Temple University Press.

Robbins, P. (general ed) (2007b) *Encyclopedia of Environment and Society*. Thousand Oaks, CA: Sage.

Robbins P. (2020) 'Is less more ... or is more less? Scaling the political ecologies of the future', *Political Geography*, 76: 1–6.

Robbins, P. and Smith, S.H. (2016) 'Baby bust: Towards political demography', *Progress in Human Geography*, 41 (2): 199–219.

Secondary Sources and References

Allen, J. (2020) 'Hope's eternal spring', *On Wisconsin*, 120 (1): 22–9.

Bassett, T.J. and Peimer, A.W. (2015) 'Political ecology perspectives on socioecological relations', *Natures, Sciences, Sociétés*, 23: 157–65.

Desvallées, L., Arnauld de Sartre, X. and Kull, C. (2022) 'Epistemic communities in political ecology: Critical deconstruction or radical advocacy?' *Journal of Political Ecology*, 29 (1): 309–40.

Heynen, N. and Robbins, P. (2005) 'The neoliberalization of nature: Governance, privatization, enclosure and valuation', *Capitalism Nature Socialism*, 16 (1): 5–8.

Hilder, (2022) 'Paul Robbins '85: Decades later, CA influences still make a difference', https://news.coloradoacademy.org/paul-robbins-85-decades-later-ca-influences-still-make-a-difference (accessed November 13, 2023).

Lawrence, F. (2021) 'Paul Robbins on how to save biodiversity in the planet', Key Conversations with Phi Beta Kappa, *The Phi Beta Kappa Society*, May 28. https://www.pbk.org/Files/Visiting-Scholar/KC-Transcript-Robbins.pdf (accessed November 13, 2023).

Peet, R., Robbins, P and Watts, M. (2011a) 'Global nature', in R. Peet, P. Robbins and M. Watts (eds), *Global Political Ecology*. Abingdon: Routledge. pp. 1–47.

Robbins, P. (2000) 'The practical politics of knowing: State environmental knowledge and local political economy', *Economic Geography*, 76 (2): 126–44.

Robbins, P. (2002) 'Letter to the editor: Obstacles to a First World political ecology? Looking near without looking up', *Environment and Planning A*, 34 (8): 1509–13.

Robbins, P. (2006) 'The politics of barstool biology: Environmental knowledge and power in greater Northern Yellowstone', *Geoforum*, 37: 185–99.

Steinberg, P. (2018) 'Introducing Paul Robbins. Is less more…or is more less? Scaling the political ecologies of the future'. Keynote lecture at the Second Biennial Conference of the Political Ecology Network (POLLEN) in Oslo, Norway. https://www.journals.elsevier.com/political-geography/videos/paul-robbins-is-less-more-or-is-more-less-scaling-the-political-ecologies-of-the-future

Sultana, F. (2021) 'Political ecology II: Conjunctures, crises, and critical publics', *Progress in Human Geography*, 45 (6): 1721–30.

Woodward King, M. (2017) 'Lawn-term relationship', *Clark Now*, Summer: 18–23.

44
Jennifer Robinson

Mary Lawhon, University of Edinburgh

Biographical Details and Theoretical Context

Jennifer Robinson grew up in Durban (now part of eThekwini), completing her Bachelor's and MA at the University of Natal (now part of the University of KwaZulu-Natal), followed by a PhD in Geography at Cambridge University. Like so many before and since, her time in the university was characterised by struggles: in South Africa, she was part of the student anti-apartheid movement; in the UK, she struggled to make sense of the relationship between the places she knew and the texts she read. For her these were neither impersonal nor ivory tower concerns: the academy was a place to contest narratives of a collective past, explore amidst new and ongoing relations, and – particularly in the South Africa of the 1990s – imagine the kinds of spaces and cities that many were trying to transform. This ongoing critical reflection on the links between theory and understudied places – and what this means for the past and future – is central to her contributions to geographical theory.

After her PhD, Robinson returned to Durban to work at the University of Natal in 1990, the same year Nelson Mandela was released from prison. Durban had always had a complicated relationship to the apartheid state, and the University of Natal was well known for its liberal activism and political scholarship. Here, her thinking continued to be informed by the rich urban, political, geographical and historical work of South African scholars such as Jeff McCarthy, Jeff Guy and Dianne Scott. She later moved to the London School of Economics in 1996 (the same year as South Africa's first democratic election), and then spent a number of years at the Open University, also in the UK (1998–2009). Many colleagues here shaped her thinking, and one can see the impact, for example, of **Doreen Massey** and John Allen on Robinson's approach to power, relationality and theory. While at the Open University, London became a key site of study, and Robinson subsequently took up a Professorship at University College London in 2009, where she extended her thinking about power, space and cities through projects such as 'Governing the Future City' and 'Making Africa Urban'.

While her collaborative and comparative work now spans global networks, South Africa remains important to Robinson's writing and thinking. Robinson is also widely recognised as an important influence in South African scholarship: she was, for example, one of four keynotes for the 100th anniversary conference of the Society of South African Geographers in 2016 (for more on her impact on South African urban geography, see Harrison and Rubin, 2016).

Those who know just a little of Robinson's work likely place her as part of the community of scholars calling for more attention to the Global South. While reasonable enough, this is a partial summary. Robinson's foundational work in calling for research on cities 'off the map' of urban studies certainly deserves attention and citation (Robinson, 2002), but her contributions to thinking about space and place go beyond this, and are important to distinguish from those of others arguing for 'southern theory'. Broadly, Robinson's work is characterised by a commitment to logic and reason and a willingness to carefully write against the grain. From her first book, *The Power of Apartheid* (1996), which pushed back against the dominant Marxist explanations amongst the South African academic left, to her insistence on the importance of including southern cities in urban theorisation in *Ordinary Cities* (2006), to the ongoing emphasis on the utility of deep engagements with the Western canon in *Comparative Urbanism* (2022), Robinson's writing insists on, theorising without over-generalising.

Spatial Contributions

Robinson's work might usefully be placed into three interwoven arcs, each of which includes a careful consideration of space and place: the spatiality of (pre- and post-apartheid) state practice, the spatiality of theory and the spatiality of comparison. While each of her three monographs focuses on a different theme, aspects of all of these can be found across the books.

Empirically and methodologically, her first book, *The Power of Apartheid* (Robinson, 1996), focused on the archives of the apartheid state, providing a deep historical examination rooted in Port Elizabeth (now Gqeberha). However, threaded throughout the book were questions of how scholars ought to understand South Africa in relation to wider scholarly debates. Herein, Robinson argued that much progressive South African scholarship at the time relied too heavily on Marxist thinking in its explanation of apartheid. She did so without rejecting the importance of economic analysis (foreshadowing her later efforts to urge non-totalising Marxian theorising, e.g., Robinson, 2022). Theoretically, the text drew extensively on the work of political scientist Michael Mann and his connection of territory and the state, providing a compelling account of how the power of the apartheid state relied on its ability to control space – and how its ultimate inability to do so led to the ultimate collapse of apartheid. Here, Robinson was not uncritical in her application of theories from elsewhere, pointing to holes and ways in which the South African case demands a more careful analysis. In this sense, her work sought to complement the burgeoning 'struggle scholarship' focused on resistance (which she noted was important) with an account of state power and its limits. In so doing, she did not position South Africa as outside of (urban) theory but instead as a 'pinnacle of modernity', needing to be understood in relation to the ways power worked elsewhere, without necessarily being explained solely through such established ideas.

While her first book had broader relevance, it was undeniably a book about South Africa, and continues to influence thinking on the history of urban apartheid (e.g., Msila, 2014). The second key arc of her work continues to be shaped by the questions that emerged in her earlier writing, drawing on South African examples, but is focused on questions of the spatiality of theory. Her 2002 paper 'Global and World Cities' urged more critical consideration of 'where' urban theory comes from. The paper was rooted in a critique of one particular line of scholarship – the 'world cities' literature as exemplified in the work of ***Peter Taylor*** and **Saskia Sassen** – and a pragmatist may wonder why this push to theorise differently mattered. Reading this in conversation with Robinson's work on Johannesburg (mentioned in the 2002 paper) makes the implications of 'theory' abundantly clear: Robinson appeared frustrated with the way theory-making happens within an Anglo-American urban geography that centres the 'global' cities of the global North. Yet it is the impact of this theory on actual city-making that spurs Robinson to write such a targeted critique. Cities like Johannesburg are not just 'outside' of theory-making, they are also actively shaped by its discourses and power, and these discourses shape decisions about material resources and built environments (see Parnell and Robinson, 2006). In short, the musings of urban scholars matter, but it is the big push by Johannesburg to adopt the language and affectations of the 'world city' concept that prompts the need for such critical reflection. This line of thinking about the links between theory and practice argues for more careful thinking about how theory travels, an argument that has shaped, and was shaped by, engagements with many other South African urbanists, including Susan Parnell, Edgar Pieterse and Vanessa Watson.

The question of the spatiality of theory became central to Robinson's subsequent writing, including a 2006 monograph that, like her 1996 book *The Power of Apartheid*, worked to consolidate arguments that had emerged across several published articles. *Ordinary Cities* is Robinson's most well-cited text, notable for its call for 'theorising from elsewhere'. Here, nuance really matters: many have clumped scholars writing in this vein into a post/decolonial/southern agglomeration. Many geographers have indeed used Robinson, Roy, Parnell, Pieterse, Chakrabarthy, the Comaroffs and Connell as interchangeable 'flags' for thinking from the south. Yet the more writing shifts from critiquing the 'provincial' nature of 'the canon' towards theorising anew, the more visible important distinctions become within this wider literature and community of authors. Understanding Jennifer Robinson's distinctive contributions means firstly insisting that this group is not a choir: it instead contains distinctive voices making different arguments (see Lawhon, 2020).

Ordinary Cities argues against the tendency to separate the study of cities into northern urban studies and southern developmental studies. Robinson insists not on the need to study 'elsewhere' separately but to bring other places into urban studies. She argues academics ought not simply apply extant theories un-reflexively – and here nuance again matters – but stops short of suggesting we start from scratch. Given frequent miscitations, it is worth stressing Robinson does not argue 'northern theory' is necessarily and inherently inapplicable to southern places. *Ordinary Cities*, and her wider work, does not reject outright the utility of the urban canon, or spatial theory more generally. Thus, while there is widespread agreement about the need to study under-researched cities and have them contribute to wider understandings of urbanism, Robinson's particular line of argument places her somewhat at odds with those advocating for 'southern urbanism'. Those using this phrase often insist on the distinctiveness of southern cities and therefore a need to study them differently, often starting with different concepts and/or scholars (e.g.,

Pieterse, 2016; Schindler, 2017; Bhan, 2019). In 2014, in a handbook explicitly about 'cities of the global south', Robinson argued for caution in the use of this label. At best, she noted, it may be useful as a temporary strategic essentialism (see also Mabin, 2014). But *Ordinary Cities* exemplifies Robinson's urging against the separation of cities into a binary North and South. As part of a group calling for more attention to 'elsewhere', Robinson's call has been widely taken up, in the broadest sense. For example, there has been a notable uptick in the number of studies of African cities published in key urban journals since her key interventions (e.g., Mercer, 2022; Lawhon and Mwaura, 2023).

How different kinds of cities might be studied together forms the central question in Robinson's third line of inquiry. Here, her writing has largely stepped outside of debates over whether to write from elsewhere and instead turned to consider comparison as a strategy through which to do so. Interestingly, this includes a shift in language from insisting on the need to research 'elsewhere' (which, while not using the terms North and South, still implies there is a 'here' and 'there') to suggesting theory can start from 'anywhere', emblematic of a longstanding push to use postcolonial thinking to arrive at 'cosmopolitan' theory (Robinson, 2003). Her writing along these lines is an effort to tease out how theory might be built starting from anywhere. Her 2022 book *Comparative Urbanism* hence provides a deep archaeology of the logic of urban comparison, what can and ought to be compared, pushing scholars to attempt new, more capacious strategies for comparison as a basis for theorisation.

The tensions noted above with the wider 'southern urban' scholarship again become apparent in reading this text. While there is a frequent nod to positionality, citation of feminists as well as scholars from Africa and the Black Radical Tradition, it is hard to not notice where and who lies at the heart of *Comparative Urbanism.* Scholars whose ideas get pages of detailed analysis include Weber, Benjamin, **Gilles Deleuze**; Marx and the 'planetary urbanists' (e.g., Neil Brenner and Christian Schmid). **Bruno Latour** and assemblage thinkers also make repeat appearances. Robinson is critical of all of these: this is no celebration but instead a critical appraisal. But Robinson does not reject these scholars as potential sources of inspiration, and instead the book is an effort to rework their thinking towards a set of usable strategies for comparison and theory-making. Here, she relies on abstract logic and a belief that scholars can separate the problematic aspects of this history and scholarship from the useful insights. This faith in reasoned debate as the foundation sits awkwardly, at times, with some contemporary post- and decolonial scholarship that finds the canon perhaps too permeated with problems to be particularly useful, and is far less confident in that scholars can adequately know how to splice their insights into those that are 'useful' and those that are 'not'. Questions of how we might distinguish 'colonial' and 'cosmopolitan' forms of comparison that were raised in her earlier work (Robinson, 2011) seem to fade into the background in this monograph.

The question of who does this work, and which places and scholars need to be more included in urban studies, returns in Robinson's conclusion to *Comparative Urbanism.* While the book is focused on the 'how' of comparison, with the literature cited to justify comparison rooted in a classic canon, the text still, in the end, insists on the need to address the many silences in urban studies. 'It is', Robinson (2022: 384–5) concludes, 'from those places previously excluded from theorisation, and those voices and scholars least presenced in the field, that the most innovative insights are anticipated'. And while much of the text is written in a fairly neutral third person, Robinson's voice appears in first person in her final lines, noting, 'I feel that some of us need to write much less, and dedicate our energies to the infrastructural and institutional transformations which are the necessary and sufficient conditions for opening urban studies to the future of the urban, and to its differences, its diversity, its divergences'. The text, then, may be read as a work focused on how urban comparison and theory-making may be done, one rooted in an appeal to logically sorting through empirics and theory. But it also suggests that there is more to consider than methodology in the pursuit of a cosmopolitan urban studies.

Key Advances and Controversies

Robinson's work has been part of a wider movement within urban studies and geography to more deeply and carefully consider under-examined places as sites for theory-making. While this movement has been influential, de-centring the production of theory, there is ongoing contestation over the utility in using the term 'southern' to describe this movement. In this sense, Robinson's writing engages her fellow scholars on two different fronts. One group whom Robinson seems to want to convince – and with whom there is an ongoing disagreement – are those making 'revanchist' efforts to assert the centrality and validity of insights from the Global North on the basis that spatial theory is universal (e.g., urbanists who regard urbanisation as having a central, generalisable logic). To this audience, Robinson relies on dispassionate appeals to logic, leaning on scholars and genealogies that will be familiar to those trained in the canon, on grounds they might well recognise and embrace. It is unclear whether this will turn the tide here: part of the power of the postcolonial critique was its insistence that reason alone is not what led modernisers to splice the world into coloniser/colonised, developed/developing, northern/southern. Whether carefully evidenced and logical arguments will really change the minds of those who believe in totalising logic is unclear. Yet Robinson's influence here goes beyond her written work, as she is urging less polarised thinking among subsequent generations of scholars and playing an important role in attempting to moderate ongoing antagonisms.

The second audience Robinson's work speaks to is those who already believe in the need to theorise from diverse places. Here, her writing is regularly cited as part of this wider movement, yet also pushes those within it to be more careful, thoughtful and reflexive in their writing. Distinct from some versions of this argument, Robinson's work provokes geographers to not simply dismiss theory from 'elsewhere', even when writing from eThekwini or Gqeberha and 'elsewhere' is London or New York. Ongoing questions remain about the relationships between where to focus work, how this work is to be done, and who is well-positioned to do it: easy as it is to suggest the need for more scholarship from the south, the long history of South African universities (including many entanglements of apartheid science) make clear that writing from the south does not necessarily make for more accurate theorisation. Post-apartheid contestation at the University of KwaZulu-Natal, for example, raised important questions about the politics of who can write even within the Global South (see also Myers, 1994; Lawhon, 2020). Building on this, one might also agree with Robinson on the analytical limits of the term 'southern' but still struggle to let go of having some kind of label. Starting from 'anywhere' and working towards 'cosmopolitanism' might well be a shared goal amongst many. At least for now, as Robinson indicates in her conclusion to *Comparative Urbanism*, starting from places that are underrepresented might well enable more innovative insights. Yet without a term like 'southern', is something lost – something affirmative, analytical and political – in the vagueness of the phrase 'places previously excluded'? How should geographers know, name and think from elsewhere without some kind of banner?

In this sense, Robinson's writing is most compelling when understood as part of an 'ecosystem' of scholars, less focused on narrating a singular right and wrong way to approach spatial theory and more focused on imbalances among many usefully different niches, strengths and strategies. There are real theoretical and strategic differences between these scholars, and there is no singular southern urbanist choir. Robinson's role in this wider conversation has been quite different from scholars who write less from, and less focused on, the canon, yet is part of a wider push for change. Her writing shows that meaningful insights can be developed in many ways, from many different places and people. These may at times be out of balance, and there will need to be ongoing inquiries into what factors matter and which axes of difference need to be acknowledged within the academy. Robinson's work encourages geographers to be mindful about the many ways to produce theory, including through engagements with established key thinkers, engagements beyond these key thinkers, and, perhaps, forging a new, unbounded and capacious canon oriented towards cosmopolitan, global ways of understanding spaces and places.

Robinson's Key Works

Robinson, J. (1996) *The Power of Apartheid: State, Power and Space in South African Cities*. Oxford: Butterworth-Heinemann.

Robinson, J. (2002) 'Global and world cities: A view from off the map', *International Journal of Urban and Regional Research*, 26 (3): 531–54.

Robinson, J. (2006) *Ordinary Cities: Between Modernity and Development*. London: Routledge.

Robinson, J. (2011) 'Cities in a world of cities: The comparative gesture', *International Journal of Urban and Regional Research*, 35 (1): 1–23.

Robinson, J. (2022) *Comparative Urbanism: Tactics for Global Urban Studies*. Chichester: John Wiley & Sons.

Secondary Sources and References

Bhan, G. (2019) 'Notes on a Southern urban practice', *Environment and Urbanization*, 31: 639–54.

Harrison, P. and Rubin, M. (2016) 'South African cities in an urban world', *South African Geographical Journal*, 98 (3): 483–94.

Lawhon, M. with contributions from Le Roux, L., Makina, A. and Truelove, Y. (2020) *Making Urban Theory: Learning and Unlearning through Southern Cities*. London: Routledge.

Lawhon, M. and Mwaura, A. (2023) 'African cities in conversation: Who are we listening and talking to?', *Urban Geography* Virtual Special Issue 44 (9). https://www.urbangeographyjournal.org/journal/virtual-special-issues/6-african-cities-in-conversation (accessed November 13, 2023).

Mabin, A. (2014) 'Grounding southern city theory in time and place', in S. Oldfield and S. Parnell (eds), *The Routledge Handbook on Cities of the Global South*. London: Routledge. pp. 43–58.

Mercer, C. (2022) 'Introduction: Africa's urban futures', https://www.ijurr.org/spotlight-on/african-futures/introduction-africas-urban-futures/ (accessed November 13, 2023).

Msila, V. (2014) *A Place to Live: Red Location and its History from 1903 to 2013*. Stellenbosch: African Sun.

Myers, G.A. (1994) 'Eurocentrism and African urbanization: The case of Zanzibar's Other Side', *Antipode*, 26 (3): 195–215.

Parnell, S. and Robinson, J. (2006) 'Development and urban policy: Johannesburg's city development strategy', *Urban Studies*, 43 (2): 337–55.

Pieterse, E. (2016) 'Epistemological practices of southern urbanism', in W. Ding, A. Graafland and A. Lu (eds), *Cities in Transition II. Power, Environment, Society*. Rotterdam: nai010 Publishers. pp. 310–25.

Robinson, J. (2003) 'Postcolonialising geography: Tactics and pitfalls', *Singapore Journal of Tropical Geography*, 24 (3): 273–89.

Robinson, J. (2011) 'Comparisons: Colonial or cosmopolitan?', *Singapore Journal of Tropical Geography*, 32 (2): 125–40.

Robinson, J. (2014) 'New geographies of theorizing the urban: Putting comparison to work for global urban studies', in S. Oldfield and S. Parnell (eds), *The Routledge Handbook on Cities of the Global South*. London: Routledge. pp. 57–70.

Schindler, S. (2017) 'Towards a paradigm of Southern urbanism', *City*, 21 (1): 47–64.

45
Gillian Rose

Sophia Maalsen, University of Sydney
Jessica McLean, Macquarie University

Biographical Details and Theoretical Context

Gillian Rose was born in 1962 in England. She completed her BA in Geography at the University of Cambridge in 1985, where she began her 'initial stumblings around feminist historians' accounts of public and private space', which would prove to be a generative starting-point for her career (Rose in WGSG, 1997: 35). Rose then went on to complete a PhD, awarded in 1990, on 'Locality, Politics and Culture: Poplar in the 1920s' at Queen Mary College (QMC), University of London. After completing her PhD, Rose taught at QMC and Edinburgh University before moving to The Open University in 1999, where she later took on the role of Associate Dean (Research) in the Faculty of Social Sciences. In 2017, Rose joined the School of Geography and the Environment at the University of Oxford and was Head of School from 2019 to 2022. She was elected a Fellow of the British Academy in 2015.

Rose acknowledges that she had it 'relatively easy in the academy', helped by her 'Cambridge connections', the 'fashionableness of a certain kind of feminism' and, as she says, her 'compulsive desire to be nice to people like me' (Rose in WGSG, 1997: 35). However, Rose reflected that she did not experience her competence and niceness unproblematically, and that it often led to a feeling of anger, with feminism being a way to 'make sense of that paradox, of the performance and its angry excess' (Rose in WGSG, 1997: 35). Feminism has helped Rose make sense of her paradoxical experience and is also one of the key theoretical framings associated with her work.

As a cultural geographer, Rose's research interests have changed throughout her career but remain true to a central theme of 'the techniques and politics of knowledge production about places' (SoGE, 2023). Within this theme, Rose is particularly known for her contribution to feminist debates and visual culture. Two key bodies of work have established her reputation in the area: her early feminist work – which she outlined in her influential book *Feminism and Geography: The Limits of Geographical Knowledge* (1993) – and her work on visual methods captured in her book, *Visual Methodologies: An Introduction to Researching with Visual Materials* (published in its fifth edition, 2022) which speaks to the continuing impact of her work in the field of Geography.

Spatial Contributions

Feminist geographers have been active in the discipline of Geography since at least the 1970s, addressing masculinist thinking and practices on multiple fronts and undermining hegemonic practices. From highlighting inequities within geographic scholarship to challenging masculinist thinking and practices including sexism within the academy, feminists have been remaking what geography is, and how it functions, with a focus on gender disparities. Rose's core contribution to this feminist scholarship within Geography is detailed in *Feminism and Geography: The Limits of Geographical Knowledge* (Rose, 1993). This work profoundly critiques the character of Geography as a discipline historically dominated by men. It also questions the masculine bias in the production of geographical knowledge with its origins in exploration and map-making. Rose particularly critiques the 'male gaze' inherent to fieldwork, and explicitly calls out the discipline's marginalisation of women, lack of engagement with feminism, and the sidelining of women geographers and the knowledge that they have produced. Rose also engages with humanistic geographers' approach to place, outlining and critiquing their framing of emotional ties to place in her chapter 'No Place for Women?' (in Rose, 1993). In her close reading of **Yi Fu Tuan**'s analysis of affectionate attachment to place as 'topophilia', she problematises humanistic geographers' emphasis on the universal experiences of connections to home as positive, warm and uncomplicated. As **Tim Cresswell** (2004: 25) later observed, Rose shows that 'many women do not share the rosy view of home/place that humanistic geographers place at the centre of the discipline'. *Feminism and Geography* is, in Rose's (1993: 5) words, 'not a book about the geography of gender, but about the gender of geography'.

Later, Rose collaborated with Alison Blunt as co-editors of *Writing Women in Space: Colonial and Post-Colonial Geographies* (1994). This work continued to progress debates on gender in geography but expanded representation and voices beyond the predominantly white and middle-class focus which had been a key critique of her earlier work. This work accounted for the complex and contested relations between gender, race, class, and sexuality, both experienced and represented as spatial differentiation (Mahtani, 2011).

Rose's critique of mainstream Geography countered the modernist and extractivist notion that space is, or ever could be, 'fully knowable'. She argued that any understanding of space can only be from a particular viewpoint rather than a totalising perspective, something now commonly accepted. From this, Rose advanced the notion of 'paradoxical space'. Paradoxical space encompasses the researchers' view of space while acknowledging and accommodating the different and multiple understandings of space as it is experienced and produced by others. To quote Rose, it refers to when spaces that 'would be mutually exclusive if charted on a two-dimensional map – centre and margin, inside and outside – are occupied simultaneously' (Rose, 1993: 140). By conceptualising space as paradoxical, Rose made room for understanding how diverse ways of being can co-exist, full of contradictions and tensions, gaps and tenuous connections. Paradoxical space is a form of hybrid thinking that shares much in common with other forms of spatial metaphors relating to in-betweenness. **Cindi Katz** described the final chapter, where Rose started to sketch out a 'politics of paradoxical space' in pursuit of an emancipatory geography, as one that offers 'a compelling means to this goal [*of emancipatory geography*] and suggest[*s*] the real promise of feminist geography' (Katz, 1997: 230).

In addition to her contributions to feminist geography, Rose has also advanced innovative methodological approaches pertaining to the visual. Rose's original engagement with visual methods was inspired by work she encountered when designing an undergraduate course on The Cultural Politics of Landscape, while simultaneously thinking through the value of feminist film theory for geographical practice (Rose, 2021: 201). At this time, Rose's understanding of the visual was primarily of images and representations (Rose, 2021: 201).

Visual Methodologies has provided a salient starting point for many students and researchers interested in visual research and who require some orientation in visual methodology and theory. As a handbook, it responds to contemporary trends and has changed shape over time. For example, in the first edition, Rose paid heed to the 'web' only through acknowledging its absence (Crouch, 2002: 708) but subsequent editions, particularly the fourth and fifth, have been largely guided by the rapid growth of digital technologies and the significant impact they have had on reconfiguring the ways we engage with the visual (Crouch, 2002: 708; Caton, 2019). However, incorporating a moving feast of visual trends with methodologies limits the extent to which these can be explained. To that end, the work provides a broad overview of the field but has been critiqued for its lack of depth and insufficient guidance on how to actually pursue the methodologies discussed without reference to other material (Zajko, 2012: 450). However, the work fulfils its purpose admirably as an orientation to a field, as attested by the publication of multiple editions.

For deeper insights on visual approaches, Rose's *Doing Family Photography: The Domestic, the Public and the Politics of Sentiment* (2010) combines her feminist geographical focus on the domestic with her visual approach. Based on a qualitative research project involving conversations with middle-class mothers, it shows family photos are less about images per se but are material objects and objects of social practice (Chalfen, 2011: 176). Here, Rose highlights the emotive and relational power of 'snaps' and their ability to be shared and travel beyond the home. She shows that this social practice is predominantly gendered with women doing most of the photographic labour, rendering family photography as a feminine space (Chalfen, 2011: 177).

Rose's understanding of the visual has developed and she now approaches the visual as part of a broader assemblage of 'bodies, discourses, technologies, practices and affects' rather than focusing on the content of the image and its impact on the subjectivity and corporeality of the viewer (Rose, 2021: 201). This has become particularly salient as the world becomes increasingly digitalised. Situating the visual as an assemblage offers a useful framework for understanding images as they are shaped and mediated by different technologies. Rose's work on visual methodologies has therefore more recently evolved to reflect this digital shift. Her visual methodology has become influenced by the ways that digital technology mediates the urban and the visualisations that smart tech can produce.

The influence of the digital on the visual also emerges in Rose's work on urban aesthetics and on seeing the city digitally. Her work with Mónica Degen, on *The New Urban Aesthetic: Digital Experiences of Urban Change*, is illustrative of the evolution of her visual approach to accommodate technological change and the affordances this offers for visual experiences (Degen and Rose, 2022). The approach taken here is of a 'relational aesthetic' in that the contribution is focused on 'new relations between digital technologies, urban materialities, and embodied experiences' (Gassner, 2023: 225). Rather than focusing on beauty, they take a political approach that looks at how space is produced and experienced through aesthetic labourers – architects, app developers and users, residents, visitors and workers – to draw attention to this labour and problematising the relations between them (Gassner, 2023: 226). Rose's work in *The New Urban Aesthetic* builds upon her contributions to smart city literature where she has predominantly interrogated the ways in which smart cities have been visualised – whether that be in the media and marketing material of the corporate software providers who see cities as a market for their products, or the visualisation of the data the smart city generates. For example, it extends her work with Willis in which they visualise digital images of smart cities uploaded to Twitter, to show the affective qualities of Twitter through its hyper connectivity, flow and glow (Rose and Willis, 2019: 416). Their analysis of Tweets in ImagePlot revealed that the smart city is visualised in various blues and oranges, which they arranged into affective engagements with the smart city through participation, learning and anticipation (Rose and Willis, 2019: 422). Having demonstrated the insights that a social media saturated engagement with the smart city can give, they argue that smart city debates should consider the affective elements of being smart that social media can generate. They also argue for methodological experimentation that takes advantage of the affordances of digital media.

Key Advances and Controversies

Rose's (1993) *Feminism and Geography* offers an important contribution to feminist geographic scholarship by targeting binary thinking that accompanied so much early geographic research and by offering new ideas for thinking differently about space. It was felt keenly at the time. Rose's work was, however, also met with critique, particularly from some feminist geographers who thought that it presented an unsophisticated and out-of-date analysis of geographical work. At this point, **Cindi Katz** (1997) noted that it ignored much of the changes to Geography made by feminist scholars, including the work involved in the creation of *Gender, Place & Culture: A Journal of Feminist Geography* and women-specific geography study groups. Katz also argued that Rose's theoretical contributions in *Feminism and Geography* perpetuated binaries along the lines of the Lacanian 'Other' (see also Rose (1995) for an in-depth application of the psychoanalytical critique of knowledge). This binary othering feeds into critiques of the work's lack of interrogation of the 'categories or relations of gender, class, ethnicity, nationality, or sexual preference', as well as the lack of engagement with 'any idea of change or struggle' (Hyndman, 1995: 197). Addressing the conceptual basis of Rose's feminist thinking, Bondi (1995) suggests that if we accept her premise that geographical knowledge is intrinsically masculine, then advocating for a feminist geography is a contradictory stance.

The lack of engagement with other feminist geographic scholarship and the book's individualisation of emancipatory spaces led Hyndman (1995: 197) to describe the contribution as a 'soloist feminist geography'.

This use of 'soloist' refers to the work's lack of interest in engaging with other feminist work, and its focus on the individual rather than a broader feminist politic. To quote Hyndman, this leads to Rose constructing her 'own brand of feminist geography, rather than build upon or engage in the work of other feminist geographers' (1995: 197). Perhaps this soloist tendency was partly a product of Rose's own relationship with feminism as a 'lifeline' which she described as a feminism that 'feels intensely personal' (WGSG, 1997: 35). Because of the personal nature of her feminism, she conceded 'that because it is so personal it might also become very individual' (WGSG, 1997: 35). As Katz and Hyndman highlighted, Rose's work potentially downplayed progress already made by feminist geographers. This could be a result of different geographies and experiences: Katz and Hyndman were speaking from North America, Rose from the UK. Nevertheless, the critique that Rose fails to engage deeply with other perspectives than white middle-class England is common (Mahtani, 2011).

Despite the critiques, Rose's landmark text inspired further reflection and responses as feminist geographic scholarship continued to build and diversify. For example, *Feminisms in Geography: Rethinking Space, Place, and Knowledges*, edited by Pamela Moss and Karen Falconer Al-Hindi (2008), talks directly to Rose's text and provides a space for geographers to pick up the epistemological contributions Rose offered and to engage with these in constructive ways. This edited collection served to highlight how feminist geographic work is inherently multiple (*feminisms* rather than feminism). As a text that challenges hegemonic practices on many levels, it provided a space for readers to determine their own trajectory through an anti-anthology format.

In terms of theoretical contributions to Geography, Rose's notion of paradoxical space has been influential, joining associated conceptualisations of hybrid realities to destabilise assumed spatial dynamics. Referring to the concept as 'tantalising' yet 'unspecified', Hyndman (1995: 206) observed that the concept would 'provoke geographical imaginations'. If not provocative, it has certainly endured, with the concept subsequently used to understand multiple examples of the complex experiences and production of spaces. From Mahtani's (2001) use of the term in her research of 'mixed race women' in Canada, to Gibb's (2022) work on the evacuation camps as paradoxical spaces for women, the concept has been applied in different contexts but retained Rose's conceptual framing of alternative spatial dynamics. Gibb (2022: 2) uses the concept to show that the dynamics of post-disaster evacuation camps compelled women within these spaces 'to adopt conflictual strategies that simultaneously exposed them to harms and resisted ideal survivor subject norms'. This situation occurred, Gibb argues, because the camp was managed in a way that 'prioritized physical safety at the expense of gendered social norms and relations' (Gibb, 2022: 2). Paradoxical space has also been used to describe the experience of digital spaces. For example, McLean and Maalsen (2013: 244) described social media as operating 'in a space between public and private domains, a paradoxical space where "intimate" confessions or conversations are forged without expectation of permanence'. The affordances of social media, to allow the existence of multiple identities at the same time in different spaces, and blur the boundaries of assumed dichotomies – public/private; 'real' world/

online – are viewed as a novel materialisation of paradoxical space (McLean and Maalsen, 2013).

Rose's visual methodologies have also been widely adopted and increasingly across disciplines as digital technologies have accelerated the production and repetition of images. From gerontology (Reyes et al., 2022) to politics and media (Mendonça and Caetano, 2021), Rose's advocacy of visual methods has had significant impact. In a powerful and evocative piece, Zaragocin and Caretta (2020) use *cuerpo-territorio* – which recognises the relationship between the body, knowledge and territory – as a distinct decolonial and visual method that can help overcome the limitations of verbal and written methods. As a participatory approach, they show how mapping feelings and emotions is powerful in understanding experiences. In this particular case, women were asked to show, on a map of the body, where they felt the effects of extractive industry. This allowed participants to create a collective narrative around the embodied experiences of mining (Zaragocin and Caretta, 2020: 1511). Rose's contributions on both gender and methods are at play here. While Zaragocin and Caretta's work stems from indigenous feminist traditions in Latin America, it contributes to broaching the impasse in Anglo-feminist methods. It is in this dialogue across visual approaches and gendered critiques that we see Rose's ideas continue to do work and create opportunities for others to further develop the approaches she has advocated.

Rose's Key Works

Rose, G. (1993) *Feminism and Geography*. Minneapolis, MN: University of Minnesota Press.

Rose, G. (1995) 'Distance, surface, elsewhere: A feminist critique of the space of phallocentric self/knowledge', *Environment and Planning D: Society and Space*, 13: 761–81.

Rose, G. (1997) 'Situating knowledges: Positionality, reflexivities and other tactics', *Progress in Human Geography*, 21 (3): 305–20.

Rose, G. (2010) *Doing Family Photography: The Domestic, the Public and the Politics of Sentiment*. London: Routledge

Rose, G. (2022) *Visual Methodologies: An Introduction to Interpreting Visual Materials* (5th edition). London: Sage.

Rose, G. and Willis, A. (2019) 'Seeing the smart city on Twitter: Colour and the affective territories of becoming smart', *Environment and Planning D: Society and Space*, 37 (3): 411–27.

Secondary Sources and References

Blunt, A. and Rose, G. (eds) (1994) *Writing Women and Space: Colonial and Postcolonial Geographies*. New York, NY: Guilford.

Bondi, L. (1995) 'Review of *Feminism and Geography*, by G. Rose', *Feminist Review*, 51: 133–5.

Caton, L. (2019) 'Gillian Rose, Visual methodologies: An introduction to researching with visual materials', *Journal of Early Childhood Literacy*. doi.org/10.1177/1468798419896175

Chalfen, R. (2011) 'Doing family photography: The domestic, the public and the politics of sentiment', *Visual Studies*, 26 (2): 176–8.

Cresswell, T. (2004) *Place: A Short Introduction*. Malden, MA: Blackwell.

Crouch, D. (2002) 'Book Review: Visual methodologies', *Progress in Human Geography*, 26 (5), 708–10.

Degen, M.M. and Rose, G. (2022) *The New Urban Aesthetic: Digital Experiences of Urban Change*. London: Bloomsbury.

Gassner, G. (2023) 'The new urban aesthetic: Digital experiences of urban change', *Planning Perspectives*, 38 (1): 225–7.

Gibb, C. (2022) 'The evacuation camp as paradoxical space for women', *Political Geography*, 93: article #102456. doi.org/10.1016/j.polgeo.2021.102546

Hyndman, J. (1995) 'Solo feminist geography: A lesson in space', *Antipode*, 27 (2): 197–207.

Katz, C. (1997) 'Feminism and geography: The limits of geographical knowledge by G. Rose', *Ecumene*, 4 (2): 227–30.

Mahtani, M. (2001) 'Racial ReMappings: The potential of paradoxical space', *Gender, Place & Culture*, 8 (3): 299–305.

Mahtani, M. (2011) 'Gillian Rose', in P. Hubbard and R. Kitchin (eds), *Key Thinkers on Space and Place* (2nd edition). Thousand Oaks, CA: Sage. pp. 329–36.

McLean, J. and Maalsen, S. (2013) 'Destroying the joint and dying of shame? A Geography of revitalised feminism', *Geographical Research*, 51: 243–56.

Mendonça, R.F. and Caetano, R.D. (2021) 'Populism as parody: The visual self-presentation of Jair Bolsonaro on Instagram', *The International Journal of Press/Politics*, 26 (1): 210–35.

Moss, P. and Falconer Al-Hindi, K. (eds) (2008) *Feminisms in Geography: Rethinking Space, Place and Knowledges*. Lanham, MD: Rowman & Littlefield.

Reyes, L., Versey, H.S. and Jarmin, Y. (2022) 'Emancipatory visions: Using visual methods to co-construct knowledge with older adults', *The Gerontologist*, 62 (10): 1402–8.

Rose, G. (2021) 'Response to "Visual studies now"', *Visual Studies,* 36 (3): 201–2.

School of Geography and Environment (SoGE) (2023) Staff Profile: Professor Gillian Rose. https://www.geog.ox.ac.uk/staff/grose.html (accessed November 13, 2023).

Women and Geography Study Group [WGSG] (1997) *Feminist Geographies: Explorations in Diversity and Difference*. London: Longman.

Zajko, M. (2012) 'Rose, Gillian, Visual Methodologies: An Introduction to Researching with Visual Materials. 3rd ed.', *Canadian Review of Sociology/Revue Canadienne de Sociologie*, 49: 449–51.

Zaragocin, S. and Caretta, M.A. (2021) '*Cuerpo-Territorio*: A decolonial feminist geographical method for the study of embodiment', *Annals of the American Association of Geographers*, 111 (5): 1503–18.

46
Edward Said

Daniel Clayton, University of St Andrews

Biographical Details and Theoretical Context

Edward W. Said (1935–2003) was a Palestinian American Professor of English and comparative literature at Columbia University, New York, where he spent his entire career. He was born in West Jerusalem (then in British-occupied Palestine) to a Lebanese mother and Palestinian father who built a thriving stationery business in Cairo. His family was Arab Christian and he was educated at elite schools (Victoria College in Cairo and Mount Hermon in Massachusetts) before entering Princeton University (BA English and History) and Harvard University (MA and PhD English 1964), and joining Columbia as an instructor in 1963 (Said, 1999b). He was a US citizen on account of his father's US military service (in World War I), was twice married, and lived in Manhattan. He battled leukaemia for the last ten years of his life, although his illness did not diminish his productivity. His papers are now archived at Columbia University, and there is an enormous literature on his work and influence (CPS-ESAC, n.d.; Iskandar and Rustom, 2010; Brennan, 2021: 411–16).

Said worked primarily in the domain of literature. Though steeped in the Western canon, he reshaped its boundaries through the pursuit of eclectic interests, including classicalmusic and opera. His work was integral to the 1970s and 1980s 'cultural turn' in the humanities and social sciences, which connected philosophical interest in how language mediates knowledge and meaning to wider consideration of how power works through 'writing, vision, and study' (discourse, representation, epistemology, narrative, text, image, and spectacle) (Said, 1978: 201). His 1978 book *Orientalism* (translated into at least 36 languages) was groundbreaking in this regard. It provided a foundational impetus to postcolonial theory, and remains a key statement about how literature, culture and politics are intertwined.

Said was an acclaimed writer, critic, and activist, and a charismatic public intellectual. He was a fearless spokesperson for Palestinian causes in the United States and critic of US foreign policy in the Middle East, with links, for a while, to the Palestine Liberation Organisation (Brennan, 2021). He was influenced by a plethora of writers (with Theodor Adorno, Eric Auerbach, Joseph Conrad, **Michel Foucault**, Lucien Goldman, Georg Lukács, Jonathan Swift, and Giambattista Vico in an idiosyncratic mix). His diverse oeuvre is replete with novel (if now household) critical terms, such as 'imaginative geography', 'secular criticism', 'travelling theory', and 'worldliness'.

Timothy Brennan (2021: 101) notes that Said had a 'taste for perverse allegiances', including a contrarian fascination with 'writers [such as Conrad and Swift] he should have disliked', and was, to those who knew him, a complex man – gracious and loyal, but irascible too. His work took a political turn following the 1967 Arab-Israeli Six-Day War, starting with an essay 'The Arab Portrayed' (Said, 1970), which grew into *Orientalism*, and with the theme of 'exile' looming ever larger in his work, and most intimately in his memoir *Out of Place* (1999b; also Said, 2000b). However, 'For all his writing on exile', Brennan (2021: xviii) continues, 'he was a rooted man – imaginatively in Palestine and actually in New York'. **Cindi Katz** and **Neil Smith** (2003: 638) extolled his perspicacity as 'an equal-opportunity indicator of oppression' and thus 'intellectual in the profoundest sense', bold in his denunciation of dogmatism, hostility, and oppression wherever he found them.

Said (2003: xiii) was concerned with both the potential and interdictions of humanistic culture, and the location of the intellectual therein. He probed the 'profound difference between the will to understand for purposes of co-existence and enlargement of horizons, and the will to dominate for the purposes of control' (Said, 2003: ix) and saw the latter in Orientalism, which he defined as a dichotomising, disparaging, and domineering 'system of representation' (or discourse) 'framed by a whole set of forces that brought the Orient [East] into Western learning, Western consciousness, and later, Western empire', and with texts and images creating 'not only knowledge but also the very reality they appear to describe' (Said, 1978: 94, 203). He deployed a theatrical understanding of representation, with the Orient 'a closed field, a theatrical stage affixed to Europe', with difference thus 'dramatised' and the Orient characterised as Europe's 'Other' ('surrogate and even underground self') (Said, 1978: 63, 3). As ***Derek Gregory*** (1999: 115) – one of the first geographers to bring Said's work to the discipline – notes, 'metaphors of "theatre" and "text" [in *Orientalism*] are not innocent [...] [but] speak to power and the production of spaces and the practices that take place within (indeed through) them'.

Said (1978: 49–73) expedited these claims through the idea of 'imaginative geography'. As ***Derek Gregory*** (2009: 369-70) captures it, imaginative geographies are 'representations of other places – of peoples and landscapes, cultures and "natures" – that articulate the desires, fantasies and fears of their authors and the grids of power between them and their "Others"'. Here, the prescriptive and performative power of Orientalism resides in 'its capacity to produce the effect it names ("the Orient")', and in binary terms: as seductive, feminine, mystical, backward, and despotic, and in opposition to a putatively civilised, rational, masculine, modern and democratic West/Occident (Gregory 2012: 150). As Said (1978: 211) observed, as both an ancient scene of Western wonder and in 'the precisely actual form in which the modern Orient was lived in, ... the *geographical space* of the Orient was penetrated, worked over, taken hold of' (Said, 1978: 55, 211). Gregory (2012: 150) suggests that Said 'identified two closely connected cultural-political performances': first, that 'the Orient' was 'an exotic and bizarre space'; and second, that it was a space that had to be 'domesticated and disciplined'.

Said (1978: 328) insisted that Orientalism prevailed over Arab and Islamic scholarship and latter-day 'decolonializing' thought and writing. Yet this did not dim his quest to 'situate these [proscriptions] in a geography of other identities, peoples, cultures, and then to study how, despite their differences, they always overlapped one another' (Said, 1993: 330). His aim was to nurture ways of knowing that do not 'get involved either in self-congratulation (when one discusses one's own [culture or discipline]) or hostility and aggression (when one discusses the "other")' (Said, 1978: 325).

Spatial Contributions

Geographers have been drawn to the overt spatial sensibilities in Said's work. The words 'geography', 'place' and 'space' abound in his writing, and his work has been seen, often in combination with that of **Michel Foucault**, as promulgating a 'spatial turn' in critical inquiry (Frank, 2009). Yet his geography was more suggestive than programmatic, partly because his direct engagement with the discipline was limited, but mainly because his geographical interests spanned different scales and problems.

Orientalism named and challenged a regionalised constellation of power, knowledge, and geography, and was not, he insisted, about getting at a real or true Orient but at a project of 'knowledgeable manipulation' that produced this part of the world 'discursively', as Europe's 'Other' (Said, 1978: 50). Imaginative geography is part and parcel of 'a tradition, or what Michel Foucault calls a discourse' of Orientalism, Said (1978: 94) pointed out, 'whose material presence or weight, not the originality of a given author, is responsible for the text [or image] produced'.

Culture and Imperialism (1993) extended Said's concern with discourse to the modern-global span of imperialism from the nineteenth century through to the present (the Iraq War of 1991). It attended more directly to 'the actual geographical underpinnings of the imperial' and 'actual geographical possession of land' (Said, 1993: 78). Said also paid close attention to the literature of post-war decolonisation and the figure of 'the insurrectionary native' (Said, 1993: 268). He portrayed geography as a force field of liberty and constraint: as a fundamental restriction to the colonised (one that needs to be usurped), and, for the coloniser, with 'geography... [being] yours to move about in more or less freely' (Said, 1993: 159). In a passage frequently cited by geographers, he wrote that 'Just as none of us is outside or beyond geography, none of us is completely free from the struggle over geography. That struggle is complex and interesting because it is not only about soldiers and cannons but also about ideas, about forms, about images and imaginings' (Said, 1993: 7). Said (1993: 51; original emphasis) sought to read the imperial archive 'not univocally but *contrapuntally*, with a simultaneous awareness both of the metropolitan history that is narrated and of those other histories against which (and together with which) the dominating discourse acts'.

These two books form the crux of Said's influence on geography and geographers. However, a string of books, from *The Question of Palestine* (1979) to *From Oslo to Iraq and the Road Map* (2004a), also brought a more localised concern with the Middle East. Said (1999a: 36) wrote, for example, of the 1990s Oslo Accords: 'they required us [Palestinians] to forget and renounce our history of loss, dispossessed by the very people who taught everyone the importance of not forgetting the past. Thus we are the victims of the victims, the refugees of the refugees'.

Finally, and as above, Said (1993: 7) had a cross-cutting interest in questions of belonging, dispossession, and exile, and urged that while 'No one can deny the persisting continuities of long traditions, sustained habitations, national languages, and cultural geographies [...] there seems no reason except fear and prejudice [...illusion or dogma] to keep insisting on their separation and distinctiveness' (Said, 1993: 59, 336). He propounded an 'outside' perspective that sprang from a post-war 'constellation of words and valences that surrounds Europe and the West [and] acquired a much more fraught and even rebarbative meaning from observers outside Europe and the West' (the critique of empire and a politics of dispossession) (Said, 2004c: 17). But he also positioned himself as a 'nomad' whose morality, in Adorno's (1994 [1951]: 39) aphorism, is to 'not be at home in one's home' (and occupy an *émigré* location). Said connected with the 'global sense of place' (Massey, 1991) thinking that imbued 1980s and 1990s postmodernism and the idea of globalisation by upholding the integrity of place and of the intellectual as someone who manifestly represents a standpoint, yet also held fast to the existential significance, in an age of flows (and both creatively, and from the vantage point of in situ trauma and dispossession), of being 'out of place' – and with all of this part of what he termed a 'democratic humanism' (Said, 2004b).

Key Advances and Controversies

'Discourse', 'empire', and 'geography' are keywords in the advances and controversies associated with Said's work. Said alighted on the issue of what ***Trevor Barnes*** and James Duncan (1992) capture as 'writing worlds': how space is produced, and geography is known and imagined, by being inscribed (written, perceived, represented). All knowledge is bound up with 'the politics, the considerations, the positions, and the strategies of power', he implored in an essay entitled 'Travelling theory', and the systemic qualities of discourse are integral to the locational politics of disciplines (Said, 1983: 242). In short, he underscored the 'discursive' nature of space, place, and inquiry.

Said's work became pivotal to a critical human geography which challenged spatial science as a foundation for the discipline and opened up the constructed, contingent, and contested nature of space, place, and geographical knowledge. The discipline's Western/exclusionary cast became noticeable and raised difficult questions about how geographers related to and represented 'other' people and places. Geography (as both discourse and discipline) could not be regarded as an impartial knowledge or project with either an innate character or a self-contained history.

Some of the discipline's most distinguishable attributes – for example, adventure, discovery, exploration, mapping, travel writing, environmental determinism and geopolitics – had Eurocentric, masculinist, and racist complexions and worked as tools of empire, with lasting influence on the image and credibility of the subject. 'Postcolonial geographies' (Blunt and McEwan, 2002) emphasised past and present relations between Geography and empire. In turn, the colonial past became a fulcrum of theoretical debate about Geography's critical remit and future (Clayton, 2003). The 'pastness' of Geography's imbroglio with empire permeated the discipline's centre and present.

The idea of 'imaginative geography' has now been applied to a great array of (Western and non-Western) situations and problems, and not least through work on other potent 'locational imaginaries' such as oceans and tropics (see Jazeel, 2019). It has also been expanded to include issues that were less of a focus for Said, such as class, gender, nature and science (see Driver, 1992; Gregory, 1999; 2012). Yet such diffuse critical ventures with this idea mostly share Said's central worry about 'the invention and construction of a geographical space called the Orient, for instance, with scant attention paid to the actuality of the geography and its inhabitants' (Said, 2000a: 181). His 'for instance' pointed to something more constitutive and proleptic in human thought and practice: how people encounter and deal with difference by simplifying, distorting, and manipulating it (see also Harley, 1992).

However, debate pertained to how Said connected knowledge and power, and related 'ground' to 'text'. Said (1978: 328) regarded Orientalism as a 'failure' of human imagination and will because it occupies and defends 'a position of irreducible opposition to a region of the world it considered alien to its own'. Or as Joanne Sharp (2023: 18) notes, Europe projected 'a single culture into the space of the "Orient" at odds with diversity of peoples, cultures and environments contained in this geographical area' and with this space 'defined by texts [of the powerful and privileged] and not by people from the Orient itself'. Yet in summoning, as such, Orientalism's determining character ('worldwide hegemony' as he put it (Said, 1978: 328)) and spatial/divisive declensions, as a way of questioning the idea of a true knowledge or authentic reality, Said also implied that his own critique somehow came from outside the field of vision and system of knowledge he named and arraigned (see Young, 1990: 120–2). Was this paradox of 'inside' and 'outside' a disabling flaw? Was Said doing to the West what it did to the Orient: providing an 'overgeneralized' and 'reductive' account (Jazeel, 2019: 58)? Was the will to power Said saw in Orientalism more an article of faith than a demonstrable fact or effect? And did what he described as 'the voyage in' from the era of decolonisation of 'peripheral off-center work that gradually enters the West and then requires acknowledgement', and its evolution into postcolonialism not involve its own inside/outside seductions of power and symbolise late capitalism's corporatisation of university/campus 'diversity' (Said, 1993: 239, 216; Ahmad, 1994)?

Second, exactly how does Orientalism have a 'material presence or weight'? How were the scholarly and literary pursuits of Orientalists translated into colonial power? Said answered some of these questions in *Culture and Imperialism* by 'rethinking geography' (Robbins et al., 1994: 21). Yet for some geographers, his frequent and elastic use of the word rang hollow, and his 'contrapuntal reading', which shuttled between the general and the particular, was hard to locate (Said, 1993: 78). For example, Cole Harris (2004) suggested that such shuttling – for example, from Austen's Antigua to Verdi's Cairo and Conrad's Congo – made it difficult to know where Said was placed, except in literature and culture. Harris reckoned that for all its 'decentred' concern with otherness, the colonised, and dispossession, much of the critical (postcolonial) geographical literature on empire tended to revolve around imperial centres and metropolitan postcolonial anxiety and not enough around 'edges of empire', such as the settler colonial space of British Columbia, where he worked (see Jacobs, 1996). In short, the critical location of postcolonial geography, and what it spoke for, needed to be examined.

Still, there was plenty in Said's work that chimed with Harris's (2004; 2020: 184–8) argument – now writ increasingly large in 'decolonial geographies' – that the study of settler colonialism should proceed 'from sites of dispossession' rather than via sweeping theories about colonial discourse, and should attend to 'colonialism's basic geographical dispossessions of the colonised'. Indeed, as Said (1993: 93) himself averred, 'the actual geographical possession of land is what empire in the final analysis is all about', continuing:

> If there is anything that distinguishes the imagination of anti-imperialism, it is the primacy of the geographical element. Imperialism after all is an act of geographical violence through which virtually every space in the world is explored, charted, and finally brought under control. For the native, the history of colonial servitude is inaugurated by the loss of the locality to the outsider; its geographical identity must thereafter be searched for and somehow restored. (Said, 1993: 222)

On the other hand, as 'more and more regions challenge the classical empires', Said (1993: 190–1) observed, 'you [in the metropole] have to reconceive [imperial] reality as something that can be held together by you the artist, in history rather than geography. Spatiality becomes, ironically, the characteristic of an aesthetic rather than of political domination'.

There is nothing 'past' about either this imagination of anti-imperialism, or the need to engage this 'spatial aesthetic' (it is still enshrined in imperial landscapes and mausoleums of the so-called 'culture wars' – statues, museums, school curricula, and public captivation with imperial derring-do). Today's postcolonial, decolonial, anti-racist, and anti-imperial geographies can also continue to learn from Said's observation in his Preface to the 25th anniversary edition of *Orientalism* that empire brought an 'immense distortion' (conceit) into the worldviews of imperial nations and cultures – a lasting bias that operates through disavowal of empire's inherent violence, and with much still hinging (as it did for Conrad – the subject of Said's first book) on how 'we' seek to demarcate, distance, and insulate ourselves from 'them' (and what 'we' did to 'them') and see 'our' problems as removed from 'theirs' (Said, 2003: xxii; and see Lester, 2022). Said's concerns with 'who speaks', 'what is seen, and from where', and how difference is produced and placed as a function of power – what might be described as his locational humanism – remain vital.

Said's Key Works

Robbins, B., Pratt, M.L., Arac, J., Radhakrishnan, R. and Said, E. (1994) 'Edward Said's Culture and Imperialism: a symposium', *Social Text*, 40: 1–21.

Said, E. (1970) 'The Arab portrayed', in I. Abu-Lughod (ed), *The Arab-Israeli Confrontation of June 1967*. Evanston, IL: Northwestern University Press. pp. 1–9.

Said, E. (1978) *Orientalism*. New York, NY: Vintage.

Said, E. (1979) *The Question of Palestine*. New York, NY: Times Books.

Said, E. (1983) *The World, the Text and the Critic*. Cambridge MA: Harvard University Press.

Said, E. (1993) *Culture and Imperialism*. New York, NY: Alfred A. Knopf.

Said, E. (1999a) 'The one-state solution', *The New York Times Magazine,* 10 January, Section 6: 36. https://www.nytimes.com/1999/01/10/magazine/the-one-state-solution.html (accessed November 13, 2023).

Said, E. (1999b) *Out of Place: A Memoir*. New York, NY: Vintage.

Said, E. (2000a) 'Invention, memory, and place', *Critical Inquiry,* 26: 175–92.

Said, E. (2000b) *Reflections on Exile and Other Essays*. Cambridge, MA: Harvard University Press.

Said, E. (2003) *Orientalism*. Preface to 25th anniversary edition. London and New York, NY: Penguin, pp. i–xxx.

Said, E. (2004a) *From Oslo to Iraq and the Road Map*. New York, NY: Pantheon Books.

Said, E. (2004b) *Humanism and Democratic Humanism*. New York, NY: Columbia University Press.

Said, E. (2004c) *Freud and the Non-European*. London: Verso.

Secondary Sources and References

Adorno, T. (1994 [1951]) *Minima Moralia* (English 8th edition). London: Verso.

Ahmad, A. (1994) 'Orientalism and after', in P. Williams and L. Chrisman (eds), *Colonial Discourse and Post-Colonial Theory*. New York, NY: Columbia University Press. pp. 162–71.

Barnes, T. and Duncan, J. (eds) (1992) *Writing Worlds: Discourse, Text and Metaphor in the Representation of Landscape*. Abingdon: Routledge.

Blunt, A. and McEwan, C. (eds) (2002) *Postcolonial Geographies*. London: Continuum.

Brennan, T. (2021) *Places of Mind: A Life of Edward Said*. London: Bloomsbury.

Clayton, D. (2003) 'Critical imperial and colonial geographies', in K. Anderson, M. Domosh, S. Pile and N. Thrift (eds), *Handbook of Cultural Geography*. London: Sage. pp. 354–68.

CPS-ESAC (n.d.) *Edward Said Archival Collection — Center for Palestine Studies | Columbia University*. http://palestine.mei.columbia.edu/said/archival-collection (accessed November 13, 2023).

Driver, F. (1992) 'Geography's empire: Histories of geographical knowledge', *Environment and Planning D: Society and Space*, 10: 23–40.

Frank, M. (2009) 'Imagining geography as a travelling concept: Foucault, Said, and the spatial turn', *European Journal of English Studies*, 13: 61–77.

Gregory, D. (1999) 'Scripting Egypt: Orientalism and the cultures of travel', in D. Gregory and J. Duncan (eds), *Writes of Passage: Reading Travel Writing*. London and New York: Routledge. pp. 114–50

Gregory, D. (2009) 'Imaginative geography', in D. Gregory, R. Johnston, M. Watts (eds), *The Dictionary of Human Geography* (5th edition). Oxford: Blackwell. pp. 369–71.

Gregory, D. (2012) 'Dis/ordering the Orient: Scopic regimes and modern war', in T. Barkawi and K. Stanski (eds), *Orientalism and War*. London: Hurst. pp. 150–70.

Harley, B. (1992) 'Rereading the maps of the Columbian encounter', *Annals of the Association of American Geographers*, 82: 522–44.

Harris, C. (2004) 'How did colonialism dispossess? Comments from an edge of empire', *Annals of the Association of American Geographers*, 94: 165–82.

Harris, C. (2020) *A Bounded Land: Reflections on Settler Colonialism in Canada*. Vancouver: UBC Press.

Iskandar, A. and Rustom, H. (eds) (2010) *Edward Said: A Legacy of Emancipation and Representation*. Berkeley, CA: University of California Press.

Jacobs, J. (1996) *Edge of Empire: Postcolonialism and the City*. Abingdon: Routledge.

Jazeel, T. (2019) *Postcolonialism*. Abingdon: Routledge.

Katz, C. and Smith, N. (2003) 'An interview with Edward Said', *Environment and Planning D: Society and Space*, 21: 635–51.

Lester, A. (2022) *Deny and Disavow: The British Empire in the Culture War*. London: Sunrise.

Massey, D. (1991) 'A global sense of place, *Marxism Today*, 38: 24–9.

Sharp, J. (2023) *Geographies of Post-Colonialism: Spaces of Power and Representation* (2nd edition). London: Sage.

Young, R. (1990) *White Mythologies: Writing History and the West*. Abingdon: Routledge.

47
Milton Santos

Lucas Melgaço, Vrije Universiteit Brussel

Biographical Details and Theoretical Context

Milton Santos is widely recognised as Brazil's most influential geographer. Santos was not only an expert in his field, but a respected thinker who developed a set of coherent and complementary concepts into a solid theoretical framework. Despite his passing more than two decades ago in 2001, the 'Miltonian School of Geography' is still hugely influential in Brazil and has recently regained wider attention thanks to the translation of some key works into English.

Milton Santos was born in 1926 in Brotas de Macaúbas, Bahia. He received his early education from his parents, both school teachers, who taught him reading, writing, maths, and French, with the latter proving useful throughout his life in terms of connecting with the French academic world. As a young man, Santos moved to the state capital, Salvador, to continue his studies, eventually earning a law degree in 1948, although he rarely practised law. Instead, he worked as a journalist and as a geography teacher. In 1956, he moved to France to pursue a doctoral degree in Geography at the University of Strasbourg. His dissertation – 'Le Centre de la Ville de Salvador: Étude de Géographie Urbaine' – was completed under the guidance of his mentor, Jean Tricart. Santos credits Tricart with teaching him the rigours of academic research and sparking his interest in Marxist theories (Mamigonian et al., 1989).

Santos returned to Brazil in 1959, taking office in the government of the state of Bahia (Chief of the Civil House of the Presidency of the Republic in the state of Bahia [1961] and President of the Economic Planning Commission Foundation of Bahia [1962–1964]; Costa et al., 2022). He also worked as a professor at the Federal University of Bahia from 1961 until 1964, when a military coup d'état forcibly removed the democratically elected government. Santos was arrested by the military police and imprisoned for three months. His incarceration was probably due to a combination of factors, including his positions when working for the government (he defended, for example, the taxation of large fortunes) and the progressive nature of his writings as a journalist, such as his report of João Goulart's visit to Cuba.

Following his three-month prison sentence, Santos was released and saw himself obliged to leave the country. This led to a 13-year period of exile, during which he pursued an academic career across various countries, including France, The United States, Canada, Venezuela and Tanzania. By Santos's own account (Yázigi, 1996), his exile prevented him from producing more empirically oriented works about his home country, thus nudging him towards a theoretical approach to geography.

After travelling to Brazil on several occasions, Santos definitively returned to the country in 1977, becoming associate professor at the University of São Paulo in 1984, where he remained until his death in 2001. The last two decades of his life saw the production of some of his most important works, for which he received several awards, including *honoris causa* distinctions from twenty universities in Europe and Latin America. In 1994, he became the first scholar from the so-called Global South to win the prestigious Vautrin Lud International Geography Prize, the most significant award in the field and considered by many to be the 'Nobel Prize of Geography'. By the end of his prolific career, he had sole-authored more than 40 books, edited 14 collections, published 58 book chapters and written about 300 scientific articles (Santos, 2001).

Most of Santos's work was first published in Portuguese, with a few exceptions in French and Spanish. Among the latter is his book *L'espace Partagé: Les Deux Circuits de L'économie Urbaine des Pays Sous-Développés*, originally published in French in 1975. It was translated into English by Chris Gerry in 1979, and remained Santos's only book available in English until 2017. Except for this translation, a special issue of *Antipode*, compiled by Santos and Richard Peet in 1977, also included a handful of Santos's articles in English. Among these articles was 'Society and Space: Social formation as theory and method' (Santos, 1977; translated by Stephan Slaner), in which Santos defended the idea that a socio-economic formation should actually be considered a socio-economic and spatial formation. This idea played a crucial role in the reaffirmation of the importance of geography among the existent social sciences.

In the twenty-first century much of his work is being rediscovered thanks to several translations. *Por Uma Outra Globalização: Do Pensamento Único à Consciência Universal* was translated by Lucas Melgaço and Tim Clarke and published in 2017 as *Toward an Other Globalization: From the Single Thought to Universal Conscience* (2017a). In 2021, *Por Uma Geografia Nova: da Crítica da Geografia à Geografia Crítica* was translated by Archie Davies as *For a New Geography* (for a discussion on Davies's choice of title, see below). Brenda Balletti's translation *The Nature of Space* (*A Natureza do Espaço: Técnica e Tempo: Razão e Emoção*) was published in that same year. Even though only a limited number of Santos's books are available in English (especially bearing in mind that the other 37 books and most of his articles are still only available in Portuguese), these three translations nevertheless provide a solid introduction to Santos's work as they cover many of his key theoretical ideas.

Toward Another Globalization was originally published in Portuguese in 2000, at the end of Santos's career. It is a book that consolidates an idea that Santos had already started developing in 1971, when he published, in French, *Le Métier de Géographe en Pays Sous-Développé* (Santos, 1971). In both books, Santos attempts to theorise the world from the perspective of 'the South' or those 'from below' (Santos, 2017). *Toward Another Globalization* was written with the idea of reaching a wider audience. As such, it serves as an accessible introduction to Santos's writing and thinking, and tackles one of the topics for which he became world-famous: globalisation. Santos reflects on the promises and discourses that surround globalisation (globalisation as fable), and the impact it creates, particularly on poor countries (globalisation as perversity). However, he concludes in an optimistic tone by focusing on the possibilities of a globalisation led by those he refers to as 'hegemonized actors' (globalisation as possibility). Indeed, optimism was one of Santos's trademarks, as was his firm belief in bottom-up transformations. In an interview for the popular talk show with Jô Soares, he claimed that '[t]he geographer is, above all, a philosopher, and philosophers are optimists, because before them lies infinity' (Santos in Soares, 1995: 14:00). It is worth mentioning, in this respect, that the Brazilian filmmaker Silvio Tendler's (2006) documentary *Encontro com Milton Santos: O Mundo Global visto do lado de cá?* is based on the central premises of *Toward another Globalization*. Subtitled in English, the documentary compiles different interviews with Santos, in which he articulates statements such as:

> Never in the history of humanity have there been such suitable technical and scientific conditions to build a world of human dignity, only that these conditions have been appropriated by a handful of companies that have chosen to construct a perverse world. (Santos in Tendler, 2006: 10:02).

For a New Geography, on the other hand, is a much more complex book than *Toward Another Globalization*. Originally published in 1978, it marked the end of Santos's long exile abroad. This dense book showcases Santos's effort to position his theories in the theoretical landscapes of the time. His voice echoes with other authors of the so-called critical and radical geography movements, who sought to oppose a blindly quantitative and computerised approach. In the book, Santos presents a new epistemology of Geography, calling attention to the fact that geographers should have a clear view of the object of geography, namely geographic space. Such clarity would prevent Geography from becoming what Santos terms a 'widow of space' (2021a: 62). *Por uma Geografia Nova* was one of Santos's first major theoretical efforts, culminating in *The Nature of Space*, which is considered by many to be his most important book.

The Nature of Space delves into the fundamental questions that were raised in *For a New Geography*, such as the definition and object of geography. However, it expands upon these concepts by incorporating theories and ideas that were developed over the two decades between both books. Essentially, *The Nature of Space* serves as a comprehensive guide to the key concepts that were utilised throughout Santos's career. With this book, he 'seeks to provide a geographic contribution to the production of a critical social theory' (Santos, 2021b: 6). In recognition of its contributions to the field of human sciences, the book received Brazil's Jabuti Award in 1997.

These three books provide valuable insights into Santos's ideas. He was heavily influenced by a variety of authors from the so-called Global North, such as ***Jean Baudrillard***, Emile Durkheim, Jacques Ellul, Jean-Paul Sartre, Gilbert Simondon, **Bruno Latour, Henri Lefebvre**, Max Sorre, Jürgen Habermas, Lewis Mumford, ***Anthony Giddens***, ***Torsten Hägerstrand***, and Ortega y Gasset. However, Santos's work is not just a repetition or mere application of foreign concepts and theories. Rather, Santos re-elaborated and reinterpreted many of these concepts to produce an original perspective. These reinterpretations, along with Santos's creativity, have resulted in the creation of his own vocabulary. Authors such as Ferreira (2007) and Souza (2009) have even organised some of his concepts into glossaries. Still, more than a set of concepts, Santos created a powerful historical and critical dialectic method for geography that is both theoretical and analytical.

Spatial Contributions

Santos's theory revolves around the importance of geographic space, which is the fundamental object of Geography. Although his definition of geographic space evolved over time, it consistently maintained that space is not merely an empty container which society occupies, but rather a content-form (Santos, 2021b: 6). In *The Nature of Space* he defines geographic space as follows:

> Space [...] is a mixture, a hybrid, formed [...] from the indissociable union of systems of objects and systems of actions. The systems of objects, the space-materiality, form territorial configurations, where the action of subjects, whether rational or not, is established to create a space. This space – geographic space – is more than the social space of sociologists because it also includes materiality. (Santos, 2021b: 291)

To Santos (2021b: 2), however, 'a discussion of space necessarily presupposes an approach to method; to speak of an object without speaking of method is simply to state a problem without truly understanding it'. Periodisation is one of the important tools in Santos's method as it allows for space and time to be approached together. He suggests that we look at space through a periodisation of technique (*técnica*, in Portuguese) or, in other words, as a periodisation of the different technical systems:

> Given that geographic space is always an indissociable set of systems of objects and systems of actions, it manifests differently in different historical moments according to the nature of the objects and actions present in each period. Because technique is also social, the system of objects and system of actions together constitute technical systems, and the succession of those systems makes up the history of geographic space. (Santos, 2021b: 230)

By periodising space and technique, Santos highlights the transformation from a natural milieu to the emergence of the current technical-scientific-informational milieu. The characteristics of such a milieu are better understood through support concepts such as technical unicity, enlargement of contexts, convergence of moments, knowability of the planet, and contemporary acceleration (see Santos, 2017a; 2021b and Melgaço, 2013 for an explanation of these concepts). Santos relies on this theoretical framework to explain spatial inequalities, as different countries, regions, and places do not necessarily share the same technical density (Santos, 2021b).

In his search for an ontology of space, Santos suggests that space should be analysed by considering its analytic categories, which, according to him, consist of structure, process, function and form (Santos, 1985). For Santos, geographic space is both an abstraction and a totality, whose 'empiricization' (Santos, 2021b: 27) occurs through other concepts such as region, networks, scale and place.

Place is incidentally a good example of how Santos re-signified existing geographical concepts. For Santos (2021b: 184), place is 'where fragments of the network take on a unique and socially concrete dimension as the result of the contiguous occurrence of socially aggregated phenomena based on solidarity events that are the outcome of diversity and of repetitive events that may include an element of surprise'. He then goes on to say that '[e]ach place is, in its way, the world' (Santos, 2021b: 216). Place is a central concept in Santos's complex theory as it combines the nodes of the networks with the banal space, 'the space of all people, all businesses, and all institutions' (Santos, 2021b: 193). Place is where both hegemonic and non-hegemonic rationalities meet, and that can lead to revolutionary change:

> [t]he recognition that concurrent rationalities exist in the face of the dominant rationality provides new perspectives on method and action, which allow for shifting perspectives on spatial and social evolution and suggest changes in the very epistemology of geography and the social sciences as a whole. (Santos, 2021b: 6)

An inexperienced reader of Santos may find these concepts and quotes hard to read at first, as a deeper understanding of their meanings and the connections between them may require an in-depth reading of Santos's other works. They are nevertheless a teaser and a testament to the richness of Santos's contributions as a thinker on space and place.

Key Advances and Controversies

During an interview on the Brazilian TV show *Roda Viva*, Santos once stated that the intellectual is there to cause discomfort. He followed his own advice to the letter, evidenced by his often provocative statements, such as: 'I consider myself an "outsider" intellectual, which is rare in Brazil. I don't belong to any political party or intellectual groups, I don't adhere to any creed, and I don't engage in any activism' (Santos in Tendler, 2006: 4:55). Despite the fact that Santos did write on the conditions of Black Brazilians (Santos, 1996; 2000) and discussed such topics in several interviews, Black movements seem to have harboured the expectation that Santos should have been more vocal on these issues and should have reflected in more detail on his unique position as a Black professor in the predominantly white environment of Brazilian academia. Santos was, however, often critical of the strategies and approaches of Black movements and always very apprehensive about speaking in the name of Afro-Brazilians. Reflecting on being a Black scholar in Brazil, he once said:

> I believe it is difficult to be Black and it is difficult to be an intellectual in Brazil. These two things, combined, result in a challenging situation, right? It is difficult to be Black because, outside of situations of visibility, daily life is very burdensome for Black people. It is difficult to be an intellectual because it is not part of the Brazilian culture to calmly listen to critical words. (Santos in Tendler, 2006: 5:22)

Santos's provocative nature certainly earned him a number of academic opponents throughout his career. He has been criticised by right-wing scholars for being too radical, and by radical Marxists for being too lenient or not committed enough. Particularly in the last two decades of his career, a rise in academic popularity was followed by more vocal criticism against his person and his work. One common critique relates to his strategy of working with dialectic pairs. Throughout his career, there are many appearances of such pairs: psychosphere and technosphere, hegemonic actors and hegemonised actors, opaque spaces and luminous spaces, territory as a rule and ruled territory, territory as a resource and as a shelter (notions borrowed from Jean Gottmann, 1975), upper circuit and lower circuit, fixes and flows, reason and emotion, horizontalities and verticalities, among others. Despite Santos clearly stating that these dialectic pairs are pedagogical strategies to facilitate the understanding of his analyses, some scholars regard them as proof of Santos's dualistic or excessively structuralist approach.

Santos has also been accused of repeating himself and occasionally contradicting his own statements throughout his many writings. It is questionable, however, that this should be the subject of criticism. His was a work in progress that gradually improved over time. This is clear, for example, when looking at the different ways he conceptualised space in his earlier and later writings. He kept reworking the concept and was still going back to it and re-signifying it in the years before his death. In his last writings, he associated the concept of geographic space to his then new and, for some, controversial concept of used territory, a concept which, unfortunately, he was never able to develop fully. The term first appears in 1994, in the text *The Return of the Territory* (also translated into English: Santos, 2017b) and crops up again in the article *Territory and Local Knowledge: Some Analytical Categories* (published in Portuguese; Santos, 1999), as well as in the manifesto that Santos wrote with some of his students at the University of São Paulo.

The manifesto was distributed as a pamphlet among the attendants of the 2000 XI National Meeting of Geographers in Florianópolis, Brazil; its English translation was published in 2017 by *Antipode* (Bernardes et al., 2017). The controversial text of the manifesto went on to ignite animated discussions. The text takes a bold stance against the reduction of geography to a mere utilitarian discipline to serve the market and eloquently points out what Santos terms 'epistemological indolence' – the lack of interest among geographers in establishing a robust theoretical framework (Melgaço, 2017).

Before delving into the captivating world of Santos's writings, English-speaking readers should be aware of two important issues. Firstly, they are reading the work of someone who passed away over two decades ago, with writings dating back over half a century. It is worth noting that Santos wrote about topics such as globalisation, networks, and the technical-scientific-information milieu but that he died in 2001, before the 9/11 attacks, and before mobile internet and social media became commonplace. Identity discussions were also less prevalent, and terms like 'woke' and 'intersectionality' were not yet common. This poses a challenge for Santos's translators. For instance, he frequently used the term 'man' to refer to individuals or to all of humanity. Secondly, readers must keep in mind that they are reading a translation, which is inherently limited and problematic. While translations are an act of generosity towards the author being translated, they can also be violent, particularly when the author is no longer alive to offer their opinion on the translation of specific terms and passages. For example, Davies (2019) suggests that Santos's term 'técnica' would be better translated as 'technics' rather than 'technique'. Despite his solid argumentation, this remains a matter of debate. Nonetheless, it serves as an example of the choices translators make and the inherent loss in choosing. It is also worth noting that Davies chose to translate Santos's term 'geografia nova' as new geography. However, Santos also used the term new geography (or *nova geografia*, in Portuguese) to refer to the approach of geographers in the 1960s and 1970s who relied on quantitative and computational methodologies, in contrast to what Santos meant by *'geografia nova'*. Translating both 'geografia nova' and 'nova geografia' as 'new geography' loses the nuanced distinction that Santos aimed to convey.

Although there have been controversies surrounding the translation of Santos, they should not discourage readers from discovering his works. In fact, these controversies highlight the depth of Santos's theories and should serve as an invitation for more texts to be translated, and for the existing translations to be revisited and improved. If we are to decolonise geography, it is crucial that more people read, discover, and quote this key thinker from the Global South, whether it be in the original language or through translations.

Santos' Key Works

Bernardes, A., Zerbini, A., Gomes, C., Bicudo, E., Almeida, E. and Betioli Contel, F. ... and de Lima Belo, V. (2017) 'The active role of Geography: A manifesto' (trans. T. Clarke and L. Melgaço), *Antipode,* 49 (4): 952–8.

Santos, M. (1977) 'Society and space: Social formation as theory and method' (trans. S. Slaner), *Antipode,* 9 (1): 3–13.

Santos, M. (2017a) *Toward an Other Globalization: From the Single Thought to Universal Conscience* trans. L. Melgaço and T. Clarke). Berlin: Springer.

Santos, M. (2017b) 'The return of the territory' (trans. by L. Melgaço, T. Clarke and C. Prouse), in L. Melgaço and C. Prouse (eds), *Milton Santos: A Pioneer in Critical Geography from the Global South.* Berlin: Springer. pp. 25–31.

Santos, M. (2021a) *For a New Geography* (trans. A. Davies). Minneapolis, MN: University of Minnesota Press.

Santos, M. (2021b) *The Nature of Space* (trans. B. Balleti). Durham, NC: Duke University Press.

Secondary Sources and References

Costa, E., Zomighani Jr., J. and Boscariol, R. (2022) 'Milton Santos, intelectual negro en la dictadura militar y sus teorías del espacio y urbanización en el Tercer Mundo', *Scripta Nova,* 26 (3): 65–93.

Davies, A. (2019) 'Milton Santos: The conceptual geographer and the philosophy of technics', *Progress in Human Geography,* 43 (3): 584–91.

Ferreira, A. (2007) 'Un penseur, par ses mots: le lexique de Milton Santos', in J. Lévy (ed), *Milton Santos: philosophe du mondial, citoyen du local.* Lausanne: Presses Polytechniques et Universitaires Romandes. pp. 27–38.

Gottmann, J. (1975) 'The evolution of the concept of territory', *Social Science Information sur les Sciences Sociales,* 1 (3/4): 29–47.

Mamigonian, A., Machado, E., Buss, M., and Pereira, R. (1989) 'Entrevista com o Prof. Milton Santos', *Geosul,* 4 (7): 116–47.

Melgaço, L. (2013) 'Security and surveillance in times of globalization: An appraisal of Milton Santos' theory', *International Journal of E-Planning Research,* 2 (4): 1–12.

Melgaço, L. (2017) 'Thinking outside the bubble of the Global North: Introducing Milton Santos and "The Active Role of Geography"', *Antipode,* 49 (4): 946–51.

Santos, M. (1971) *Le Métier du Géographe en Pays Sous-Développés*. Paris: Ophrys.

Santos, M. (1979) *The Shared Space: The Two Circuits of the Urban Economy in Underdeveloped Countries* (trans. C. Gerry). London: Routledge.

Santos, M. (1985) *Espaço e Método*. São Paulo: Nobel.

Santos, M. (1996) 'As cidadanias mutiladas', in J. Lerner (ed), *O Preconceito*. São Paulo: Imprensa Oficial do Estado de São Paulo. pp. 133–44.

Santos, M. (1999) 'O território e o saber local: Algumas categorias de análise', *Cadernos IPPUR*, 2: 7–12.

Santos, M. (2000, May 7) 'Ser negro no Brasil hoje', *Folha de São Paulo*. São Paulo: Caderno Mais.

Santos, M. (2001) 'Curriculum Vitae. *Milton Santos*' http://www.miltonsantos.com.br/site/miltonsantos_curriculum.pdf (accessed November 13, 2023).

Soares, J. (1995) Entrevista Milton Santos, 10 July on the television programme *Jô Soares Onze e Meia*. www.youtube.com/watch?v=jzUIHAAilSM

Souza, M.A. (2009) 'Santos, M.', in R. Kitchin and N. Thrift (eds), *International Encyclopedia of Human Geography*, vol. 10. Amsterdam: Elsevier. pp. 11–14.

Tendler, S. (2006) *Encontro com Milton Santos ou o mundo global visto do lado de cá*. Rio de Janeiro: Caliban Produções, DVD (89 min).

Yázigi, E. (1996) 'Milton Santos e a criatividade', in M.A. de Souza (ed), *O mundo do cidadão, um cidadão do mundo*. São Paulo: Hucitec. pp. 396–425.

48
Saskia Sassen

James Faulconbridge, Lancaster University

Biographical Details and Theoretical Context

Saskia Sassen was born in The Hague, The Netherlands, in 1947 but moved with her parents to Buenos Aires, Argentina shortly after. This move from the Netherlands to Argentina was to be one of many international relocations experienced by Sassen, moving to Italy as a child and then studying as a university student at the Universita degli Studi di Roma, before periods at the Universidad Nacional de Buenos Aires and the Université de Poitiers. She obtained her MA and PhD at the University of Notre Dame in Indiana, US. As she points out, this constant mobility made her somebody who is 'always a foreigner, always at home' (Sassen, 2005: 222) and this transnational perspective continues to influence her work.

One way that her mobility has influenced her work relates to her use of language. Growing up competent in five languages gave her an awareness of particular 'gaps' in a given language compared to another, and vice versa. This awareness projected itself on the larger stage of research and explanation because, for Sassen, it made theorising important and meaningful as part of an attempt to compensate for existing language deficiencies. Another influence of mobility was the experience of acting in ways that were 'out of place' due to a lack of situated knowledge. Sassen found that small actions, particularly when 'out of place', could have an effect on even the most powerful events or processes. This led her eventually to develop the proposition that, under certain conditions, powerlessness can be complex and thereby be a factor in the making of a history and a politics (see Sassen, 2007a, and Gane, 2004 for a more detailed discussion of these points). In addition, Sassen's time at universities in different countries facilitated encounters with a range of philosophies and theoretical perspectives as diverse as Marxist political economy, the work of Berger, Luckman, Kuhn, **Deleuze** and Guattari, **Foucault** and Althuser. This would influence her in a variety of ways, and most importantly would encourage her to develop an approach to research focused on what she sees as strategic empirical domains – that is, conditions that are heuristic in that they produce knowledge about more than themselves: cities, immigration, and nation-states are three such subjects.

In the context of her nomadic early years, it is perhaps unsurprising that migration acted as the focus for the research which led to Sassen's first major publication – *The Mobility of Labor and Capital* (1988 and continuously reprinted since then). In this book, her original contribution is to contest the notion that foreign direct investment will actually develop countries and keep them from becoming labour exporters; her research suggests quite the opposite. In *Guests and Aliens* (2000) Sassen follows up on this work and uses 200 years of intra-European migrations to explore two common notions about migration: a) if richer countries do not control their borders, they will be invaded by poor migrants; and b) when the outsider is from the same religion and race, discrimination does not occur. She finds that Europe's history shows that both of these commonly held notions are wrong.

Sassen is also a prolific writer on the sociology of globalisation, developing her arguments through several key books. In *Globalization and its Discontents* (1999) she covers a number of key issues associated with globalisation: immigration, cities, the feminisation of survival, the nation-state, and networked technologies. In doing this she highlights the often negative consequences for many processes of globalisation. In *A Sociology of Globalization* (2007b) Sassen covers a vast array of scholarship in the social sciences and shows that such work provides significant resources – empirical, methodological and theoretical – for studying the mix of processes, conditions, and subjectivities that are commonly grouped under the term globalisation. *Territory, Authority, Rights* (published originally in 2006 and in an updated form in 2008) is particularly focused on the global spaces that inhabit national institutional and territorial framings. Sassen finds that much of the transformation we call global is better described as a *denationalising* of what was historically constructed as national. It is, however, the books *The Global City* (first published in 1991 and revised in 2001) and *Cities in a World Economy* (first published in 1994 and revised in 2000, 2006, 2012 and 2019) that gave Sassen's work on globalisation most prominence. The books, which set out Sassen's understanding of the role of global cities such as London and New York in the global economy of the late twentieth and early twenty-first centuries, were acclaimed worldwide in academic and public policy circles. Both books offer a complete rethinking of the processes that make cities and define their role in globalisation.

Sassen has also turned her attention to the 'brutality and complexity' that results from the knowledge and technology that underlie globalisation. Her 2014 book *Expulsions* received critical acclaim for its charting of the predatory ways that systems of knowledge expropriate livelihoods, homes, spaces and rights. At the core of Sassen's argument is the idea that contemporary technologies and knowledge create systems that make the tracing of responsibility for expulsions difficult if not impossible, protecting those who benefit and inhibiting resistance by those experiencing brutality. Expulsions are also unpredictable and unforeseen – whether they be individuals from their homes as a result of sub-prime mortgages that years later propagated a global financial crisis; or the disfiguring and destroying of lands and waters as a result of engineering innovations that enabled fracking. Sassen's prowess at unpicking the complexity that underlies expulsions opened up a new approach to unmasking globalisation processes and effects.

The above only begins to summarise Sassen's impressive list of publications. In addition, she has also authored numerous journal articles and has published interviews, book chapters and edited books. She regularly contributes articles to *OpenDemocracy* and *The Huffington Post*. Her publications have been translated into twenty-one different languages.

Sassen has held academic positions at numerous institutions including at Harvard University, where she had her first appointment in the form of a postdoctoral fellowship, and the University of Chicago, Columbia University and the London School of Economics and Political Science. In 2009, Sassen was appointed the Robert S. Lynd Professor of Sociology and co-chairs the Committee on Global Thought at Columbia University in New York. In addition, Sassen has held various positions on editorial boards, has been awarded numerous honorary degrees and has secured herself the position as one of the leading figures in the study of the sociology of globalisation.

Spatial Contributions

It is difficult to capture all of Sassen's contributions to spatial theory but three stand out as being especially important. First, Sassen has provided inspired analysis of the nature, causes and consequences of migration in the global economy in the late twentieth and early twenty-first century, challenging many assumptions about this process. Her work has helped place migration at the centre of debates about globalisation and has shown the impossibility of explaining contemporary geographies of migration without reference to geographies of the global economy. The importance of this theme in Sassen's work is unsurprising considering her biography: in addition to mobility during the early years of her life, Sassen now frequently travels between London and New York and has homes in both cities which she shares with her husband, the renowned sociologist Richard Sennett.

Initially through *The Mobility of Labor and Capital,* which was inspired by the experience of living in multiple countries and witnessing a range of social, economic and political tensions (see Sassen, 2007a), Sassen redirected thinking about labour migration and revealed it to be a phenomenon encouraged by and an integral part of economic globalisation. Developing this idea, *Globalization and its Discontents* and *Guests and Aliens* both highlight how contemporary economic migration, especially to the US in the latter part of the twentieth century, mirrored processes that had been experienced worldwide and, as such, was not a new phenomenon but a particular manifestation of the leading role of the US in the global economy. Sassen also uses this work on migration to highlight the inequality and hardship associated with many forms of migration. Those fulfilling demand for maids, nannies and cleaners, entering the informal economy such as in clothing sweatshops or becoming part of the underground economy such as the sex industry in major cities have been shown to be the 'losers' of global capitalism, being exploited and suffering inequality, particularly in the context of migration to global cities (see below). Her participation in the accompanying documentary to the film *Children of Men,* directed by Alfonso Cuarón (2006), is just one example of how Sassen has conveyed these arguments to the public using diverse media.

Secondly, in relation to the theorisation of globalisation, Sassen's work comprehensively documents the value of studying not the outcomes of globalisation (e.g., transnational corporations) but the underlying infrastructures and practices that make up the global economy. As she puts it:

> A key proposition that has long guided my research is that we cannot understand the x – in this case globalization – by confining our study to the characteristics of x – i.e., global processes and

> institutions. This type of confinement is a kind of endogeneity trap, one all too common in the social sciences and spectacularly so in the globalization literature [...] There are consequences to a type of analytics that posits that an explanation of x needs to be configured in terms of non-x. For one, it demands a focus on the work that produced the new condition – in this case, globalization. (Sassen, 2006: 4)

By focusing on the non-x – i.e., the institutions, technologies and infrastructures that evolved during the twentieth century to produce globalisation – Sassen has been able to engage both in debates about the distinctiveness of globalisation in the late twentieth and early twenty-first century (see Held et al., 1999, on such debates) and about the spatial architecture of the global economy (see Castells, 2000, in particular). This has led her to highlight the way that the global and national are interdependent and, therefore, that globalisation as a set of processes interacts with the national scale in more ways than many have recognised. It also underlies the arguments in her book *Expulsions* about the importance of deciphering the unexpected, often hidden yet powerful effects of the knowledge and technology that underlie contemporary global economic practices.

Sassen's third major contribution has been to develop ideas about the importance of place and local and national scales in globalisation through theorisation of the role of cities in globalisation. Through careful analysis of cities including London, New York and Tokyo, the books *Cities in a World Economy* and *The Global City* show that cities – as particular types of place – have actually become more important in recent global times. As she puts it:

> Overlooking the spatial dimension of economic globalization and overemphasizing the virtual information dimensions have distorted the role played by major cities in the current phase of economic globalization. (Sassen, 2019: 3)

In particular, Sassen argues that cities are strategic places for the making of the organisational instruments needed to redeploy economic activities across the globe. The instruments produced in cities are both informational and directly related to global finance but are also political and infrastructural, ranging from technologies to regulations. In particular, Sassen reveals through her work on global cities that the deep economic histories of complex cities have produced the key components of the so-called knowledge economy. This also explains why as corporate economic globalisation expanded in the 1990s and onwards, the number of major and minor global cities continued to grow. Sassen's work on world cities has, therefore, provided the foundations for further investigations of the uneven geographies of the global service economy in the same way that **Doreen Massey** and others had earlier laid the foundations for work on the geography of global manufacturing (see Massey, 1995). In addition, her work has made her a leading member of public policy groups such as the Mastercard Worldwide Centers of Commerce programme and more recently the Committee on Global Thought at Columbia University.

Key Advances and Controversies

Sassen's aim in her work has been to unsettle existing modes of thinking and theorising.

As a result, each of the three major contributions discussed above have advanced theory and understanding of globalisation and set new agendas for research. In particular her theorisation of the sociology of globalisation has redirected thinking by highlighting the importance of a range of *practices* that constitute the global – ranging from the construction of monetary and fiscal policies to the exercising of political imaginaries and production of international law. As a result she has been able to highlight that:

> When the social sciences focus on globalization – still rare enough deep in the academy – it is typically not on these types of practices and dynamics but rather on the self-evidently global scale. And although the social sciences have made important contributions to the study of this self-evidently global scale [...] there is much work left. At least some of this work entails distinguishing (a) the various scales that global processes constitute, ranging from supranational and global to sub-national, and (b) the specific contents and institutional locations of the multi-scalar globalization. (Sassen, 2007a: 91)

Through her work, and especially through her books *Territory, Authority, Rights*, and *Expulsions*, she has, therefore, forced scholars of globalisation to take seriously the ways in which the global is constructed through multiple institutions, technologies and infrastructures rather than existing as an independent, singular process. In doing this Sassen has argued that the national scale is far more important in globalisation than many have realised, in particular because globalisation is actually a process of denationalisation in which the apparatus constructed to support the role of the nation-state in the past is reconfigured to support global institutions and frameworks in the present. This has informed her arguments about territory that appear across multiple works and will inform her next book, *When Territory Exits Existing Frameworks*.

In addition to advancing theoretical debates about the process of globalisation, Sassen has also forced a rethinking of the way migration and its role in the global economy is theorised. *The Mobility of Labor and Capital* countered claims that growing levels of foreign direct investment into developing countries would prevent worker out-migration and showed that investment actually drives emigration rather than encouraging citizens to stay and work in the newly established branch plants. By highlighting what at the time was a relatively novel concept – *transnational space* as created by foreign direct investment – Sassen showed how the US had become connected to developing countries in new ways, not least because of the presence of US firms overseas and the growing awareness of the ideologies and supposed ideals of US life this created. When coupled with increasing demand for low-paid labour in the US, this spurred flows of migrant workers. By developing these points in her subsequent work, Sassen has shown how migration is intimately tied to processes of globalisation, not just in the form of the transnational elites conceptualised by ***Manuel Castells*** and Sklair, but also in the form of poorly paid labourers who flock to global cities. She has shown that these workers are an integral part of the global economy (what Sassen, 2019, calls the 'other workers in the advanced corporate economy') and, as a result, it has become clear that low-skilled as well as high-skilled migration has to be central in any theorisation of the 'human side' of the global economy.

Finally, Sassen's work on global cities has emphasised the fact that the information economy is *made*, that it has a production process, and hence needs a range of specific places and diverse types of workers, from experts to cleaners. This suggestion that the informational economy is produced, contextually, is in line with her suggestion that the national and global are not mutually exclusive and allows her to highlight how production and innovation in global cities produces the technologies, infrastructures and regulations that constitute globalisation. As a result, Sassen has shown that cities need to be conceptualised as sites of global economic productivity as well as sites of struggle, thus making them a crucible for economic, social and political processes in the context of contemporary globalisation.

As a leading light in the study of globalisation, Sassen has both laid the groundwork for and developed work by a range of other leading scholars of space and place, including Manuel Castells and ***Immanuel Wallerstein,*** and in particular work on cities and the urban, including **Jennifer Robinson's** work on comparative urbanism and ***Peter Taylor's*** work on globalisation and world cities (as manifest in the Globalization and World Cities network and project). At times she has disagreed with some of her fellow theorists. For example, Rantanen (2006) notes that for a long time Manuel Castells disagreed with one of her first published articles where she argued that low-paid migrants are an integral part of the global economy. Likewise, Sassen prefers to use the term global city when referring to the production capitals of the global economy, in line with her belief that a fundamental reconstitution of the economy has occurred as part of globalisation in the late twentieth and early twenty-first century, while others (e.g., Taylor, 2003) prefer the term world city to reflect a subtly different conceptualisation based on world systems analyses. Nonetheless, such differences are overshadowed by Sassen's achievement in gaining widespread recognition for the idea that globalisation is a socio-technical achievement, with the socio rather than the technical being emphasised. As a result, she has been at the heart of successful attempts to counter hyperbolic assertions of globalisation as the 'end of geography' (O'Brien, 1992) and the emergence of a 'borderless world' (Ohmae, 1992) and has promoted a social analysis of technology and globalisation which parallels the work of Karin Knorr-Cetina (see, for example, Knorr-Cetina and Bruegger, 2002) but which is unique and distinctive because of its emphasis on the importance of social place in the production of globally networked economic space.

Sassen's Key Works

Sassen, S. (1988) *The Mobility of Labor and Capital: A Study in International Investment and Labor Flow.* Cambridge: Cambridge University Press.

Sassen, S. (1991) *The Global City: New York, London, Tokyo.* Princeton, NJ: Princeton University Press (2nd edition, 2001)

Sassen, S. (1996) *Losing Control? Sovereignty in an Age of Globalization.* New York, NY: Columbia University Press.

Sassen, S. (1999) *Globalization and its Discontents.* New York, NY: The New Press.

Sassen, S. (2000) *Guests and Aliens.* New York, NY: New Press.

Sassen, S. (2005) 'Always a foreigner, always at home', in A. Sica and S. Turner (eds), *The Disobedient Generation: Social Theorists in the Sixties*. Chicago, IL: University of Chicago Press. pp. 221–51.

Sassen, S. (2006) *Territory, Authority, Rights: From Medieval to Global Assemblages*. Princeton, NJ: Princeton University Press (2nd edition, 2008).

Sassen, S. (2007a) 'Going digging in the shadow of master categories', in M. Deflem (ed), *Sociologists in a Global Age: Biographical Perspectives*. Aldershot: Ashgate. pp. 85–98.

Sassen, S. (2007b) *A Sociology of Globalization*. New York, NY: W.W. Norton.

Sassen, S. (2014) *Expulsions: Brutality and Complexity in the Global Economy*. Cambridge, MA: Harvard University Press.

Sassen, S. (2019) *Cities in a World Economy* (5th edition). London: Sage.

Secondary Sources and References

Castells, M. (2000) *The Rise of the Network Society*. Malden, MA: Blackwell.

Cuarón, A. (dir.) (2006) *Children of Men* [film]. Strike Entertainment/Hit and Run Productions.

Gane, N. (2004) 'Saskia Sassen: Space and power', in *The Future of Social Theory*. London: Continuum. pp. 125–42.

Held, D., McGrew, A., Goldblatt, D. and Perraton, J. (1999) *Global Transformations: Politics, Economics and Culture*. Cambridge: Polity.

Knorr-Cetina, K. and Bruegger, U. (2002) 'Global microstructures: The virtual societies of financial markets', *American Journal of Sociology*, 107 (4): 905–50.

Massey, D. (1995) *Spatial Divisions of Labour: Social Structures and the Geography of Production*. Basingstoke: Macmillan.

O'Brien, R. (1992) *Global Financial Integration: The End of Geography*. London: Royal Institute of International Affairs.

Ohmae, K. (1992) *The Borderless World: Power and Strategy in the Interlinked Economy*. New York, NY: Harper Business.

Rantanen, T. (2006) 'Digging the global mine: An interview with Saskia Sassen', *Global Media and Communication*, 2 (2): 142–159.

Sassen, S. (forthcoming) *When Territory Exits Existing Frameworks*. Cambridge, MA: Harvard University Press.

Taylor, P.J. (2003) *World City Network: A Global Urban Analysis*. London: Routledge.

49
Amartya Sen

Simon P.J. Batterbury, University of Melbourne
Jude L. Fernando, Clark University

Biographical Details and Theoretical Context

Amartya Kumar Sen is an intellectual of global stature. He was born in 1933 and grew up in Dhaka (then India, now Bangladesh) in a middle-class academic family, and in Santiniketan (in present-day West Bengal), attending a school founded by the Bengali literary figure, Tagore (Sen, 2021). He has remained in academic institutions his entire life. He first attended the elite Presidency College in Calcutta, graduating rapidly with a degree in economics (during which time he also survived self-diagnosed throat cancer), and moved to England in 1953 to read economics at Trinity College, University of Cambridge. After more undergraduate work he then completed a PhD in one year (Sen, 1960), before returning to India to a professorship at the new Jadavpur University, Calcutta at the age of just 22.

He has held positions in economics, philosophy or both, at the University of Delhi from 1963 to 1971, then the London School of Economics (1971–1977), Oxford (1977–1987), and Harvard (1987–1998). He was then Master of Trinity College, Cambridge (1998–2003, the first Asian to chair an Oxbridge College) before returning to Harvard in 2004. From 2012–2015 he was Chancellor of the newly established Nalanda University in Bihar, but his comments about the policies of the ruling Bharatiya Janata Party, which he believed was prejudicial to unity and well-being, led him to withdraw. Among over 90 honorary doctorates and 25 major accolades, Sen was awarded the Senator Giovanni Agnelli International Prize in Ethics, the Bharat Ratna (the President of India's highest honour), the National Humanities Medal from President Obama in 2012, and in 1998 The Nobel Prize for Economics. The latter was overdue, perhaps because Sen was never 'captured by economics imperialism' (Fine, 2001: 13). He has repeatedly criticised some of the core tenets of neoclassical economics, and published widely in other fields, without forsaking the discipline.

Over the decades he has contributed to social choice theory (a technical field in economics) and welfare economics, the understanding and measurement of poverty, explanations of famine and hunger, agrarian change and rural development issues, as well as engaging with ethics and political philosophy and co-writing a sweeping history of India (Drèze and Sen, 2013). His empirical investigations involve wide-ranging international comparisons, but with a primary focus on South Asia. Perhaps the central theme that links his writings – which span over 60 years and 30 books – has been his passionate concern to redress economic and social inequality and to improve human wellbeing.

In his early years Sen was preoccupied with welfare economics and made important contributions to the formalistic (mathematical) expression of social choices (Sen, 2018 [1970]). His early technical work resulted in significant challenges to Kenneth Arrow's 'impossibility theorem', specifically the notion that individuals act *purely* according to self-interest or utilitarian criteria when making welfare or income decisions (Sen, 1977). Insisting that non-utility concerns like equity, class position, and family influence matter, Sen developed an interpersonal comparison of utility (the value of a commodity achieved in consumption), beyond the established utilitarian technique of 'revealed preferences'. Social choice theory continued to figure in Sen's work, particularly when assessing how aggregative judgements may be arrived at, given the 'diversity of preferences, concerns, and predicaments of the different individuals within society' (Sen, 1998: n.p.).

Sen progressed his arguments into considerations of the ethical foundations of justice and development. He became interested in human 'capability', defining it as 'a set of functioning bundles representing the various alternative "beings and doings" that a person can achieve with his or her economic, social, and personal characteristics' (Drèze and Sen, 1989: 18; Sen, 1985). Capability is 'tantamount to the freedom of a person to lead one kind of life rather than another', in an Aristotelian sense (Nussbaum and Sen, 1993: 2). It follows, therefore, that an individual's capability can be *repressed* (for example, by denying a person access to basic services including food, education, land, freedom of expression, or health care), or *enabled* (for example, through supportive actions by other individuals and institutions including local communities, or even a 'developmentalist' state).

Development as Freedom (Sen, 1999) followed from influential work on inequality and welfare (Sen, 1992a; 1992b). In the book, wellbeing was linked to the enhancement of capabilities, and the core of development itself is the enjoyment of freedom – not just freedom of speech in a narrow sense, but freedom (positive and negative) for individuals to lead valuable lives: 'development consists of the removal of various types of unfreedoms that leave people with little choice and little opportunity of exercising their reasoned agency' (Sen, 1999: xii). For Sen, freedom is not just important in and of itself; it is also *instrumental* since it enables people to attain desired ends, and *constitutive* in the making of collective moral decisions through dialogue (Gasper, 2000). Democracy has the greatest capacity to expand basic freedoms, and expanding human capabilities enhances freedom. Sen stresses the importance of recognising cultural values in this process (Sen, 2004; 2009).

In the 1990s Sen applied these ideas in improvements to the indicators used to measure poverty rates and human development (Sen, 1992b; 1997). The Human Development Index (HDI) and the United Nations Development Programme's *Human Development Report* provided an alternative to the neoliberal 'Washington Consensus' on income-related poverty measures. He and researchers at the UNDP argued that development priorities should be geared towards improving human development (capabilities) – assessed through wide-ranging multivariate indices – rather than through growth-centred economic policy. He also argued quite strongly for the extension of freedom and independence to women and children in developing countries, which he said has demonstrable positive effects on wellbeing, poverty alleviation, and mortality (Agarwal et al., 2005). He regarded the neglect of women's nutrition and health (not least among poor African-Americans in the US), and sex-selective abortion in developing countries, as criminal (Sen, 1990; 1999).

Despite these applications of his work, Sen has largely abstained from collective protest or activism, although he was involved in left-wing Bengali politics as a student (Barsamian, 2011 [2001]; Sen, 2021). He was always cautious about giving policy advice directly, but he used funds from his Nobel to set up the Pratichi Trust, concerned with literacy and schooling in India and Bangladesh (issues explored in Drèze and Sen, 2002). In Indian affairs he has remained a strong supporter of gender equality, pluralism, and pro-poor policies; he also opposes the excesses of Hindu nationalism. *The Argumentative Indian* (2005) highlights positive features of Indian traditional culture. Secularism is a part of Indian cultural roots, he argues. In the book he reflects on the importance of 'Voice and Heterodoxy', and 'Reason and Identity' in India among other sweeping themes, foregoing narrower and ahistorical characterisation based on religion and culture. The cosmopolitan, democratic and secular qualities of Indian culture have very deep roots, according to Sen.

Equality and justice, and the rights of the poor not to suffer 'unfreedoms', are abiding concepts in his work (Barsamian, 2011: n.p.). *The Idea of Justice* (2009) cross-references most of the major arguments. It is a major work of political philosophy, which controversially dismisses a search for instrumental social justice delivered through strong institutions as John Rawls proposed, in favour of reasoned, pluralist, and comparative conceptions of what is 'right', focusing on delivering acceptable outcomes for how well we live and ought to live. The reader is taken full circle back to his early work on social choice; justice is a function of its definition and expression, and simple criteria and rules can never define it. The book does not offer a theory of justice, making a distinction between freedom and unfreedom difficult, especially where inequalities and injustices co-exist with ethno-religious nationalism.

Spatial Contributions

Geographers have debated and applied Sen's pleas for the realisation of equality and justice, although Sen has never contributed directly to the discipline. Much of his work does involve inter-regional and inter-state comparisons (e.g., Drèze and Sen, 2002) and makes detailed reference to places and historical events. Sen's cosmopolitan positionality and belief in universal rights leads him to argue that culture transcends place, rather than, following **Arturo Escobar**'s pluriversal concepts, that culture 'resides in places'. Again distancing himself from the pluriverse, he says 'culture' is not impermeable to rational considerations and rational choices (Sen, 2004).

Sen's work on poverty and famine has attracted great interest. He witnessed the direct effects of the 1943–5 famine in Bengal. The central argument of his acclaimed *Poverty and Famines* (Sen, 1981; see also Devereux, 2001) was that famine is not caused by a negative Malthusian relationship between population and food supply, but by the inability of famine-prone individuals to *access* food in times of great need, even when food supplies were adequate and the rich were still eating well. The implication is that famine is often a food *demand* problem, not a *supply* problem. 'Endowments' are the assets and resources that people may theoretically access. 'Entitlements' are those that are available, and are therefore cognate with 'acquisition power'. 'It is the totality of entitlement relations that governs whether a person will have the ability to acquire enough food to avoid starvation' (Sen, 1981: 155). To survive a famine, individuals need to grow food (production-based entitlements), buy it (trade-based entitlements), sell their labour (own labour entitlement), or receive food from others (inheritance and transfer entitlement). There are also 'extended entitlements' (including looting – Drèze and Sen, 1989). *Poverty and Famines* was highly influential, and his 'food entitlement decline' (FED) explanation was extensively elaborated, empirically and theoretically, with the gifted Indian economist Jean Drèze (Drèze and Sen, 1989; 1990). They argue India's avoidance of full-blown famine since 1947 has been due to voting, a moderately free press, and food distribution and work programmes. These ideas have been influential among geographers and anthropologists in South Asia and beyond, influencing theories of risk (Bohle, 1993; Blaikie et al., 2003), and underlying critique of colonial food policy and governance (Watts, 1983). Endowments and risks also feature in sustainable livelihood research (Bebbington, 1999). 'Livelihoods thinking' shows how different entitlements are juggled by individuals and households, guiding work in several international development agencies. Sen's theories have also informed famine early-warning systems that scrutinise price signals in markets for signs of impending food stress. Famine relief today targets access to food, as well as basic food relief.

Key Advances and Controversies

Sen's work is particularly challenging to critical scholars because it is difficult to stereotype. He exhibits 'cautious boldness', raising issues dear to the critical left, while retaining faith in some aspects of free market economics (Gasper, 2000). The radical implications that follow from his support for human emancipation, equality, justice and poverty alleviation have been taken up by others, although less so over the last decade. Calls to extend his frameworks were, for a long time, more common than pleas to reject them (Desai, 2001; Fine, 2001; Alkire, 2002). For example, Ben Fine (a former student) found an 'unresolved tension between micro-foundations of entitlement analytic and the broader recognition of famine as irreducibly macro, not least because famine is more than the sum of its individual parts' (Fine, 2001: 8). Sen has of course been aware of political-economic forces, pointing out that 'famine is dependent upon 'the exercise of power and authority [...] alienation of the rulers from those who ruled [...] the social and political distance between the governors and the governed' (Sen, 1999: 170) but Fine suggested Sen's understanding of power and structures is 'superimposed, not built, upon the micro-foundations'.

Geographer ***Michael Watts***, who provided the most detailed Marxist argument that famine in West Africa had, at its roots, colonial origins (1983), was broadly sympathetic to Sen. He argued that if poverty (and famine) is a result of 'capability failure' and a lack of entitlements, the prime response must be to overcome vulnerability and to address demand for food. But processes of oppression can lie at the roots of famine, and the development of 'critical autonomy' and a strong 'sense of society' is needed to enhance freedoms (Watts, 2000: 57). Following Marx, Michael Watts suggests entitlements are 'both constituted and reproduced through conflict, negotiation and struggle' (Watts, 2000: 62).

The instrumentality of 'complex emergencies' and 'war famines', where 'freedom to choose' is clearly constrained by the powerful, were not always recognised by Sen. Famine can be used as a political weapon. Armed conflict and violence in famine situations serve certain elite interests (De Waal, 1989; Keen, 1994; Middleton and O'Keefe, 1998). Other important criticisms include the fact that many famine deaths occur through ill health, not starvation (De Waal, 1989), and that endowments – like communal livestock herds, or common property – lack clear property rights, making them 'fuzzy' and thus hard to catalogue (Devereux, 2001: 253).

Only in the last few years has Sen glanced toward a postcolonial perspective, exploring Indian historical failures and linking unfreedoms in some degree to the neoliberal capitalist system (Drèze and Sen, 2013). But he is wary of tackling European colonisation head on, or its responsibility for underdevelopment, dependency or violence (Comiling and Sanchez, 2014: 9). His work has never supported Gandhi's belief in local self-sufficiency and modern technological scepticism, and he saw an 'ugly side' to many forms of localist, communitarian, and sectarian politics.

Here, his support for civil society movements is qualified by a strong faith in reason: 'One's concern for equity and justice in the world must not carry one into the alien territory of unreasoned belief' (Barsamian, 2011: n.p.). For Sen, economic globalisation can tackle poverty and inequality, even though it 'doesn't always work' (Barsamian, 2011: n.p.)and should not be reduced to the unfettered expansion of trade and markets. His point here is that globalisation is nothing new, and that the poor should be able to share in its gains. Corbridge argued that 'Sen's liberalism leaves him poorly equipped to deal with questions of entrenched power and the politics of conflict and social mobilization' (2002: 203), and hence his failure (or refusal) to elaborate a strong account of the political economy of capitalism. This is surprising given the political and ethically charged issues that Sen has addressed, and the fact that he is well aware of the debates (Fine, 2001: 12; Barsamian, 2011: n.p.; Sen, 2003; 2006). In Sen's work there are few links made between the production of 'opulence' resulting from growth and from wealth, and the reduction of entitlement for the poor (Cameron and Gasper, 2000: 986). Other critics call for a richer conception of 'self' that sees human agents as much more than beings seeking to attain capabilities (Giri, 2000: 1018). Selfhood and care, nor spirituality, are not always a full part of Sen's analysis despite his call to insert value into welfare analysis and in his accounts of social conflict (Alkire, 2002; Sen, 2006).

There is also a certain level of ambiguity in Sen's notion of 'development as freedom'. Freedom is, more often than not, accompanied by its foreclosure. Sen is less than clear about tackling the conflicts that result from people's pursuit of freedom: development seems to be about opening opportunity and 'choice', to which there must surely be limitations (Gasper, 2000: 999, Chandler, 2014), particularly in a world far more constrained by limits to consumption, and economic, political, and environmental crises today than when he began his writing in the 1960s. Carbon emissions reductions will surely limit the forms of development he envisages, but were not a feature of debates during much of his professional life.

In *The Argumentative Indian* Sen contends that secularism necessitates the separation of religion and state, through 'equidistance' (2005). This is because 'as the state must deal with different religions and members of different religious communities, there must be a basic symmetry of treatment', with 'no violation of secularism for a state to protect everyone's right to worship as he or she chooses, even though in doing this, the state has to work with – and for – religious community' (Sen, 2005: 296). The difficulty is that religion continues to be significant to people, and religious fundamentalist politics are rising. Hinduism seen as a rational and peaceful faith, without central leadership, has been overtaken by state-supported religious nationalism that also deploys neoliberal economic policies. Sen's vision is threatened.

Lastly, how exactly are freedoms achieved? By what political means? How may the poor marshal power to challenge vested interests other than through votes? The 2022 People's Assemblies in Sri Lanka were one concerted action. Does Sen support such protest? Critics see difficulties and ambiguities here, and a failure to recognise the vitality of 'concerted struggles against the powers of vested interests, at all spatial scales' (Corbridge, 2002: 209). As Chandler (2014: 78) says, 'Sen's framing takes the understanding of socio-economic and political processes out of the framing of liberal modernity'. In *The Idea of Justice* (2009) the *quality* of governance, and the opportunities it affords for rational debate, was proposed as most important for equitable development and human wellbeing. In *Identity and Violence* he rejected simplistic explanations for violence like Samuel Huntingdon's *Clash of Civilizations* (1996) for more complex and multifaceted ones, rejecting the reductionist idea that inequality and poverty always generate violent conflict (Sen, 2006). In these and other works, it is clear that Sen extends but does not overturn neoclassical economics (Fine, 2001).

Amartya Sen has been a beacon of common sense in the interdisciplinary terrain he occupies. His work on entitlements and capability stress human agency rather more than constraint, and they carry 'some sense of worth and of real people's lives' (Gasper, 2000: 996). His insistence on understanding and resolving inequality, tackling justice and expanding basic freedoms has placed many mainstream economists and utilitarians in a heavily compromised position. As he approaches his ninth decade, social scientists in general should be grateful that this important ambassador for 'humane economics' (Desai, 2001) has reached the ears of policymakers and publics worldwide, whether directly or indirectly, with his beguiling ideas.

Sen's Key Works

Drèze, J. and Sen, A.K. (1989) *Hunger and Public Action*. Oxford: Clarendon.

Drèze, J. and Sen, A.K. (2013) *An Uncertain Glory: India and its Contradictions*. Princeton, NJ: Princeton University Press.

Nussbaum, M. and Sen, A.K. (1993) *The Quality of Life*. Oxford: Clarendon.

Sen, A.K. (1981) *Poverty and Famines: An Essay on Entitlements and Deprivation*. Oxford: Clarendon Press.

Sen, A.K. (1997 [1973]) *On Economic Inequality*. An Expanded Edition with a Substantial Annexe by James Foster and Amartya Sen. Oxford: Clarendon.

Sen, A.K. (1999) *Development as Freedom*. Oxford: Oxford University Press.

Sen, A.K. (2005) *The Argumentative Indian*. London: Penguin.

Sen, A.K. (2006) *Identity and Violence: The Illusion of Destiny*. London: Penguin.

Sen, A.K. (2009) *The Idea of Justice.* Cambridge, MA: Harvard University Press.

Sen, A.K. (2018 [1970]) *Collective Choice and Social Welfare: An Expanded Edition.* Cambridge, MA: Harvard University Press.

Secondary Sources and References

Agarwal, B., Humphries, J. and Robeyns, I. (eds) (2005) *Amartya Sen's Work and Ideas: A Gender Perspective.* London: Routledge.

Alkire, S. (2002) *Valuing Freedoms: Sen's Capability Approach and Poverty Reduction.* Oxford: Oxford University Press.

Barsamian, D. (2011 [2001]) 'Amartya Sen. Interview with Sen by David Barsamian'. *The Progressive Magazine,* April 2001 issue. https://progressive.org/latest/amartya-sen-barsamian-110929/ (accessed November 13, 2023).

Bebbington, A.J. (1999) 'Capitals and capabilities: A framework for analyzing peasant viability, rural livelihoods and poverty', *World Development,* 27: 2021–44.

Blaikie P. M., Cannon, T., Davis, I. and Wisner, B. (eds) (2003) *At Risk: Natural Hazards, People's Vulnerability and Disasters.* London: Routledge.

Bohle, H-G (ed) (1993) *Worlds of Pain and Hunger: Geographical Perspectives on Disaster Vulnerability and Food Security.* Saarbrücken: Breitenbach.

Cameron, J. and Gasper, D. (2000) 'Introduction: Assessing and extending the work of Amartya Sen', *Journal of International Development,* 12 (7): 985–8.

Chandler, D. (2014) 'Where is the human in human-centred approaches to development? A critique of Amartya Sen's "Development as Freedom"', in S. Mezzadra, J. Reid and R. Samaddar (eds), *The Biopolitics of Development: Reading Michel Foucault in the Postcolonial Present.* New Delhi: Springer. pp. 67–86.

Comiling, R. and Sanchez, J.M.O. (2014) 'A postcolonial critique of Amartya Sen's capability framework', *Perspective in the Arts & Humanities Asia,* 4 (1): 1–26.

Corbridge S.E. (2002) 'Development as freedom: The spaces of Amartya Sen', *Progress in Development Studies,* 2: 183–217.

Desai, M. (2001) 'Amartya Sen's contribution to development economics', *Oxford Development Studies,* 29 (3): 213–23.

Devereux, S. (2001) 'Sen's entitlement approach: Critiques and counter-critiques', *Oxford Development Studies, 29* (3): 244–63.

Drèze, J. and Sen, A.K. (eds) (1990) *The Political Economy of Hunger* (3 volumes). Oxford: Clarendon Press.

Drèze, J. and Sen, A.K. (2002) *India: Development and Participation.* Oxford: Oxford University Press.

Fine, B. (2001) 'Amartya Sen: A partial and personal appreciation'. *CDPR Discussion Paper 1601.* London: Centre for Development and Policy Research. SOAS. https://eprints.soas.ac.uk/2338/1/Amartya%20Sen-%20A%20Partial%20and%20Personal%20Appreciation.pdf (accessed November 13, 2023).

Gasper, D. (2000) 'Development as freedom: Taking economics beyond commodities – the cautious boldness of Amartya Sen', *Journal of International Development,* 12: 989–1001.

Giri, A.K. (2000) 'Rethinking human well-being: A dialogue with Amartya Sen', *Journal of International Development,* 12: 1003–18.

Huntington, S.P. (1996) *The Clash of Civilizations and the Remaking of World Order.* New York, NY: Simon and Schuster.

Keen, D. (1994) *The Benefits of Famine. A Political Economy of Famine and Relief in*

Southwestern Sudan, 1983–1989. Princeton, NJ: Princeton University Press.

Middleton, N. and O'Keefe, P. (1998) *Disaster and Development: The Politics of Humanitarian Aid*. London: Pluto.

Sen, A.K. (1960) *Choice of Techniques*. Oxford: Blackwell.

Sen, A.K. (1977) 'Rational fools: A critique of the behavioral foundations of economic theory', *Philosophy and Public Affairs*, 6 (4): 317–44.

Sen, A.K. (1985) *Commodities and Capabilities*. Amsterdam: North-Holland.

Sen, A.K. (1990) 'More than 100 million women are missing', *New York Review of Books*, December 20: 61–6. https://www.nybooks.com/articles/1990/12/20/more-than-100-million-women-are-missing/ (accessed November 13, 2023).

Sen, A.K. (1992a) *Choice, Welfare and Measurement*. Oxford: Blackwell.

Sen, A.K. (1992b) *Inequality Re-examined*. Oxford: Oxford University Press.

Sen, A.K. (1998) 'Autobiography', *Swedish Academy of Sciences Nobel Prize website*. (Published in *Les Prix Nobel*, 1999) https://www.nobelprize.org/prizes/economic-sciences/1998/sen/biographical/ (accessed November 13, 2023).

Sen, A.K. (2003) *Rationality and Freedom*. Cambridge, MA: Harvard University Press.

Sen, A.K. (2004) 'How does culture matter?', in V. Rao and M. Walton (eds), *Culture and Public Action*. Stanford, CA: Stanford University Press. pp. 38–50.

Sen, A.K. (2021) *Home in the World: A Memoir*. London: Penguin.

de Waal, A. (1989) *Famine That Kills: Darfur, Sudan, 1984–1985*. Oxford: Clarendon.

Watts, M.J. (1983) *Silent Violence: Food, Famine and Peasantry in Northern Nigeria*. Berkeley, CA: University of California Press.

Watts, M.J. (2000) 'Struggles over geography: Violence, freedom and development at the millennium', *Hettner Lectures No. 3*. Heidelberg: Department of Geography, University of Heidelberg.

50
AbdouMaliq Simone

Suzanne Hall, London School of Economics

Biographical Details and Theoretical Context

Improvisations, inventions, fall backs and the capricious life of everyday commitments and innate precarities are constitutive of the ever-evolving urban curatorial practice of AbdouMaliq Simone. To read Simone's work, to hear his talks, is to embrace a coalescence of images, vocabularies and oscillations, cognisant of but never curtailed by conventions. It would be contradictory to locate Simone's work in anything as definitive as a 'school' or even movement, and equally, it would be a misrepresentation of his embodied and varied intellectual journey to represent this in the all-too-linear constraints of a formal biography. Rather, Simone's work presents an opportunity to think with a kind of solidarity that emerges from attuned ethico-political affinities and shared experiments that necessarily transcend disciplines, geographies and forms. Such practices of solidarity sustain lively ways of understanding the immensely complex and multifarious conditions of urban life, captured in Simone's simultaneously elemental and erudite concern with 'how stuff gets done' (as he describes it on his personal website, http://abdoumaliqsimone.com/).

Born in the US, AbdouMaliq Simone spent much of his childhood in Freetown, Sierra Leone, recalling a situation which was:

> fairly folded in a parochial community, a network of people who didn't have a lot of money. They couldn't spread themselves out into the surrounding urban area. So there was this kind of mixed existence; being part of a complicated neighborhood, but also being apart and withdrawn from the city and folded in a network of schools and institutions, a kind of expatriate community that didn't have a lot of money and had precarious status. In some ways this is how I grew up and later on in late adolescence, early adulthood, I escaped back to West Africa. It was at that moment that I revisited these memories and began to have an interest in African cities. (cited in Burga, 2009: n.p.)

Working in the 1970s and 1980s as an activist in communities in Accra, Abidjan and Khartoum, whilst taking University teaching jobs, he was to obtain his PhD (in social psychology) from the Wright Institute, California in 1979, moving subsequently to positions in Chicago, Göttingen and Sheffield (where he was appointed Senior Professorial

Fellow in 2018), at the same time as maintaining connections in Bangkok, Jakarta, London, and Turin.

Simone's work hence emerges from a vital range of conversations and collaborations across continents, as much as his inhabitation of varied working-class districts within a multifarious 'Urban South' (Simone, 2018). Such located experiences are generative of a visceral culture of knowledge as much as a contestation of an over-generalised 'Global South' that eludes planetary interdependencies, international circuits of extraction and cultural specificity. Similarly, Simone's work pushes against the categoric periphery described via normative tropes of 'urbicide' (i.e., systemic urban destruction) or expectations of assimilation to 'global' norms. His epistemic imagination turns to the pulse of life and place shaped inside and against the durable structures of harm driven by hierarchy, speculation and extraction:

> What if the uninhabitable enabled a kind of thinking that challenged or refused what it means to viably inhabit a place? What if it was a method to more fully understand the *rhythms of endurance*, the surges of life that carry bodies forward and back between destinations that are altered in each approach, each retreat? (Simone, 2018: 10; original emphasis)

The underlying fascination with rhythms of endurance is evident in his long-lived collaborations with Edgar Pieterse and the extended family at the African Centre for Cities at the University of Cape Town in South Africa. Their intellectual riff has generated provocations on city-making and the possibilities of political action lodged in the intricacies of the everyday (see Pieterse and Simone, 2013; Simone and Pieterse, 2017). This collaborative impulse is also sustained in Simone's close engagement with the Rujak Center for Urban Studies in Jakarta and his work on the plural formations of resourcefulness in the context of the mega city (see *Jakarta, Drawing the City Near*, 2014, *City Life from Jakarta to Dakar*, 2010 and his own website http://abdoumaliqsimone.com/). The practice of thinking through collaboration is extended in invited lecture series and close intellectual connections, as with Michele Lancione and the Beyond Inhabitation collective (Lancione and Simone, 2021), and the Urban Institute at Sheffield where Simone participated in the Sheffield Urbanism Lecture Series, which was generative of his monograph *The Surrounds: Urban Life Within and Beyond Capture* (2022). The commitment of conversing together is echoed in Ash Amin's curation of Simone's Smuts Memorial lectures on Postcolonialism at Cambridge University, leading to *Improvised Lives* (Simone, 2018). Simone's engaged, trans-local practice is further enriched amongst fellow travellers invested in the vital intricacies of collective life (see, for example, Bhan et al., 2020).

The breadth and depth of these spaces and explorations allude to a vital thinker on space and place whose imaginative capacity is shaped adjacent to but also outside of institutional academic space. Simone's intellectual range spans the craft of an extraordinary writer, the quiet observer and an attuned collaborator. Here, the empirical context of life and the possibilities of understanding the urban are immersive, emotive and bodily.

Spatial Contributions

Simone scrutinises the spatial life of precarious and creative practice that occurs within and against the relentless political economy of 'harm' orchestrated by states and corporations. His attentive engagement unfolds through varied avenues, and has contributed, inter alia, to the conceptualisation of people as infrastructure, understanding of southern urbanisms and the exploration of global Blackness. Across these realms, Simone provides lively methodological engagements as much as conceptual ones, captured in capricious and irrepressible forms where the reader is enrolled in the experience of thinking with feeling, of being anchored while also being unsettled. This way of pursuing our understanding of the complexities of urban life engages with the varied practices of openness that operate within the confines of capitalism. The quotidian arts of improvisation that are core to Simone's spatial consciousness are revealed through the liminal repertoires that make their way through the labyrinthine geographies of uncertainty. Quotidian practices of the 'hustle', for example, reveal the simultaneity of generative and restricted possibilities of how inhabitants of precarity craft a leeway. In making the hustle, the vital aesthetics of performance and the rudimentary requirements of making do and getting by shape space (Matlon, 2011; Thieme, 2018). Importantly, improvisation, whether articulated through refusal or recalibration, is a making of space and making of knowledge that supplements and recomposes the uneven and often violent relations of distribution. These modes of shaping and understanding are neither retrievable via 'a plausible statistical mean', 'an average story', nor a normative commitment to 'progress' (Simone, 2018: 14; see also Harney and Moten, 2013 and **Katherine McKittrick**, 2020).

In Simone's (2004b) influential essay 'People as infrastructure: Intersecting fragments in Johannesburg', he entwines the economic-cultural-social nexus of urban life, foregrounding the ensembles of inhabitation shaped in the enrolment of objects, ideas, technologies and collaborations. These energetic infrastructures of urban life are far from fixed; they emerge through regular interactions and unpredictable encounters, constantly calibrated in relation to durable precarity. The vital yet tenuous assemblage of people as infrastructure is also a commitment that requires careful looking. It is a recognition of an inhabited politics, a piecing together of fragments to compose a viable platform for urban life, a condition of possibility, 'which attempts to derive maximal outcomes from a minimal set of elements' (Simone, 2004b: 411). Although Simone subsequently reflects on the origin of his almost '"throw-away" formulation' of 'people as infrastructure' he places particular emphasis on his 'deep-seated concern with how to consider urban collective life as something beyond the capture of individuals, households, communities and institutions' (Simone, 2021: 1341). Rather than the recurring sociological engagement with social ties inscribed in the knowledge architectures of community and the enclave, Simone draws our attention to the possibilities of collaboration and coping in the emergent working-class districts of extensive urbanisation.

The simultaneity of 'structured structures', of actors and policies and encounters and events, of the unanticipated and the eventual, is evoked by the notion of a spatial composition of 'the surrounds' (Simone, 2021; 2022). The idea serves to emphasise how space is shaped through a 'lived resonance' that exceeds the actual and explanatory limits of fixed relations, or the guarantee of planned outcomes. Simone's emphasis is not based on a theoretical premise about the organisation of capital; his empirical and anthropological explorations are located in lifeworlds where the prospects of large-scale (re)distribution are absent. The loose analytic of 'the surrounds' focuses our attention on the visceral realities of how space, and the capacity to interact with evasive spatial conditions, is composed through unpredictable operations:

> Across these urban majorities, there was the conviction that messy interchanges among different sentiments and actors would eventually produce something better than existed now, even when people might be hard pressed to demonstrate real results. In the rough and tumble everyday worlds of trying to make the most from what was available, of putting together people and things that didn't seem to really go together, residents tended to be convinced to let many different scenarios play out. They demonstrated a basic tolerance for ways of doing things that didn't necessarily correspond to their values or beliefs. What was important was an experience of forward momentum, even in small increments; the feeling that one wasn't stuck in place, that there were ways of 'working' the situation, playing the field, that allowed one a sense of agency. (Simone, 2021: 1342–43)

The connections Simone draws between the innate plight of urban majorities across the planet and their persistent retrieval of resources in order to live life and make livelihood, is evoked in the spatial relation and specific contexts of southern urbanisms. Rather than a geographic designation, the idea and vocabulary of the urban south offers us the possibility of thinking in and across space to engage with the associations that are assembled in order to hold onto already limited resources. Our attention turns to the insurgent efforts and ceaseless ensemble work that operate within and manoeuvre against the extensive reach of the political economy of harm. The tenet is not limited by the determined expectations of resilience or even the commons, but is shaped through the everyday mobilisations and cultures of resourcing. Crucially, this provisioning is invariably forged outside of the conventions of what we might assume or designate as primary architectures of (mal)distribution: the state and the market. What the vocabulary of southern urbanisms opens out is simultaneously the syntax of quotidian resourcing, and how these modes of composition configure an alternative periphery or margin (Lancione and Simone, 2021). This locus is a diminutive territory; it is a space from which to anticipate precarity, a place to be at risk and to take risk.

From here, we see productive oscillations amongst emergent vocabularies outside of Anglo-American or Eurocentric canons. For example, Caldeira (2017: 3) has identified 'autoconstruction' as a means of city-making by residents who obliquely interact with the state, additionally bypassing the dominant logics of capitalist distribution. Such claims to space are generative of political possibility; they open out our comprehension of citizen and citizenship beyond the excluding limits of legislated designations. Bhan (2019: 645) similarly proposes the vocabularies of *squat*, *repair* and *consolidate*, as 'terms that institutions of authority need to consider', thereby inverting the assumptions of expertise and 'informality' in processes of provisioning. In this critical inversion, Bhan foregrounds a 'staying power' sustained 'more through a claim to legitimacy than legality', pushing us to consider what it means to inhabit and exceed the pre-established designations of illegality on the part of sovereignty.

Designations of legitimacy and legality, of citizen and denizen, are lodged in the enduring structures of extraction and harm underpinned by the racial order of capitalism, and the 'ruling calculus of what counts as lives worth living' (Simone, 2023: 339). A vital part of our current terrain of thinking about 'race', space and place is enriched by postcolonial, decolonial and anti-racist commitments that attend to the ever-present axes of imperial economies, racial geographies and their knowledge formations. Within these explorations that seek to redress structural and epistemological forms of dominance, are wide-ranging engagements with practices of refusal and abolition. The attention to political and cultural imagination opens out the possibilities of dismantling the racial systems that procure premature death, alongside the reconstituting of spaces that sustain life (**Gilmore**, 2022). Such liberatory imagination extends to an embrace of life-affirming embodiments that pulsate through music, poetry, dance, fiction and more (McKittrick, 2016). This too is generative of alternative vocabularies that provoke us to think with the spatial formations that resonate behind, below or beyond the segregating imperatives of hierarchy disguised as progress. Conceptions of the 'undercommons' (Harney and Moten, 2013), or 'rogue urbanisms' (Pieterse and Simone, 2013), intentionally evade the neoliberal logics of the purified public realm and the promises of urban regeneration, fully aware that these terms cloak the base realities of exclusion and displacement.

Simone's work on 'Blackness' engages directly with the untenable order of racial violence that is carried forward in the authorised and condoned political economies of dispossession. But rather than focus on the sustained punitive dimensions of systemic racial violence, Simone's work considers the generative solidarities and dexterous experiments that are embodied in urban citizenships 'at the edge'. The relation of collective life to power is arguably explored in his work less as a definitive *insurgency* that might deploy a variety of formalised infrastuctures, from legal claims to interventions in planning processes. Simone highlights *endurance* as a resolute and sustained mode of contestation and survival:

> Yet endurance for large numbers of urban residents remains predicated on indifference to and acts of detachment from prevailing modes of urban power, in operations of stealth and supplement long embodied by blackness, and Black Power. (Simone, 2016: 183)

This quotidian attunement to making life liveable, while not in the first instance conditioned by the commitment to liberation, nonetheless holds close and refines practices of care and recalibration, that surpass the capture of domination. In this embrace, Simone brings in the base conditions of working-class life, and the range of Black urban experience anchored in the enforcement of racist segregation and violence. He also opens up thinking about Blackness by considering the possibility of contestation through the ordinary spaces of our cities, extending it to varied situations and places in order to think closely and expansively about rhythms of endurance:

> This is not a story of survival against the odds or the resilience of the oppressed. Rather, it is about the persistence of potentiality within any format of urbanization – its black side. Something that is not reduced to metaphor but rather a reiteration of political contestation played out day in and day out across the ordinary spaces of city life [...] I want to talk about a force that perhaps never comes to the fore but, nevertheless, endures. (Simone, 2016: 184–5)

Simone thus recognises the requirement of 'calculated indifference' as integral to the innate context of uncertainty, incompletion and harm:

> Whether you pass through the roadblock, manage to mobilize a life savings into something tangible, or manage to turn that corner just before the police or debt collectors arrive, this time of the maybe, more than simply a wager or speculation, is a continuous rebellion against what is on offer, against how one is regarded. But it is a rebellion without redemption. (Simone, 2022: 102)

His open lexicon evokes relations and affinities between and across the 'unruly majority' working at 'ongoing rebelliousness' and, at the same time, emphasises a sense of *temporariness*.

Key Advances and Controversies

The language of advances and controversies, or paradigmatic schools of urban and spatial theory, is not strictly useful for unpacking the resonance of Simone's work. It is important to focus on where his work has had a particular reach in offering alternative frames of understanding. Here, the significance of Simone's work needs to be considered alongside extended debates in urban theory which are often organised in relation to structures of order, both critical and methodological, that are also invariably positional (although less frequently articulated as such). A crucial starting point is the question of how to understand the extant condition of an urban majority living outside of waged labour, housing contracts, and recourse to public services and the assumed promise of libertarian rights. One vital perspective orientates around the ability of urban political economy to grasp and comprehend the distinctive forms of institutional and corporate establishments that produce the extended urban space of capitalism. Critical urban theory seeks to analyse the urban as a system oversaturated by uninhibited iterations of speculation, accumulation, debt, eviction and premature death. Core to the systemic production of profound harm is the neoliberal state, shown to be actively shaping social relations through which the systems of durable inequality are maintained.

Crucial here is the question of the extent to which the ubiquity of planetary exploitation and the specificity of making of life without expectation of authorised provision can be readily analysed through established explanatory frameworks. Simone's work is important in the push to take urban theory beyond 'inherited concepts and methods [...] and oft-repeated mantra' (Brenner et al., 2011: 226), from an emphasis on 'wide-reaching engagements' and 'broader strategies' (Brenner et al., 2011: 227) to 'a rethinking of agency' and engagement with a 'critical imaginary' (McFarlane, 2011: 205–8). The elemental concerns surely converge (if divergently) in questions of the enduring production of crisis and the organised forms of power and survival that support, contest or occupy the urban condition. In the contexts of capitalism Simone is immersed within, 'the "crisis" exceeds being a matter of repair' (Lancione and Simone, 2021: 969). The possibilities of transformation or amelioration forged by unionised collective action and socialist distribution, or by constitutional and institutional democratic recourse preordained by liberalism, are out of place.

To embrace improvisation as a political practice is to recognise it as a constitutive mode of spatial production, one that is neither reducible to a minor detail nor an ethnographic imperative. Through this conceptual and inevitably *grounded* position, Simone advances how to think critically, but not conventionally, about the prospects of citizenry in and beyond the circuits of dispossession. Thinking about the kinds of citizens and citizenships that are socially produced in space and that produce space, requires thinking about multifarious claims that operate in and rearticulate extant realities of displacement, whether we analyse these as casualisation, border control, or gentrification (Hall, 2021). Simone's attention on improvised lives allows us to see co-constitutive reckonings of power, sustained in ineluctable denigrations and punishments, the organisational capacities of the sidestep, and the quotidian yet ubiquitous cultures of 'push back'.

Simone's Key Works

Pieterse, E.A. and Simone, A. (eds) (2013) *Rogue Urbanism: Emergent African Cities*. Cape Town: Jacana Media and the African Centre for Cities.

Simone, A. (2004a) *For the City Yet to Come: Changing African Life in Four Cities*. Durham, NC: Duke University Press.

Simone, A. (2004b) 'People as infrastructure: Intersecting fragments in Johannesburg', *Public Culture*, 16 (3): 407–429.

Simone, A. (2010) *City Life from Jakarta to Dakar: Movements at the Crossroads*. Abingdon: Routledge.

Simone, A.M. (2014) *Jakarta: Drawing the City Near*. Minneapolis, MN: University of Minnesota Press.

Simone, A. (2018) *Improvised Lives: Rhythms of Endurance in an Urban South*. Hoboken, NJ: John Wiley & Sons.

Simone, A. (2022) *The Surrounds: Urban Life Within and Beyond Capture*. Durham, NC: Duke University Press.

Simone, A. and Pieterse, E. (2017) *New Urban Worlds: Inhabiting Dissonant Times*. Cambridge: Polity.

Secondary Sources and References

Bhan, G. (2019) 'Notes on a Southern urban practice', *Environment and Urbanization*, 31 (2): 639–54.

Bhan, G., Caldeira, T., Gillespie, K. and Simone, A. (2020) 'The pandemic, southern urbanisms and collective life', *Society and Space*, August 3. https://www.societyandspace.org/articles/the-pandemic-southern-urbanisms-and-collective-life (accessed November 14, 2023).

Brenner, N., Madden, D.J. and Wachsmuth, D. (2011) 'Assemblage urbanism and the challenges of critical urban theory', *City*, 15 (2): 225–40.

Burga, H.F. (2009) 'An interview with AbdouMaliq Simone'. *Berkeley Planning Journal*, 22 (1). doi.org/10.5070/BP32215367

Caldeira, T.P. (2017) 'Peripheral urbanization: Autoconstruction, transversal logics, and politics in cities of the global south', *Environment and Planning D: Society and Space*, 35 (1): 3–20.

Hall, S.M. (2021) *The Migrant's Paradox: Street Livelihoods and Marginal Citizenship in Britain*. Minneapolis, MN: University of Minnesota Press.

Harney, S. and Moten, F. (2013) *The Undercommons: Fugitive Planning and Black Study*. Wivenhoe: Minor Compositions.

https://www.minorcompositions.info/wp-content/uploads/2013/04/undercommons-web.pdf (accessed November 14, 2023).

Gilmore, R.W. (2022) *Abolition Geography: Essays Towards Liberation*. London: Verso.

Lancione, M. and Simone, A. (2021) 'Dwelling in liminalities, thinking beyond inhabitation', *Environment and Planning D: Society and Space*, 39 (6): 969–75.

Matlon, J. (2011) 'Il est garçon: Marginal Abidjanais masculinity and the politics of representation', *Poetics*, 39 (5): 380–406.

McFarlane, C. (2011) 'Assemblage and critical urbanism', *City*, 15 (2): 204–24.

McKittrick, K. (2016) 'Rebellion/invention/groove', *Small Axe: A Caribbean Journal of Criticism*, 20 (1): 79–91.

McKittrick, K. (2020) *Dear Science and Other Stories*. Durham: Duke University Press.

Simone, A. (2016) 'Urbanity and generic blackness', *Theory, Culture & Society*, 33 (7–8): 183–203.

Simone, A. (2021) 'Ritornello: "People as infrastructure"', *Urban Geography*, 42 (9): 1341–8.

Simone, A. (2023) 'Making use of everything: Tangier and its Southern, peripheral practices', *South Atlantic Quarterly*, 122 (2): 339–59.

Thieme, T.A. (2018) 'The hustle economy: Informality, uncertainty and the geographies of getting by', *Progress in Human Geography*, 42 (4): 529–48.

51
Neil Smith

Hamish Kallin, University of Edinburgh

Biographical Details and Theoretical Context

Neil Smith (1954–2012) was a prominent Marxist geographer, best known for his ground-breaking work on gentrification, uneven development, and the production of space. Born in Leith, working-class docklands attached to Edinburgh, Scotland, he spent most of his childhood in Dalkeith, a small town south-east of the city. In the early 1970s, he studied Geography at St. Andrews University. An international exchange to the University of Pennsylvania in 1974/75 took him to Society Hill, an inner-city neighbourhood of Philadelphia built for the gentry, who had abandoned it as a 'slum' by the 1940s, but which was starting to be gentrified (it is now one of the most affluent and expensive neighbourhoods in the city). In his undergraduate dissertation (under the supervision of Joe Doherty), Smith began to think through the forces shaping this dramatic 'regeneration', emphasising the role of state power, land values, rent, and the interests of property developers. The idea of a 'back to the city movement' was, he argued, a myth that naturalised the process. Thus began a lifelong intellectual pursuit of theorising, understanding, and fighting against gentrification.

Moving to Baltimore to undertake a PhD supervised by **David Harvey**, he developed this work into two trailblazing interventions in the gentrification debates (Smith, 1979a; 1979b), which utilised a Marxist analysis to explain the process as part of the uneven circulation of capital within cities. Both papers sought to explain gentrification with respect to the wider spatial dynamics of power, ideology, and profit derived from rent. His doctoral thesis followed this trail of uneven capital circulation so far that it left gentrification in the rear-view mirror. Widely hailed as a masterpiece on publication, *Uneven Development: Nature, Capital and the Production of Space* (Smith, 1984) became a highly influential text developing Marxist thought in Geography and vice versa. It foregrounds the uneven processes of capitalism as essential to its functionality (rather than an unfortunate side-effect) and develops a captivating understanding of nature, scale, and place as socially produced (where the 'social' is riven with inequality, shaped thoroughly by capitalist relations).

He secured his first academic job (in the Geography Department of Columbia University) shortly before submitting his thesis. There he pursued a third strand of research related to the history of the discipline itself.

Fascinated by Isaiah Bowman (1878–1950) – a prominent American geographer whose work was deeply intertwined with imperial statecraft – Smith's (1987) discussion of Bowman's role in the closure of Geography at Harvard touched a nerve in the climate of uncertainty facing the discipline. Shortly before Columbia University closed their own Geography Department (alongside all but one of the Ivy League universities), Smith moved to Rutgers University in 1986, where he was quickly tenured. The Bowman project would slowly mature into *American Empire* (Smith, 2004a), a remarkable book intertwining the history of US Geography as an institutionalised discipline with the geopolitics of US imperialism.

In 1988 he was promoted to Associate Professor and in 1990 to Professor. In the 1990s, he re-defined the gentrification debates on multiple fronts through a series of publications on New York in particular. His work in this era kept capital and class in clear view, but entwined them with discussions of race, police brutality, homelessness, culture, representation, and more, as exemplified in his influential book, *The New Urban Frontier: Gentrification and the Revanchist City* (Smith, 1996). Increasingly, he became a well-known figure on the public left, for this work was politically energising in the context of a city undergoing violent shifts that exposed its stark lines of inequality and dispossession. What defined his impact was a clear grounding of his interventions in specific debates (around urban space and gentrification, for example) to much wider conceptual and intellectual concerns. As a case in point: reflecting on the homelessness crisis in New York, Smith (1992) further politicised the notion of scale as socially produced, showing how power and resistance construct, contain and break beyond different spheres of thought and action. In 1997, working with Graciela Uribe (and many others), he co-organised the first International Conference of Critical Geography in Vancouver, which has survived as a bastion of radical thinking within the discipline.

At the turn of the millennium, he took up a post as Distinguished Professor of Anthropology and Geography at the City University of New York's Graduate Center, where he founded and became Director of the interdisciplinary Center for Place, Culture, and Politics, a position he would hold for the rest of his life. The first decade of the millennium was a turbulent time in the US, with the September 11 attacks on the World Trade Center followed swiftly by the 'War on Terror' and the hardening of global security and military policy this provoked. Smith's response to this illustrates how sharply his theory evolved with political observation, and how readily it sought to shape that reality: the war, in his eyes, existed *between* scales, merging a galvanised US nationalism with global geopolitical intent, all the while intertwined with the interests of capital (Smith, 2002b; 2004b). In this conjecture, any theory of scale fit for purpose needed to grasp the duplicity, liquidity and falsity of so many powerful scalar motifs, themes he would pull together in *The Endgame of Globalization* (Smith, 2005a). When Hurricane Katrina hit New Orleans in August 2005, Smith penned a short blog post that typified what was so influential about his writing. Echoing the work of his long-time friend and collaborator, Phil O'Keefe (O'Keefe et al., 1976), 'There's No Such Thing as a Natural Disaster' (Smith, 2006) was timely, political, easy to engage with, and conceptually eye-opening for those within and beyond the academy.

Several of his contributions in the early 2000s – on the continued importance of class analysis in geography (Smith, 2000), the generalisation of gentrification as a global strategy of capital accumulation (Smith, 2002a), rethinking nature as accumulation strategy (Smith, 2007) – confirmed his status as one of the most prominent radical geographers. In the aftermath of the 2007/08 global financial crisis, Smith grasped at the apparent fragility of neoliberalism and the political possibilities this engendered. Capital's global order looked like it was 'dead but dominant' (Smith, 2008) and the 'revolutionary imperative' remained essential (Smith, 2010). Smith passed away on September 29, 2012, aged just 58. Alcohol had taken its toll on his body, and Geography lost one of its most inspiring and creative radical intellectuals (for tributes and critical reflections on his life see Cowen et al., 2012; Slater, 2012; Mitchell, 2014).

Spatial Contributions

As a theoretician of gentrification, Smith's contributions were myriad and highly significant in shaping the way subsequent scholars and activists understood, contested, and theorised the changing role of urban space in the global economy. The initial challenge was to denaturalise neoclassical models of land rent theory inherited from the Chicago School, models that presupposed a 'rationality' behind the economic evolution of cities powered by consumer sovereignty. Smith argued that consumer preference always takes place in a world shaped by more significant relations of power and profit. His conceptual arsenal was rich and particularly generative. The *rent gap* (Smith, 1979b) juxtaposes capitalised ground rent (how much money is currently being made from any given parcel of land) against potential ground rent (how much money could be made if the land was put to a different use). This was an ingeniously simple way of spatialising and historicising the profit motive behind gentrification, showing that the relationship between disinvestment and reinvestment made urban redevelopment more profitable. The link to his work on *uneven development* (Smith, 1982) showed gentrification to be but one contemporary symptom of a wider dynamic within global capitalism. It was this inherent unevenness, in perpetual motion, that made gentrification possible. The *urban frontier* (Smith, 1986; 1996) as a framing mechanism tied the emerging mythology of 'civilising' the inner city to the predatory nature of capital. The 'frontier' in Smith's work was both a line of profitability (beyond which reinvestment was still too risky) and a territorial mentality, tightly intertwined with a cultural politics bearing the legacy of American settler colonialism. By urbanising the idea of a vengeful turn through his work on *revanchism*, Smith (1996, 2001) showed how the disembodied power of finance capital marched in step with the brutally embodied power of the police, situating gentrification within a broader reactionary turn that was at once racist, classist and sexist, and demonstrated how a 'new' political culture of vindictiveness had profoundly spatial impacts. Connected to all these concepts was the importance of *public space* as a realm of political possibility, democracy, and repression.

The majority of this work was focused, in the first instance, on New York City, where the transition from deindustrialised bankruptcy to global real estate hotspot contained within it all the processes Smith so vividly explained. The theories were never designed to stay in place, however. Intellectually (as scholars elsewhere began to use, test, critique, and develop the ideas) and materially (as the globalisation of finance capital and real estate speculation picked up pace), the relevance of Smith's ideas became evident across the globe. By the early 2000s, he responded to this diffusion of ideas and financialisation by reframing gentrification as a *global* strategy of capital accumulation (Smith, 2002a), where the state's fixation on urban regeneration combined with corporate investors' fixation on the 'security' of real estate assets to push gentrification-like processes wherever they might be profitable. At the same time, revanchism was similarly upscaled and globalised (Smith, 2009). Smith's invocation of revanchism hence became a captivating invitation to think through the class dynamic of gentrification as a spatial struggle that was impossible to explain without attending to histories of disinvestment, ideologies of reinvestment, state power, inequality and unevenness in both global North and South. Smith's conception of revanchism offered a *political* and *politicising* conception of what gentrification is and how it happens.

More widely, Smith's work on uneven development began with the assertion that there is an intractable contradiction between the tendency towards *equalisation* (the overcoming of obstacles and differences; the imposition of exchangeability in the value of all places, things, and human relations, etc.) and the tendency towards *differentiation* (rich and poor, boss and worker, places of production and consumption, etc.) within global capitalism. The dialectical relationship between the two (never reconciled, always in motion) meant that uneven development was not some unfortunate side-effect of capitalist development but essential to its working and constantly reproduced (Smith, 1984). This fed into the wider assertion that space was itself socially produced, shot through with power imbalances, never a pre-existing abstract terrain but a contested, ideological, and material realm. Whilst the theoretical importance and influence of this text has obtained classic status, its insistence on the dynamism of uneven development means the empirical reality it describes is always, to a certain degree, in want of an update (Smith, 2011). A new preface for the second edition (1990) and a new afterword for the third edition (2008) went some way towards this.

Central to *Uneven Development* was a rethinking of the relationship between nature, society, and capitalism. Smith (1984) rejected the separation of (non-human) nature from (human) society as an ideological framing. Instead, he sought to show how capitalism seeks the 'unity' of the human and non-human to incorporate as much as possible within the sphere of exchange. This was a provocative and influential thesis that significantly reshaped the way geographers considered nature and how nature was theorised in relation to Marxism, economics, and capitalism (Castree, 2003).

Smith's engagement with scale began alongside his early research on gentrification (Smith, 1982). At this stage, scale appeared as a hierarchy of three spheres (urban, region, nation) within which capital and power acted, and could act, differently. In subsequent work, scale became something much more political and much less fixed: it was also, crucially, a categorisation that could be, and was, changed. The idea of 'scale-jumping' (Smith, 1992) – intentionally shifting between scales – and 'scale-bending' (Smith, 2004b) – transgressing the norms of what kind of social activities are appropriate at what scale – were both simple phrases that carried within them a complex rethinking of the ways in which capital, power, labour, ideas, and institutions remake, contest, and exploit scale. In this later work, scale begins to function a bit like David Harvey's notion of a 'spatial fix', i.e., jumping or bending scale provides a (temporary) solution to capitalism's intractable crises, just as it also provides potentially fruitful opportunities for dissent, organising, and resistance. Throughout this work – influenced by Sallie Marston and **Cindi Katz** – Smith began, more and more, to connect his Marxist-inspired political economy to insights from feminist scholarship on the importance of the body as a political scale, recognising the value of social reproduction to the survival of capitalism, and identity to the culture of capitalism (Jones et al., 2017).

What united Smith's work was a passionate and evolving attempt to think through what it might mean to be a Marxist or, more broadly, a radical geographer in the era of 'post-socialism' (historically speaking) and 'post-structuralism' (intellectually speaking). The leitmotif of all his work was a connection back to the spaces produced by inequality, class, and capitalist power. Not one to dabble in fatalistic or fixed theorisations, his ideas were engineered to inspire hope.

Key Advances and Controversies

Smith's work has been most influential in Geography and Urban Studies, where his key texts continue to be regularly cited. His impact cemented the possibility, and increased the popularity, of using Marxist concepts in Geography, and more broadly sharpened the radical understanding of how space and scale are made and remade under capitalism. He helped to foreground the importance of thinking politically about the intellectual ideas we inherit and create.

All of Smith's headline concepts are still being used and developed. The rent gap, for example, has evolved significantly. Whilst it has been criticised for being difficult to empirically measure, and is often mistaken for a measure solely of land values (as opposed to the forces shaping those values), its ability to fuel debate shows no signs of waning. For example, recent work expands the concept to consider how short-term lets and touristification impact local housing markets (Wachsmuth and Weisler, 2018), state governance capitalises on territorial stigmatisation (Risager, 2022), and the frontier of gentrification is powered by debt (Kallin, 2020). Gentrification scholarship more generally has continued to evolve, and Smith's (2002a) assertion of the *global* relevance of the concept has been reaffirmed whilst being challenged by an emphasis on comparative gentrification studies across a wide range of contexts (Lees et al., 2016).

Yet Smith's work is certainly no stranger to criticism. Whilst his Marxism was increasingly open and flexible, this did not inoculate it against accusations of economic determinism. This was evident in the 'production' vs. 'consumption' debates in gentrification theory, where his 'supply-side' explanation (concerned with the forces producing urban space for profit) was pitted against 'demand-side' explanations (where the changing desires of the new middle classes took precedence), as typified at the time through the work of ***David Ley*** (1987). A huge amount of ink was spilled debating this supposed dichotomy, before many gentrification scholars began to acknowledge that it was somewhat odd to consider one without the other (Lees et al., 2008). Smith often adopted a combative tone in debate, pulling no punches in his critique. Perhaps most controversially, his biting criticism of the 'neo-critical geography' championed by Ash Amin and **Nigel Thrift** (Smith, 2005b) sought to demarcate a genuinely radical future for the discipline against what he saw as the increasing dominance of depoliticising concepts and the business-like mentality of universities. His insistence on the importance of class as a (if not *the*) fault line of exploitation (Smith, 2000) also marked him out as an avid defender of views that others were quick to discard. He had a tendency to reach for larger and larger scales in his theoretical work, which was arguably at odds with the broader intellectual rejection of 'totalising' theory and the more recent shift towards emphasising difference (though the centrality of unevenness to all his work means it is certainly possible to join these dots rather than oppose them).

Neil Smith was a passionate advocate for a more radical understanding of how geography is made, one which should be forever attentive to challenging the ways capitalism inscribes inequalities and unevenness across space. As Don Mitchell (2014) – one of Neil's doctoral students – puts it in a beautifully balanced obituary, he was undoubtedly a contradictory man, sometimes in ways that were troubling, and ultimately destructive of his own life. He was an intellectual heavyweight, a brilliant writer, a fearless intellectual, an inspiration to many, and undoubtedly one of Geography's key radical thinkers.

Smith's Key Works

Smith, N. (1979b) 'Toward a theory of gentrification: A back to the city movement by capital, not people', *Journal of the American Planning Association*, 45 (4): 538–48.

Smith, N. (1984) *Uneven Development: Nature, Capital, and the Production of Space*. Oxford: Blackwell.

Smith, N. (1996) *The New Urban Frontier: Gentrification and the Revanchist City*. New York, NY: Routledge.

Smith, N. (2002a) 'New urbanism, new globalism: Gentrification as global urban strategy', *Antipode*, 34 (3): 427–50.

Smith, N. (2004a) *American Empire: Roosevelt's Geographer and the Prelude to Globalization*. Berkeley, CA: University of California Press.

Smith, N. (2005a) *The Endgame of Globalization*. New York, NY: Routledge.

Secondary Sources and References

Castree, N. (2003) 'The production of nature', in E. Sheppard and T.J. Barnes (eds), *A Companion to Economic Geography*. Oxford: Blackwell. pp. 275–89.

Cowen, D., Harvey, D., Haraway, D., Rameau, M., Bacon, N., Bissen, M. [...] and Jaar, A. (2012) 'Neil Smith: A critical geographer', *Environment and Planning D: Society and Space*, 30: 947–962.

Jones, J.P. III, Leitner, H., Marston, S.A. and Sheppard, E. (2017) 'Neil Smith's scale', *Antipode*, 49 (S1): 138–52.

Kallin, H. (2021) 'In debt to the rent gap: Gentrification generalized and the frontier of the future', *Journal of Urban Affairs*, 43 (10): 1393–404.

Lees, L., Shin, H.B. and López-Morales, E. (2016) *Planetary Gentrification*. Cambridge: Polity.

Ley, D. (1987) 'The rent-gap revisited', *Annals of the Association of American Geographers*, 77: 465–8.

Mitchell, D. (2014) 'Neil Smith, 1954–2012: Marxist Geographer', *Annals of the Association of American Geographers*, 104 (1): 215–22.

O'Keefe, P., Westgate, K. and Wisner, B. (1976) 'Taking the naturalness out of natural disasters', *Nature*, 260: 566–7.

Risager, B.S. (2022) 'Rent gap governance: State-led gentrification in the Danish "ghetto"', *City*, 26 (2–3): 199–223.

Slater, T. (2012) 'Rose Street and revolution: A tribute to Neil Smith', *ACME: An International E-Journal for Critical Geographies*, 11 (3): 533–546.

Smith, N. (1979a) 'Gentrification and capital: Practice and ideology in Society Hill', *Antipode*, 11 (3): 24–35.

Smith, N. (1982) 'Gentrification and uneven development', *Economic Geography*, 58 (2): 139–55.

Smith, N. (1986) 'Gentrification, the frontier, and the restructuring of urban space', in N. Smith and P. Williams (eds), *Gentrification of the City*. Boston, MA: Allen and Unwin. pp. 15–34.

Smith, N. (1987) '"Academic war over the field of Geography": The elimination of Geography at Harvard, 1947–51', *Annals of the Association of American Geographers*, 77 (2): 155–72.

Smith, N. (1992) 'Contours of a spatialized politics: Homeless vehicles and the production of geographical scale', *Social Text*, 33: 54–81.

Smith, N. (2000) 'What happened to class?' *Environment and Planning A*, 32: 1011–32.

Smith, N. (2001) 'Global social cleansing: Postliberal revanchism and the export of zero tolerance', *Social Justice*, 28 (3): 68–74.

Smith, N. (2002b) 'Scales of terror: The manufacturing of nationalism and the war for U.S. globalism', in M. Sorkin and S. Zukin (eds), *After the World Trade Center: Rethinking New York City*. New York, NY: Routledge. pp. 97–108.

Smith, N. (2004b) 'Scale bending and the fate of the national', in E. Sheppard and R. McMaster (eds), *Scale and Geographic Inquiry: Nature, Society, and Method*. Oxford: Blackwell. pp. 192–212.

Smith, N. (2005b) 'Neo-critical geography, or, The flat pluralist world of business class', *Antipode*, 37: 887–99.

Smith, N. (2006) 'There's no such thing as a natural disaster', in *Items: Insights from the Social Sciences*. https://items.ssrc.org/understanding-katrina/theres-no-such-thing-as-a-natural-disaster/ (accessed November 14, 2023).

Smith, N. (2007) 'Nature as accumulation strategy', *Socialist Register*, 43: 19–41.

Smith, N. (2008) 'Neoliberalism is dead, dominant, defeatable – then what?' *Human Geography*, 1 (2): 1–3.

Smith, N. (2009) 'Revanchist planet', *The Urban Reinventors Online Journal*, 3: 2–18.

Smith, N. (2010) 'The revolutionary imperative', *Antipode*, 41 (1): 50–65.

Smith, N. (2011) 'Uneven development redux', *New Political Economy*, 16 (2): 261–5.

Wachsmuth, D. and Weisler, A. (2018) 'Airbnb and the rent gap: Gentrification through the sharing economy', *Environment and Planning A: Economy and Space*, 50 (6): 1147–70.

52
Gayatri Chakravorty Spivak

Robina Mohammad, Robina-therapies.com, Kuala Lumpur and Singapore
James D. Sidaway, National University of Singapore

Biographical Details and Theoretical Context

Gayatri Chakravorty Spivak was born in Kolkata, in colonial India, in 1942 to a middle-class family. She graduated from Presidency College (now Presidency University), Kolkata in 1959 with first-class honours in English having received gold medals for English and Bengali literature. The College had been established in 1817 and was a constituent institution of the University of Calcutta when Spivak attended. Through her studies, Spivak was highly conscious that the teaching of English Literature in colonial India was a central part of the 'civilising mission' of imperialism, and the legacies of colonisation, struggle for independence and 'postcolonialism' have remained central to her work throughout her career.

Spivak continued her education in the US where she pursued a Master's degree (1962) in English at Cornell University. She also spent a year in Britain at Cambridge University as a fellow at Girton College before returning to the US to take up an instructors' post at the University of Iowa. At the same time, she worked on a PhD. Her doctoral dissertation (later published as Spivak, 1974) was on the work of Irish poet and politician W.B. Yeats (1865–1939, recipient of the Nobel Prize for literature in 1923), undertaken again at Cornell University, supervised by Paul de Man (1919–1983). Spivak was appointed Avalon Foundation Professor in the Humanities at the Department of English and Comparative Literature at Columbia University (New York) in 1992, where she worked alongside **Edward Said** (1935–2003). She has received numerous honorary degrees and awards. A frequent traveller and engaging presence, Spivak is often invited to speak across many disciplines as well as outside the academy in activist circles and at literary events. Selected recordings are easily found online and may usefully supplement engagement with her writings. They can also offer an introduction to her frequently elliptical but engrossing and extensively referenced modes of presentation and thought.

For many geographers and other scholars, the name of Gayatri Chakravorty Spivak is synonymous with deconstruction and postcolonial feminism. It is no secret, however, that Spivak's speaking and writing, which often has a dense style and complexity, has sometimes yielded bewilderment or even invoked dismissal. It is certainly challenging. Perhaps too there is a tendency of invoking her work without close engagement. However, many find her words and thought to be compelling and richly suggestive. Amongst her wide range of book-length published work is a translation into English and critical introduction to Jacques Derrida's *De la Grammatologie* (1976); *In Other Worlds: Essays in Cultural Politics* (1987); *Outside in the Teaching Machine* (1993); *A Critique of Post-Colonial Reason* (1999); *Death of a Discipline* (2003); *Other Asias* (2005), and *An Aesthetic Education in the Era of Globalization* (2013). Many of these books comprise sets of essays, sometimes retaining the tone of the original spoken delivery. She has also authored dozens of articles, some of the most influential being: *Three Women's Texts and a Critique of Imperialism* (1985); *Can the Subaltern Speak* (1988a) and *Who Claims Alterity* (1989); *Poststructuralism, Marginality, Postcoloniality and Value* (1990a). The wide range of her interests encompass 19th and 20th century literature, Marxism, Feminism, deconstruction, general French theory, globalisation and the postcolonial, political theory, and critical social theory. In spoken presentations, she often describes herself as 'a Europeanist', reflecting her immersion in comparative literature debates and social and literary theory that have a European provenance.

Spivak's introduction to deconstruction was via Paul de Man who was its most prominent advocate in the US at a time when Derrida was making similar arguments in Europe. Both de Man and Derrida questioned the stability and transparency of textual meaning. Deconstruction developed as a critique of 'logocentrism', wary of the privileging of an authoritative presence held to be characteristic of much Western thought. It is – in part – an approach to reading. Yet it is not so much a method as a strategy for de-linking, destabilising and undermining these oppositions by revealing their mutual constitution. As Spivak argues:

> Deconstruction does not say there is no subject, there is no truth. It simply questions the privileging of identity so that someone is believed to have the truth. It is not the exposure of error. It is constantly and persistently looking into how truths are produced. That's why deconstruction doesn't say logocentrism is a pathology, or metaphysical enclosures are something you can escape. Deconstruction, if one wants a formula, is among other things, a persistent critique of what one cannot not want. (Spivak, 1994: 280)

Barbara Johnson (1980: 5) thus notes 'the deconstruction of a text does not proceed by random doubt or arbitrary subversion, but the careful teasing out of warring forces of signification within the text'.

Spatial Contributions

For Matthew Sparke, Spivak's 'mode of deconstructive rereading' offers 'a spur to respond ethically to the erasures represented by any particular geography':

> Every geography, whether assumed or explicitly elaborated as such, every mapping, picturing, visualization, landscaping, theorization, and metaphorization of space becomes readable in this sense not just for what it includes, but also for what it overwrites and covers up in the moment of representing spatially. (Sparke, 2005: xv–xvi)

However, the frequent density of Spivak's prose means that the geographical applications of her work are not necessarily immediately obvious and her own 'scattered speculations on geography' (Spivak, 2014) may not offer an easy guide. Moreover, it is difficult to separate her fully from Derrida's impacts. However, one contribution to Geography relates to the so-called 'crisis of representation' and the vexed issue of 'positionality' that emerged in the 1990s and which continues to be present in geographical writing, increasingly influenced by discussions of the decolonial and the growing influence of Black geographies. In this context, Spivak's work was drawn on to highlight the ethnocentricity of hegemonic knowledge and consider the exclusions, occlusions and marginalisations found in colonial and many modernist epistemologies. Spivak's (1990a: 224) concern with 'the imbrication of techniques of knowledge with strategies of power' is that representations do not just describe reality but are constitutive of it. Spivak (1988a) elaborates on this when she defines representation as both 'a speaking of' that describes or depicts a perceived reality, through language and communication and a 'speaking for' which is the same as the representation of a particular constituency (Spivak, 1988a).

In her essay *'Can the Subaltern Speak?'* (Spivak, 1988a), she deals directly with such questions, and takes issue with claims by intellectuals to speak for the disempowered, placing them alongside the benevolent colonialists who saved Hindu women from widow sacrifice in the nineteenth century. In doing so, Spivak argues that such interlocutors (even as they claim to represent and speak for the disempowered) have silenced subordinated subalterns (where the latter are distinguished in their class, gender and caste). The essay considers wider historiographical themes about who and what is excluded from or overwritten in archives and historical narratives. The subaltern may seek to speak or find other ways to make a statement, but she is frequently misheard, ignored or her words and actions are misinterpreted by a dominant culture whose terms of meaning, truth and understanding exclude subalterns by positioning them and interpreting their actions through dominant frames of reference. She examines how, in Morton's (2003: 57) terms; 'the historical and structural conditions of political representations do not guarantee that the interests of particular subaltern groups will be recognised or that their voices will be heard'. A 2010 reprinting of variants of the essay, together with a reflection by Spivak and eight engaged commentaries (Morris, 2010), reflect its enduring influence. The category of the subaltern has been subject to wider historical analysis, in particular in South Asian histories that sought to move beyond both colonial and canonised nationalist narratives. Spivak's work is therefore part of a wider arc of Subaltern Studies, which included the Indian historian Ranajit Guha (1923–2023), who, like Spivak, was educated at Presidency College. Amongst geographers, the implications of theoretical insights have been explored in a set of essays on contemporary and historical *Subaltern Geographies* (Jazeel and Legg, 2019) and in considerations of geopolitical categories and discourses (Cheong, 2019). Spivak's reflections on the subaltern are also points of departure in recent textbooks by geographers on postcolonialism (Jazeel, 2019) and decolonising geographical knowledge (Radcliffe, 2022).

At the same time, Spivak has argued that Western feminisms have sometimes propagated the 'lie of a global sisterhood', failing to recognise the differences between women

(Spivak, 1986: 226). In response to such critiques, the questions of differences *between* women have occupied feminist geographers, particularly in terms of the implications for the production of geographical knowledge (**McDowell**, 1993). This saw feminist geographers at the forefront of debates on the implications of the politics of representation for the production of geography both in terms of fieldwork and writing (Katz, 1992; Nast, 1994; Mohammed, 2001), leading to ongoing reflections on these dilemmas (Asher, 2019).

For Spivak, an 'insider' can simultaneously be an 'outsider'. She argues that authorial claims based on one-dimensional experience of marginality serve to negate multiple differences which distinguish and mark the distance of those being represented from the researcher/writer. Thus 'Other' or 'Third World' intellectuals may be regarded by some as the 'authentic inhabitants of the margin' (Spivak, 1990a: 224), positioned to legitimately speak on behalf of their group, constituting and adjudicating both the 'limits of knowledge about the Other and the graphematic space the latter occupies/inscribes' (Spivak, 1990b: 175). But it is precisely the Third World intellectual's distance from the wholly Other – produced by the uneven experience of oppression – that positions some 'Others' closer to the 'Same', from which they derive the power to speak for the subaltern.

Another way in which questions of representation have been registered in Geography is in the history of the discipline. Spivak offers inspiration to those who would seek alternative histories and lineages of geographical writing in the age of formal European imperialism. To McEwan, therefore, Spivak offers some pathways to the recovery of subaltern agency and presence in the critical re-examination of colonial geography and exploration:

> Deconstructing the colonial archive also allows for a more subtle analysis of the production of geographical knowledge, and the possibility of decentring a putatively western tradition by viewing the production of those knowledges as a complex process of cultural exchange and negotiation, in which colonized populations played an active part. (McEwan, 1998: 374)

Similarly, Spivak offers some departures for re-thinking geographies of development by reconceptualising development as a project of worlding, whose origins and trajectory are imbued with both the potential for liberation and the reality of enduring forms of imperial, class and gender inequality (Radcliffe, 1994; Roy, 2020).

Geographers have also seized the deconstructive impulse in Spivak and others to problematise the discourse of 'globalisation' and the assumptions about its pervasiveness and putative power (**Gibson-Graham**, 1996). This also yielded engagement by geographers with ways that Spivak proposes 'planetary' as an alternative frame instead of globalisation (Sidaway et al., 2014). Spivak (2003) proposes that 'the planet [...] overwrite the globe', for:

> Globalization is the imposition of the same system of exchange everywhere. In the gridwork of electronic capital, we achieve that abstract ball covered in latitudes and longitudes, cut by virtual lines [...] The globe is on our computers. The planet is the species of alterity, belonging to another system. (Spivak, 2003: 72)

Key Advances and Controversies

For some, Spivak's density of writing and her mode of deconstruction may be off-putting – or worse. Terry Eagleton (1999), for example, called Spivak's *A Critique of Post-Colonial Reason* 'obscurantist'. **Edward Said** (1983) had earlier criticised the style of some literary theory that 'alienates' rather than communicates with the reader, something that for many is confirmed by Spivak's work. For Said, the dense jargon [of literary theory] 'obscures' social reality and 'encourage[s] a scholarship of "modes of excellence" very far from daily life' (Said, 1983: 4). Moreover, Spivak's avowed political aim is to understand the oppressed, the marginalised, the excluded, the subaltern. How can this be pursued through the act of reading, of deconstruction? How can Spivak reconcile such political aims with the charges of textual idealism that have sometimes been levelled at deconstruction, apparently confirmed in the contention that 'there is nothing outside of the text' (Derrida, 1976: 158)?

Yet Spivak (1988a) deploys deconstruction to contend with such problematic and limited understanding of texts. Her deconstruction follows Derrida's understanding of the text:

> The text is not the book; it is not confined in a volume, itself confined to the library. It does not suspend reference – to history, to the world, to reality [...] (Derrida, 1988: 137)

Rather it is:

> [...] differential network, a fabric of traces referring endlessly to something other than itself, to other differential traces. Thus the text over runs all the limits assigned to it so far (not submerging or drowning them in an undifferentiated homogeneity, but rather making them more complex, dividing and multiplying strokes and lines) – all the limits, everything that was set up in opposition to writing (speech, life, the world, the real, history, [...] every field of reference – to body or mind, to conscious or unconscious, politics, economics, and so forth). (Derrida, 1979: 84–5)

'There is nothing outside the text' might therefore mean texts contain multiple traces of the wider world, and/or that the world is rendered through a series of 'texts'; not simply literary texts and codes, but perhaps other inscriptions of difference and meaning (such as accounts and systems of value like money). Following Derrida, Spivak's work aims to make visible the constitution of the world in and through the fabric and fabrication of texts; for example, those that elaborate government policies or produce legal apparatus.

Two key areas of Spivak's contribution have been on postcolonial issues relating to class and gender. In relation to the former, Spivak reads Marx creatively. Her deconstructive approach to texts (including that of *Das Kapital* itself) sympathises with Marx's determination to expose the categories of value, economy and class to critical scrutiny (Castree, 1997). In her essay on *'Supplementing Marxism'*, Spivak rethinks the meaning and definition of socialism, which becomes a project that:

> [...] is not in opposition to the form of the capitalist mode of production. It is rather a constant pushing away – a differing and a deferral – of the *capitalist* harnessing of the social mobility of capital. (Spivak, 1995: 119)

In specific terms, Spivak (1989) illuminates the continued disempowerment of the 'subaltern' in postcolonial India. With reference to internal colonisation, transnational capitalism and new forms of imperial power, she explains how the national independence movement left the existing highly polarised class structure in South Asia more or less intact. The Subaltern Studies group of radical historians of India (see Chari, 2009) sought to rewrite a history from below, but Spivak engages with some of the limits to this. She has drawn on the work of Subaltern Studies to argue that Western Marxism has proven unable to represent the complexity of subaltern histories of insurgency and resistance. Spivak thus questions the ability of both conventional nationalist histories or of western Marxism (even as developed by the Subaltern Studies group) to recover and represent the historical roles of disenfranchised Indians (see Morton, 2003).

For Spivak, Marxism and feminism are vital political projects, but she argues these should be supplemented with deconstruction to provide those conceptual tools or vocabulary that would do justice to the everyday lives of marginalised (subaltern) women – both historically and in the present day – for such lives are complex, contradictory, and frequently obscured from view. They are written out of history, or invoked as passive masses or victims without their subjectivity intact and therefore represented minus their capacity to make history and geography. Moreover, the geopolitical category of the 'Third World' is – for Spivak – to be questioned in terms that regard it as not simply as a place or space that is a mere supplement to the Western history of capitalism. Instead, Spivak sees it as necessary (but neglected) constituent of that wider history; without which Western hegemony and capitalism could not have taken their exploitative course.

Rather than attempting to represent 'Third World' women, Spivak has thus sought to disrupt the codes and conventions of western knowledge that have sought to represent the Third World. In 'Literary representation of the subaltern' (1988b) Spivak provides a critical reading of 'Breast-giver' by Mahasweti Devi (1995). 'Breast-giver' follows the life of the protagonist Jashoda who is employed as a wet nurse in an upper-class, high-caste Bengali household. Spivak highlights the limits of the bourgeois 'Mother India' ideal, marking the failure of the Indian nationalist project to eliminate gendered-classed oppressions. She also draws on this work to point out to Western feminists that women's reproductive labour is not always unwaged and domestic. In 'Three women's texts and a critique of imperialism' (1985) Spivak reads Brontë's (1847) *Jane Eyre* against the hegemonic western feminist readings which she argues point to the complicity of Western feminism and imperialism. Such readings celebrate *Jane Eyre* as a text that privileges the narrative of its female protagonist who is read as a liberated western female individual. Spivak's reading finds Jane's liberation as an individual rests on the representation of Bertha Mason (the Jamaican Creole, who is the 'mad' first wife of Jane's fiancé Rochester), as 'Other' and 'not yet human' (Spivak, 1985: 247).

Spivak's (2003) *Death of a Discipline* returns to the issue of the production of disciplinary knowledge in the context of translation, globalisation and diversity. It draws attention to the ways in which the business (flows of money, bottom lines and profits) of knowledge production shapes fields such as comparative literature. Here, the US becomes a centre of publication and academic authority which then travels back to influence the sites where 'foreign' literatures originate. A similar interrogation of categories is sustained in *Other Asias* (2005); an eclectic set of essays about comparative areas studies, disciplinary knowledge and what is signified by the term 'Asia' in the context of geopolitical and geoeconomic shifts. In related terms, Spivak (2013: 468) asks what it means to compare nations and continents when:

> Comparison assumes a level playing field and the field is never level, if only in terms of the interest implicit in the perspective. It is, in other words, never a question of compare and contrast, but rather a matter of judging and choosing. (Spivak, 2013: 468)

Likewise, Spivak (as in her 1998 essay on 'Revisiting the "Global Village"' and subsequently in her 2013 book *An Aesthetic Education in the Era of Globalization*) interprets education as 'cultural talk' that recodes the disjuncture and difference whose parameters derive from capitalism and imperialism.

Throughout these and her extensive breadth of writings, teaching, talks and interviews, Spivak's attention to the 'margins' and the occlusion of difference, presence and power will continue to be instructive for those who take the time and trouble to engage carefully with her works and to tease out their mappings of enduring colonial legacies and indictments of neo-colonial complicities.

Spivak's Key Works

Spivak, G.C. (1974) *Myself Must I Remake: The Life and Poetry of W.B. Yeats*. New York, NY: Crowell.

Spivak, G.C. (1985) 'Three women's texts and a critique of imperialism', *Critical Inquiry*, 12: 243–61.

Spivak, G.C. (1987) *In Other Worlds: Essays in Cultural Politics*. New York, NY: Methuen.

Spivak, G.C. (1988a) 'Can the subaltern speak? Speculations on widow sacrifice', in C. Nelson and L. Grossberg (eds), *Marxism and the Interpretation of Culture*. London: Macmillan. pp. 271–313.

Spivak, G.C. (1988b) 'Literary representations of the subaltern', in R. Guha (ed), *Subaltern Studies*. New Delhi: Oxford University Press. pp. 86–111.

Spivak, G.C. (1989) 'Who claims alterity?', in B. Kruger and P. Mariani (eds), *Remaking History*. Seattle, WA: Bay Press. pp. 269–92.

Spivak, G.C. (1990a) 'Poststructuralism, marginality, postcoloniality and value', in P. Collier and H. Geyer-Ryan (eds), *Literary Theory Today*. Cambridge: Polity. pp. 219–44.

Spivak, G.C. (1993) *Outside in the Teaching Machine*. New York, NY: Routledge.

Spivak, G.C. (1994) 'Bonding in difference', in A. Arteaga (ed), *An Other Tongue: Nation and Ethnicity in the Linguistic Borderlands*. Durham, NC: Duke University Press. pp. 273–83.

Spivak, G.C. (1995) 'Supplementing Marxism', in C. Cullenberg and B. Magbus (eds), *Wither Marxism: Global Crises in International Perspective*. New York, NY: Routledge. pp. 109–19.

Spivak, G.C. (1999) *A Critique of Post-Colonial Reason: Toward a History of the Vanishing Present*. Cambridge, MA: Harvard University Press.

Spivak, G.C. (2003) *Death of a Discipline*. New York, NY: Columbia University Press.

Spivak, G.C. (2008) *Other Asias*. Malden, MA: Blackwell.

Spivak, G.C. (2013) *An Aesthetic Education in the Era of Globalization*. Cambridge, MA: Harvard University Press.

Secondary Sources and References

Asher, K. (2019) 'Fieldwork', in Antipode Editorial Collective, T. Jazeel, A. Kent, K. McKittrick, N. Theodore, S. Chari [...] and M.W. Wright (eds), *Keywords in Radical Geography: Antipode at 50*. Chichester: Wiley. Available from https://onlinelibrary.wiley.com/doi/book/10.1002/9781119558071 (accessed November 14, 2023).

Castree, N. (1997) 'Invisible Leviathan: Speculations on Marx, Spivak and the question of value', *Rethinking Marxism*, 9: 45–78.

Chari, S. (2009) 'Subaltern Studies', in D. Gregory, R.J. Johnston, G. Pratt, M. Watts and S. Whatmore (eds), *The Dictionary of Human Geography* (fifth edition). Oxford: Blackwell-Wiley. pp. 727–28.

Cheong, S. (2019) 'A subaltern geopolitics?', *Geopolitics*, 24: 989–94.

Derrida, J. (1976) *Of Grammatology* (trans. G.C. Spivak). Baltimore, MD: Johns Hopkins University Press.

Derrida, J. (1979) *Spurs: Nietzsche's Styles/Eperons: Les Styles de Nietzsche* (trans. B. Harlow). Chicago, IL: University of Chicago Press.

Derrida, J. (1988) *Limited Inc.* (trans. S. Weber). Evanston, IL: Northwestern University Press.

Devi, M. (1995) 'Breast-giver', in S. Thames and M. Gazzaniga (eds), *The Breast: An Anthology*. New York, NY: Global City Press. pp. 86–111.

Eagleton, T. (1999) 'In the gaudy supermarket', *London Review of Books*, 21: 10.

Gibson-Graham, J.K. (1996) *The End of Capitalism (As We Knew It): A Feminist Critique of Political Economy*. Oxford: Blackwell.

Jazeel, T. (2019) *Postcolonialism*. London: Routledge.

Jazeel, T. and Legg, S. (eds) (2019) *Subaltern Geographies*. Athens, GA: University of Georgia Press.

Johnson, B. (1980) *The Critical Difference*. Baltimore, MD: Johns Hopkins University Press.

Katz, C. (1992) 'All the world is staged: Intellectuals and the projects of ethnography', *Environment and Planning D: Society and Space*, 10: 495–510.

McDowell, L. (1993) 'Space, place and gender relations, part II: Identity, difference, feminist geometries and geographies', *Progress in Human Geography*, 17: 305–18.

McEwan, C. (1998) 'Cutting power lines within the palace? Countering paternity and Eurocentrism in the "Geographical tradition"', *Transactions of the Institute of British Geographers*, 23: 371–84.

Mohammad, R. (2001) 'Insiders' and/or 'outsiders': Positionality theory and praxis', in M. Limb and C. Dwyer (eds), *Qualitative Methodologies for Geographers*. London: Arnold. pp. 101–17.

Morris, R. (ed) (2010) *Can the Subaltern Speak? Reflections on the History of an Idea*. New York, NY: Columbia University Press.

Morton, S. (2003) *Gayatri Chakravorty Spivak*. London: Routledge.

Nast, H.J. (1994) 'Women in the field: Critical feminist methodologies and theoretical perspectives', *Professional Geographer*, 46: 54–60.

Radcliffe, S.A. (1994) '(Representing) post-colonial women: Authority, difference and feminisms', *Area*, 26: 25–32.

Radcliffe, S. (2022) *Decolonizing Geography: An Introduction*. Chichester: Wiley.

Roy, A. (2020) "The shadow of her wings": Respectability politics and the self-narration of geography', *Dialogues in Human Geography*, 10 (1): 19–22.

Said, E. (1983) *The World, the Text, the Critic*. Cambridge, MA: Harvard University Press.

Sidaway, J.D., Woon, C.Y. and Jacobs, J.M. (2014) 'Planetary postcolonialism', *Singapore Journal of Tropical Geography*, 35: 4–21.

Sparke, M. (2005) *In the Space of Theory: Postfoundational Geographies of the Nation-State*. Minneapolis, MN: University of Minnesota Press.

Spivak, G.C. (1986) 'Imperialism and sexual difference', *Oxford Literary Review*, 7 (1–2): 225–40.

Spivak, G.C. (1990b) 'Versions of the margin: Coetzee's Foe reading Defoe's Crusoe/ Roxana', in J. Arac and B. Johnson (eds), *Consequences of Theory*. Baltimore, MD: Johns Hopkins University Press. pp. 154–80.

Spivak, G.C. (1998) 'Cultural talks in the hot peace: Revisiting the "global village"', in P. Cheah and B. Robbins (eds), *Cosmopolitics: Thinking and Feeling Beyond the Nation*. Minneapolis, MN: University of Minnesota Press. pp. 329–50.

Spivak, G.C. (2014) 'Scattered speculations on geography', *Antipode*, 46: 1–12.

53
Nigel Thrift

Barney Warf, University of Kansas

Biographical Details and Theoretical Context

Working at the intersections of economic, urban, and cultural geography, Nigel Thrift has long been recognised as one of the discipline's most imaginative, articulate, and productive scholars, and he is one of human geography's most heavily-cited figures. In an exceptionally lengthy series of papers and books, Thrift has made major contributions to geographical thought, initially through his focus on time and time geographies, then through his interest in critical human geography; post-Fordism and geographies of finance and banking; and latterly, his focus on urban life, actor-network theory and non-representational theory. It is the latter that he is perhaps most associated with, and both he and many of his former PhD students have been crucial in articulating the contours of a 'non-representational' geography.

Born in Bath in 1949, Thrift earned his BA in Aberystwyth, Wales, and a PhD from the University of Bristol in 1979, supervised by ***Peter Haggett***. He was stationed briefly in Leeds (1976–1977), spent four years (1979–1983) as a Research Fellow at the Australian National University, and three years (1984–1987) as a Lecturer and Reader at Saint David's University College, Lampeter, Wales. In 1987 he returned to Bristol as Professor in the School of Geographical Sciences before moving to administrative roles: between 2003 and 2006 he was Pro-Vice-Chancellor for Research at Oxford University after which he served as Vice-Chancellor at the University of Warwick. Nonetheless, while undertaking these sizeable leadership roles, he continued to publish, and also had visiting Professorships at institutions including Tsinghua University in Beijing, the University of Vienna, Macquarie University, the Institutes of Advanced Study in Sweden and the Netherlands. He was the Executive Director of Schwarzman Scholars 2016–17. Thrift is also an Academician of the Association of Learned Societies in the Social Sciences and became a Fellow of the British Academy in 2003. He is the recipient of numerous awards, including the Scottish Geographical Society's Gold Medal in 2008, the Royal Geographical Society's Victoria Medal in 2003, the Medal of the University of Helsinki, and Distinguished Scholarship Honors from the Association of American Geographers in 2007, an Honorary LLD from Bristol in 2010 and another Honorary LLD from Monash University in 2013. In 2015 he was knighted for his services to higher education. He also served for many years as managing editor of *Environment and Planning A*, and in 1982 co-founded *Environment and Planning*

D: Society and Space, one of the most important journals for promoting interdisciplinary thinking on space and place.

Thrift's earliest publications saw him working with Australian geographer Don Parkes on time in the city (Parkes and Thrift, 1975), extending the work of ***Torsten Hägerstrand*** and the Lund group to develop time-geography. This interest in time and temporality was sustained for many years (e.g., May and Thrift, 2001; Glennie and Thrift, 2009), and included studies of the restructuring of 'time consciousness' in contexts ranging from the mediaeval to the contemporary. The latter included consideration of the imperatives of commodity production and capital accumulation (Thrift, 1988), a notion that closely echoed E.P. Thompson's famous portrayal of the social dynamics of factory time. It was during the fluorescence of social theory in the 1980s and 1990s, however, that he came into his own as a leading theoretician of geographies as situated social practices. Throughout, Thrift has displayed a Marxist inspiration of the non-structuralist variety, including an abiding concern for class, a position that he later extended to a variety of non-class based forms of social determination (Thrift and Williams, 1987). Thrift has long insisted that geographies are produced by people in pre-discursive, practical ways. Crucially, he played a key role in introducing ***Anthony Giddens***' notion of structuration into geography (Thrift, 1985a): his paper 'On the determination of social action in space and time' (Thrift, 1983) was a landmark essay that both situated structuration theory within the broader intellectual context of micro-macro divisions in the social sciences and pointed toward the need for a reconceptualisation of regional geography.

The publication of the two-volume *New Models in Geography* (Peet and Thrift, 1989) heralded the ascendancy of this new brand of political economy to a central and pivotal status within geography, ostensibly supplanting the 'old models' of location theory, including those associated with his former supervisor ***Peter Haggett***, but also offering a rejoinder to the Marxist geography that appeared fixated with capital and class at the expense of practice and agency. Extending interpretations concerned with class into new domains, the essays in these two influential volumes offered comprehensive and insightful interrogations of the relations of society and space via studies of the production of nature, the state, gender, uneven development, the nation-state, urbanisation, race, culture, and postmodernism.

Thrift's interest in the differential regional effects of globalisation generated a series of case studies. He has worked extensively on the restructuring of the British space-economy in the face of globalisation and mounting social and regional polarisation, including the rise of the greater London region (Thrift, 1987). He also examined other European regions in various contexts (Amin and Thrift, 1992), including the rising role of dense human networks in leading innovative regions characteristic of advanced post-Fordist production systems. Thrift also worked extensively on Pacific Asia, particularly Vietnam (Thrift and Forbes, 1986; Forbes and Thrift, 1987). These empirical analyses complemented his insistence that regions mattered not only ontologically, i.e., as units of empirical analysis, but also epistemologically, i.e., that no social process unfolds in the same way in different places. The latter line of thought would prove to hold wide appeal to the growing legions of geographers who insisted that local uniqueness mattered.

Throughout the 1980s, the new critical regional geography developed linkages between top-down global political economy and bottom-up concerns for living actors and everyday life. Thrift played a key role in articulating this vision of geography, defending it against both positivist and Marxist critics for whom regional analysis could, by definition, scarcely amount to more than reconstituted empiricism. Unlike chorology, however, the reconstituted regionalism was both theoretically informed and paid great attention to the connections among regions, their linkages to the global division of labour, and the sticky issue of the production of spatial scale. The new regionalism made local uniqueness a valid object of scholarly scrutiny, demonstrating that the local mattered and that no social process played out exactly in the same way in different time–space contexts. The recovery of regions for social theory obligated the discipline to delve into the intricate dynamics of local actors who populated them and made them work: thus the localities school and structuration theory's focus on everyday life were mutually complementary. Yet this relational materialist perspective maintains that the global space of flows does not float in a disembodied space: rather, global flows and connections are constructed by human beings who are always embedded in networks of power and knowledge that are themselves part of an ever-changing structural context.

This concern with the intersections between the local and global extended into his work on the most abstract and globalised of economic sectors, finance. On this topic, Thrift entered into a fecund relationship with Andrew Leyshon that generated a series of works concerning the international financial system, British financial capital, and banking networks in London (Thrift and Leyshon, 1992; 1994; Leyshon and Thrift, 1997). These works both elucidated the impacts of electronic money and national deregulation and simultaneously demystified money by stressing the key roles played by interpersonal relations predicated on trust and face-to-face contact, including the critical but often overlooked roles played by local milieu, trust, and institutional thickness, i.e., the cultural context in which financial transactions are always and everywhere embedded. Money, even in its most rarefied electronic form, is thus much more than an abstract economic relation, for its real power lies in its symbolic value and links to national and global distributions of social power (Thrift, 2001). Like all social relations, financial systems are inevitably discursively constituted through particular social-cultural practices in different times and places. Such an approach brought a badly needed human quality to a topic dominated by ethereal understandings devoid of actors, and is vital if the new forms of soft managerial capitalism are to be understood in all their complexity.

Throughout his career, Thrift has been fascinated by cities as dense nuclei of economic and social transactions, on the changing possibilities for economic development afforded by globalisation (Thrift, 1987), and the mutations in urban form and life that such transformations entailed (Pile and Thrift, 2000). Cities, he has argued, particularly in the knowledge-driven economies of soft capitalism, both reflect and constitute the changing dynamics of social life as contingent constellations of meaning (Amin and Thrift, 2002; 2017). Yet he also delved into issues of big data and smart technologies which appear to generate 'sentient' cities which possess agency (Thrift, 2014). Along with economic geographer Ash Amin, his two books *Cities: Reimagining the Urban*

(2002) and *Seeing Like a City* (2017) set out an agenda for urban studies which acknowledges this dispersed agency and tentatively provides an introduction to a de-centred urban theory. In *Killer Cities* he (Thrift, 2021) delivers on this agenda by examining how cities are organised to facilitate mass animal deaths, bringing the non-human to the forefront of urban studies. Here, his interest in slaughter and extinction is not surprising given he has long exhibited an interest in violence (Thrift and Forbes 1986; Thrift, 2007).

Alongside these important interventions in economic, regional and urban geography, Thrift is perhaps best known in the twenty-first century as the progenitor of a non-representational geography. Through his work in the 1980s and 1990s, Thrift had played an influential role in getting geographers acquainted with various forms of post-structuralism, particularly as it pertains to subjectivity, identity, language, representation, discourse, and performativity (Thrift, 2000; 2008). Although he has never referred to himself as a postmodernist per se, his theoretical orientation reveals a distrust of broad meta-theories, an appreciation for local uniqueness, and an emphasis on social structures and change as inherently open-ended and contingent. In the twenty-first century this consolidated into a putative non-representational geography – sometimes termed a more-than-representational geography – which considers the pre-cognitive aspects of social practice as crucial in the making of the world. Here, Thrift drew on Massumi's interpretations of **Gilles Deleuze**'s thought, the work of German philosopher Sloterdijk and Spinoza's ethics, to articulate a geography where questions of affect and atmosphere are considered vital in the making of post-humanist worlds.

Spatial Contributions

If there is any theme that permeates Thrift's multitudinous works, perhaps it is his sustained interest in connectivity of one sort or another. His papers have, in different ways and contexts, focused on the diffusion of information (Thrift, 1985b), the reciprocal relations between knowledge and power, and increasingly globalised exchanges of human, symbolic, and financial capital. Thus, he exhibits a consistent concern for the flows of information and knowledge that shape individuals' conceptions of themselves and one another and that play a major role in the routinised reproduction of social relations. Invoking actor-network theory, Thrift extended the insights of **Bruno Latour** and actor network theory, which effectively abandons the distinction between agents and structures altogether. This view focuses on the exercise of power by actors rather than just the embeddedness of power in networks. Throughout his career, Thrift has worked assiduously to portray geographies as embodied, embedded, contingent, and ever changing, harnessing the fluidity of spatial relations to demonstrate how they are embodied within changing relations of power.

Thrift's writings paint a thorough portrait of contemporary, globalised, neoliberal capitalism, dominated by finance and digital information technology. This historically-specific constellation of discourses and practices includes novel forms of production and consumption that reflect, and drive, its pronounced penetration of culture and consciousness. The symbolic economy calls for new forms of performativity for actors in order to function and succeed (Thrift, 2000). In analysing systems of signification, including discourses, language, texts, and representations, Thrift pioneered the use and adoption of a non-representational theory of action that stresses performative, embodied knowledges (Thrift, 1997). Non-representational theory, which has been widely debated, stresses the pre-conscious dimensions of capitalism, its expression through worlds of affect and emotion that underlie, and guide, the intentions and behaviours of actors (Thrift, 2004; 2007). Thrift's interpretation of affect situates it within the political role of technologies in Western democracies.

Central to this process was the adoption of information technology and electronic communications, key indicators of post-Fordist soft capitalism and its attendant digitised, virtual dimensions (Thrift, 1995; 1997). The massive time–space compression unleashed by digital globalisation has thus altered both the field of possibilities of nation-states and of individual actors in everyday life: the renewed regionalism and reconstituted rhythms of daily life are two sides of one coin. Digital capitalism operates, he maintains, in a manner different from its predecessors: it is more mobile, more flexible, more insistently symbolic in nature, and lodges itself into the innermost recesses of the human psyche. At the level of the corporation and everyday life, he has also taken an interest in the possibilities extended by the widespread use of computational technology and mobile telecommunications. In one particularly original paper, for example, he argued that new developments in toy technology were producing commodities that were capable of engaging with, and producing, the world (Thrift, 2003). In sum, he argued the contingent, simultaneously determinant interactions between culture and technology form a long-standing theme threaded throughout his papers.

Most recently, given his administrative experience, Thrift has turned his attention to the social role of British research universities (Thrift, 2016a; 2022). Adopting a Latourian perspective, he views institutions of higher education as assemblages of people, buildings, equipment, and power. He has also criticised dominant narratives that portray universities in a state of perpetual decline.

Key Advances and Controversies

Geographers' engagements with Thrift's ideas have taken a variety of forms, of which five are suggested here. First, the popularity of structuration theory in the discipline owed no small part to his writings, which sensitised the discipline to the socially constructed and contingent nature of space. Thus, human geographers often came to frame their topics, questions, and responses within the terms of intended actions and unintended results, an approach that elevated everyday life to a central role in social theory.

Second, his repeated calls for an historical geography of time were important both in moving the field toward an emphatically dynamic approach to theoretical and empirical issues and underscored the social construction of time, which to all appearances appears natural. By denaturalising time, Thrift's work pointed toward the broader denaturalisation of other phenomena, including language, power, poverty, representations, and the body.

Third, economic geographers have been motivated by his books and papers to emphasise the role of human actors in varying historical contexts, demystifying even money, the most abstract of social relations. For example, international banking networks were shown to be the very human products of business executives enmeshed in culturally-specific webs of meaning. Thrift's corpus of writings also advanced the field beyond Marxism, with its frequent teleological social predetermination and reification of class.

Fourth, his sensitivity to regions, not simply as receivers of broader changes but as generators in their own right, has made an appreciation for local uniqueness de rigueur within the field. Thrift and others illustrated that no social process unfolds in the same way across different places, raising the significance of context in social explanation to a central position. In this reading, the local was far from simply a handmaiden to the global, and the critical question of spatial scale was problematised across several intellectual perspectives.

Fifth, geographers have been inspired by his post-structuralism, and Thrift's work fostered a critical sense of discourse and representation within the field, contrary to much conventional thinking in cultural geography, generating marked interest in questions of practice and agency and how it is sutured to space in ways that are pre-cognitive and embodied. In large part due to Thrift's efforts, actor-network theory became normalised in human geography, and his book *Non-Representational Theory: Space, Politics, Affect* (Thrift, 2008) is a much-cited touchstone.

Yet for all this, Thrift's work is not without controversy. Despite a clear interest in ethics and the politics of the left (see, for example, Amin and Thrift, 2013), as well as his writing on emancipatory cities (Thrift, 2007), the turn to non-representational theory has sometimes been seen as politically problematic, with affect granted too deterministic a role in the making in the world (Barnett, 2008). His book *Killer Cities* (2021) was also criticised as overly anthropocentric given the rising concern in geography for non-human actors (see, for example, **Kathryn Yusoff**).

Despite this, Thrift's contributions to human geography have left it a far more sophisticated, subtle, and nuanced field. Here, his engagement with continental philosophy, and willingness to engage with interdisciplinary understandings of embodiment, agency and the performative, mark him out as a major influence on twenty-first century geography.

Thrift's Key Works

Amin, A. and Thrift, N. (2013) *Arts of the Political: New Openings for the Left.* Durham, NC: Duke University Press.

Crang, M. and Thrift, N. (eds) (2000) *Thinking Space*. London: Routledge.

Glennie, P. and Thrift, N. (2009) *Shaping the Day: A History of Timekeeping in England and Wales 1300–1800*. Oxford: Oxford University Press.

May, J. and Thrift, N. (eds) (2001) *TimeSpace: Geographies of Temporality*. London: Routledge.

Parkes, D.N. and Thrift, N. (1975) 'Timing space and spacing time', *Environment and Planning A*, 7 (6): 651–70.

Thrift, N. (1981) 'Owners' time and own time: The making of a capitalist time consciousness, 1300–1880', in A. Pred (ed), *Space and Time in Geography: Essays Dedicated to Torsten Hägerstrand*. Lund Studies in Geography Series B, No. 48.

Thrift, N. (1983) 'On the determination of social action in space and time', *Environment and Planning D: Society and Space*, 1 (1): 23–57.

Thrift, N. (1996b) *Spatial Formations.* Thousand Oaks, CA: Sage.

Thrift, N. (2000) 'Performing cultures in the new economy', *Annals of the Association of American Geographers*, 90 (4): 674–92.

Thrift, N. (2005) *Knowing Capitalism.* Thousand Oaks, CA: Sage.

Thrift, N. (2008) *Non-representational Theory: Space, Politics, Affect*. London: Routledge.

Thrift, N. (2011) 'Lifeworld, Inc–and what to do about it', *Environment and Planning D: Society and Space*, 29 (1): 5–26.

Thrift, N. (2021) *Killer Cities*. Thousand Oaks, CA: Sage.

Thrift, N. (2022) *The Pursuit of Possibility: Redesigning Research Universities*. Bristol: Policy.

Thrift, N. and Olds, K. (1996) 'Refiguring the economic in economic geography', *Progress in Human Geography*, 20 (3): 311–37.

Secondary Sources and References

Amin, A. and Thrift, N. (1992) 'Neo-Marshallian nodes in global networks', *International Journal of Urban and Regional Studies*, 16 (4): 571–87.

Amin, A. and Thrift, N. (2002) *Cities: Reimagining the Urban*. Cambridge: Polity.

Amin, A. and Thrift, N. (2017) *Seeing Like a City*. Cambridge: Polity.

Barnett, C. (2008) 'Political affects in public space: Normative blind-spots in non-representational ontologies', *Transactions of the Institute of British Geographers*, 33 (2): 186–200.

Forbes, D. and Thrift, N. (eds) (1987) *The Socialist Third World: Urban Development and Territorial Planning*. Oxford: Blackwell.

Leyshon, A. and Thrift, N. (eds) (1997) *Money/Space: Geographies of Monetary Transformation*. London: Routledge.

Peet, R. and Thrift, N. (eds) (1989) *New Models in Geography: The Political-Economy Perspective*. Boston, MA: Unwin-Hyman.

Pile, S. and Thrift, N. (eds) (2000) *City A–Z*. London: Routledge.

Thrift, N. (1985a) 'Bear and mouse or bear and tree? Anthony Giddens's reconstitution of social theory', *Sociology*, 19: 609–23.

Thrift, N. (1985b) 'Flies and germs: A geography of knowledge', in D. Gregory and J. Urry (eds), *Social Relations and Spatial Structures*. Basingstoke: Macmillan. pp. 366–402.

Thrift, N. (1987) 'The fixers: The urban geography of international commercial capital', in M. Castells and J. Henderson (eds), *Global Restructuring and Territorial Development*. London: Sage. pp. 203–33.

Thrift, N. (1988) 'Vivos voco: Ringing the changes in the historical geography of time consciousness', in T. Schuller and M. Young (eds), *The Rhythms of Society*. London: Routledge. pp. 169–212.

Thrift, N. (1995) 'A hyperactive world', in R.J. Johnston, P. Taylor and M. Watts (eds), *Geographies of Global Change*. Oxford: Blackwell. pp. 82–95.

Thrift, N. (1997) 'The rise of soft capitalism', in A. Herod, S. Roberts and G. Toal (eds), *An Unruly World? Globalisation and Space*. London: Routledge. pp. 25–71.

Thrift, N. (2001) '"It's the romance, not the finance, that makes the business worth pursuing": Disclosing a new market culture', *Economy and Society*, 4: 412–32.

Thrift, N. (2003) 'Closer to the machine? Intelligent environments, new forms of possession and the rise of the supertoy', *cultural geographies*, 10 (4): 389–407.

Thrift, N. (2004) 'Intensities of feeling: Towards a spatial politics of affect', *Geografiska Annaler*, 86: 57–78.

Thrift, N. (2007) 'Immaculate warfare: The spatial practices of extreme violence', in D. Gregory and A. Pred (eds), *Violent Geographies: Fear, Terror and Political Violence*. London: Routledge. pp. 279–300.

Thrift, N. (2014) 'The "sentient" city and what it may portend', *Big Data & Society* 1 (1): 1–21.

Thrift, N. (2016a) 'The university of life', *New Literary History*, 47 (2): 399–417.

Thrift, N. and Forbes, D. (1986) *The Price of War: Urbanization in Vietnam, 1954–1985*. Boston, MA: Allen and Unwin.

Thrift, N. and Leyshon, A. (1992) 'Liberalisation and consolidation: The single European Market and the remaking of European financial capital', *Environment and Planning A*, 24: 49–81.

Thrift, N. and Leyshon, A. (1994) 'A phantom state: The de-traditionalization of money, the international financial system and international financial centers', *Political Geography*, 13 (4): 299–327.

Thrift, N. and Williams, P. (eds) (2014 [1987]) *Class and Space: The Making of Urban Society* (2nd edition). New York, NY: Routledge.

54
Anna Tsing

Jamie Lorimer and Eben Kirksey, University of Oxford

Biographical Details and Theoretical Context

Anna Lowenhaupt Tsing is Professor of Anthropology at the University of California, Santa Cruz. She received her BA from Yale University and completed her MA (1976) and PhD (1984) at Stanford University. Her initial ethnographic field research was in the Meratus Mountains of South Kalimantan, part of Borneo, which is a province of Indonesia. Her first book, *In the Realm of the Diamond Queen* (1993), was a path-breaking account of cultural hybridity, transculturation, and marginality in Kalimantan. It spoke to the concerns of feminism and post-structuralist theory, and is considered emblematic of the 'writing culture' turn in anthropology, with its attention to the poetics and politics of ethnography and representation (Clifford and Marcus, 1986).

Tsing's second book, *Friction* (2004), accounted for the rapid deforestation of Kalimantan in the context of spatial, discursive, and metaphoric constructions of the local and the global. In it she tracked how capital, scientific knowledge, and political ideas produce 'friction' at sites where people often profoundly misunderstand one another, talking of the 'grip of worldly encounter' where global flows coagulate and congeal with local ways of life (Tsing, 2005: 1). Her work was informed by long-term relationships with Meratus Dayak elders and Indonesian environmentalists, which underpinned a measured affirmation of indigenous approaches. In an essay entitled 'Indigenous voice' (2007: 33), Tsing argues that the global Indigenous movement that was emerging in the 1990s 'endorses authenticity and invention, subsistence and wealth, traditional knowledge and new technologies, territory and diaspora'.

In 2010 Tsing received a Guggenheim Fellowship to support an ethnography of the global trade in the matsutake mushroom, a highly prized gift in Japan. In 2015 she published *The Mushroom at the End of the World*, a widely-read book that helped pioneer multispecies and new materialist approaches in the Environmental Humanities. She suggests that we should understand that 'human nature is an interspecies relationship' and explore nature as the outcome of environmental disturbance and the messy dynamics of capitalism on resource frontiers. Her focus on a mushroom that thrives in 'blasted landscapes' – e.g., the post-atomic ruins of Hiroshima, Japan or formerly clear-cut industrial forests – helps her consider the 'possibilities of coexistence' within the ruined landscapes of the Anthropocene.

In 2013, Tsing was granted the Niels Bohr Professorship at Aarhus University in Denmark, where she helped run a transdisciplinary program exploring the Anthropocene, named the Aarhus University Research on the Anthropocene (AURA) research centre. The project ran for a five-year period until 2018 and proved highly productive leading to the publication of a host of collaborative articles, an edited collection *Arts of Living on a Damaged Planet: Ghosts and Monsters of the Anthropocene* (2017), and the *Feral Atlas* (2020), a collaborative digital humanities project. Since the end of AURA, Tsing has focused on the social and ecological problems associated with plantation agriculture and has become a member of the high-profile set of social scientists giving commentary on the Anthropocene and Plantationocene.

Spatial Contributions

Anna Tsing has made a series of important contributions to spatial theory, including innovative accounts of globalisation and the development of more-than-human approaches to place. In her early work, Tsing was part of the wider postcolonial project aiming to provincialise Western thought. Gazing through the wrong end of a telescope from the standpoint of Indonesia, the luminary scholar of postcolonial nationalism **Benedict Anderson** (1998) imagines that he can see an image of Europe that is smaller, miniaturised, distant, and diminished. Tsing offers a contrasting vision of Europe and global landscapes of power that diffracts conventional geographic images with funhouse mirrors. Her key interlocutors in Indonesia told stories about distant lands that reflected the words of foreigners as if through 'a wall of broken mirrors fastened at odd angles, overlapping and askew' (Tsing, 1993: 253).

In the Realm of the Diamond Queen features the voices of Meratus Dyak elders, who are masters of fabulation and parody, understanding global geography from the margins. Here, Tsing decodes jumbled Indigenous accounts of spiritual landscapes. She is attentive to how Indigenous peoples, who are rooted in a particular place, creatively borrow from other places and peoples as well as from regional, national, and international sources of power (Tsing, 1993: 260). She relates both bizarre and serious accounts of new local and global histories that displace conventional colonial, national, and postcolonial stories about territory, sovereignty, and belonging.

The central ideas of *Friction* (2005) emerge from the 'zones of awkward engagement' in the contact zones of Indonesia's forested hinterlands, where the Meratus Dyak encounter (often illegal) logging ventures, environmentalists and activists who oppose deforestation, as well as soldiers who defend legitimate and illegitimate capitalist enterprises. She traces here how global connections come to life in 'friction', as rubber meets the road to enable and exclude particular trajectories, processes, and dreams. She writes that 'friction is not a synonym for resistance [...] Hegemony is made as well as unmade with friction' (Tsing, 2004: 6). The frontiers of 'salvage accumulation' – which she defines as the process through which firms amass capital without controlling the conditions under which commodities are produced – are key spaces and places in *Friction*. Tsing (2005: 28) describes the frontier as being on the 'edge of space and time: a zone of not yet – not yet mapped, not yet regulated. It is a zone of unmapping: even in its planning, a frontier is imagined as unplanned'.

This interest in salvage economies and their global connections comes to the fore in *The Mushroom at the End of World*, which pioneers a multi-sited research design. Anthropologists have been conducting multi-sited ethnographic research since 1995, when George Marcus introduced new strategies for following people, conflicts, and 'things'. Tsing pioneered multi-sited methods for following a *living* thing: a species of mushroom, tracing the global chains of connection that bring matsutake mushrooms from dynamic multispecies landscapes in diverse places – from Europe, to the hinterlands of China, and the Pacific Northwest of the US – to consumers in Japan.

Tsing's final set of spatial contributions comes from current work on the social and ecological consequences of the rise of global plantation agriculture. Working with other anthropologists and historians (including her close colleague **Donna Haraway**) she developed the concept of the Plantationocene (Mitman et al., 2019) as an alternative label for the Anthropocene, situating planetary rupture in the histories of colonialism and the Columbian exchange of people and organisms between the 'old' and 'new' worlds. She builds from her analysis of matsutake forest ecologies to map the 'patchy Anthropocene' (Tsing et al., 2019), contrasting the 'Holocene refugia' created by sustainable agroforestry with the Plantationocene ecologies common to industrial forestry and the production of commodity crops like palm oil, soy, and bananas. She argues that agroforestry helps protect biodiversity and indigenous livelihoods, while the plantation model tends towards violent 'modular simplification', enclosing lands, excluding people, and creating hotspots for pathogenesis (the emergence of weeds, pests, and zoonotic diseases) and enabling their 'proliferation' along the networks of globalisation (Tsing, 2017).

In her later work, Tsing has also been integral in pioneering approaches to multispecies ethnography and more-than-human conceptualisations of space and place. Multispecies ethnography conjoins approaches from ethnography, ethology, and ecology to trace the entangled relations between humans and a wide range of nonhumans (Kirksey and Helmreich, 2010). It delivers rich descriptions of place that afford agency to animals, plants, and microbes – as well as human infrastructure – recognising their ability to shape spatial patterns and processes. It develops in part from the Actor-Network Theory of **Bruno Latour** (and others), **Gilles Deleuze's** concept of the assemblage, and the synthesis of their work by geographers including Sarah Whatmore. A multispecies approach also attends to the practical, embodied, and affective dimensions of place, connecting with parallel developments in non-representational theory pioneered by figures such as **Nigel Thrift** and feminist scholars including **Sara Ahmed** and Elizabeth Grosz.

Tsing elaborates her approach to multispecies ethnography in her mushroom book, offering a relational conception of situated human–environmental interaction and a musical 'polyphonic' (Tsing, 2015) conception of place. By following the mushroom through time and space she contrasts the temporalities that enable both the flourishing of mushrooms and the resurgence of agroforestry relations, with the plantation logics that lead to the 'feral proliferation' of weeds in the Anthropocene (Tsing, 2017). Here Tsing is careful not to vilify any individual organism, but focuses on the relations and intensities that make them pathological.

In *The Mushroom at the End of the World*, Tsing (2015: 238) takes readers to today's out-of-the-way places, to the edges of capitalist encroachment, to blasted landscapes where she finds things that are 'strange and wonderful'. She draws these far-flung places together to provide an account of 'the end of the world', and in writing about the bleak contemporary moment she is attentive to the ways that capitalist extraction, industrialisation, and consumer practices have had calamitous effects on the planet. Nonetheless, she finds 'contaminated diversity' in the ruins of capitalism, tracing how people are eking out an existence in precarious landscapes. Cautious optimism is emerging as people forage and garden in the ruins – creating conditions of life for cherished fungi. Even as worlds end, foragers find joy and multisensory delight with the aroma and taste of freshly picked mushrooms. While writing about mushrooms as companion species (see Haraway, 2008), she insisted that anthropologists start paying attention to the 'unruly edges' of multispecies landscapes (Tsing, 2012). The seams of empire constitute disordered but productive edges, she suggests, that offer spaces to forage for edible and medicinal mushrooms.

In her later work Tsing (2019) has developed interdisciplinary methodologies for mapping proliferation and resurgence through collaborations with natural scientists, developing novel 'arts of noticing' (Tsing, 2015). In one compelling speculative example she explores how humanist geographical imaginaries of territories are scrambled by tracing the global aerial circulation of the fungal spores that drift across oceans, riding air currents in a networked biogeography that troubles clear-cut distinctions between native and alien species (Tsing, 2014). Alongside Haraway and Latour, she argues that social scientists should make new alliances with the biological sciences in the aftermath of the science wars and in the face of contemporary environmental harms. She finds inspiration in post-genomic conceptions of the human and in symbiotic models of human–nonhuman collaboration, affirming a scientific ontology that stands in contrast to the selfish bounded individual of the modern synthesis in biology and the social theory founded upon it.

Key Advances and Controversies

Tsing has been at a forefront of three significant conceptual developments in the social sciences in the twenty-first century. First, Tsing's focus on reading the global from the margins and on tracing global connections paralleled wider developments in Geography and Anthropology concerned with mapping patterns of globalisation and 'following the things' (Cook, 2004) along heterogeneous commodity chains. Tsing's work resonates with feminist and heterodox approaches to economic geography, including the writings of **Doreen Massey** and **J.K. Gibson-Graham** on the importance of 'reading for difference', pushing up against prevalent modes of critique offered by their neo-Marxist contemporaries (including **Neil Smith** and **David Harvey**). Tsing has developed a distinct, playful, and affirmative style, interweaving critique with comedy and parody, and experimenting with different genres of representation including visual arts, science fiction, and opera. She deliberately eschews what Latour terms the 'negative' mode of critique that fails to make an alliance with the worlds it describes.

Second, in her work in Indonesia and then on the matsutake and plantation ecologies, Tsing centres the perspectives of Indigenous peoples, alongside marginal workers. She has been at the forefront of scholarship that centres Indigenous perspectives without essentialising them as premodern, disconnected or necessarily ecologically benign. Instead, her accounts start from the margins and amidst the ruins of colonial capitalism, exploring diverse knowledge practices, survival strategies, and modes of multispecies co-existence that emergence in spite of exploitation and degradation. In contrast to the powerful models of geo-engineering, planetary stewardship, and 'nature-based solutions' offered by Anthropocene scientists, Tsing suggests that we will learn more from the 'unruly edges' and the worlds already shattered by social and environmental change. Her ruins thinking has come to exemplify a minoritarian strand of Anthropocene social science that is evidenced in different ways in the writings of others, including Elizabeth Povinelli, Nigel Clark (2011), and **Kathryn Yusoff**.

Third, Tsing and Donna Haraway have led the wider flourishing of multispecies studies, with their two books *When Species Meet* (2008) and *The Mushroom at the End of the World* (2015) serving as key texts across a range of undergraduate and graduate programmes. Tsing in particular has been instrumental in shifting the focus of this work away from charismatic animals and proximal encounters, to explore the lively agencies of fungi, bacteria and weedy plants as they proliferate within the global geographies of industrial agriculture. Tsing's work connects to the wider interest in Geography in topological conceptualisations of disease, health and biosecurity that maps the configuration of the networks and the hotspots that drive the emergence and spread of modern pathogens, often from within the centres of modern agriculture – e.g., in the writings of Steve Hinchliffe et al. (2016), Julie Guthman (2019), and Maan Barua (2022). Likewise Tsing's focus on plantations and her articulation of the Plantationocene give expression to wider anxieties about the unequal social and environmental risks that critics felt were masked by the term Anthropocene, which blames humanity as an undifferentiated species (Malm and Hornborg, 2014).

Tsing's critics come from those approaching these topics from different strands of spatial theory. For those who ground their politics in a critique of capitalism, Tsing, Haraway, Latour and others are 'dithering while the planet burns' (Hornborg, 2017). Andreas Malm (2020: 116) argues that the posthuman turn in critical theory occludes scientific reports about 'accelerated risk of species extinction from climate change'. A connected but somewhat different strand of criticisms comes from theorists concerned about Tsing's seeming celebration of ruins and those who live amongst them. According to David Chandler and Julian Reid, the Anthropocene is celebrated by Tsing, as an era which might be 'grasped as a world that is fuller, more lively and more entangled than the soulless, simplified and atomised world of modernity' (Chandler and Reid, 2019: 12). Writers including Clive Hamilton (2017) and Stephanie Wakefield (2018) join Chandler and Reid in suggesting that Tsing goes too far in rejecting the modern Enlightenment promise of progress and human emancipation. For Chandler and others this stems from the alliance Tsing is seen to be making with Indigenous peoples and their knowledge practices and forms of social organisation. They argue that in grounding their politics in the margins and in the local, they lack the organising power required for the necessary politics of the planetary, undermining environmental projects that necessitate an amplified anthropocentrism (Hamilton, 2017).

Meanwhile a different set of criticisms are offered by Zoe Todd (2016), a Métis scholar who is concerned that scholars involved in the 'more-than-human' turn have not been citing

Indigenous thinkers who have been writing about legal theory and human–animal relations. Broadly, there have been concerns that scholars in the United States are complicit in what **Eve Tuck** and Wayne Yang (2012: 2) term 'settler moves to innocence', trading on Indigenous scholarship without sufficient moves to decolonise the academy. The bibliography of Tsing's important books is replete with citations of Indigenous activists from Indonesia (i.e., Tsing 2005). While Tsing has not been at the forefront of the movement to decolonise the academy, her work stands as an example of how to think with Indigenous intellectuals in out-of-the-way places. A parallel set of criticisms come from anti-racist scholars and Black geographers concerned that the model of the Plantationocene offered by Tsing and Haraway does not pay sufficient attention to violent histories of slavery and how they manifest in the colonial present (Davis et al., 2019).

These accusations are subject to rich debate and should be understood in context as internal discussions within a body of scholarship that shares broad interests in mapping and critiquing the geographies of the Anthropocene: a field to which Anna Tsing has made varied and original contributions, with the potential for future development as her current work emerges and travels. Tsing works from the assumption that the world will not be 'saved' by activism, environmentalism, or political policy (Anonymous, 2011). She is less revelling in the wreckage of the Anthropocene, but noticing small patches where multispecies justice might be co-created by feral forms of life and precarious people (Ihar, 2022). Through long-term relationships and collaborations with Indigenous peoples, she is working to understand how contemporary forms of colonialism and nationalism create experiences of marginality in out-of-the-way places.

Tsing's Key Works

Tsing, A.L. (1993) *In the Realm of the Diamond Queen: Marginality in an Out-of-the-way Place.* Princeton, NJ: Princeton University Press.

Tsing, A.L. (2005) *Friction: An Ethnography of Global Connection.* Princeton, NJ: Princeton University Press.

Tsing, A.L. (2007) 'Indigenous voice', in M. Cadena and O. Starn (eds), *Indigenous Experience Today*. Oxford: Berg. pp. 33–68.

Tsing, A.L. (2012) 'Unruly edges: Mushrooms as companion species', *Environmental Humanities*, 1: 141–54.

Tsing, A.L. (2014) 'Strathern beyond the human: Testimony of a spore', *Theory, Culture & Society*, 31 (2–3): 221–41.

Tsing, A.L. (2015) *The Mushroom at the End of the World: On the Possibility of Life in Capitalist Ruins*. Princeton, NJ: Princeton University Press.

Tsing, A.L. (2017) 'A threat to Holocene resurgence is a threat to liveability', in M. Brightman and J. Lewis (eds), *The Anthropology of Sustainability: Beyond Development and Progress*. New York, NY: Palgrave Macmillan. pp. 51–65.

Tsing, A.L. and Bubandt, N. (2017) *Arts of Living on a Damaged Planet: Ghosts and Monsters of the Anthropocene*. Minneapolis, MN: University of Minnesota Press.

Tsing, A.L., Mathews, A.S. and Bubandt, N. (2019) 'Patchy Anthropocene: Landscape structure, multispecies history, and the retooling of Anthropology', *Current Anthropology*, 60 (S20): S186–S197.

Tsing, A.L., Deger, J., Saxena, A.K., and Zhou, F. (2021) *The Feral Atlas*. Stanford, CA: Stanford University Press. Available at https://feralatlas.org/ (accessed November 15, 2023).

Secondary Sources and References

Anderson, B. (1998) *The Spectre of Comparisons: Nationalism, Southeast Asia, and the World.* New York, NY: Verso.

Anonymous (2011) *Desert.* St. Kilda: Stac an Armin Press.

Barua, M. (2022) 'Plantationocene: A vegetal geography', *Annals of the American Association of Geographers*: 1–17. doi.org/10.1080/24694452.2022.2094326.

Chandler, D. and Reid, J. (2019) *Becoming Indigenous: Governing Imaginaries in the Anthropocene.* Lanham, MD: Rowman & Littlefield.

Cook, I. (2004) 'Follow the thing: Papaya', *Antipode*, 36 (4): 642–64.

Clark, N. (2011) *Inhuman Nature: Sociable Living on a Dynamic Planet.* London: Sage.

Clifford, J. and Marcus, G.E. (1986) *Writing Culture: The Poetics and Politics of Ethnography.* Berkeley, CA: University of California Press.

Davis, J., Moulton, A.A., Van Sant, L., and Williams, B. (2019) 'Anthropocene, Capitalocene, ... Plantationocene?: A Manifesto for ecological justice in an age of global crises', *Geography Compass*, 13 (5): article #e12438.

Guthman, J. (2019) *Wilted: Pathogens, Chemicals, and the Fragile Future of the Strawberry Industry.* Berkeley, CA: University of California Press.

Hamilton, C. (2017) *Defiant Earth: The Fate of Humans in the Anthropocene.* Chichester: Wiley.

Haraway, D.J. (2008) *When Species Meet.* Minneapolis, MN: University of Minnesota Press.

Hinchliffe, S., Bingham, N., Allen, J. and Carter, S. (2016) *Pathological Lives: Disease, Space and Biopolitics.* London: Blackwell.

Hornborg, A. (2017) 'Dithering while the planet burns: Anthropologists' approaches to the Anthropocene', *Reviews in Anthropology*, 46 (2–3): 61–77.

Ihar, Z. (2022) 'Multispecies mediations in a post-extractive zone', in S. Chao, K. Bolender and E. Kirksey (ed), *The Promise of Multispecies Justice.* Charlotte, NC: Duke University Press. pp. 205–26.

Kirksey, S.E. and Helmreich, S. (2010) 'The emergence of multispecies ethnography'. *Cultural Anthropology*, 25 (4): 54576.

Malm, A. (2020) *The Progress of This Storm: Nature and Society in a Warming World.* London: Verso.

Malm, A. and Hornborg, A. (2014) 'The geology of mankind? A critique of the Anthropocene narrative', *The Anthropocene Review*, 1 (1): 62–9.

Marcus, G.E. (1995) 'Ethnography in/of the world system: The emergence of multi-sited ethnography', *Annual Review of Anthropology*, 24: 95–117.

Mitman, G., Haraway, D. and Tsing, A.L. (2019) 'Reflections on the Plantationocene: A conversation with Donna Haraway and Anna Tsing', *Edge Effects,* October 12. Available at https://edgeeffects.net/haraway-tsing-plantationocene/ (acccessed November 15, 2023).

Todd, Z. (2016) 'An Indigenous feminist's take on the ontological turn: "Ontology" is just another word for colonialism', *Journal of Historical Sociology*, 29 (1): 4–22.

Tuck, E. and Yang, K.W. (2012) 'Decolonization is not a metaphor', *Decolonization: Indigeneity, Education & Society*, 1 (1): 1–40.

Wakefield, S. (2018) 'Infrastructures of liberal life: From modernity and progress to resilience and ruins', *Geography Compass*, 12 (7): article #e12377.

55
Yi-Fu Tuan

Tim Cresswell, University of Edinburgh

Biographical Details and Theoretical Context

Yi-Fu Tuan was a leading figure in the theoretical tradition of humanistic geography that emerged in the 1970s as a critique of the dehumanised traditions of the quantitative revolution and an affirmation of the importance of meaning in the human relationship to environments. Tuan's work (along with that of other humanists such as ***Anne Buttimer***, David Lowenthal, and David Seamon) became one of the cornerstones for the subsequent revival of cultural geography and, later, the geohumanities. His articulation of ideas of space and place as key components of human experience have been foundational within geography and across other disciplines ranging from literary studies to landscape architecture. By his death in 2022, Tuan was one of the most influential and highly cited geographers globally.

Tuan was born in Tianjin, China, in 1930. His father, Mao-Lan Tuan worked for the Foreign Office of the Chinese Nationalist government and the family moved frequently due to the civil war and various postings. At the age of ten, Yi-Fu moved with his family to Australia via Hong Kong and in 1946 the Tuan family moved to the Philippines and then London as Mao-Lan Tuan transferred between embassies in the years leading up to the founding of the People's Republic of China in 1949. After attending school in London for two years, Yi-Fu briefly studied at University College London before moving to the University of Oxford to study geography. While his intent was to focus on the humanities and human geography, it was the geomorphology taught by Robert Beckinsale that inspired him and he subsequently moved to Berkeley to study for his Doctorate with the Berkeley geomorphologist John E. Kesseli. His thesis focused on the pediment landforms in Southeast Arizona. He completed his doctorate in 1957 and took up his first position at the University of Indiana before moving to the University of Chicago for a post-doctoral position in statistics. In 1959 he moved to the University of New Mexico as one of a two-member department, and it was there that he had the freedom to explore the big humanities-based questions that would become central to humanistic geography in the 1970s. During his time in New Mexico, Tuan continued his geomorphological research but began to publish papers concerning human attitudes to the environment that would eventually become the work he was best known for, including papers in J.B. Jackson's journal, *Landscape* – an important venue for geographers grappling with questions of human experience of the world around them. In a sign of things to come thirteen years later, one of these contributions was called 'Topophilia: or, Sudden Encounter with the Landscape' (Tuan, 1961).

In 1966, Tuan moved to the University of Toronto for a joint appointment between Geography and Landscape Architecture before moving to the University of Minnesota as Professor of Geography and East Asian Studies where he spent fifteen years between 1968 and 1983. It was while he was at Minnesota that Tuan wrote his best-known works, *Topophilia: A Study of Environmental Perception, Attitudes, and Values* (1974a) and *Space and Place: The Perspective of Experience* (1977). In 1984 Tuan moved to the University of Wisconsin, Madison and became both a Vilas Professor (a university honor given to exceptional scholars) and the John Kirkland Wright Professor (a personal position he chose to name after one of his main inspirations in geography). Tuan retired in 1998 but continued to publish books on a regular basis. By the time of his death in 2022, Tuan had been publishing contributions to academic geography for 68 years. He was the author of 23 books in addition to a series of self-published pamphlets he continued to write and send to friends and colleagues in his later years. His work has been translated into Chinese, German, Italian, Japanese, Polish, Portuguese, Spanish and Swedish. While he is best known as an author of books, Tuan also published over 150 papers and book chapters. He received multiple awards including a John Simon Guggenheim Fellowship, the Callum Geographical Medal of the American Geographical Society, the Lauréat d'Honneur of the International Geographical Union, and the Vaudrin-Lud Prize of the International Festival of Geography. In 2013 he received the inaugural Award in Creativity in Geography from the Association of American Geographers. He was a Fellow of the British Academy, the American Academy of Arts and Sciences, and the American Association for the Advancement of Science.

Spatial Contributions

Tuan's main contribution to thinking about space and place is the development of humanistic geography in the 1970s. Tuan was never intentionally a leader, but a set of ideas coalesced around, or coincided with, his work in the 1970s that became recognised as a major school of thought within the discipline. Although Tuan had used the term 'humanistic' to describe his interpretive method as early as 1963, it was in an article published in the *New Mexico Quarterly* and unlikely to have been read by many geographers (Tuan, 1963). Humanistic Geography, as such, had not yet been named. In 1974 Tuan published his paper 'Space and Place: Humanistic Perspective' (Tuan, 1974b) and over the rest of the 1970s humanism gained momentum as both a critique of positivist spatial science and an affirmation of the importance of meaning, experience, and subjectivity in our understanding of the world around us (Ley and Samuels, 1978; Buttimer and Seamon, 1980; Ley, 1980). Some of the groundwork for humanistic geography could be found in the work of J.B. Jackson and his magazine, *Landscape,* as well as the writings of David Lowenthal, John Kirkland Wright, Carl Sauer, and even, Tuan reminded us, Alexander von Humboldt, who all variously called for attention to worlds of meaning, imaginative geographies, and human perception (Wright,

1947; Jackson, 1957; Sauer, 1965). By 1971, Tuan was introducing geographers to some of the key components of the philosophy of phenomenology, showing how focusing on human experience of the world might provide a rich seam of research that had, with a few exceptions, been largely ignored up to that point (Tuan, 1971). Tuan was not primarily interested in the exegesis of other people's work and he refused to get into the weeds of phenomenological, or any other, philosophical approach – instead he gently introduced geographers to a way of thinking about the world that would remain important to geographers up to the present day. At the heart of humanistic geography was the recognition of the importance of what it is to be human and to pay attention to the 'human' in human geography. This involved resisting the impulse of quantitative spatial science to think of humans in the same way as a physical scientist might think of rocks or atoms. Humans, according to Tuan and other humanists, had capacities – reason, imagination, emotion, sensorial experience, subjectivity – that made them stand out from other living and non-living elements of the world around us. A properly human geography, then, would be interested in these aspects of being human and the role they play in what Tuan saw as the heart of the geography – the way we make the earth into a home (Tuan, 1991a).

It is worth pausing to consider the title of this collection – *Key Thinkers on Space and Place*. Such a title would not have been possible in 1970 as geographers did not really deal with concepts of space and place in a reflective manner. It is true that spatial scientists of the 1950s and 1960s used the word space, but they did not really think about it as a concept. It was in Tuan's work, and particularly his 1977 book *Space and Place: The Perspective of Experience* that the twinned concepts of space and place became central to the discipline. If space was taken for granted in spatial science, place was hardly considered at all beyond the bare-bones version of 'location'. Tuan often used seemingly opposed pairs to think with – dominance and affection (Tuan, 1984), morality and imagination (Tuan, 1989), cosmos and hearth (Tuan, 1996)' among others. But *Space and Place* is the work he remains best known for and, despite subsequent critique and revision, these ideas have been central to the development of the discipline ever since. Tuan's focus in 1977 was explicitly on the phenomenological experience of space and place, not on, for instance, the structural production of space and place within a capitalist economy (see **David Harvey, Doreen Massey** and **Neil Smith**). From the perspective of experience, he argued, space was an arena of action while place was a 'center of meaning' and 'field of care'. While Tuan's definition of space (in relation to place) was not widely commented upon, his account of place as a locus of meaning and care has remained a starting point for many accounts of the role of place in human and more-than-human life (Cresswell, 2015). This view of place as a center of meaning was already prefigured in both an early paper and his earliest landmark text where he introduced the term *Topophilia* – the love of place – as a way of thinking about the way we value the environments around us (Tuan, 1961; 1974b). While his work in the 1970s has been the most influential, he continued to work on the core question of how humans invested the world around them with meaning throughout his career, exploring themes including morality (Tuan, 1989), aesthetics (Tuan, 1993) and fantasy (Tuan, 1998) and their role in human relations to the environment.

Key Advances and Controversies

While Tuan's account of space and place within a humanist framework remains his most significant contribution to geography there are other areas where Tuan's work represents some of the earliest geographical reflections on ideas that are now more commonplace in the discipline. His book *Dominance and Affection: The Making of Pets* (Tuan, 1984) was one of the first examples of what might now be called 'animal geography' (Wolch and Emel, 1998), taking the human relationship to other animals as an important topic of research that illuminates aspects of the ways humans attempt to mould the world into their own image. The book's title also, and unusually for Tuan, centers the idea of power in a way that gently mirrors **Foucault**'s work on discipline. *Landscapes of Fear* (Tuan, 1979) has at its heart a specific emotion, or affect – fear – that was an early example of what we might now call emotional geography, or geography of affect. Tuan's consistent use of the scale of the body as a site for our engagement with space and place prefigured work on geographies of the body and embodiment that became more central to the discipline in the 1990s (Duncan, 1996; Teather, 1999). One important aspect of this was Tuan's focus on the senses in *Space and Place* and elsewhere that gave permission for later geographers to explore the worlds of sensory geographies (Rodaway, 1994; Thrift, 2002; Cosgrove, 2008). In all these cases, of course, it was not Tuan alone that opened up these fields and many later geographers would be critical of his humanistic approaches to these themes. However, in 1970, when Tuan started exploring the possibilities that came with a fully human geography that recognised all the capacities that come with being human, all these themes were more or less absent from the mainstream of geographical research and writing.

In a wider sense, Tuan's work formed one of the bedrocks of what became the 'new cultural geography' in the 1980s as geographers worked on fusing the insights of humanistic geography with those of broadly Marxist approaches to culture (**Denis Cosgrove**, 1983; **Peter Jackson**, 1989; Daniels, 1990). Advocates of the new cultural geography took from Tuan and others, such as ***Anne Buttimer***, the importance of worlds of meaning but recognised that Tuan and others paid too little attention to the structural conditions that meaning-making is conditioned by. Tuan was a committed universalist – looking for general statements about the human condition in the great humanist tradition. This was the main critique made of the work of Tuan and other humanists – that with all his capacity for intuition, he never seemed to intuit the difference that asymmetrical power hierarchies make. He was very reluctant, for instance, to speak or write as a Chinese American man or for Chinese Americans. It was not until the latter years of his life that he addressed his attraction to other men in his moving autobiography, *Who Am I? An Autobiography of Emotion, Mind, and Spirit* (Tuan, 1999). The humans in Tuan's work, for the most part, were seemingly universal humans. As Tuan was writing through some of the most tumultuous times in American and world history, including the Vietnam War, the Civil Rights Movement, second wave feminism, and the emergence of Gay Pride, it is strange that these hardly make an appearance in his work which was otherwise extraordinarily wide-ranging in the sources he drew from. Rather than turning attention to these political movements Tuan was focused on themes such as human goodness

(Tuan, 1986; 2008) and the roles of beauty and fantasy in the constitution of place and landscape (Tuan, 1993; 1998). The new cultural geographers, on the other hand, began to explore 'maps of meaning' in relation to social hierarchies based on, for instance, class, gender, and race – all aspects of human life that Tuan more or less ignored in his quest for a generalised sense of the human condition. Geographers beyond cultural geography were also merging the insights of humanism with a more structural, critical approach to power in edited collections such as *Remaking Human Geography* – a book that explicitly sought to marry humanism with historical materialism (Kobayashi and Mackenzie, 1989). By the 1990s, cultural geography (no longer called 'new') was one of the most innovative areas of the discipline and was where substantive work on Tuanian themes such as place, meaning, the emotions, bodily experience, and perception could be found. Within cultural geography we have seen the rise of 'literary geographies' where geographers and literary scholars explore the worlds of literature through the lens of a concern for space, place and other geographical themes (Hones, 2022). The use of literature in geographical scholarship was promoted early on in his work (Tuan, 1978) and became a central theme in the wider oeuvre of humanistic geography (Pocock, 1981). The field of literary geography has recently expanded into an embrace of creative writing in the form of 'geopoetics' (Magrane et al., 2020). Most recently, cultural geographers have been central to a more interdisciplinary field of geohumanities (Dear et al., 2011; Cresswell and Dixon, 2017) – a field whose journal, *GeoHumanities*, is subtitled, 'Space, Place, and the Humanities' – a nod to Tuan and the other humanistic geographers who made such a field possible.

As I write, in 2023, the influence of Tuan's ground-breaking work can be seen across the broad sweep of human geography. In addition to the specific themes I have mentioned, it should be noted that the very way he wrote was different from the conventions of the social sciences and humanities – and that geographers since Tuan have gradually begun to write in more diverse and interesting ways – especially in cultural geography and the geohumanities. At the same time, however, Tuan's influence is often so pervasive that it is also diluted; there is not a body of work that is obviously Tuanian. Tuan did not set himself up to be a leader or creator of paradigms. He did not have a large body of acolytes. His doctoral students were few and far between, given the length of his career. There were, however, books dedicated to his influence written since his retirement (Adams et al., 2001; Nogué, 2018). It is also the case that humanism in its entirety has been subject to significant critique as an approach to the world born of the Renaissance in Europe that seemingly adopts a universal standpoint while actually being a specific view of the world most often embodied in white European masculinity (Rose, 1993). To many critical geographers, humanism is part of the problem in its inability to diagnose the difference that difference makes (Anderson, 2007). It is for this reason, and others, that recent years have seen the rise of 'post-humanism' – a body of work that not only recognises the difference within humanity but also insists that the human qualities of reason, imagination, desire, and subjectivity, have been overly celebrated in humanism and that other actors in the world, both living and inert, might be equally important (Castree and Nash, 2006).

Tuan's Key Works

Tuan, Y-F. (1971) 'Geography, phenomenology, and the study of human nature', *Canadian Geographer*, 15: 181–92.

Tuan, Y-F. (1974a) *Topophilia: A Study of Environmental Perception, Attitudes, and Values*. Englewood Cliffs, NJ: Prentice Hall.

Tuan, Y-F. (1974b) 'Space and place: Humanistic perspective', *Progress in Human Geography*, 6: 211–52.

Tuan, Y-F. (1977) *Space and Place: The Perspective of Experience*. Minneapolis, MN: University of Minnesota Press.

Tuan, Y-F. (1979) *Landscapes of Fear*. New York, NY: Pantheon.

Tuan, Y-F. (1984) *Dominance and Affection: The Making of Pets*. New Haven, CT: Yale University Press.

Tuan, Y-F. (1991a) 'A view of Geography', *Geographical Review*, 81 (1): 99–107.

Tuan, Y-F. (1991b) 'Language and the making of place: A narrative-descriptive approach', *Annals of the Association of American Geographers*, 81 (4): 684–96.

Tuan, Y-F. (1998) *Escapism*. Baltimore, MD: Johns Hopkins University Press.

Tuan, Y-F. (1999) *Who am I? An Autobiography of Emotion, Mind, and Spirit*. Madison, WI: University of Wisconsin Press.

Secondary Sources and References

Adams, P.C., Hoelscher, S.D. and Till, K.E. (eds) (2001) *Textures of Place: Exploring Humanist Geographies*. Minneapolis, MN: University of Minnesota Press.

Anderson, K. (2007) *Race and the Crisis of Humanism*. London: Routledge.

Buttimer, A. and Seamon, D. (eds) (1980) *The Human Experience of Space and Place*. New York, NY: St. Martin's.

Castree, N. and Nash, C. (2006) 'Posthuman geographies', *Social & Cultural Geography*, 7 (4): 501–4.

Cosgrove, D.E. (1983) 'Towards a radical cultural geography: Problems of theory', *Antipode*, 15 (1): 1–11.

Cosgrove, D.E. (2008) *Geography and Vision: Seeing, Imagining and Representing the World*. London: I.B. Tauris.

Cresswell, T. (2015) *Place: An Introduction*. Chichester: Wiley.

Cresswell, T. and Dixon, D.P. (2017) 'GeoHumanities', in D. Richardson, N. Castree, M.F. Goodchild, A. Kobayashi, W. Liu, W. and Marston, R.A. (eds), *International Encyclopedia of Geography*. Oxford: Wiley Blackwell. pp. 1–9.

Daniels, S. (1990) 'Marxism, culture and the duplicity of landscape', in N. Thrift and R. Peet (eds), *New Models in Geography*. Boston: Allen and Unwin. pp. 177–220.

Dear, M.J., Ketchum, J., Luria, S. and Richardson, D. (eds) (2011) *GeoHumanities: Art, History and Text at the Edge of Place*. London: Routledge.

Duncan, N. (ed) (1996) *Bodyspace: Destabilizing Geographies of Gender and Sexuality*. London: Routledge.

Hones, S. (2022) *Literary Geography*. New York, NY: Routledge.

Jackson, J.B. (1957) 'The abstract world of the hot-rodder', *Landscape: Magazine of Human Geography*, 7: 22–7.

Jackson, P. (1989) *Maps of Meaning*. London: Unwin Hyman.

Kobayashi, A. and Mackenzie, S. (eds) (1989) *Remaking Human Geography*. London: Unwin Hyman.

Ley, D. (1980) 'Geography without man: A humanistic critique'. *University of Oxford, Research Paper* 24.

Ley, D. and Samuels, M. (eds) (1978) *Humanistic Geography: Prospects and Problems*. London: Croom Helm.

Magrane, E., De Leeuw, S., Russo, L. and Perez, C.S. (eds) (2020) *Geopoetics in Practice*. London: Routledge.

Nogué, J. (2018) *Yi-Fu Tuan: El Arte de la Geografia*. Vilassar de Dalt: Icaria Edition.

Pocock, D.C.D. (eds) (1981) *Humanistic Geography and Literature: Essays on the Experience of Place*. London: Croom Helm.

Rodaway, P. (1994) *Sensuous Geographies: Body, Sense, and Place*. London: Routledge.

Rose, G. (1993) *Feminism and Geography: The Limits of Geographical Knowledge*. Cambridge: Polity.

Sauer, C. (1965) 'The morphology of landscape', in J. Leighly (ed), *Land and Life*. Berkeley, CA: University of California Press. pp. 315–50.

Teather, E.K. (1999) *Embodied Geographies: Spaces, Bodies and Rites of Passage*. London: Routledge.

Thrift, N. (2002) 'All nose', in K. Anderson, M. Domosh, S. Pile, et al. (eds), *Handbook of Cultural Geography*. London: Sage. pp. 9–13.

Tuan, Y-F. (1961) 'Topophilia: Or, sudden encounter with the landscape', *Landscape*, 11 (Fall): 29–32.

Tuan, Y-F. (1963) 'The desert and the sea: A humanistic interpretation', *New Mexico Quarterly* (Autumn): 329–31.

Tuan, Y-F. (1978) 'Literature and Geography: Implications for geographical research', in D. Ley and M. Samuels (eds), *Humanistic Geography: Prospects and Problems*. London: Croom Helm. pp. 194–206.

Tuan, Y-F. (1986) *The Good Life*. Madison, WI: University of Wisconsin Press.

Tuan, Y-F. (1989) *Morality and Imagination: Paradoxes of Progress*. Madison, WI: University of Wisconsin Press.

Tuan, Y-F. (1993) *Passing Strange and Wonderful: Aesthetics, Nature, and Culture*. Washington, DC: Island Press.

Tuan, Y-F. (1996) *Cosmos & Hearth: A Cosmopolite's Viewpoint*. Minneapolis, MN: University of Minnesota Press.

Tuan, Y-F. (2008) *Human Goodness*. Madison, WI: University of Wisconsin Press.

Wolch, J.R. and Emel, J. (1998) *Animal Geographies: Place, Politics, and Identity in the Nature–Culture Borderlands*. London: Verso.

Wright, J.K. (1947) 'Terrae incognitae: The place of the imagination in geography', *Annals of the Association of American Geographers*, 37: 1–15.

56
Eve Tuck

Jen Rose Smith, University of Washington
Sara Smith, University of North Carolina at Chapel Hill

Biographical Details and Theoretical Context

Eve Tuck's scholarship, which works across disciplines, genres, and communities in solidarity, exemplifies socially engaged research that is conducted with and for the benefit of working-class, racialised, and Indigenous communities. Tuck is an Unangax̂ scholar who is an enrolled member of the Aleut community of St Paul Island, Alaska. She spent her formative years in Pennsylvania and New York, and graduated from The City University of New York in 2008, with a PhD in Urban Education. She is Professor of Indigenous Studies and James Weldon Johnson Professor at NYU Steinhardt, and is director of NYU's Center for Indigenous Studies. Tuck is internationally regarded for her contributions to educational research and Indigenous Studies and has worked closely with diverse urban and rural communities to conduct research that improves education and social policy, particularly for disadvantaged and marginalised youth.

Tuck was the founding director of the Tkaronto Collaborative Indigenous Research for Communities, Land and Education (CIRCLE) Lab, which is dedicated to social change and collaborative writing using arts- and materials-based research, participatory research with youth and communities, and visual and audio research. Under Tuck's leadership, in 2022 the Tkaronto CIRCLE Lab released the Collaborative Indigenous Research Digital Garden – a virtual visiting place to bring together conversations of Indigenous and participatory research – and the Land Education Dreambook, a guidebook for organisations working with and developing land education programmes for Indigenous youth (Tkaronto CIRCLE Lab, 2023). Tuck also makes a podcast with graduate students called The Henceforward (http://www.thehenceforward.com), discussing relationships between Indigenous and Black communities in what is for now known as the US and Canada.

Tuck's work contributes to collaborative Indigenous research, Indigenous feminisms, and land education. Her research focuses on how Indigenous social thought can be engaged to create more fair and just social policy, more meaningful social movements, and robust approaches to decolonisation. Much of Tuck's work is collaborative, and she regularly co-authors with K. Wayne Yang among other interlocutors. Within her many collaborations, she emphasises the act of intentional curation, an attentiveness toward the politics of citation, and the uplifting of new and junior scholars. Moreover, Tuck's research not only theorises collaborative Indigenous research, but also models what forms that scholarship might take, such as co-creating multiple edited volumes with a range of collaborators, utilising omissions and refusals of information strategically, and playing with form, genre, and use of figures. These choices of craft destabilise the idea of sole authorship as most important. Her methods spring from expansive collaborative endeavours that are creative, generative, and politically-focused.

Spatial Contributions

Tuck has written extensively on land, land education, place, and researching place. She engages widely with geographic theories and carefully distinguishes between European universalist and Indigenous theorisations. Tuck's writing on land evokes the *where* of scholarly work and the politics of place. Together with McCoy and McKenzie, Tuck spells out how Indigenous scholarship engages with land (including water, air, and subterranean earth) as a set of relationships that cannot be generalised or made planar (Tuck et al., 2016). Land is not irreducibly material: land has emotional, spiritual, and intellectual dimensions and contours. Land is both teacher and site of memory, intergenerational and instructive. Tuck and her collaborators' approach to land is therefore incommensurable with settler visions of land that justify or encourage the replacement of Indigenous people and turn land into a set of property relations. Crucially, the conceptualisation of Indigenous land relations offered by Tuck and colleagues is not simply about untouched wilderness, or static and fixed in a pre-contact temporality. For Tuck, the urban, the rural, and all places among and in between are sites of Indigenous land relations.

Tuck describes one part of her methodology as 'visiting', which she categorises as an Indigenous feminist practice. Visiting is impermanent, 'a passing through of space and time', that emphasises relationality and care and 'does not assume entitlement to space, or assert control, or ownership over land [...] it suggests a responsibility to the past and future of a place through the impermanence of our presence' (Tuck et al., 2023: 1). Elsewhere, Tuck and Recollet (2018: 1) describe a visitation as 'a remembrance of an old futuristic way of relating to place, non-human persons, and each other'. Understanding visiting as part of a spatial methodology that is temporal, temporary, and a practice of consent and ethics, Tuck and her collaborators offer visiting not as a corrective but as empowerment of Indigenous practices of being in community. Visiting articulates an understanding of the relationship to the land where and with whom we conduct our research, a method that provides different ethical and intellectual possibilities to those we are in conversation with – human and more-than-human.

Tuck crucially asks us to reflect on the limits of research, particularly for settler researchers. This intervention includes the foundational text 'Decolonization is not a metaphor' (Tuck and Yang, 2012) and is a thread throughout her oeuvre. Tuck and Yang write against academic appropriation of decolonisation, clarifying, 'decolonization is not a metaphor. When metaphor invades decolonization, it kills the very possibility of decolonization; it recenters whiteness, it resettles theory, it extends innocence to the settler, it entertains a settler future' (Tuck and Yang, 2012: 3). Against this damaging misuse, they insist, 'decolonization in the settler colonial context must involve the repatriation of land simultaneous to the recognition of how land and relations to land have always already been differently understood and enacted; that is, all of the land, and not just symbolically' (Tuck and Yang, 2012: 7). The piece argues for the fundamentally related nature of decolonisation and abolition and warns against false uses of decolonisation language as a settler claim to innocence. Put another way, decolonisation cannot happen without social justice, without land return, or without an assessment of the ongoing material consequences of colonisation for racialised and dispossessed communities.

Tuck also makes important contributions to the understanding of academic spaces. She describes the overwhelming number of scholars who wish to understand political complexity only by engaging Western or continental theory, and insists on contemplating Indigenous and Black frameworks of thought. However, Tuck remarks: 'I find myself less willing to do this now. I am weary after so many conference presentations in which Indigenous scholars present work and then someone in the audience asks them a question that expects them to do more work' (cited in Smith et al., 2019: 14). Tuck has shared a template of how she organises Q&A sessions to anticipate and refuse inappropriate audience reflections. She argues, 'Indigenous and decolonial theories are unfairly, inappropriately expected to answer to whiteness and to settler relationships to land in the future [...] a settler future is preoccupied by questions of, *What will decolonization look like? What will happen after abolition? What will be the consequences of decolonization for the future?*' (cited in Smith et al., 2019: 15). Tuck understands this as a distraction: 'Decolonization is not the endgame, not the final outcome of a long process, but the next now, the now that is chasing at our heels. I am lucky to come from the long view' (cited in Smith et al., 2019: 16). Further, Tuck is wary of the settler tendency to absorb the insights of Native scholarships and incorporate them as aesthetic or metaphor without affecting real or lasting change. With Gaztambide-Fernández, Tuck writes about this as an additional form of settler colonialism, in which settlers extract the ideas of Native Scholarship and continue exclusionary hiring practices in their predominantly white universities instead of ceding authority to Native scholars and Native communities (Tuck and Gaztambide-Fernández, 2013).

Running through all of Tuck's work is an attention to temporality – both orientations toward the future and the simultaneity of the past in the present. She disrupts tendencies that centre stories of Indigenous damage and trauma and asks instead for desire-centred research (Tuck, 2009; 2012). For Tuck, desire is a motivating force that pulls people into the future. It does not erase trauma, and still makes space for the realities of trauma, but insists on not limiting our understanding of life to pain alone (Tuck, 2012: 15). Tuck makes a necessary intervention into a proliferation of trauma narratives in the academy to ask researchers to 'suspend damage' (Tuck, 2009). Damage-centred research assumes that if we can effectively *show harm* that will initiate change or reparations. When taken up by settler researchers, this theory of change positions outsiders as changemakers (community members themselves already know the harm they face and do not need it proved to their own communities) and treats these outsiders as 'judge and jury'. As a salve to this approach, Tuck urges us to take up a desire-based framework. This framework offers a way of storytelling and relating to land, ourselves, our stories, and communities. For Tuck (2009: 416), 'desire is an antidote, a medicine to damage narratives'.

Key Advances and Controversies

Tuck provides a blueprint for research and scholarship that is deeply geographic. She asks us to begin from the land on which we move, speak with the people to whom we are accountable, find flexible and creative methods that match our aims, define and centre decolonisation as an objective and measure for our work, and to invent and invest in a theory of change emerging from the political specificities driving our work. Her work has widely influenced the field of education, Native American and Indigenous Studies, and the humanities and social sciences more broadly, but we also find that geographers can more productively take up her contributions to help reshape our discipline's knowledge production practices. In our readings across Geography, Tuck is often cited in relation to debates on decolonisation, and we urge that geographic scholarship take up her engagement with place.

The importance of Tuck and Yang's insistence that 'decolonization is not a metaphor' cannot be overstated. One of Tuck's key advances is to insist on the *specificity* of decolonisation and draw attention to the work that metaphor can do when wielded carelessly. They call for a moratorium on the appropriation of 'decolonization' as a synonym for diversity, equity, or inclusion work. Reflecting on the intervention seven years later, Tuck writes, 'To say that decolonization is not a metaphor is to resist using decolonization as a trendy term, and in settler colonial contexts, to resist delinking decolonial projects from the rematriation of Indigenous land and life. In settler states that are also antiblack states founded through the violence of chattel slavery, decolonization also must involve abolition' (in Smith et al., 2019: 13). This work is a *radical* intervention into decolonisation scholarship, as scholars now are forced to grapple with the question: whose land am I on? Is my work furthering the material return of Indigenous land? Additionally, Tuck's insistence on attending to the histories of place pushes especially settler scholars conducting fieldwork 'elsewhere' to be critically accountable to the ongoing consequences of violence and dispossession in the spaces where they live, teach, or write. This troubles the undergirding idea that 'real' ethnography is done somewhere outside North America, and destabilises any universalised, homogenised understanding of the Global North.

Tuck's scholarship also works against the dangerous forms of metaphorising place. In their commentary on a conference session on 'demetaphorizing complex ecologies', Tuck and McKenzie (2014: xv) reflect on how the environmental studies participants wished to connect in new ways to the same territory that Indigenous participants already spoke of having intact relations with for generations. Later in the same text, Tuck and McKenzie caution on how space- and place-based metaphors can sidestep responsibility to material places and, 'invoke place superficially, too easily', rather than in a politically grounded manner (Tuck and McKenzie, 2014: 18). Tuck and McKenzie provide us with a means to centre Indigenous theories, methodologies, and methods that centre accountability to material land and community.

Geographers seeking to outline what decolonisation means for Geography frequently incorporate Tuck and Yang's cautions on research and refusal. Tuck's framing on the need for *material* decolonisation and cautions on extractive settler research is also central in recent interventions. Notably, when the 2017 Royal Geographical Society–Institute of British Geographers' annual conference declared

'Decolonising geographical knowledges' as the theme, a flurry of conversations invoked Tuck and Yang's guidance on the meaning of the term. Esson et al. (2017: 385) cite their piece in reference to *material* decolonisation to argue that, 'The focus on decolonizing knowledges should be positioned as a means to this end, not as an end in itself'. They argue that the conference risked doing exactly what Tuck and Yang warn against.

Geographers and other scholars use Tuck's work to consider the tangled connections of desire and haunting in shaping our present and future, and how we consider the ways that the past remains with us. Building from Avery Gordon, Tuck and Ree write that haunting is 'the relentless remembering and reminding that will not be appeased by settler society's assurances of innocence and reconciliation [...] individuals are haunted, but so are societies' (2013: 642). Tuck has inspired some geographers to centre desire and move away from damage-based narratives (Daigle, 2019; Dodds and Smith, 2022; Rivera, 2023). Others make extensive use of the concept of haunting. For example, Best and Ramirez discuss how Black women artists trouble urban space by refusing to forget the connections between Black women and property (Best and Ramírez, 2021). Haunting has also been crucial to several recent pieces working through how we might understand settler legacies and Indigenous, Latinx, and Black futures.

While Tuck's work has been extremely influential, traditional and disciplinary geographers have yet to take up Tuck's scholarship in ways that scholars of land and Native politics in Native Studies have for decades. Tuck is impatient with academia's tendency to extract knowledge from land and people and render it into an abstract currency, and her work challenges geographers to transform our relationships with the lands we write about and from. She insists that we forge more ethical relationships with an intentional theory of change, recognising that we are not entitled to all spaces, processes, and knowledges. Her theorisations of land, her methodologies and strategies of engagement and relentless attention to cultivating futures across the situated relations to land and life held by Indigenous and Black communities have much to offer our discipline if we attend to them.

Tuck's Key Works

Tuck, E. (2009) 'Suspending damage: A letter to communities', *Harvard Educational Review*, 79 (3): 409–28.

Tuck, E. and Gaztambide-Fernández, R.A. (2013) 'Curriculum, replacement, and settler futurity', *Journal of Curriculum Theorizing*, 29 (1): 72–89.

Tuck, E. and McKenzie, M. (2014) *Place in Research: Theory, Methodology, and Methods*. New York, NY: Routledge.

Tuck, E. and Ree, C. (2013) 'A glossary of haunting', in S. Holman Jones, T.E. Adams and C. Ellis (eds), *Handbook of Autoethnography*. New York, NY: Routledge. pp. 639–58.

Tuck, E. and Yang, K.W. (2012) 'Decolonization is not a metaphor', *Decolonization: Indigeneity, Education & Society*, 1 (1): 1–40.

Tuck, E. and Yang, K.W. (2014) 'R-words: Refusing research', in D. Paris and M.T. Winn (eds), *Humanizing Research: Decolonizing Qualitative Inquiry with Youth and Communities*. Thousand Oaks, CA: Sage. pp. 223–47.

Tuck, E., McCoy, K. and McKenzie, M. (2016) *Land Education: Rethinking Pedagogies of Place from Indigenous, Postcolonial, and Decolonizing Perspectives*. New York, NY: Routledge.

Tuck, E., Stepetin, H., Beaulne-Stuebing, R. and Billows, J. (2023) 'Visiting as an

Indigenous feminist practice', *Gender and Education*, 35 (2): 144–55.

Secondary Sources and References

Best, A. and Ramírez, M.M. (2021) 'Urban specters', *Environment and Planning D: Society and Space*, 39 (6): 1043–54.

Daigle, M. (2019) 'The spectacle of reconciliation: On (the) unsettling responsibilities to Indigenous peoples in the academy', *Environment and Planning D: Society and Space*, 40 (2): 276–91.

Dodds, K. and Smith, J.R. (2022) 'Against decline? The geographies and temporalities of the Arctic cryosphere', *The Geographical Journal*, 189 (3): 388–97.

Esson, J., Noxolo, P., Baxter, R., Daley, P. and Byron, M. (2017) 'The 2017 RGS-IBG chair's theme: Decolonising geographical knowledges, or reproducing coloniality?', *Area*, 49 (3): 384–8.

Rivera, I. (2023) 'Undoing settler imaginaries: (Re)imagining digital knowledge politics', *Progress in Human Geography*, 47 (2): 298–316.

Smith, L.T., Tuck, E., and Yang, K.W. (2019) *Indigenous and Decolonizing Studies in Education: Mapping the Long View*. New York, NY: Routledge.

Tkaronto CIRCLE Lab (2023) 'Tkaronto CIRCLE Lab'. https://www.tkarontocirclelab.com/ (accessed November 15, 2023).

Tuck, E. (2012) *Urban Youth and School Pushout: Gateways, Get-Aways, and the GED*. New York, NY: Routledge.

Tuck, E. and Recollet, K. (2018) *Visitations: You are Not Alone*. Catalog of the exhibition #callresponse, by C. Belcourt, I.V. Castellanos, M. Crosby, M. Hupfield, U. Johnson, C. L'Hirondelle, I. Murdoch, E. Neff, T. Tagaq, T. Willard and L. Williamson-Bathory. Vancouver, BC: Grunt Gallery.

57
John Urry

Peter Adey, Royal Holloway University of London

Biographical Details and Theoretical Context

John Urry was born in London in 1946. He gained his first degrees from Christ's College, Cambridge in 1967, a 'double first' BA and MA in Economics, before completing his PhD in Sociology from the same institution in 1972. He joined Lancaster University as Lecturer in Sociology, guiding its early development, and becoming Head of Department in 1983. He was made Professor in 1985 and later Dean of Faculty and Distinguished Professor of Sociology. With Mimi Sheller – supported by centre administrator Pennie Drinkall – Urry co-founded the Centre for Mobilities Research (CEMORE) in 2003, and from 2003 to 2015 was its Director. In 2015, with Linda Woodhead, he co-founded the Centre for Social Futures, also at Lancaster. He remained at Lancaster until his sudden death in 2016, completing 44 years of service there in total. A peaceful tribute to John was created at the university in the form of a purposefully-designed memorial garden on the Bowland North campus.

Urry's research interests extended from his PhD on the emergence of capitalist societies to his later examinations of mobility, energy, climate change, and financial and social futures. In over 40 books and hundreds of articles, his work charted broad changes in the nature of capitalist society, but always with an eye to the material infrastructures and conditions that underpinned these transitions. A sociologist by trade, Urry was a dominant force in international sociological debates for more than four decades, but worked extensively with human geographers, principally in British and European contexts, and voiced opinion on themes which cut across neat disciplinary barriers. Urry was integral to international and interdisciplinary research through the editorship of the international journal *Mobilities* and the creation of spaces such as the Cosmobilities network – a cross European-Network of mobility researchers – and actively collaborated in international projects, such as with researchers in China.

Writing in the aftermath of May 1968, Urry's doctoral work concerned the sociology of power in Britain and an interest in changes in the philosophy of the social sciences. In the manner of his early research, Urry consistently sought to depart from any kind of *status quo*, seeking new avenues for research and adopting innovative conceptual and methodological tools to interrogate them. Urry's work was marked by its insistence to 'turn': his arguments swerved, changed direction and departed from established positions. At the same time, however, they were grounded and fixed, but not in the sense that he understood his objects of enquiry as immobile: indeed, this is the exact opposite of his most general points on mobilities. Rather, Urry's investigations of tourism, business interactions, talk, and new forms of economic life were always bound to concrete material contexts and activities even when studying the footloose forms of financial offshoring and elite lifestyles reliant on hypermobility but also apartness and withdrawal from public scrutiny. Urry's work drove apparently superficial actions down into sub-surface layers, contexts and structures of economies, labour relations, forms of governance, materialities, and the minutiae of practices, somewhat in spite of the mammoth scales his writing often alluded to.

In the late 1970's Urry became increasingly interested and influential in the burgeoning study of the region and locality. The Conference of Socialist Economists (CSE) and Lancaster University's own Regionalism Group were to play a particularly significant role. Drawing together research of this type, the groups benefited from an emerging turn within Geography towards 'locality studies' or a new kind of 'regional geography'. At that time this research was inspired by **Doreen Massey's** (1979) path-breaking work along with Phillip Cooke and others. In this context, Urry fostered productive intellectual forays into spatial thought and theory as he sought to understand the role of locality upon the social and spatial structures of Britain's emerging service economy, its declining manufacturing industries and the social and material transformations of Thatcherite ideology.

Urry's interdisciplinarity was shaped by processes that influenced many social scientists at work in Britain and the US in the early 1980s. Urry felt a distinctive need to surpass the disciplines, to make them more 'ill-disciplined' (see Adey et al., 2010), particularly as departments were forced into smaller and more open-looking units. In 1985 Urry published an influential joint-edited collection with ***Derek Gregory*** (1985). Along with several other collections at the time (Allen and Massey, 1988; **Thrift** and Williams, 1987), it attempted to map out the relations between societal and spatial structures thereby forging new connections between geographers and sociologists coming to terms with the local distinctiveness of state-invoked restructuring.

With Scott Lash, *The End of Organized Capitalism* (Lash and Urry, 1987) marked a major development in Urry's study of capitalist economies and societies. Lash and Urry's approach did not follow a traditional political economy path. In their analysis of the restructuring of labour relations, new forms of leisure time and practices of consumption, Lash and Urry helped set in motion a path towards a culturally-nuanced 'postmodern' theory of capitalism (Lovering, 1990: 212). Bubbling away in *The End of Organised Capitalism* we see evidence of the evolution of Urry's thinking – specifically in his conception of mobility – then-taken as an albeit interesting by-product of the transition of extensive economic and political processes.

Mobility (or 'new mobilities') became probably Urry's most important contribution to spatialised social science by putting the social *into* mobility. Developed in *Sociology Beyond Societies* (Urry, 2000) and *Mobilities* (Urry, 2007), following *The Tourist Gaze* (1990) and *Economies of Signs and Space* (Lash and Urry, 1994), Urry gradually sought to re-imagine what we mean by 'the social' (see also Axhausen et al., 2006). Urry argued that modern, Western societies are far from solid but composed of complex mobilities and interactions (Sheller and Urry, 2006; 2016). With the influence of Lancaster colleagues such as Mimi Sheller, Anne-Marie Fortier and Sylvia Walby, mobility took on the horizon of both gendered material practices and a metaphorical movement eliciting more supple subject positions than a singular and masculine male point-of-view. Clearly inspired by Actor-Network Theory and its proponents – such as John Law and Lucy Suchman – mobility was taken by Urry to encompass the movements not only of people but also those *things* whose orbits comprise complex socio-material relationships. Urry explained how 'material transformations' are 'remaking the 'social'. The mobilities of travel, the movements of images and especially information, were 'reconstructing the "social as society" into the "social as mobility"' (Urry, 2000: 2).

The mobility agenda was pushed even further as international circuits of knowledge from clusters emerging between the Netherlands, Scandinavia and Germany aligned ***Ulrich Beck's*** 'risk society' thesis, 'cosmopolitanism' (see, for example, Axhausen et al., 2006; Canzler et al., 2008), and new understandings of mobility and technology. In this sense, Urry was foundational to the making of a 'new mobilities paradigm' (Sheller and Urry, 2016) which had implications for studies not just of transport and travel per se but understandings of society's reliance on particular forms of energy and oil-based economies and even international financial systems (Urry, 2014). In his later years Urry became increasingly concerned with questions of climate change and the necessary transition to a post-oil zero-carbon society, arguing that profound socio-technical reorganisation was needed to avoid 'negative system effects including greater levels of ungovernability, major losses of income and population, and large increases in personal and system surveillance' (Urry, 2014: 57). Focusing on the importance of social science (contra science) in understanding and adapting to climate change, Urry outlined the importance of systems-level change in dislodging engrained social habits, and in his final book, on social futures, explored how potential futures are closed down by lock-ins and complex interdependencies which often make low-carbon futures appear unfashionable and unachievable (Urry, 2016).

Spatial Contributions

After his death in 2016, many memorial events, an edited collection, special issues and appreciative write-ups celebrated Urry's many contributions to academia. Several of these alighted on his advancements of Sociology and the interdisciplinary social sciences, and especially his work within the 'spatial turn', acknowledging his long-term contributions to geography. Urry was always vocal in his articulation of the disciplinary relation between Sociology and Geography, tracing out the two discipline's entwined paths in a contribution to the collection *New Models in Geography* (Urry, 1989). Therein, Urry (1989: 295) argued they had not come together simply through a process of convergence whereby 'sociologists "add" space and geographers "add society" to their respective analyses' but that this involved the cross-fertilisation of theories, concepts and methods in quite uneven ways.

Urry's (1981) interventions into the geographies of localities were especially prominent within Sociology given the local was endowed with the properties of a powerful social object. The investigations of Lancaster's Regional group into the consumption of specific locations and geographies proved influential for the spatial analysis of gender, class and region (Murgatroyd et al., 1985; Bagguley et al., 1990). Charting changes to the capitalist economy, the growth of the middle-classes and increased opportunities to travel, alongside other developments in transportation technology such as the railroad and the aeroplane, Urry conceptualised tourism as one of the defining characteristics of being 'modern'. Urry's turn to consumption, tourism and mobility departed from accepted 'academic prejudices'. As set out in *The End of Organized Capitalism,* (Lash and Urry, 1987) and later in *Economies of Signs and Space* (Lash and Urry, 1994), Urry refused to accept the prioritisation of manufacturing over services, 'production rather than consumption, of "work" rather than "leisure", of structure rather than mobility, and of work-related mobility rather than leisure mobility' (Lash and Urry, 1994: 254). His work showed particular kinds of movement, tourism and consumption could be historicised into *stages* or timeframes identifiable in capitalist societies: for example, from the organised exploration of pre-capitalism, individual travel of the wealthy in liberal capitalism, mass tourism under organised capitalism through to the 'end of tourism' under disorganised capitalism. This laid the groundwork for important geographical analyses of consumption by geographers such as ***Peter Jackson***, Phil Crang and Ian Cook.

The dichotomy of production and consumption so engrained in social science was also befuddled by Urry's illumination of the new economies of images, ideas, data, and other material objects. Geographically speaking, the implication was that any space, terrain and landscape could be commodified and therefore circulated – places could be *consumed.* Representations, images and the gaze of the tourist were central to this expedition into the visual, sensuous and embodied dimensions of new experiences of capitalism. The *tourist gaze* was a 'systemised' and 'socially organised' way of looking. This could mean the images and visions of a tourist destination – stimulated in anticipation through 'daydreaming and fantasy' (Urry, 1990: 3) – or passed on through other mobile media such as tourist brochures and television broadcasts.

Within these frameworks, the tendency to see mobility as 'derived' from more substantial processes made it a particularly 'difficult task for social scientists' to grapple with (Urry, 1990: 7). Mobility appeared fleeting, ephemeral. But Urry's 'new mobilities paradigm' – developed with Mimi Sheller, Kevin Hannam and others – demanded the re-conceptualisation of processes such as social exclusion. Those unable to move could be excluded from job markets, access to social services such as education, health and other benefits; it could prevent one from voting or evacuating a disaster area (**Cresswell**, 2006). **Stephen Graham** and Simon Marvin's (2001) exploration of networked 'technological mobilities' in *Splintering Urbanism* is perhaps the most comprehensive example of this kind of work, exploring the unequal social relations caught up in infrastructural provision and access within built material environments and sunk physical infrastructures. Although mobilities might transform political spaces, composing new sites of political engagement and action, in concert with Doreen Massey's (1994) 'politics of mobility', Urry's 'new mobilities' was to inspire a body of research invested in the inequalities and politics of mobility and movement, including much work considering the possibilities of a low-carbon future (see Adey et al., 2023).

But it would be wrong to suggest that Urry's attention to the mobile meant he paid too much attention to social and spatial life. To the contrary, Urry's mobilities work involved precise spatial thinking exploring how capacities for mobility were structured within economies, systems of production, and other spatial formations. Urry's cluster of research projects and writings around the end of the car, oil and financial offshoring, expressed these conceits through a range of spatial metaphors of 'containers', 'confinement', 'moorings', 'lock-in', 'exit', and 'islanding'. At the heart of these approaches were commitments to relationality, interdependent networks, and especially 'dependency'. His attention to automobile-centric, oil-based, high-carbon societies in a context of climate change systematically highlighted the difficulties of doing mobility otherwise. Urry saw society's choices confined by a reliance on systems of production, oil resources and manufacturing structures, as well as the inefficient design of cities, and the (ironically) rigid yet fluid social practices reliant upon them. He even posited the 2008–2009 US financial crisis to be, in part, a product of a high-carbon automobile-based society which overvalued risky mortgage products fuelling forms of suburban housing growth that were precariously reliant on cheap energy (Urry, 2014). Further, *Offshoring* (Urry, 2014) sought to expose the dependence of financial systems, elite actors, and their lifestyles on spaces outside of tax authorities, legal orders and public scrutiny, describing the withdrawn spaces, enclaves, islands, and secrecy that society's elites depend upon.

The later stages of Urry's research developed a major theme in much of his writing and thinking: attention to the shape and form of the future (Urry, 2016). We might say that Urry's approach to futures was also highly spatialised, revealing the alternative spaces that are hinted at in the present, that might be born, that might take different forms. Several of his later works – even *Mobilities* (Urry, 2007) – considered alternative futures that could be imagined, experimented with, and considered as potential societal pathways we might want to live with or avoid (Tyfield et al., 2016).

Key Advances and Controversies

Urry's research was pivotal for the way geographers came to understand mobility.

Without mobility there would be no social world. Consumption would fail without the means to get at services or products or have them brought to us. Friendship would be hard without physical proximity or mediated contact by communications like the telephone or email. Leisure activities would be difficult without the means to get to a destination. Work, too, is increasingly done on the move; from the aircraft cabin to driving on the motorway (Laurier, 2004). It is just these sorts of mobilities which Urry argued remade the social into societies constituted by mobiles – by people, objects and things. Re-imagining the social away from the closed worlds of immobile face-to-face encounters and rigid networks of relationships (Urry, 2004a), Urry's 'new mobilities paradigm' helped unify a disparate body of research for whom mobility is at their heart.

The impact of these styles of thinking brought mobilities to the forefront of diverse sets of research interests and disciplinary agendas in the twenty-first century, and 'mobility' as a concept into the general parlance of social scientific research (Adey, 2009). For example, scholars of travel and tourism have sought to place mobility somewhere at the centre of their understandings of tourism, while in criminology the issue of mobility is similarly taken as a fundamental subject to tackle (Aas, 2007). Moreover, while policy and regulation enable and shape mobilities, policy itself is now understood to travel, being shared, copied and 'in motion' (McCann, 2008).

Within geography, Urry's influence on mobility was patently felt within a plethora of research. Developments in transport geography (Knowles et al., 2008) accommodated the 'new mobilities' paradigm to better theorise more socially sophisticated concepts of mobility, although some argued that transport geography's engagement with more social, political and culturally inscribed approaches to transport has come about not just because of mobilities research (Schwanen, 2016).

Urry's mobilities are particularly invested with significance for Tim Cresswell. Through his humanistic reading, Cresswell argues that mobilities are frequently bound up in ideological codings and relations of power, dislocating the investiture of fixed points, pauses or places as centres of social significance (**Tuan**, 1977; Cresswell, 2004). Alternatively, for Nigel Thrift (1996), mobility captures something more-than meaningful experiences but what he determines as the 'structures of feeling' of western modernity. Modernity itself struck Urry as thoroughly mobile: 'modern society is a society on the move' (Urry and Lash, 1994: 252). Harnessing the inner and minor worlds of performance, habit and pre-cognition, with some help from phenomenology, mobility helped establish non-representational theory as a prominent conceptual and empirical concern which can deal with motion, action and process (McCormack, 2002; Adey, 2009). More considered accounts of technology and movement considered the unfolding 'cyborg assemblages' of travel from the 'driver-car' (Dant, 2004; Laurier, 2004; Merriman, 2004) to the train (Bissell, 2009b) and the airport.

Urry's apparent prioritisation of the visual in his 'new mobilities' and the 'tourist gaze' idea did, however, provoke some criticism. Urry argued that the 'gaze' and the way it is constructed via varied practices of seeing worked to secure tourism mobilities as a 'field apart' from other kinds of embodied leisure. Critics contended that it is the whole of the body and not just the eyes that see, experience and compose tourist activity and other styles of mobile practice. Tourist scholar Pau Obrador-Pons (2007) puts this point simply but effectively when he argued tourist dwelling involves the whole body. Others extended this work towards phenomenological accounts of the feeling and sensation of cycling (Spinney, 2006) or the sensations of travelling on a bus at rush hour (Harada et al., 2023).

The prioritisation of mobility as the basis of social action has also had its critics. Post-phenomenological accounts have criticised Urry and others for emphasising the vitalism of perpetual action and excitation. For instance, Bissell (2009a; 2009b; see also Rose and Wylie, 2006) sought to account for the mobile body as less-than-active. In his reworking of mobile experience, Bissell explores moments of passivity, boredom, inactivity and silence as a 'turning-away' from the world during very mobile encounters. Harrison (2008) and Kraftl and Horton (2008) similarly addressed the relation between passivity and productivity arguing for a less excitable account of the everyday but towards more deadened moments of acquiescence and stillness.

Another question has arisen from criticisms levelled at the un-economic character of the 'new mobilities' paradigm in general and the failure of some to ground mobility in economic processes and relations. There is a concern here that newer examinations of the mobile are rather 'light'. However, these arguments do not necessarily map that well directly onto Urry's research which, time and again, considered mobility in relation to the signature of the temporal, geographic and economic. For example, his writings on automobiles, oil and offshoring always saw mobility as embedded within wider political, economic, industrial and financial systems which constrained and straightjacketed change. Urry's focus upon the weighty relationships between mobilities and resources ultimately helped geographers appreciate mobility as embedded deep within the political and economic contexts from which it emerged.

Urry's Key Works

Lash, S. and Urry, J. (1987) *The End of Organized Capitalism*. Cambridge: Polity.

Lash, S. and Urry, J. (1994) *Economies of Signs and Space*. London: Sage.

Urry, J. (1973) *Reference Groups and the Theory of Revolution*. London: Routledge.

Urry, J. (1990) *The Tourist Gaze: Leisure and Travel in Contemporary Societies*. London: Sage.

Urry, J. (1995) *Consuming Places*. London: Routledge.

Urry, J. (2000) *Sociology Beyond Societies: Mobilities for the Twenty-First Century*. London: Routledge.

Urry, J. (2003) *Global Complexities*. Cambridge: Polity.

Urry, J. (2007) *Mobilities*. Cambridge: Polity.

Urry, J. (2013) *Societies Beyond Oil*. London: Bloomsbury.

Urry, J. (2014) *Offshoring*. Chichester: John Wiley.

Urry, J. (2016) *What is the Future?* Chichester: John Wiley.

Secondary Sources and References

Aas, K.F. (2007) 'Analysing a world in motion – Global flows meet "criminology of the other"', *Theoretical Criminology*, 11 (2): 283–303.

Adey, P. (2009) *Mobility*. London: Routledge.

Adey, P., Bissell, D. and Urry, J. (2010) 'Mobilities, meetings and futures: An interview with John Urry', *Environment and Planning D: Society and Space*, 28 (1): 1–16.

Adey, P., Cresswell, T., Lee, J.Y., Nikolaeva, A., Nóvoa, A. and Temenos, C. (2023) *Moving Towards Transition: Commoning Mobility for a Low-Carbon Future*. London: Zed.

Allen, J. and Massey, D. (1988) *The Economy in Question*. London: Sage.

Axhausen, K., Larsen, J., and Urry, J. (2006) *Tracing Mobilities: Towards a Cosmopolitan Perspective*. Aldershot: Ashgate.

Bagguley, P., Lawson, J.M., Shapiro, D., Urry, J., Walby, S. and Warde, A. (1990) *Restructuring Place, Class and Gender*. London: Sage.

Bissell, D. (2009a) 'Conceptualising differently-mobile passengers: Geographies of everyday encumbrance in the railway station', *Social & Cultural Geography*, 10 (2): 173–95.

Bissell, D. (2009b) 'Visualising everyday geographies: Practices of vision through travel-time', *Transactions of the Institute of British Geographers*, 34 (1): 42–60.

Canzler, W., Kaufmann, K. and Kesselring, S. (2008) *Tracing Mobilities: Towards a Cosmopolitan Perspective*. Aldershot: Ashgate.

Cresswell, T. (2004) *Place: A Short Introduction*. Oxford: Blackwell.

Cresswell, T. (2006) *On the Move: Mobility in the Modern Western World*. New York, NY: Routledge.

Dant, T. (2004) 'The driver-car', *Theory, Culture & Society*, 21 (4–5): 61–75.

Graham, S. and Marvin, S. (2001) *Splintering Urbanism: Networked Infrastructures, Technological Mobilities and the Urban Condition*. London: Routledge.

Gregory, D. and Urry, J. (eds) (1985) *Social Relations and Spatial Structures*. London: Macmillan.

Harada, T., Birtchnell, T. and Du, B. (2023) 'The rush of the rush hour: Mobility justice for seniors on public transport in Sydney, Australia', *Social & Cultural Geography*, 24 (2): 212–31.

Harrison, P. (2008) 'Corporeal remains: Vulnerability, proximity, and living on after the end of the world', *Environment and Planning A*, 40 (2): 423–45.

Knowles, R., Docherty, I. and Shaw, J. (2008) *Transport Geographies: Mobilities, Flows, and Spaces*. Oxford: Blackwell.

Kraftl, P. and Horton, J. (2008) 'Spaces of every-night life: For geographies of sleep, sleeping and sleepiness', *Progress in Human Geography*, 32 (4): 509–24.

Laurier, E. (2004) 'Doing office work on the motorway', *Theory, Culture & Society*, 21 (4–5): 261–79.

Lovering, J. (1989) 'The restructuring debate', in R. Peet and N. Thrift (eds), *New Models in Geography: The Political-Economy Perspective, Vol. One*. London: Unwin Hyman. pp. 198–223.

Massey, D. (1979) 'In what sense a regional problem', *Regional Studies*, 13 (2): 233–43.

Massey, D.B. (1994) *Space, Place and Gender*. Cambridge: Polity.

McCann, E.J. (2008) 'Expertise, truth, and urban policy mobilities: Global circuits of knowledge in the development of Vancouver, Canada's "four pillar" drug strategy', *Environment and Planning A*, 40 (4): 885–904.

McCormack, D.P. (2002) 'A paper with an interest in rhythm', *Geoforum*, 33: 469–85.

Merriman, P. (2004) 'Driving places – Marc Auge, non-places, and the geographies of England's M1 motorway', *Theory, Culture & Society*, 21 (4–5): 145–67.

Murgatroyd, L., Savage, M., Shapiro, D., Urry, J. and Walby, S. (1985) *Localities, Class and Gender*. London: Pion.

Obrador-Pons, P. (2007) 'A haptic geography of the beach: Naked bodies, vision and touch', *Social & Cultural Geography*, 8 (1): 123–41.

Rose, M. and Wylie, J. (2006) 'Animating landscape', *Environment and Planning D: Society and Space*, 24 (4): 475–9.

Schwanen, T. (2016) 'Geographies of transport I: Reinventing a field?', *Progress in Human Geography*, 40 (1): 126–37.

Sheller, M. and Urry, J. (2006) 'The new mobilities paradigm', *Environment and Planning A*, 38 (2): 207–26.

Sheller, M. and Urry, J. (2016) 'Mobilizing the new mobilities paradigm', *Applied Mobilities*, 1 (1): 10–25.

Spinney, J. (2006) 'A place of sense: A kinaesthetic ethnography of cyclists on Mont Ventoux', *Environment and Planning D: Society and Space*, 24 (5): 709–32.

Thrift, N. (1996) *Spatial Formations*. London: Sage.

Thrift, N. and Williams, P. (1987) *Class and Space: The Making of Urban Society*. London: Routledge and Kegan Paul.

Tuan, Y. (1977) *Space and Place: The Perspective of Experience*. London: Edward Arnold.

Tyfield, D.P., Zuev, D., Li, P. and Urry, J. (2016) *The Politics and Practices of Low-Carbon Urban Mobility in China: 4 Future Scenarios*. A CEMORE report, https://eprints.lancs.ac.uk/id/eprint/80292/1/Low_Carbon_China_Mobilities_Futures_Scenarios_CeMoRe_Report_Final.pdf

Urry, J. (1981) 'Localities, regions and social class', *International Journal of Urban and Regional Research*, 5: 455–73.

Urry, J. (1989) 'Sociology and Geography', in R. Peet and N. Thrift (eds), *New Models in Geography: The Political-Economy Perspective, Vol Two*. London: Unwin Hyman. pp. 295–317.

Urry, J. (2004a) 'The "system" of automobility', *Theory, Culture & Society*, 21 (4–5): 25–39.

Urry, J. (2014) 'Epilogue: Oil, the American suburbs and the future of mobility', in P. Adey, D. Bissell, K. Hannam, P. Merriman and M. Sheller (eds), *The Routledge Handbook of Mobilities*. London: Routledge. pp. 585–93.

58
Gill Valentine

Rik Huizinga, Utrecht University
Bettina van Hoven, University of Groningen

Biographical Details and Theoretical Context

Gill Valentine has been an important feminist voice in contemporary human geography and has inspired geographers' thinking for well over three decades. Valentine has published over 170 journal articles and edited book chapters as well as 17 books contributing to social and cultural geography, feminist geography, children's geographies and the geographies of encounter and difference. Through her strong empirical work on gender, sexuality, childhood and difference, Valentine pioneered research examining the relational production of places highlighting how social difference matters. In so doing, she has made key contributions to developments in Geography, deepening and pluralising spatial thinking.

Valentine attended Durham University, graduating with a first-class BA (Hons) in Geography in 1986. In 1989 she was awarded her PhD from the University of Reading. Between 1989 and 1991, Valentine worked as a journalist for Haymarket Publications and BBC Wales. Although not completely satisfying as a professional direction, she appreciated that being a journalist made her 'follow [her] nose for interesting trends and patterns in society' (Valentine, personal communication, March 10, 2023). Valentine came back to academia as an ESRC Research Fellow at the University of Reading. After a lecturer position at the University of Manchester, she worked between 1994 and 2004 at the University of Sheffield, before joining the University of Leeds as a Professor of Geography and Director of Leeds Social Science Institute in 2005. In 2012 she returned to the University of Sheffield as a Professor of Geography and, in 2017, she was appointed Provost and Deputy Vice Chancellor for Social Sciences.

Throughout her academic career, Valentine has drawn heavily from personal observations and experiences-in-place to identify and spatialise under-researched lived experiences and social phenomena. Her interest in spatial thinking originated at secondary school where a passionate geography teacher sparked an interest in observing and understanding spatial patterns. Although originally more drawn towards physical geography, it was during her undergraduate studies at Durham University that Valentine became interested in human geography to explain the issues that she wanted to understand or that she felt needed to be researched. Intrigued by humanistic geography and environmental perception, Valentine became interested in the ways she and those around her engaged with their surroundings and the imaginative bond people maintain with their environments. Valentine herself recalls feeling uncomfortable walking in the dark between Durham's city centre and the university campus due to reported sightings of men exposing themselves in front of female students (Valentine, personal communication, March 10, 2023). Recognising how her own experience of the environment, her perception of space and her conduct within particular places was related to feelings of fear and anxiety she took inspiration from **Yi-Fu Tuan**'s *Landscapes of Fear* (1979) for her doctoral research, entitled *Women's Fear of Violence in Public Space: A Spatial Expression of Patriarchy*.

Valentine's academic work covers interrelated and timely thematic areas such as social identities, citizenship and belonging, children and parenting, lifestyle and consumption cultures, and encounter and difference. Despite her wide research interests, throughout her work Valentine maintains a tenacious consistency in how she views and seeks to understand the world, how she believes complex social phenomena are measurable through spatial thinking and how to report on it. Her research projects are characterised by being both theoretically informed and strongly rooted in empirical investigation. Valentine also known for using innovative methodologies and for her engagement with policy and public debates.

By carving out a space in a discipline that has a very masculinist, heterosexist, ableist and racist history (McDowell, 1990: see entries on **Audrey Kobayashi**, **Linda McDowell**, **Laura Pulido**), Valentine played a brave and prominent role in the emergence of women in geography, not only in the UK but also across non-Anglophone contexts (see Drooglever Fortuijn, 2019). Throughout her career, she performed a pioneering role in creating circumstances that enabled more diverse generations of scholars to find their place in academia. This is reflected in the various collaborations on research projects that tend to include early career researchers. Valentine also shaped the discipline by being one of the founding editors of *Social & Cultural Geography* and an editor of *Gender, Place and Culture* and *Children's Geographies*. Additionally, Valentine's talent for communicating her ideas and thoughts across disciplinary boundaries in a compelling and highly readable way, possibly reflecting her training as a journalist, led to a wide readership.

Valentine's many original contributions have been honoured. She is a Fellow of the British Academy, a Fellow of the Academy of Social Sciences and a Fellow of the Royal Geographical Society. Her scholarly work as well as her efforts to shape governmental and academic institutional policy are widely appreciated and recognised; amongst others, she has been distinguished with the prestigious Royal Geographical Society/Institute of British Geographies Murchison Award (2015) for her publications relating to the geography of difference, equality and diversity, the RGS/IBG Gill Memorial Award (2000) for contributions to Geography and Gender, the Philip Leverhulme Prize (2001), and the Sage Prize for Innovation and Excellence (2017).

Spatial Contributions

Together with other feminist scholars such as **Doreen Massey** and **bell hooks**, Valentine's work is characterised by the key premise that space is never free of power relations and dominant structures. Inspired by scholars such as Janet Townsend who were teaching about contemporary debates in feminist geography in the 1980s, Valentine refused the idea that space is neutral and considered places to be constituted relationally. The introduction of more postmodernist discourses in cognate disciplines opened up space in Human Geography to explore and critically think about knowledge production that claims validity across different groups, cultures, and races; theoretical developments that influenced Valentine's academic oeuvre throughout. However, while the 'cultural turn' in Human Geography was influential, and geographers turned to Cultural Studies to understand relations of dominance and subordination (Bell, 2009), Valentine did not engage with these debates directly, as for her these discussions could be overly abstract: as she stresses, 'for me theory doesn't work if it's not empirically informed and adapted' (Valentine, personal communication, March 10, 2023).

Nevertheless, the cultural turn opened up new spaces of inquiry in Geography to research unequal relations between individuals and societies such as the body. Intellectual curiosity about geographical contributions to cultural studies themes caught the attention of publishers, notably Tristan Palmer, the commissioning editor of Routledge at the time, who published *Mapping Desire* (Bell and Valentine, 1995) and who permitted experimentation with the 'unconventional' physicality of the book (see also Peake, 2016). Despite the new wind blowing in human geography, Valentine warned of the danger of 'being caught up in its own pretensions' (Valentine, 2001: 167) preferring to centre on the ongoingness of everyday practice and the force of norms related to everyday life rather than on more abstract concerns, such as the organisation of meaning through representation.

Valentine's key contribution has been to problematise, pluralise and refine hegemonic understandings of space and place in context and establish credibility to work on social issues around young people and exclusion. By introducing ideas and concepts from other disciplines to Geography, Valentine's work centres around spatialising often taken-for-granted facets of everyday life, thereby contributing to an understanding of how place influences the various ways in which identities are ascribed to people. By playing a critical part in getting different marginalised groups onto the geographical agenda, Valentine's contributions have had a significant impact on Geography and have been agenda-setting for fields such as feminist, sexuality and childhood research. Most of her work is concerned with embodied experiences and issues around social belonging, politics of difference and exclusion, and how these are expressed, reinforced and contested spatially. Through her methodological focus on the body as a spatial scale and her interest in the 'ideal body' in socio-spatial relations, Valentine is partly responsible for introducing a new scale of inquiry to the discipline.

Arguably, Valentine's most significant contribution to thinking through space and place is her work on the study of gender and sexuality. Geographies of gender and sexuality as a field of scholarly inquiry have contributed to understanding differences in sexualities and the spatial organisation of these differences as a key feature of human geography. Valentine's work on lesbian geographies highlighted the heteronormativity of space by exploring the production and regulation of heterosexual space. The work was revolutionary as Valentine showed how heteronormativity was not merely attributed to a specific location or locale but is operational everywhere. *Mapping Desire: Geographies of Sexualities* (1995), co-edited with David Bell, was an early effort to bring together a group of researchers to investigate experiences around sexuality from a geographical perspective, many of these invoking queer theories which were little discussed or even acknowledged in geographical scholarship at the time. The idea for this edited volume originated from the 'Sexuality and Space Network' designed to discuss the fluidity of sexual identities and identity politics. Although studies on the geographies of sexualities were not new per se, the book was immediately well received and cited, and stimulated the emergence of the Sexuality and Space Specialty Group of the AAG in 1996 (Peake, 2016) and further geographical forays into queer theorisation. The book was not only well received in the academic press but was also widely reviewed in the mainstream, non-academic gay press.

In addition to Valentine's contributions to gender, LGBT communities and sexualities research, she played an important role in developing work on the geographies of childhood. *Cyberkids: Children in the Information Age* (2002), co-authored with Sarah Holloway, was the fruit of a project that developed alongside the emergence of the internet, considering the emergence of new and unfamiliar concerns about children's online safety. Two other edited book projects, *Cool Places: Geography of Youth Cultures* (1998), with Tracey Skelton, and *Children's Geographies: Playing, Living, Learning* (2000), with Sarah Holloway, showed how geography matters in the social construction of childhood and youth, exposing the moral panics (e.g. stranger danger) that encouraged parents to restrict younger children to occupying domestic and commercial spaces, while condemning older teenagers to occupy the liminal spaces of the streets.

Although Valentine picks up research ideas from other disciplines such as Anthropology and Sociology and introduces new theories in the field of Geography, this relationship is not unidirectional. Although perhaps not as distinctive as the works on gender, sexuality and childhood, the 'Living with Difference' research project illustrates her commitment to put forward the role of place in other disciplines. Interested in the question of how to develop the capacity to live with difference, Valentine critically engages with work in Psychology such as Gordon Allport's (1954) idea of the 'contact hypothesis' and sociological work on cosmopolitanism. Thinking through difference, encounter, and prejudice from a spatial perspective, she illustrates the mundane experiences, positions, and activities in place that shape people's experience of difference and organise everyday life around perceptions of difference. In her authoritative reflection on geographies of encounter (Valentine, 2008), Valentine critically engages with the debates around 'meaningful contact', conviviality, and multiculturalism. She highlights the role of power relations in place to redefine theoretical thinking in other disciplines. Moreover, in her article on intersectionality (Valentine, 2007) she builds on Kimberlé Crenshaw's (1993) theorisation to redefine intersectionality within feminist geography. To understand how identities and social relations are shaped through power in place, she points at the usefulness of an intersectional lens for geographers to move beyond looking at singular identity markers.

Key Advances and Controversies

Valentine embodies a commitment to the relevance of geography in a changing world in which new and sometimes problematic relations are altering the dynamics of everyday life and ordinary practices. Valentine's work exposes the ways in which ordinary practices are patterned by norms that become lived as common sense. These themes appear throughout her work, be it on the geographies of everyday fear, heteronormativity or consumption. Valentine has contributed significantly to theoretical debates on understanding how space and place permeate social, cultural, economic and political life and relations, and how inequality and exclusion are articulated in these processes. As such, by making under-researched ways of thinking visible through spatiality, her work offers complementary perspectives on existing theories on how space and place come into being. Moreover, her work recognises the notion of place in its content, approach, methodological decision-making and outputs. Valentine's oeuvre comprises theoretically informed and empirically based innovative research with attention to novel and creative research methods in order to investigate everyday life patterns previously deemed unimportant in Geography.

Valentine's work has sometimes been viewed as controversial, since despite currents contributing to exciting new and emancipatory developments, Geography as a discipline and institution can be rather traditional and rigid (Drooglever Fortuijn, 2019). Although acknowledging the existence of caring social spaces that allowed her to thrive intellectually, Valentine's (1998) 'Sticks and stones may break my bones' paper reveals that researching marginalised identities came at a personal cost. At the time little was written about women's experiences or experiences that represented gay and lesbian lives. Valentine reports that her academic skills as well as her place in the academy were questioned up to a point where 'colleagues' threatened her and insisted she should quit her career. Moreover, the work on lesbian geographies received some press coverage, which was used to rouse a campaign that accused her of misusing tax money (see Valentine, 1998). Valentine further recalls an incident after a computer disaster whilst working to a deadline. Having lost important files, Valentine and co-author David Bell asked an administrative assistant to help re-type the work. After having read the content, the person refused to type up the manuscript given its 'abnormal' and 'shocking' content. Although these experiences left her bruised, Valentine received various forms of support across the discipline and her work on sexualities is considered ground-breaking. Commenting on the 1998 paper, she suggests:

> It is still the most enduring paper as people still contact me about it. It created a more positive and receptive environment as it mobilised people to recognize issues in the discipline. There was no quality legislation to protect peoples' employment and there was a lot more blatant homophobia in public life and discrimination in workplaces. (Valentine, personal communication, March 10, 2023)

Valentine's broad interest in the notions of space and place and the spatialities of often marginalised or absent groups in Human Geography resonates with those working in other theoretical traditions. Although not appreciated equally across the discipline, she was keen to look beyond Geography and to bring in exciting ideas, trends and methods into the field from other disciplines. This has led some researchers to accuse her of being a 'magpie geographer' (Valentine, personal communication, March 10, 2023). Yet, in the context of the 'spatial turn', a paradigm shift in thought across the social sciences and humanities pointing at the negligence of the spatial dimension in the investigation of social phenomena, she continues to be an influential figure to facilitate the intellectual exchange between geographers and other researchers.

The strong empiricism in Valentine's work has been key to advancing the social geography of the everyday and its quality has led the discipline to appreciate qualitatively informed research. Deeply connected to Valentine's orientation towards everyday life and the importance of ordinariness in her work lies a political commitment to listen and give voice to individuals and groups who experience marginalisation. Inseparable from her work is an underlying methodology that allows her to flesh out how everyday practice is, on the one hand, shaped through patterned norms, but simultaneously avoids the methodological trap of assuming individuals are just reproductive of that same set of norms. Consequently, despite the strong focus on everyday life, her work is not reductive and she avoids treating everyday life merely as a mechanism for the reproduction of everyday life.

As a strong believer in qualitative research, it was Valentine's belief that human geography needed to capture insights grounded in people's actual lived experiences and to put forward their voices. The detailed and lively manner in which Valentine described her research processes and methodological considerations was pioneering. Throughout, her work remains transparent about knowledge production by describing carefully how she arrived at her insights. For example, her use of the long interview quote which is systematically unpacked in detail has been influential and has been taken up by many other researchers in various fields (see Valentine, 2013).

Despite these key advances to understanding space and place, the empirical and theoretical insights her articles and chapters produce remain largely aimed at the UK – or at the least the Anglo-American geographical tradition – reasserting a particular view of geography as self-evidently dominant (see also Boyle, 2004). This, at least in part, might be explained through the type of funding Valentine received and the scope of these funding bodies aimed at timely issues and challenges in the UK. For example, the geographies of gender and sexuality have become established fields of inquiry in part because of her work but in particular centred within the British and North American scholarly traditions. Despite the 'now widespread acceptance of the need for a global and racialized take', it remains a field predominantly covering studies of white queers in the Global North that perpetuate Western hierarchies (Peake, 2016: 575). The work on geographies of food, too, is probably most recognised by more 'affluent populations of Western societies' (Kodras, 1998).

Overall, Valentine's material traces in Geography and related disciplines show the value of subjective modes of knowing in understanding the spatialities of young people's everyday lives and experiences of exclusion. Throughout her academic work, she has engaged productively with geographies' various 'turns' as well as newly emerging forms of marginalisation and inequalities due to societal change and development. Since her diverse research output and methodological approach are firmly anchored in empiricism and theory, her spatial contributions continue to be meaningful and inspire people across various academic fields. Moreover, her brave and pioneering work has paved the way for scholars who desire to pursue similar research interests that still remain positioned in the margins of geography.

Valentine's Key Works

Bell, D. and Valentine, G. (eds) (1995) *Mapping Desire: Geographies of Sexualities*. London: Routledge.

Bell, D. and Valentine, G. (1997) *Consuming Geographies: We Are Where We Eat*. London: Routledge.

Holloway, S.L. and Valentine, G. (eds) (2000) *Children's Geographies: Playing, Living, Learning*. London: Routledge.

Skelton, T. and Valentine, G. (eds) (1998) *Cool Places: Geographies of Youth Cultures*. London: Routledge.

Valentine, G. (1989) 'The geography of women's fear', *Area*, 21 (4): 385-390.

Valentine, G. (1993) '(Hetero)sexing space: Lesbian perceptions and experiences of everyday spaces', *Environment and Planning D: Society and Space*, 11: 395–413.

Valentine, G. (2007) 'Theorizing and researching intersectionality: A challenge for feminist geography', *The Professional Geographer*, 59 (1): 10–21.

Valentine, G. (2008) 'Living with difference: Reflection on geographies of encounter', *Progress in Human Geography*, 32 (3): 323–37.

Valentine, G. (2013) 'Tell me about … : Using interviews as a research methodology', in R. Flowerdew and D. Martin (eds), *Methods in Human Geography: A Guide for Students Doing a Research Project*. London: Longman. pp. 110–127.

Secondary Sources and References

Allport, G.W. (1954) *The Nature of Prejudice*. Boston, MA: Addison-Wesley.

Bell, D. (2009) 'Cultural studies and human geography', in R. Kitchin and N. Thrift (eds), *International Encyclopedia of Human Geography*. Amsterdam: Elsevier. pp. 437–41.

Boyle, M. (2004) 'Critical readings of key thinkers on space and place', *Environment and Planning A*, 37: 161–87.

Crenshaw, K.W. (1993) 'Mapping the margins: Intersectionality, identity politics and violence against women of color', in M. Albertson Fineman and R. Mykitiuk (eds), *The Public Nature of Private Violence*. New York, NY: Routledge. pp. 93–118.

Droogleever Fortuijn, J. (2019) 'Women geographers and the International Geographical Union', *Boletín de la Asociación de Geógrafos Españoles*, 81, 2819: 1–16.

Holloway, S. and Valentine, G. (2002) *Cyberkids: Children in the Information Age*. London: Routledge.

Kodras, J.E. (1998) 'Review of *Consuming Geographies: We are Where We Eat* by David Bell and Gill Valentine', *Annals of the Association of American Geographers*, 88 (4): 733–6.

McDowell, L. (1990) 'Sex and power in academia', *Area*, 22 (4): 323–32.

Peake, L. (2016) 'David Bell and Gill Valentine (eds) (1995). *Mapping Desire: Geographies of Sexualities*', *Progress in Human Geography*, 40 (4): 574–8.

Tuan, Y-F. (1979) *Landscapes of Fear*. New York, NY: Pantheon.

Valentine, G. (1998) '"Sticks and stones may break my bones": A personal geography of harassment', *Antipode*, 30 (4): 305–32.

Valentine, G. (2001) 'Whatever happened to the social? Reflections on the "cultural turn" in British human geography', *Norsk Geografisk Tidsskrift – Norwegian Journal of Geography*, 55 (3): 166–72.

59
Eyal Weizman

Mikko Joronen, Tampere University

Biographical Details and Theoretical Context

Eyal Weizman is a British-Israeli architect known especially for his investigative (forensic) approach to the built environment and war. Weizman, born in Haifa, 1970, graduated in architecture at the Architectural Association in London (1998) and completed his PhD at London Consortium/Birkbeck College (2006), after which he has been a Professor of Architecture at the Academy of Fine Arts in Vienna and professeur invité at EHESS Paris, and is currently working as a professor of Spatial and Visual Cultures at Goldsmiths, University of London. Weizman's commitment to the elaboration of the effects of war, especially as they are evidenced through built environments, material ruins, and debris, has played a crucial role in tracing the damage of contemporary military/state violence at various sites around the world, particularly Palestine. Weizman's early works in particular, such as *Hollow Land: Israel's Architecture of Occupation* (2007), *The Least of All Possible Evils: Humanitarian Violence from Arendt to Gaza* (2011), and *A Civilian Occupation: The Politics of Israeli Architecture* (2003, edited with Rafi Segal), have offered manifold insights into architecture's role in Israel's colonialism, vertical politics, and humanitarian thanatopolitics (i.e., the politics of death) in Palestine (see also Weizman, 2009).

After the establishment of *Forensic Architecture* research group at Goldsmiths in 2010, a forensic investigative approach has become a central trademark of Weizman's work, evidenced particularly at *Mengele's Skull: The Advent of Forensic Aesthetics* (2012, with Thomas Keenan), *The Conflict Shoreline: Colonization as Climate Change in the Negev Desert* (2015, with Fazal Sheikh), *Forensic Architecture: Violence at the Threshold of Detectability* (2017), and *Investigative Aesthetics: Conflicts and Commons in the Politics of Truth* (2021, with Matthew Fuller). In these books, Weizman (co)develops forensic analysis that mobilises architecture as an investigative tool for detecting and reconstructing violence related to state oppression, military aggression, human rights abuses, environmental degradation, police brutality, and state silences worldwide – from gold mining violence in the Amazon to the Beirut port explosion and Russian airstrikes on Kyiv and Mariupol. Weizman has also been influential outside academia, including engagements with the International Criminal Court (ICC), the Centre for Investigative Journalism (CIJ), Israeli human rights NGO B'Tselem, and the Decolonizing Architecture Art Research (DAAR) collective he co-established in 2007. In addition to several personal awards, the Forensic Architecture group was also a partner of the New York Times' EMMY-award-winning film *One Building, One Bomb* that reconstructed a state-induced chemical attack in Syria.

Spatial Contributions

In his most acknowledged book, *Hollow Land* (2007), Weizman approaches the complexity of Israeli frontier practices in occupied Palestinian territories (oPt) through the geographies of *elastic space*. In such spaces, obfuscation and structured chaos become key organising principles that constantly blur the clear border between the inside and the outside. Instead of 'borders' Weizman refers to 'frontier zones' that, as elastic, are constantly moving and transforming, and the status of which hence remains unclear. At such frontier zones power comes from multiple sources and remains diffused to many actors, including Israeli state bodies, petty sovereigns (single military commanders and soldiers), and importantly, non-state actors. Among the last are the settlers, civilians living in settlements built on dispossessed Palestinian lands, who for Weizman provide one of the key forces of colonisation and violence in Palestine. Weizman shows how often the state of Israel lets the settlers and settler groups do its 'dirty job' through various vigilante forms of violence, the military intervening merely to prevent further escalation, most commonly by protecting settlers from those Palestinians who choose to protest.

While the vigilante tactics of settlers consist of various forms of intimidation and sabotage (see Shalhoub-Kevorkian, 2015), elasticity comes to the fore perhaps most clearly in 'outpost' establishment. Outposts are unauthorised additions to settlements (in themselves illegal according to Fourth Geneva Convention) built without state-authorised permits, most often by bringing caravans and trailers to Palestinian lands close to existing settlements. While Israeli authorities do at times evict outposts, they systematically legalise them afterwards, for instance under the pretext of security or as a punitive measure responding to Palestinian violence. Far from being organised through a state-machinated top-down master plan, much of the colonisation in Palestine thus takes place through bottom-up organisation, *ad hoc* decisions, and retrospective formalisation that flexibly operate through state–settler nexuses.

In elastic spaces power thus becomes unpredictable, something appearing to the outside as chaotic and messy. This, Weizman argues, is the key mechanism of control at use in oPt, one where colonial violence is obfuscated with a systematical promotion of *complexity*. The constant remaking and rearrangement, sometimes expanding the frontier zone, at times withdrawing it, thus create complexly varying rhythms of spatial control that blur the brutal treatment of Palestinians, the systematic appropriation of their lands, and the destruction of their spaces of dwelling (see Ghantous and Joronen, 2022). Weizman, however, does not just approach elastic spaces horizontally – as territorial frontiers – but also *vertically* (see also Weizman, 2003). The chapters of *Hollow Land* are organised to pinpoint the importance of this *volumetric* aspect, and the way it further complicates the notion of territorial border through the vertical layering of power, colonisation, and claims of sovereignty. The chapters move from the control of subterranean elements – groundwater, tunnels, sewage, eliminations of archaeological strata, etc. – to Israel's use of the built environment in demographic divides, control of movement, and 'altitude colonisation' implemented through hilltop settlements, eventually discussing the militarisation of aerial space.

In his early 2010s work, especially in *The Least of All Possible Evils* (2011), Weizman's focus turns to humanitarian violence, particularly to the use of the principle of 'lesser evil' in justifying moral and political means of war and extra-judicial killing. Weizman sees the principle of 'lesser evil' as a liberal version of eighteenth-century philosopher Gottfried Wilhelm Leibniz's thesis of the world as the 'best of all possible worlds' created by a God who, as a divine economist, has calculated the optimal balance between good and evil. In its contemporary liberal form, this theological thesis (with roots going all the way back to 4th century theology) has turned into a secularised political calculus for legitimising military violence as the 'least of possible evil' necessary for conducting the common good.

A crucial point regarding the contemporary mobilisation of the principle of 'lesser evil' is the way international humanitarian law, humanitarianism, and human rights have become tools abused by states, militaries, and supra-state actors in their calculated economies of violence. An example par excellence is the blockade of Gaza, where humanitarianism has become a way to continue scarce conditions of war and occupation, something that found perhaps its most brutal expression in the Israeli calculation of the minimum number of calories needed to keep the Gazan civilian population physically alive, but on the 'verge of death' (see Smith, 2016; Joronen, 2021a). Such instrumentalisation also finds other forms, for instance, (i) when 'collateral damage' to civilian areas is justified as a 'proportionate' side-effect of doing the common good; (ii) when humanitarian organisations are allowed to do the work occupying powers are obligated to under the international law; (iii) when humanitarian law is used to authorise rather than prevent violence ('lawfare'); or (iv) when targeted assassinations, torture, and deportations, even the use of chemical weapons, are defended via 'economy of evil', always calculating the 'least evil' option out of the limited choices available (on 'ticking timebomb' arguments, see Hannah, 2006). It is precisely by pushing discussion to such closed options that violence appears necessary – an unfortunate evil that is tolerated – with the military's ethical calculator always ensuring the minimum necessary amount of violence is conducted.

In *The Least of All Possible Evils* Weizman also conducts forensic investigations that have over the decades grown into a broader program with exhibitions, seminars, publications, and reconstructions often stunningly visualised at the website forensic-architecture.org. By examining the materials that remain – rubble, ruins, debris, shrapnel – in various sites (Ethiopia, Kosovo, Iraq, Afghanistan, Sudan and Gaza), Weizman uses the same analytic methods of building collapsibility, satellite imagery, geospatial data, field investigations, and so on as militaries do in their 'proportionality assessments', but to retrospectively reveal the damage done to civilian spaces. Importantly, however, forensic investigations are not only about collecting *scientific* evidence by analysing 'failing points of structures', 'weapons and ammunition remnants' and 'medical and autopsy reports'. Rather, such investigations are also connected to what lays at the heart of the Latin word *forensis*: the placing of objects as part of larger narratives to persuade audiences at different *forums*. This is never a straightforward process, as the narration of evidence needs to be mobilised in various forums for varying purposes.

This was the case, for instance, in Human Rights Watch's 2009 Gaza war report based on forensic analysis of battle damage assessments. The report was heavily criticised by Israeli authorities on the basis that the main investigator, a former US Defense Intelligence Agency analyst Marc Garlasco, was collecting Nazi-era memorabilia. Rubble and blast analysis might provide scientific evidence to a story that nevertheless needs to be interpreted, constructed, and narrated in broader political and spatial contexts. Another example is the May 2022 killing of Al Jazeera's Palestinian-American journalist Shireen Abu Akleh at West Bank: forensic architecture investigations reconstructed the shooting through a 3D model based on audio, geolocation, field, and image complex analyses, to identify, contra Israeli military claims, the only possible direction that the bullet killing Abu Akleh could have come from (Forensic Architecture, 2022). Every investigation thus contains a political struggle to reverse the way states, police, and military monopolise the use of forensic analysis, and so to counter prevalent interpretations, silences, histories trajectories, and geometries of power.

In his more recent works, particularly in *Forensic Architecture* (2017), Weizman offers a further guide for conducting forensic investigations and exhibiting the outcomes in different forms and forums. A key threshold of investigation becomes the detectability of the material damage, and the signatures left on ruins, rubble and built environment (see also Boyce and Chambers, 2021). Due to structural inequalities in gaining access to high-resolution satellite images, data from various sources is used, including images taken with cell phones, clips of security cameras, social media metadata, and personal memories. Through triangulation of materials, events and their patterns can be reconstructed (for instance, with analyses of bomb cloud formation and shadow angles) and the specific ammunition in use detected (for instance, with hit angle analysis, AI categorisations, and blast pattern-radius-height analysis).

But the proximate relationship to events also means investigations will always have an angle and an ethical stance that, despite the scientific evidence, produce situated knowledge. In *Investigative Aesthetics* (Fuller and Weizman, 2021) the question is no more of one simply investigating the world with rigorous scientific methods, but also more broadly epistemological (how to gain knowledge of the world) and ontological (how we think of the world we aim to gain knowledge from). Here investigations might seek to establish claims of truth (on how events took place) but remain committed to critical theory, the authors underline, when questioning ways in which knowledge and power become entangled to serve certain political mechanisms of truth-making. 'Investigative aesthetics' is importantly not 'post-truth', but rather seeks to inquire both, the 'making of' and the 'making up' of institutionalised and normalised forms of truth, through evidence-based (forensic) truth-telling.

Key Advances and Controversies

Weizman's work on vertical and volumetric politics has had a noticeable influence on Geography, especially on works around aerial violence (Adey, 2010), cartography (Crampton, 2010), and urban geography (Graham, 2016; Harris, 2015) (see also entry on **Stephen Graham**). Adey (2010), for instance, draws on Weizman in his work on aerial governance when arguing the analysis of space needs to be turned away from two-dimensional territoriality of a *surface* to a three-dimensional *volume* (see Adey, 2013; Adey et al., 2013; see also Gregory, 2011; Graham and Hewitt, 2012; Woon and Dodds, 2021). **Stuart Elden** (2013) has further acknowledged the way volumetric is entangled with the *calculability* of the 'metric', and thus with the ways of controlling – or 'securing' – the volume (on how calculative computing also counters the dimensionality implied by the 'volume', see **Louise** Amoore, 2018). Elden recognises Weizman as a 'key thinker of verticality', though further underlining the importance of looking, not only down from above (or up from below), but also of understanding the interplay between the vertical and horizontal through *angles* and inclines always beyond mere perpendicular.

More recently, the scope of verticality has been expanded to cover various other geographical debates, for instance, on political ecology (Mostafanezhad and Dressler, 2021), material sovereignty (Billé, 2020), and the long-term effects of bombings on earth and life (Griffiths, 2022a). Crucially, these takes underline that there is a growing need to avoid the anthropocentric gaze that ignores the soil, water, and the earth, with the focus on built environments – tunnels, bunkers, houses, and so on – in examining the verticalities of military violence (see also Bridge, 2013; Chambliss, 2020; Gordillo, 2020). However, as Mark Griffiths (2022a: 287) acknowledges, such is not merely a question of starting from the earthly 'view from below' (Adey et al., 2011), but of avoiding the distancing techno- and Eurocentric remoteness always at danger of 'expunging' the lives of those to whom the war is nothing but remote (Harker, 2014).

These concerns further resonate with the lack of focus on bodily intimacy and the intimate politics many feminist geographers have addressed as key for understanding politics (e.g., Dowler and Sharp, 2001; Hyndman, 2001; Koopman, 2011; Harker, 2014; Fregonese, 2020; see also Nieuwenhuis, 2016). A good example of this is *Hollow Land*'s perhaps unintentional intellectualisation of IDF's micro-tactics, such as troops breaking through the walls in urban counterinsurgency operations, for which ideas might have been adapted from post-structuralist philosophers like **Gilles Deleuze** and Felix Guattari, but which look significantly less intellectual for Palestinians living with their consequences. As Christopher Harker (2009: 324) points out, in its aim to think of strategic, vertical, and architectural dimensions as the main arena of conflict and violence, *Hollow Land* is 'largely devoid of the Palestinian people who continue to make their lives there', consequently recreating 'Palestine as a space that is only known through practices of occupation and the violence they entail' (Harker, 2009: 328). Here spaces of everyday, cultures of living, and the intimate ways of knowing and being, become seen only through tactical spaces of occupation, such framing ignoring the complex spatialities that might bear relation to, but importantly remain *beyond*, the grid of occupation. Similar concerns have been aired by many (e.g., Marshall 2014; Hammami, 2015; Joronen, 2017) addressing the problems of recognising (if being recognised at all) subaltern/colonised/occupied populations merely as passive targets of power or victims of trauma and suffering, who can at best claim agency by becoming 'heroes of resistance' (Harker, 2011) capable of no more than simple mirroring of prevalent matrixes of power (see Kotef and Amir, 2011; Leshem, 2015; Griffiths, 2022b; see also Petti et al., 2013). What is so set aside is not only the complex embodiment of power *in* everyday life, and the unspectacular and less visible forms of violence they reveal (Golańska,

2022; Joronen, 2021b; Shalhoub-Kevorkian, 2016), but also the 'ungovernability of living' (Joronen and Griffiths, 2022), always irreducible to strategic orchestrations of power and domination.

Weizman's work on humanitarian and legal violence has similarly gained a welcoming but critical reception in geographical literature (e.g., Lopez et al., 2015; Jones, 2016; Pallister-Wilkins, 2020; MacGormack and Gilbert, 2022). By following Weizman's discussions of 'lawfare' Craig Jones (2015, 2016) has shown how late modern war operates in *spaces* that mobilise law through plural scales and overlapping legal regimes. The grey zones of International Humanitarian Law, Jones argues, are certainly used to justify interpretations of proportionality, necessity, and collateral damage of military violence, but in the 'heat of the battle' they often play a less significant role. During operations military commanders instead adapt 'checklist' approaches to IHL, use real-time legal advice, and require well-defined 'rules of engagement' (ROE). Here, Jones (2015: 691) argues 'international', 'domestic' and 'operational' legal regimes become intersected, interpreted and operationalised through changing protocols, legal memos, military directives, court rulings, and so on. Such spatial entanglement of violence and humanitarianism can be further seen, for instance, in governance of refugee camps (Brankamp, 2022; Newhouse, 2015), humanitarian logistics (Ziadah, 2019), infrastructural developments of settler colonial states (Bhungalia, 2015; Salamanca, 2016), and in ways in which humanitarian law is used to make civilians 'responsible for their own killing' (Joronen, 2016). In this regard, humanitarian violence maintains 'parasitic relationships' between colonial conquest, genocide, racism, and biopolitical care for saving lives (Pallister-Wilkins, 2022), thus being entwined with structural forms of imperial order and racial violence (Stoler, 2016).

Importantly, Weizman's own forensic approach places the proximate relationship to events at the centre of conducting investigations. This trait of analysis has been more recently taken up by a number of geographers. Alex Jeffrey (2021), for instance, reviews the wide scope of legal geographies that focus on ways in which legal evidence is constituted through interpretative aspects of authority and expertise; silences, silencing and absence of voice; and the formal barriers of admissibility. Others have stretched forensic analyses to cover a wide spectrum of fields, from 'forensic ecology' (Gandy, 2019) to 'negative methods' (MacFarlane, 2020). Jo Sharp (2021), in turn, has drawn upon the notion of forensics to develop feminist understanding on the intimate entwinement of bodies, materialities and representations to geopolitics. Sharp, while pointing out the problems related to ways in which forensic tools replicate, not only certain forms of structural racism, but crucially the very techniques of violence they seek to uncover, also sees value in looking at forensic analyses though feminist geopolitics focusing more extensively on precarities (and abilities) of the body and the materials attached to it. This again underlines the importance of the *forum* – of how things and bodies are made political and materialities *represented*. Sharp exemplifies this through the work of forensic anthropologist Clea Koff (2004), underlining the micro-geographies of material meaning-giving in former Yugoslavia, where local women recognised bodies deceased during the battles from their personal ways of mending or the placing of stitching at old garments. Again, geographers are calling for analyses capable of recognising more nuanced geographies behind what can easily become a way of reproducing spectacular capacities of military aggression through a masculine and distancing technological gaze.

Weizman's Key Works

Fuller, M. and Weizman, E. (2021) *Investigative Aesthetics: Conflicts and Commons in the Politics of Truth*. London: Verso.

Keenan, T. and Weizman, E. (2012) *Mengele's Skull: The Advent of Forensic Aesthetics*. London: Sternberg.

Petti, A., Hilal, S. and Weizman, E. (2013) *Architecture after Revolution*. London: Sternberg.

Segal, R. and Weizman, E. (eds) (2003) *A Civilian Occupation: The Politics of Israeli Architecture*. London and New York: Verso.

Weizman, E. (2003) 'The politics of verticality: The West Bank as an architectural construction', in K. Biesenbach (ed), *Territories: Islands, Camps, and other States of Utopia*. Köln: Walther König. pp. 65–149.

Weizman, E. (2007) *Hollow Land: Israel's Architecture of Occupation*. London: Verso.

Weizman, E. (2009) 'Thanato-tactics', in A. Ophir, M. Givoni and S. Hanafi (eds), *The Power of Inclusive Exclusion. Anatomy of Israeli Rule in the Occupied Palestinian Territories*. New York, NY: Zone. pp. 543–74.

Weizman, E. (2011) *The Least of All Possible Evils: Humanitarian Violence from Arendt to Gaza*. London: Verso.

Weizman, E. (2017) *Forensic Architecture: Violence at the Threshold of Detectability*. New York, NY: Zone Books.

Weizman, E. and Sheikh, F. (2015) *The Conflict Shoreline: Colonization as Climate Change in the Negev Desert*. Göttingen: Steidl.

Secondary Sources and References

Adey, P. (2010) *Aerial Life: Spaces, Mobilities, Affects*. Chichester: Wiley-Blackwell.

Adey, P. (2013) 'Securing the volume/volumen: Comments on Stuart Elden's Plenary paper "Secure the volume"', *Political Geography*, 34: 52–4.

Adey, P., Whitehead, M. and Williams, A.J. (2011) 'Introduction: Air-target: Distance, reach and the politics of verticality', *Theory, Culture & Society*, 28 (7–8): 173–87.

Adey, P., Whitehead, M., and Williams, A.J. (2013) *From Above: War, Violence and Verticality*. London: Hurst.

Amoore, L. (2018) 'Cloud geographies: Computing, data, sovereignty', *Progress in Human Geography*, 42 (1): 4–24.

Bhungalia, L. (2015) 'Managing violence: Aid, counterinsurgency, and the humanitarian present in Palestine', *Environment and Planning A: Economy and Space*, 47 (11): 2308–23.

Billé, F. (ed) (2020) *Voluminous States: Sovereignty, Materiality, and the Territorial Imagination*. Durham, NC: Duke University Press.

Boyce, G.A. and Chambers, S.N. (2021) 'The corral apparatus: Counterinsurgency and the architecture of death and deterrence along the Mexico/United States border', *Geoforum*, 120: 1–13.

Brankamp, H. (2022) 'Camp Abolition: Ending carceral humanitarianism in Kenya (and beyond)', *Antipode*, 54: 106–29.

Bridge, G. (2013) 'Territory, now in 3D!', *Political Geography*, 34: 55–7.

Chambliss, W. (2020) 'Spoofing. The geophysics of not being governed', in F. Billé (ed), *Voluminous States Sovereignty, Materiality, and the Territorial Imagination*. Durham, NC: Duke University Press. pp. 64–51.

Crampton, J.W. (2011) 'Cartographic calculations of territory', *Progress in Human Geography*, 35 (1): 92–103.

Dowler, L. and Sharp, J. (2001) 'A feminist geopolitics?', *Space and Polity*, 5 (3): 165–76.

Elden, S. (2013) 'Secure the volume: Vertical geopolitics and the depth of power', *Political Geography*, 34: 35–51.

Forensic Architecture (2022) *Shireen Abu Akleh: The Extrajuridical Killing of a Journalist.* Forensic architecture and Al-Haq. Available from: https://forensic-architecture.org/investigation/shireen-abu-akleh-the-targeted-killing-of-a-journalist (accessed November 15, 2023).

Fregonese, S. (2020) *War and the City: Urban Geopolitics in Lebanon*. New York, NY: Bloomsbury.

Gandy, M. (2019) 'The fly that tried to save the world: Saproxylic geographies and other-than-human ecologies', *Transactions of the Institute of British Geographers*, 44 (2): 392–406.

Ghantous, W. and Joronen, M. (2022) 'Dromoelimination: Accelerating settler colonialism in Palestine', *Environment and Planning D: Society and Space*, 40 (3): 393–412.

Golańska, D. (2022) 'Slow urbicide: Accounting for the shifting temporalities of political violence in the West Bank', *Geoforum*, 132: 125–34.

Gordillo, G (2020) 'Gravity. On the primacy of terrain', in F. Billé (ed), *Voluminous States: Sovereignty, Materiality, and the Territorial Imagination*. Durham, NC: Duke University Press. pp. 159–74.

Graham, S. (2016) *Vertical: The City from Satellites to Bunkers*. London: Verso.

Graham, S. and Hewitt, L. (2013) 'Getting off the ground: On the politics of urban verticality', *Progress in Human Geography*, 37 (1): 72–92.

Gregory, D. (2011) 'From a view to a kill: Drones and late modern war', *Theory, Culture & Society*, 28 (7–8): 188–215.

Griffiths, M. (2022a) 'Thanato-geographies of Palestine and the possibility of politics', *Environment and Planning C: Politics and Space*, 40 (8): 1643–58.

Griffiths, M. (2022b) 'The geontological time-spaces of late modern war', *Progress in Human Geography*, 46 (2): 282–98.

Hammami, R. (2015) 'On (not) suffering at the checkpoint; Palestinian narrative strategies of surviving Israel's carceral geography', *borderlands* 14 (1): 1–17.

Hannah, M. (2006) 'Torture and the ticking bomb: The War on Terrorism as a geographical imagination of power/knowledge', *Annals of the Association of American Geographers*, 96 (3): 622–40.

Harker, C. (2009) 'Spacing Palestine through the home', *Transactions of the Institute of British Geographers*, 34 (3): 320–32.

Harker, C. (2011) 'Geopolitics and family in Palestine', *Geoforum*, 42 (3): 306–15.

Harker, C. (2014) 'The only way is up? Ordinary topologies of Ramallah', *International Journal of Urban and Regional Research*, 38 (1): 318–35.

Harris, A. (2015) 'Vertical urbanisms: Opening up geographies of the three-dimensional city', *Progress in Human Geography*, 39 (5): 601–20.

Hyndman, J. (2001) 'Towards a feminist geopolitics', *Canadian Geographer*, 45: 210–22.

Jeffrey, A. (2021) 'Legal geography III: Evidence', *Progress in Human Geography*, 45 (4): 902–13.

Jones, C.A. (2015) 'Frames of law: Targeting advice and operational law in the Israeli military', *Environment and Planning D: Society and Space*, 33 (4): 676–96.

Jones, C.A. (2016) 'Lawfare and the juridification of late modern war', *Progress in Human Geography*, 40 (2): 221–39.

Joronen, M. (2016) '"Death comes knocking on the roof": Thanatopolitics of ethical killing during Operation Protective Edge in Gaza, *Antipode*, 48 (2): 336–54.

Joronen, M. (2017) '"Refusing to be a victim, refusing to be an enemy". Form-of-life as resistance in the Palestinian struggle against settler colonialism', *Political Geography*, 56: 91–100.

Joronen, M. (2021a) 'To wound life, to prevent its recovery: Enforcing vulnerability in Gaza', in D. Bissell, P. Harrison and M. Rose (eds), *Negative Geographies. Exploring the Politics of Limits*. Lincoln, NE: University of Nebraska Press. pp. 206–33.

Joronen, M. (2021b) 'Unspectacular spaces of slow wounding in Palestine', *Transactions of the Institute of British Geographers*, 46 (4): 995–1007.

Joronen, M. and Griffiths, M. (2022) 'Ungovernability and ungovernable life in Palestine', *Political Geography*, 98: article #102734.

Koff, C. (2004) *The Bone Woman: Among the Dead in Rwanda, Bosnia, Croatia and Kosovo*. London: Atlantic.

Koopman, S. (2011) 'Alter-geopolitics: Other securities are happening', *Geoforum*, 42 (3): 274–84.

Kotef, H. and Amir, M. (2011) 'Between imaginary lines: Violence and its justifications at the military checkpoints in Occupied Palestine', *Theory, Culture & Society*, 28 (1): 55–80.

Leshem, N. (2015) '"Over our dead bodies": Placing necropolitical activism', *Political Geography*, 45: 34–44.

Lopez, P.J., Bhungalia, L., and Newhouse, L.S. (2015) 'Geographies of humanitarian violence', *Environment and Planning A: Economy and Space*, 47 (11): 2232–9.

MacFarlane, K. (2020) 'Negative research: Sonic methods in Geography and their limits', *The Professional Geographer*, 72 (2): 297–308

Marshall, D.J. (2014) 'Save (us from) the children: Trauma, Palestinian childhood, and the production of governable subjects', *Children's Geographies*, 12 (3): 281–96.

McCormack, K. and Gilbert, E. (2022) 'The geopolitics of militarism and humanitarianism', *Progress in Human Geography*, 46 (1): 179–97.

Mostafanezhad, M. and Dressler, W. (2021) 'Violent atmospheres: Political ecologies of livelihoods and crises in Southeast Asia', *Geoforum*, 124: 343–7.

Newhouse, L.S. (2015) 'More than mere survival: Violence, humanitarian governance, and practical material politics in a Kenyan refugee camp', *Environment and Planning A*, 47: 2292–307.

Nieuwenhuis, M. (2016) 'Breathing materiality: Aerial violence at a time of atmospheric politics', *Critical Studies on Terrorism*, 9 (3): 499–521.

Pallister-Wilkins, P. (2020) 'Hotspots and the geographies of humanitarianism', *Environment and Planning D: Society and Space*, 38 (6): 991–1008.

Pallister-Wilkins, P. (2022) 'HuManitarianism: Race and the overrepresentation of "Man"', *Transactions of the Institute of British Geographers*, 47 (3): 695–708.

Salamanca, O.J. (2016) 'Assembling the fabric of life: When settler colonialism becomes development', *Journal of Palestine Studies*, 45 (4): 64–80.

Shalhoub-Kevorkian, N. (2015) *Security Theology, Surveillance and the Politics of Fear*. Cambridge: Cambridge University Press.

Shalhoub-Kevorkian, N. (2016) 'The occupation of the senses: The prosthetic and aesthetic of state terror', *British Journal of Criminology*, 57: 1279–300.

Sharp, J. (2021) 'Materials, forensics and feminist geopolitics', *Progress in Human Geography*, 45 (5): 990–1002.

Smith, R.J. (2016) 'Isolation through humanitarianism: Subaltern geopolitics of the siege on Gaza', *Antipode*, 48 (3): 750–69.

Stoler, L.A. (2016) *Duress. Imperial Durabilities in Our Times*. Durham, NC: Duke University Press.

Woon, C.Y. and Dodds, K. (2021) 'Subterranean geopolitics: Designing, digging, excavating and living', *Geoforum*, 127: 349–55.

Ziadah, R. (2019) 'Circulating power: Humanitarian logistics, militarism, and the United Arab Emirates', *Antipode*, 51 (5): 1684–1702.

60
Brenda Yeoh

Margaret Walton-Roberts, Wilfrid Laurier University

Biographical Details and Theoretical Context

Brenda S.A. Yeoh, born in 1963 in Singapore, is Raffles Professor of Social Sciences at the National University of Singapore (NUS) and Research Leader of the Asian Migration Cluster at the Asia Research Institute (ARI). She received a first-class BA honours degree in Geography from the University of Cambridge in 1985. She then did a Diploma in Education from the Institute of Education in Singapore before working as a teacher at Victoria Junior College, one of the top three junior colleges in Singapore. After leaving the teaching service, Yeoh completed a DPhil in Geography from the University of Oxford in 1991 and joined NUS in the same year as a Lecturer, being promoted to full Professor in 2005. She has held various administrative leadership roles at NUS, including Dean of the Faculty of Arts and Social Sciences from 2010–2016, Vice Provost graduate education in 2017, and Director, Humanities and Social Science Research at the Office of Deputy President from 2018–2021. She was awarded the Public Administration Medal (Silver) in 2015 for her contributions and public service in Singapore, as well as the Public Service Medal, National Day Award, in 2022. Yeoh has been a visiting scholar at numerous institutions, including in 2000 when she received a Fulbright Scholar award to study at the University of California, Berkeley, and in 2022, when she was a Scholar of Excellence at Toronto Metropolitan University, Canada.

Yeoh has held numerous editorial and editorial board positions since 1996, the range of which indicate her broad multidisciplinary interests. This includes geography (*Progress in Human Geography, Geography Compass*), gender (*Gender, Place and Culture, NAIS Gendering Asia book series*), population (*Population, Place and Space; Asian Population Studies*), geographies of Asia (*Asian Studies Review, Singapore Journal of Tropical Geography*) and migration (*Asian and Pacific Migration Journal, International Migration*). She also holds numerous positions on international boards and committees, is a member of the International Geographical Union's Population Geography Commission and a full member of the Commission on Geography and Gender. She is a Corresponding Fellow of The British Academy (2021) and was awarded the 2021 Vautrin Lud Prize, one of the highest honours presented for developments in the discipline, and widely considered the 'Nobel Prize in Geography'. Brenda Yeoh is one of only four women awarded this honour since 1991, and Yeoh saw it as a significant acknowledgement that Asia-focused research can have global influence (in Qing, 2021: n.p.).

Yeoh's research interests include colonial and postcolonial cities, migration, gender, and transnationalism. Her own career trajectory is arguably reflective of, and deeply entangled with, postcolonial state power and its development strategy through the sponsoring of global elite student mobility (Kenway and Koh, 2013). Singapore's educational policies guide students from top junior colleges through the application process for elite universities, including the Oxbridge system, for which the Singaporean state provides scholarships (Ye and Nylander, 2015). The scholarships effectively reproduce state power and global aspirations by embedding these elite Oxbridge graduates into the Singaporean civil service once their studies are complete. Brenda Yeoh attended Hwa Chong Junior, which is one of the top two junior colleges engaged in these elite educational trajectories (Ye and Nylander, 2015).

Spatial Contributions

Yeoh is a prolific scholar of Asia, with wide ranging research interests. An emphasis on Asia within a globalising world frames much of her career and intellectual interests and marks the significant contribution her work has made to research on urban space under colonialism. Her PhD dissertation, published as *Contesting Space in Colonial Singapore: Power Relations and the Urban Built Environment* (Yeoh, 1996), was the first to review the history of Singapore's development as a colonial city where the ethnic pluralism of Singaporean society contested colonial power and control of urban development. Her careful historical analysis of the management of colonial urban space and sanitation led Yeoh to argue that 'all too often the colonial city is not treated in its own terms, and the contemporary significance of the spatial configurations of the city for the inhabitants who live out their habitual lives within its confines remains uninterrogated' (Yeoh, 1996: 9). Yeoh examined the experiences of those inhabitants, and the resources and tactics they used to counter colonial authority, concluding that 'such acts were often too dispersed, too anonymous, and far too commonplace to allow immediate effective actions against the actual culprits' (Yeoh, 1996: 15). The resulting urban spatial form 'embodied and expressed the tensions and negotiations, conflicts and compromises between different groups' (Yeoh, 1996: 18). Her spatial contribution is informed by her careful examination of subaltern communities and how their everyday forms of resistance are embedded in the resulting urban spatial forms. Her focus on the politics of space in colonial cities, starting with her graduate work at Oxford and developed in subsequent publications, provided essential and pioneering resources for understanding Singapore's development and its growing global status (Yeoh, 1992; 1993; 1996). Her later interests continued this engagement through analysis of postcolonial cities more broadly (Yeoh, 2001) and Singapore's development as a global city-state (Kong and Yeoh, 2003). Yeoh's spatial contribution is the serious and widely cited analysis of how 'plebeian' or local communities can shape the space of the colonial city, thereby challenging previous analysis that tended to privilege readings of the colonisers' unchecked power and authority.

Yeoh's early contributions to the field of historical geography through her analysis of the spatial development of the colonial city of Singapore have been surpassed by her wide-ranging, contemporary, and deeply interdisciplinary focus on the intersections of gender, migration, and transnationalism across the

Asia-Pacific region. She has published, edited or co-edited over 30 books, and hundreds of other publications. An important aspect of her research is that it is highly collaborative, indicating a wealth of global connections with scholars and institutions. Her extensive co-author network exceeds 50 people, many of whom she has multiple publications with, including Shirlena Huang, Theodora Lam, Elspeth Graham, Elaine Ho, Mika Toyota, Peggy Teo, Nicola Piper, Francis Collins, and Katie Willis.

Over her scholarly career Yeoh's spatial focus through gendered analysis has moved from the scale of the city to the intimate lives of migrant domestic workers and the older care recipients they interact with in Singapore. Some of this work is published with Shirlena Huang and others over several years (Yeoh and Huang, 1998a; 1998b; Yeoh et al., 2005; Huang et al., 2012). Over this time the issues of gender, migration, care, and the body have become more central in Yeoh's conceptualisation of migration policy and migrant subjectivities, resulting in a move away from Yeoh's initial temporal historical analysis of colonial spatial change, toward the contemporary period and deeper analysis of the gendered power geometries that migrant workers are entangled in. For example, Yeoh and Huang (1998b) explore the spatial construction of domestic workers as 'the other' in Singaporean public space and examine how the public space where workers socialise on their days off is physically and discursively monitored by employers, but how domestic workers engage in tactics of resistance through spatial negotiation and wider social practice.

The focus on Singapore also contributes to important work relating to migration policy, by showing how policy operates in its temporary migration regime (Yeoh, 2021). This analysis of the Asian migration context offers important spatial variation regarding comparative migration policy (Liu-Farrer and Yeoh, 2018). It has also allowed for nuance in terms of the spatial consequences of this temporary migration policy in the city, and the resulting spatial-temporal processes of enclosure (worker dormitories) and enclavism (e.g., weekend spaces where migrant socialisation occurs). This spatial framing was developed by Yeoh and Lam (2022), who expand analysis of urban segregation and consider how these spatial-temporal forms of enclosure and enclavism are deployed to prevent the integration of migrant workers in Singapore. Yeoh and Lam (2022) conceptualise these spatial formations not as a means of integration and claiming citizenship, as they have been employed in western urban contexts, but as disciplining measures to enforce non-integration and limited rights in Asian spaces marked by temporary migration regimes.

Yeoh's research focus on the state and the migrant allows her to engage in a double movement in the development trajectory of her research, since while the intimate and *intensive* space of the body has come more into focus in elements of her research, the spatially *extensive* analysis of state control and transnational connections between sending, receiving and transit migrant locations has also informed her analysis of international migration. In her research this transnational analysis spreads across different occupational and family groups, from high to low-paid (skilled) workers, international students (Sidhu et al., 2019), marriage migrants and the children of migrants and the role of the migration intermediaries (Lindquist et al., 2012). In the process the relationship between agency and structure is always in tension in Yeoh's analysis. This tension is comprehensively reflected upon in Yeoh's co-authored chapter with Kellynn Wee and Charmian Goh (Yeoh et al., 2020), which explores how the scale of the body negotiates key binaries (structure/agency, mind/body, etc.) revealing the mutual constitution of these binaries, but also destabilising and challenging them. Encounters between bodies and their circulations make places emotional and affective and create visceral geographies. Yeoh et al. (2020) highlight how corporeal experiences of migration are structured by, and react to, power geometries based on the intersecting axes of class, gender, race, religion, nationality, citizenship, and so on. This includes examples such as migrant bodies and border control, migrant encounters between those with different histories and geographies and transnational family life. This results in 'modes of temporariness that are institutionalised as a fundamental principle in neoliberal labour migration regimes' (Yeoh et al., 2020: 101).

Key Advances and Controversies

Brenda Yeoh's work highlights the significance of Asia in contemporary reflections on migration regimes and policy approaches, challenging migration researchers to think beyond European and North American migration policy landscapes. This includes work on the international mobility of students with a focus on transnational family formations and livelihoods. Yeoh's influential and relatively early work on international students, in collaboration with Francis Collins et al. (2014), explores tertiary education as critical infrastructure in a competitive economy and traces the experiences of international students moving through these transnational educational spaces. These themes have been extended through a focus on China's Belt and Road Initiative (BRI) and changing patterns of student mobilities in China and Southeast Asia to understand how the BRI is reshaping academic linkages across the region, as well as a deeper reflection on the possible shape global higher education might take in the wake of pandemic transformations (Sidhu et al., 2021).

Yeoh's other key advance is regarding the adoption and execution of research using a transnational perspective and methodology. This is clear in the edited collections *State/Nation/Transnation: Perspectives on Transnationalism in the Asia-Pacific* (2004) with Katie Willis, *The Cultural Politics of Talent Migration in East Asia* (2013) with Shirlena Huang, *Return: Nationalizing Transnational Mobility in Asia* (2013) with Biao Xiang and Mika Toyota (2013), *Transnational Labour Migration, Remittances and the Changing Family in Asia* (2015) with Lan Anh Hoang, and the *Handbook on Transnationalism* (2022) with Francis Collins. One of the most enduring of Yeoh's transnational research efforts and the most impressive in terms of its methodological rigour is the Transnational Migration in Southeast Asia and the Health of Children Left Behind project (CHAMPSEA). This started with a migrating out of poverty project where the analysis was aimed at informing policy development, leading to the CHAMPSEA project, which was continued through several rounds of funding and revealed the intergenerational effect of migration in terms of the experiences of the children of migrants, and their perspective as they grow up and engage with migration themselves. The project has consisted of a longitudinal examination of transnational migration/householding, familial care politics and left-behind children by following up on the children in the first wave and contributing to wider debates about intergenerational effects of migration, the feminisation of migration and the organisation of reproductive labour. The research contributed to the analysis of international migration in contexts of temporary migration, where family unification is precluded, thus forcing the creation of transnational households. This type of migration is also highly feminised, raising gendered debates about social reproduction responsibilities and the longer-term effects of international migration on children. The research focus on children counteracts the tendency to centre adults' perspectives in social analysis of migration and exploration of the health and wellbeing of children living in transnational households (Graham and Yeoh, 2013). In terms of key advances the longevity of this analysis offers examples of how to sustain transnational research inquiries, but also demonstrates why it is important to bring a gendered focus to transnational studies (Pratt and Yeoh, 2003), how to methodologically achieve this using mixed methods (Graham and Yeoh, 2013), and how findings can contribute to policy debates about temporary migration policy regimes and the effect on the family (Lam et al., 2013).

The rescaling from the body to policy and state regimes is also a focus for Yeoh's work on the Singaporean state's migration policy and its embrace of cosmopolitan discourse (i.e., Yeoh, 2004), an area where she has contributed to key advances in Migration and Urban Studies and is widely cited beyond the Asian context. This research combines sophisticated theoretical reflection with the invitation to consider migration theory and policy from an Asian perspective. This includes her critical analysis of cosmopolitanism and the Singaporean state's use of this ideal for purposes of global boosterism. Yeoh's critical interest in the Singaporean state's growth and development strategy has informed a vibrant stream of her research on city development, migration policies and ideas of urban cosmopolitanism. Alluding to a universal vision of a diverse globalised multi-ethnic, affluent, and equal society, Yeoh (2004: 2442) argues that, 'The cosmopolis, as erected in state visions and pronouncements in Singapore, however, provides little room – both conceptual and material – for the majority of migrant others called "foreign workers"'. Yeoh suggests that cosmopolitan discourse used by the Singaporean state results in an exclusionary and elitist political and social hegemony that justifies the social exclusion of the lowest paid migrant workers who are contained and controlled through state migration regulations that limit their rights and the length of their stay in Singapore, and that seeks to enclose them in migrant dorms, camps or in the homes of their employers. As an alternative to this exclusionary cosmopolitanism, Brenda Yeoh suggests 'sideways cosmopolitanism' where civil society actors play greater roles in co-creating new approaches to societal organisation and governance (Yeoh, 2004; 2013). These ideas are further developed in several other publications, including Yeoh and Soco (2014) which highlights the masculinity of cosmopolitan thinking, and considers how to forge new ways of thinking about cosmopolitan ideals in terms of the feminised subjectivities of domestic workers, and Ye (2016) who examines how cosmopolitan aspirations conflict with low-waged migration regimes structured around class.

The tension between structure and agency in the context of temporary migration has also been a key preoccupation of Yeoh more recently, looking at the Asian method and regime of international migration, and understanding how the Singaporean state, and Asian countries more widely, embrace a temporary migration regime. Yeoh has been reflecting on policy choices that can humanise this temporary migrant regime, which may be seen as problematic since it accepts the right of the state to limit mobility and impose a permanent temporariness. However, Yeoh's sensitivity to the realities of state management of migration results in more pragmatic solutions, and this has become more urgent considering the COVID-19 pandemic and the longer-term sustainability of temporary migration in Asia and beyond (Yeoh, 2020; 2022). Yeoh's reflections include policy suggestions such as longer visas promoting less churn and reduced costs, including temporary migrants in national safety nets, and integrating migrants more effectively into the receiving society as a bulwark to xenophobia. Yeoh proposes these short-term responses to build momentum toward longer-term, more sustainable migration policy approaches. She has also set an agenda for critically assessing the sustainability and resilience of migrant systems more broadly, taken up in a special issue co-edited by Triandafyllidou and Yeoh (2023), which promises to inspire further research and dialogue regarding global migration regimes in the wake of the COVID-19 pandemic in Asia and beyond.

Yeoh's Key Works

Graham, E. and Yeoh, B.S. (2013) 'Child health and migrant parents in South-East Asia: Risk and resilience among primary school-aged children', *Asian and Pacific Migration Journal*, 22 (3): 297–314.

Huang, S., Yeoh, B.S., and Toyota, M. (2012) 'Caring for the elderly: The embodied labour of migrant care workers in Singapore', *Global Networks*, 12 (2): 195–215.

Sidhu, R.K., Chong, H.K. and Yeoh, B.S. (2019) *Student Mobilities and International Education in Asia: Emotional Geographies of Knowledge Spaces*. Cham: Springer Nature.

Willis, K. and Yeoh, B. (eds) (2000) *Gender and Migration*. Cheltenham: Edward Elgar.

Yeoh, B.S. (1996) *Contesting Space: Power Relations and the Urban Built Environment in Colonial Singapore*. Oxford: Oxford University Press.

Yeoh, B.S. (2013) '"Upwards" or "sideways" cosmopolitanism? Talent/labour/marriage migrations in the globalizing city-state of Singapore', *Migration Studies*, 1 (1): 96–116.

Yeoh, B.S. (2022) 'Is the temporary migration regime in Asia future-ready?', *Asian Population Studies*, 18 (1): 1–5.

Yeoh, B.S. and Lam, T. (2022) 'Managing the non-integration of transient migrant workers: Urban strategies of enclavisation and enclosure in Singapore', *Urban Studies*, 59 (16): 3292–3311.

Yeoh, B.S., Huang, S. and Lam, T. (2005) 'Transnationalizing the "Asian" family: Imaginaries, intimacies and strategic intents', *Global Networks*, 5 (4): 307–315.

Yeoh, B.S, Wee, K. and Goh, C. (2020) 'Corporeal geographies of labor migration in Asia', in K. Mitchell, R. Jones and J. Fluri (eds), *Handbook on Critical Geographies of Migration*. Cheltenham: Edward Elgar. pp. 92–105.

Secondary Sources and References

Collins, F.L., Sidhu, R., Lewis, N. and Yeoh, B.S. (2014) 'Mobility and desire: International students and Asian regionalism in aspirational Singapore',

Discourse: Studies in the Cultural Politics of Education, 35 (5): 661–76.

Hoang, L.A. and Yeoh, B.S. (2015) *Transnational Labour Migration, Remittances and the Changing Family in Asia*. London: Palgrave Macmillan.

Kenway, J. and Koh, A. (2013) 'The elite school as "cognitive machine" and "social paradise": Developing transnational capitals for the national "field of power"', *Journal of Sociology*, 49 (2–3): 272–90.

Kong, L. and Yeoh, B.S. (eds) (2003) *The Politics of Landscapes in Singapore: Constructions of 'Nation'*. Syracuse, NY: Syracuse University Press.

Lam, T., Ee, M., Anh, H.L. and Yeoh, B.S. (2013) 'Securing a better living environment for left-behind children: Implications and challenges for policies', *Asian and Pacific Migration Journal*, 22 (3): 421–45.

Lindquist, J., Xiang, B. and Yeoh, B.S. (2012) 'Opening the black box of migration: Brokers, the organization of transnational mobility and the changing political economy in Asia', *Pacific Affairs*, 85 (1): 7–19.

Liu-Farrer, G. and Yeoh, B.S. (eds) (2018) *Routledge Handbook of Asian Migrations*. Abingdon: Routledge.

Pratt, G. and Yeoh, B.S. (2003) 'Transnational (counter) topographies', *Gender, Place and Culture: A journal of feminist geography*, 10 (2): 159–66.

Qing, A. (2021) 'NUS professor awarded prestigious "Nobel Prize" for geography', *Straits Times*, 13 October. https://www.straitstimes.com/singapore/parenting-education/nus-professor-awarded-prestigious-geography-prize (accessed November 15, 2023).

Sidhu, R., Cheng, Y.E., Collins, F., Ho, K.C. and Yeoh, B.S. (2021) 'International student mobilities in a contagion: (Im)obilizing higher education?', *Geographical Research*, 59 (3): 313–23.

Triandafyllidou, A. and Yeoh, B.S. (2023) 'Sustainability and resilience in migration governance for a post-pandemic world', *Journal of Immigrant & Refugee Studies*, 21 (1): 1–14.

Willis, K. and Yeoh, B.S. (eds) (2004) *State/Nation/Transnation: Perspectives on Transnationalism in the Asia Pacific*. London: Routledge.

Xiang, B., Yeoh, B.S. and Toyota, M. (eds) (2013) *Return: Nationalizing Transnational Mobility in Asia*. Durham, NC: Duke University Press.

Ye, J. (2016) *Class Inequality in the Global City: Migrants, Workers and Cosmopolitanism in Singapore*. London: Palgrave Macmillan.

Ye, R. and Nylander, E. (2015) 'The transnational track: State sponsorship and Singapore's Oxbridge elite', *British Journal of Sociology of Education*, 36 (1): 11–33.

Yeoh, B.S. (1992) 'Street names in colonial Singapore', *Geographical Review*, 82 (3): 313–22.

Yeoh, B.S. (1993) 'Urban sanitation, health and water supply in late nineteenth and early twentieth century colonial Singapore', *South East Asia Research*, 1 (2): 143–172.

Yeoh, B.S. (2001) 'Postcolonial cities', *Progress in Human Geography*, 25 (3): 456–68.

Yeoh, B.S. (2004) 'Cosmopolitanism and its exclusions in Singapore', *Urban Studies*, 41 (12): 2431–45.

Yeoh, B.S. (2020) *Temporary Migration Regimes and their Sustainability in Times of COVID-19*. Geneva: International Organization for Migration (IOM). https://publications.iom.int/system/files/pdf/temporary-migration.pdf (accessed November 16, 2023).

Yeoh, B.S. (2021) 'Conceptual contours of migration studies in and from Asia', *International Migration*, 59 (6): 225–33.

Yeoh, B.S. and Collins, F. (eds) (2022) *Handbook on Transnationalism*. Cheltenham: Edward Elgar.

Yeoh, B.S. and Huang, S. (1998a) 'Strategic intents', *Global Networks*, 5 (4): 307–15.
Yeoh, B.S. and Huang, S. (1998b) 'Negotiating public space: Strategies and styles of migrant female domestic workers in Singapore', *Urban Studies*, 35 (3): 583–602.
Yeoh, B.S. and Huang, S. (eds) (2013) *The Cultural Politics of Talent Migration in East Asia*. New York, NY: Routledge.
Yeoh, B.S. and Soco, M.A. (2014) 'The cosmopolis and the migrant domestic worker', *cultural geographies*, 21 (2): 171–87.

61
Oren Yiftachel

Nir Cohen, Bar Ilan University Ramat Gan
Haim Yacobi, University College London

Biographical Details and Theoretical Context

Oren Yiftachel was born in Haifa, Israel in 1956. As a child, he grew up in Kibbutz Matzuva, located in the northwestern Galilee, a mile from the Lebanese border. It may have been his formative years in the egalitarian, though socially conservative, Kibbutz at the heart of the mixed (Arab-Jewish) Galilee which inspired him to develop a critical outlook on his surroundings, most notably the relations between space and society in Israel/Palestine. As he recently recalled, 'Kibbutzim in those days abided by one, big, top-down Truth, which frustrated an adolescent with rebellious genes like me' (Yiftachel, 2017: 137).

Yiftachel left Israel in the late 1970s to travel the world as a backpacker and – later – to pursue academic education. In 1983, he completed an undergraduate degree in Urban Studies at the Western Australian Institute of Technology in Perth, followed by an MA in Urban and Regional Planning (1986) at Curtin University in Perth. After working as an urban planner in Perth for several years, he enrolled in a PhD program in Regional Development, and graduated jointly from the University of Western Australia and the Technion: Israel Institute of Technology in Haifa in 1990. He then took up a post-doctoral fellowship at the Technion (1991–1992) and worked as a planner in several professional outlets in Israel. In 1994 he joined the Department of Geography and Environmental Development at Ben-Gurion University of the Negev and has since risen through the ranks from lecturer to full professor.

A prolific writer, Yiftachel has authored or edited a dozen books and more than 120 journal articles and book chapters over the years. All have originally been written in Hebrew and English but translated into several languages, including Arabic, Spanish, Italian, German, Russian, Turkish and Japanese. A self-professed Neo-Gramscian, Yiftachel's work aims to decipher the nexus between space and power. Thematically, his work spans several fields, namely political and legal geography, as well as urban studies and planning theory. Theoretically, he has been inspired by critical scholars, including ***Manuel Castells***, **David Harvey**, ***Iris Marion Young***, Peter Marcuse, Brian McLoughlin, Margo Huxley, Ian Alexander, Leonie Sandercock, Frantz Fanon, **Edward Said**, and **Gayatri Spivak**.

His early publications, authored while still in Australia, centred on urban planning in Australian cities, primarily the Perth metropolitan area. It was already at this early stage, that he made significant contributions to the field of urban planning by focusing on disciplinary-wide, theoretical issues, including oft-cited papers on planning typologies and repression of ethnic minorities (Yiftachel, 1989; 1995). The latter has no doubt resulted from the turbulent reality he encountered upon his return to Israel in the late 1980s. Against the backdrop of the lingering Israeli occupation and the first Palestinian Intifada (1987–1991), beginning with his PhD dissertation, *Spatial Policy and Ethnic Relations: The Political Geography of Arab-Jewish Relations* (1991), Yiftachel (re)focused his scholarly attention on Israel/Palestine, exploring the unequal socio-spatial relations between Jews and Arabs and their territorial expressions.

Subsequently, writing against the grain of the mostly celebratory accounts characterising Israeli – and international – literature of the mid-1990s Oslo era, Yiftachel judiciously uncovered the repressive logic and internal mechanisms that make up Israel's land regime. Specifically, his work highlighted a wide range of discriminatory planning practices – from seizing and nationalising minority lands, through quenching their town expansion plans, to fencing their communities. His publications in this period not only marked a paradigmatic shift in the discipline of geography in Israel, spearheading its critical turn, but contributed significantly to the theorisation of political-geographic regimes, spatial relations in deeply divided societies, and territorial control of ethnic minorities. An important aspect of the work was the use of systematic *comparative analysis* of settler and ethnic states embroiled in protracted conflicts, such as Northern Ireland, Sri Lanka, Malaysia, and Australia (Yiftachel, 1991).

Though his primary focus was – and remains – the adverse effects of Israel's land regime on non-Jewish minorities, predominantly Palestinian (non) citizens, Bedouins, and Druze, by the late 1990s Yiftachel had come to acknowledge its systematic oppression of ethnic and class minorities within the Jewish majority. It was in the context of his emerging social activism in redistributive justice in land allocation that Yiftachel explored the systematic spatial oppression directed by the Ashkenazi-dominated Israeli state toward Mizrahi Jews. His landmark publications, alone and with Alexander Kedar and Erez Tzfadia, about the salient role of Israel's planning regime in the relegation of Mizrahim to the socio-economic and spatial margins of Israeli society (Yiftachel and Kedar, 2000; Yiftachel and Tzfadia, 2004) have been instrumental to the understanding of the inextricably strong ties between space and ethnic domination in Israel.

By the end of the 1990s, Yiftachel had developed his highly influential model of 'ethnocracy' in a series of articles. In one, published in the same year as the failed Peace Conference at Camp David and the outbreak of the Second Intifada, Yiftachel (2000) theoretically conceptualised the underpinnings of this unique political-geographical formation. Though he used Israel/Palestine as a primary test case, his shrewd scrutiny drew on examples from countries as different as Sri Lanka, Cyprus, and Estonia, making his model of global relevance. In collaboration with Haim Yacobi, he applied the model to cities in a highly influential paper on *Urban Ethnocracy* (2003). His book, *Ethnocracy: The Politics of Land and Identity in Israel and Palestine* (2006, and subsequently translated into several languages), was the culmination of his work on the topic and fixed his status as a leading spatial thinker outside the Western world.

By the late 2000s, together with renowned writers, like Ananya Roy, **AbdulMaliq Simone**, Gautam Bhan and Venessa Watson, Yiftachel was spearheading the so-called 'southeastern' school of urban theory (Yiftachel, 2006), which called for a more context-sensitive scholarship that accounts for the distinct planning dynamics shaping post- and neo-colonial cities in the developing world. Much of his work in this period has been inspired by the precarious spatialities characterising everyday lives of the Negev Bedouins, a short distance from his university office, and African asylum seekers in South Tel Aviv. Among his influential conceptualisations in the past fifteen years are 'grey space' – that which exists between the fully planned and legal and the unplanned and illegal spheres, 'displaceability' – the process by which precarious groups are forced to live in permanent fear of forced removal by the authorities, and more recently 'defensive urban citizenship' (DUC), the mobilisation of disenfranchised groups to thwart that which threatens to rob their meagre bundle of city-based rights (Yiftachel, 2009, 2020; Yiftachel and Cohen, 2021).

Spatial Contributions

Yiftachel's work dialogues with that of critical geographers, planning theorists and urbanists, suggesting that space is a source and product of power relations, a politically constructed entity, which is neither given nor neutral. Rather, as Yiftachel develops empirically and theoretically, space is constantly reshaped by power relations and produces social hierarchies, structural inequality, and marginality, with the ever-present potential to enhance equality and inclusiveness. His exploration of space and power is evident already in his early work, which focused on Australian metropolitan planning and its impact on 'social sustainability' – a concept he developed already in the early 1990s (Yiftachel and Hedgecock, 1993). This early work was the foundation of a couple of influential articles identifying and conceptualising the 'dark side' of planning (Yiftachel, 1994; 1998) and joined the growing body of critical planning theory led by Leonie Sandercock, John Friedmann, Brian McLoughlin, Margo Huxley, and others. Inspired by eminent urbanists ***Manuel Castells*** and **David Harvey**, Yiftachel's early work engages class analysis with wider aspects of power relations including ethnicity, minority politics, identity, and migration:

> These have exposed the regressive and oppressive impacts of spatial policies such as public housing, inner-city development, gentrification, the location of employment centres and environmental hazards or freeways. The regressive aspect of these policies often appears along a range of ethnic, racial, class, gender, and sexuality cleavages. (Yiftachel, 1998: 397)

The above should be seen as the intellectual foundations for Yiftachel's most significant contribution to the study of space and place, namely his theorisation and comparison of ethnocratic societies, their land regimes, and their effect on spatial inequalities. In a set of articles that were published in journals in the fields of public policy and politics (Yiftachel, 1999), political and urban geography, as well as in planning (Yiftachel and Yacobi, 2003), and later in a book (Yiftachel, 2006), he developed the concept of 'ethnocracy' as a distinct colonial-type regime established to enhance the expansion and control of a dominant ethnonational group in multiethnic territories.

In these regimes, prevalent in Israel/Palestine, but also other (post)colonial countries, it is ethnicity rather than citizenship that constitutes the main criterion for distributing power and resources. Consequently, ethnocratic regimes typically display highly uneven levels of ethnic segregation as well as a process of polarising ethnic politics. Ethnocratic regimes combine a degree of political openness and formal-democratic representation with political structures that facilitate the seizure of contested territory by a dominant ethno-nation. Simultaneously, the dominant group appropriates the state apparatus and controls capital flows, marginalising peripheral 'ethno-classes' and various types of minorities.

While in the earlier stage of forming the notion of ethnocracy, Yiftachel focused on state and regional scales, during the first decade of the 2000s he shifted the discussion into the urban context in ethnocratic societies. In his co-authored article on urban ethnocracy, cities are identified as key actors in the determination of local conflicts and resource allocation which are operating in the larger political fields that bind the city, especially through the discourses of ethnicisation and development (Yiftachel and Yacobi, 2003). Urban ethnocracy, as a distinct political and spatial regime, is the tangible manifestation of the wider ethnonational tensions and conflicts rooted in the ethnocratic context. This discussion was further developed in a co-edited special issue (Porter and Yiftachel, 2019), which urged scholars to consider the city as a strategic locus of colonial intent and decolonising struggle, and to explore the nexus between Indigenous lives, colonial politics, and urbanisation. Indeed, local urban politics usually revolves around struggles for space, economic resources, and political power, most commonly along ethnonational lines:

> [U]rban ethnocracy is likely to become increasingly conspicuous in the landscapes of future cities. Structural tensions are embedded in the powerful city-shaping forces [...] urbanization, globalization, commodification, and liberalization, and nationalism, ethnicization, and the containment of 'rights to the city' exercised by peripheral ethnonational groups. The struggles emanating from the interaction of these forces, and the grids of powers and identities they reflect and produce, are set to continue and to shape most cities well into the 21st century. (Yiftachel and Yacobi, 2003: 690)

The theory of ethnocracy was an entry point for Yiftachel's later urban research which has also shifted towards understanding urban informality. This led to the development of the notion of 'grey spaces' which focuses on the unequal foundations of urban processes that resulted in a rapid expansion in informal and temporary urban development forms (Yiftachel, 2009). While Yiftachel joined an extensive critical discussion on urban informality (e.g., Roy, 2005; Watson, 2012), his contribution lay in the theorisation of informality as a structural component of contemporary cities, and in the understanding that through a process of 'grey spacing', cities are governed under the logic of fragmentation, social stratification, and extensive displacement. The vast expansion of grey spaces in contemporary cities reflects the emergence of new types of colonial relations, which are managed by urban regimes facilitating a process of 'creeping apartheid' (Yiftachel, 2009). Planning, he suggests, is a lynchpin of this urban order, providing tools and technologies to classify, contain and manage deeply unequal urban societies. Importantly, Yiftachel's grey space concept is an example of 'engaged theories', which stems from his deep commitment to advocacy, legal and planning work with disenfranchised communities in Israel/Palestine, predominantly the Bedouins of the Negev (see also Kedar et al., 2018):

> Grey spacing is a power-laden process. Therefore, the concrete emergence of 'stubborn' informalities is typically handled not through corrective or equalizing policy, but by employing a range of delegitimizing and criminalizing discourses, regulations, and violence. This creates boundaries that divide urban groups according to their status; a process of 'separating urban incorporation' and 'creeping apartheid' whereby the meaning of urban citizenship depends on arbitrary features such as ethnicity, place of birth or class. The double-edged move of 'separating incorporation' preserves grey spaces in a state of 'permanent temporariness'; concurrently tolerated and condemned, perpetually waiting 'to be corrected'. (Yiftachel, 2009: 250)

Indeed, grey spaces are produced by the assemblage of those which lie literally 'in the shadow' of the formal, planned city, polity, and economy. They exist partially outside the gaze of state authorities and city plans and are unevenly incorporated into the everyday of today's globalising urban societies under the hegemonic forces of neoliberal economy and associated identity regimes, such as nationalism, ethnicity, race, and religion.

In a similar vein, Yiftachel (2020) notes that the discussion of displacement has mainly occurred in Global North contexts, with a dominant focus on class dynamics. Against this backdrop, he develops the term 'displaceability' to describe how urban dwellers are differentially affected by the depth of the threats of displacement which is primarily a spatial-political concept. Importantly, and related to his earlier work on place and space, he argues that the further unpacking of oppressive powers operating in urban territories 'provides a major agenda for critical urban researchers in the coming years' (Yiftachel, 2020: 162). He further suggests:

> Space does not permit a serious discussion of the mobilization and resistance required to combat the expansion of urban displaceability. Suffice it to note, finally, that informing and facilitating the struggle against this ominous phenomenon is surely one of the most urgent tasks for scholars and activists committed to more just urbanization in the global north, west, east, and south. (Yiftachel, 2020: 162)

Key Advances and Controversies

A prominent figure in Israeli critical social sciences, Yiftachel's work broke new ground in several fields – from urban planning to geopolitics and legal geography. Praised for his eclectic and daring approach, theoretically informed analysis, and insightful writing (Abowd, 2010: 341–2; see also Anderson, 2016), Yiftachel's work has simultaneously drawn significant criticism over the years. Not surprisingly, it was his conceptualisation of ethnocracy, born out of his dissatisfaction with the traditional scholarship that invariably depicted Israel as an ethnic democracy (Smooha, 1997), that elicited the strongest critique. In a detailed account, Yiftachel and colleagues critiqued the ethnic democracy model on both empirical and conceptual grounds. Not only, they argued, does the oxymoronic term 'ethnic democracy' – which they likened to 'hot ice' – constitute 'a serious conceptual deficiency' (Ghanem et al., 1998: 264), but given Israel's breaching of some fundamental democratic characteristics (e.g., equal citizenship, free and inclusive elections and territorial borders), the term falls short of making 'a scholarly exploration guided by empirical accuracy or conceptual coherence' (Ghanem et al., 1998: 265).

Subsequently, Yiftachel's concept of ethnocracy has been subjected to conceptual, normative, and political criticisms. While largely accepting his overall thesis concerning the ethnocratic nature of the Israeli regime, critics typically took issue with three aspects of the model. Conceptually, the very term ethnocracy was questioned by many who, like Smooha, insisted that Israel's formal political structure merits its conceptualisation as a *special type* of democracy (see also Gavison, 1999). Others found his 'continuum' or 'tipping-point' approach to distinguishing between democracies and ethnocracies too dangerous, due to its normalising potential of the latter. The 'clear break between the two', it was argued, must entail 'a move from a fundamental inequality of stake to a critical model that enshrines equality of stake and that entails nothing short of a fundamental legal revolution rather than a mere evolution' (Ben-Dor, 2007: 180).

Yiftachel was sometimes accused of adopting both a reductionist approach, which conceives of every state practice as 'deriv[ing] from or subordinated to, the fundamental territorial logic' (Rosenhek, 2008: 135) and a bird's eye view, downplaying important local contextual factors, both contemporary and historical. His innovative theoretical contribution was compromised, it was argued, by his tendency to de-contextualise matters, ignoring Israel's geopolitical and demographic situation, including the rival interests of its main ethnic groups and viable political alternatives. Normatively, critics pointed to the absence of a convincing political philosophical defence of his case. Some questioned the validity of his key premise 'that all states at all times [...] ought to elevate the demos over any ethnos as a principle of political organization', challenging his assertion that the avowedly Jewish character of the Israeli state is to blame for 'all of the evils against which he protests' (Arkush, 2007: 162).

However, it was undoubtedly his practical, political solution for the dismantling of Israel's ethnocratic regime that drew the strongest criticism. It is worth recalling that Yiftachel opposed the two-state program, long the preferred solution of the Zionist left, prescribing instead a vision of 'phased bi-nationalism' as a way out of the century-long Israeli/Palestinian conundrum. Despite his assertion that such a utopian scenario ought to progress slowly and gradually, critics maintained that his vision '[i]s not as realistic as he would like his readers to believe that it is' (Arkush, 2007: 165). To some, his solution, premised on underestimation of the future demographic reality in the territory between the (Jordan) River and the (Mediterranean) Sea, would doom Israeli Jews to living in 'a state where they can only expect to become a steadily shrinking minority' (Arkush, 2007: 166). To others, Yiftachel's political program does not go nearly far enough; by shying away from the one-state solution, 'where Jews will live as a group under condition of equality', his otherwise 'absolutely brilliant diagnosis' essentially 'becomes the most sophisticated tool for silencing Palestinian solidarity and the possibility of addressing past and present injustices caused to them' (Ben-Dor, 2007: 185). In a poignant response, Yiftachel (2007: 192–5) repudiated his critics' support of the one-state solution, noting that political strategies that are 'grounded in political realities' and not 'abstract principles' would lead to the dismantling of Israel's ethnocracy and find 'a path for the coexistence, peace, and justice' in the two-nations' bleeding homeland.

Yiftachel's Key Works

Yiftachel, O. (1989) 'Towards a new typology of urban planning theories', *Environment and Planning B: Planning and Design*, 16 (1): 23–39.

Yiftachel, O. (1998) 'Planning and social control: Exploring the dark side', *Journal of Planning Literature*, 12 (4): 395–406.

Yiftachel, O. (1999) "Ethnocracy': The politics of Judaizing Israel/Palestine', *Constellations*, 6 (3): 364–90.

Yiftachel, O. (2000) 'Social control, urban planning, and ethno-class relations: Mizrahi Jews in Israel's development towns', *International Journal of Urban and Regional Research*, 24 (2): 418–38.

Yiftachel, O. (2006) 'Re-engaging planning theory? Towards "south-eastern" perspectives', *Planning Theory*, 5 (3): 211–22.

Yiftachel, O. (2006a) *Ethnocracy: Land and Identity Politics in Israel/Palestine*. Philadelphia, PA: University of Pennsylvania Press.

Yiftachel, O. (2007) 'From ethnocracy to peace through gradual bi-nationalism: A response to Oren Ben-Dor', *Holy Land Studies*, 6 (2): 187–95.

Yiftachel, O. (2009) 'Theoretical notes on "grey cities": The coming of urban apartheid?' *Planning Theory*, 8 (1): 88–100.

Yiftachel, O. and Cohen, N. (2021) 'Defensive urban citizenship: A view from southeastern Tel Aviv', in A.K. Mohan, S. Pellissery and J. Gómez Aristizábal (eds), *Theorising Urban Development From the Global South*. Cham: Palgrave Macmillan. pp. 149–74.

Yiftachel, O. and Tzfadia, E. (2004) 'Between periphery and "third space": Mizrahim in Israel's development towns', in A. Kemp, O. Yiftachel, D. Newman and U. Ram (eds), *Israelis in Conflict: Hegemonies, Identities and Challenges*. Brighton: Sussex Academic Press. pp. 203–35.

Yiftachel, O. and Yacobi, H. (2003) 'Urban ethnocracy: Ethnicization and the production of space in an Israeli "mixed city"', *Environment and Planning D: Society and Space*, 21 (6): 673–93.

Secondary Sources and References

Abowd, T. (2010) 'Oren Yiftachel, Ethnocracy: Land and identity Politics in Israel/ Palestine', *International Journal of Middle East Studies*, 42 (2): 340–2.

Anderson, J. (2016) 'Ethnocracy: Exploring and extending the concept', *Cosmopolitan Civil Societies: An Interdisciplinary Journal*, 8 (3): 1–29.

Arkush, A. (2007) 'Ethnocracy, land and identity politics in Israel/Palestine', *Israel Studies*, 12 (3): 161–7.

Ben-Dor, O. (2007) 'Debating Israeli ethnocracy and the challenge of secular democracy: A critique of Oren Yiftachel', *Holy Land Studies*, 6 (2): 177–86.

Gavison, R. (1999) 'Jewish and democratic? A rejoinder to the "ethnic democracy" debate', *Israel Studies*, 4 (1): 44–72.

Ghanem, A.A., Rouhana, N. and Yiftachel, O. (1998) 'Questioning "ethnic democracy":

A response to Sammy Smooha', *Israel Studies*, 3 (2): 253–67.

Kedar, A., Amara, A. and Yiftachel, O. (2018) *Emptied Lands*. Stanford, CA: Stanford University Press.

Porter, L. and Yiftachel, O. (2019) 'Urbanizing settler-colonial studies: Introduction to the special issue', *Settler Colonial Studies*, 9 (2): 177–86.

Rosenhek, Z. (2009) 'Ethnocracy: Land and identity politics in Israel/Palestine', *Israel Studies Forum*, 24 (1): 133–5.

Roy, A. (2005) 'Urban informality: Towards an epistemology of planning', *Journal of the American Planning Association*, 71 (2): 147–58

Smooha, S. (1997) 'Ethnic democracy: Israel as an archetype', *Israel Studies*, 2 (2): 198–41.

Watson, V. (2013) 'Planning and the "stubborn realities" of global south-east cities: Some emerging ideas', *Planning Theory*, 12 (1): 81–100.

Yiftachel, O. (1991) 'State policies, land control, and an ethnic minority: The Arabs in the Galilee region, Israel', *Environment and Planning D: Society and Space*, 9 (3): 32962.

Yiftachel, O. (1995) 'Planning as control: Policy and resistance in a deeply divided society', *Progress in Planning*, 2 (44): 115–84.

Yiftachel, O. (1998) 'Planning and social control: Exploring the dark side', *Journal of Planning Literature*, 12 (4): 395–406.

Yiftachel, O. (2017) *Spatial Power: Critiquing and Fixing*. Beer Sheva: Ben Gurion University, Department of Geography and Environmental Development. [in Hebrew].

Yiftachel, O. (2020) 'From displacement to displaceability: A southeastern perspective on the new metropolis', *City*, 24 (1–2): 151–65.

Yiftachel, O. and Hedgcock, D. (1993) 'Urban social sustainability: The planning of an Australian city, *Cities*, 10 (2): 139–157.

Yiftachel, O. and Kedar, S. (2000) 'Landed power: The making of the Israeli land regime', *Theory and Criticism*, 16 (1): 67–100. [in Hebrew]

62
Kathryn Yusoff

Rory Rowan, Trinity College Dublin

Bibliographical Details and Theoretical Context

Kathryn Yusoff trained as a sculptor at the University of Northumbria and completed an MA in Visual Cultures at Bath Spa University before embarking on a career as an academic geographer. Her PhD, carried out in Royal Holloway, University of London, explored the visual culture of Antarctic studies and attests to a founding interest in climate change, geographic knowledge, and aesthetics. Following the completion of her PhD, she was a Fellow at The Open University, then a Lecturer at the University of Exeter, the Lancaster Environment Centre and Queen Mary, University of London, where she became Professor of Inhuman Geography in 2017. Yusoff established herself as a major intellectual figure through the introduction of new theoretical approaches and philosophical traditions, work she pursued as a founding member of the Race, Culture and Equality Working Group (RACE) at the RGS-IBG and as editor of the *Society and Space* Open Site (2012–2016).

Yusoff's early writing explored the role of art practices and visual culture in constructing – and limiting – the geographic imaginaries through which climate change is apprehended, and the tensions within arts–science collaborations in environmental change research (Yusoff, 2005; 2009; 2010; 2012; Gabrys and Yusoff, 2006; 2011; 2012). This early work established two core features of Yusoff's later work: a concern with the geographic imaginaries of biodiversity loss, extinction, and climate catastrophe; and a deeply collaborative, and frequently interdisciplinary, ethos expressed in close intellectual alliances with co-authors.

Yusoff's work has developed in conversation with the broader turn towards exploring non-human agency, more-than-human ontologies, and hybrid ecologies. Yusoff, with her collaborators, is engaged in radicalising the work of earlier 'social–nature' geographies by expanding its analysis to include non-human agency, particularly the geophysical forces of the earth. However, she is sceptical of attempts to recuperate the non-human into comfortably ordered, albeit more-than-human, ecologies in a manner that excludes the disruptive potential of indeterminate matter and reinscribes anthropocentric norms through the backdoor of hybridity (Yusoff, 2015a; Hird and Yusoff, 2016). She has drawn on work in queer ecology to give conceptual priority to thinking through the indeterminacy of bodies as key to the project of 'denaturing' dominant ideologies of Nature (Yusoff, 2017b; Yusoff and Waterton, 2017; Clark and Yusoff, 2018). In addition,

Yusoff has consistently taken up feminist Science and Technology Studies' concern with how different forms of matter come to *matter*, both epistemologically and ontologically (Hird and Yusoff, 2016; Johnson et al., 2019). Yusoff's concern with material indeterminacy has taken her beyond a concern with 'life' and its regulation into deep engagements with the non-living and the inhuman, specifically the geological. Yusoff's interest in the geologic has also put her work in dialogue with recent attempts to rethink the spatial relations, arenas, and materials relevant to political geography (see, for example, Elden, 2013; Peters et al., 2018).

Yusoff's theoretical concerns have always developed in a dialectic relation to the social and political context in which she works, and they are shaped by a deep commitment to environmental and racial justice. The anti-racist thrust of Yusoff's work is a critical response to the re-mainstreaming of explicitly white supremacist ideologies and their capture of state power. It has grown in the intellectual space opened by the Black Lives Matter movement and the Standing Rock protests. These movements have profoundly altered the intellectual environment in the UK, the US, and elsewhere, where critical debates about the legacies of empire, colonialism and slavery have become structural features of mainstream cultural and political discourse. Yusoff's work has served to bring these intellectual currents out of marginalised sub-disciplinary silos and into areas of debate typically sheltered from their challenge.

Spatial Contributions

Yusoff is best known for a decade's worth of striking interventions around the philosophical and political implications of the Anthropocene. Much of her work has taken shape within – and *against* – the broad interdisciplinary debates that have emerged around the Anthropocene and the 'new epoch of thought' it announced (Yusoff, 2016). This new geological 'age of man' has revealed that the Earth is not a 'stable platform upon which dynamic social processes play out' and Yusoff contends that it requires social thinkers to make a much 'closer engagement with the earth sciences' (Clark and Yusoff, 2017). Yusoff takes seriously its incitement to rethink the relationship between the human and the geologic, embarking on a 'geological turn' (Yusoff, 2013) that follows the Earth Sciences into the sphere of social relations and human subjectification.

What is most alluring and disturbing about the Anthropocene for Yusoff is the novel mode of 'geological subjectification' it supposes (Yusoff, 2013): the Anthropos, a 'new collective geologic identity' (Yusoff, 2018b) for the human as 'geologic world-maker and destroyer of worlds' (Yusoff, 2016: 5). Yusoff is critical of the 'homogenous geomorphizing' inherent in the Anthropos as a universal geological subject (Yusoff, 2016: 3). Within the supposed planetary novelty of 'Anthropocene Man', she identifies the restoration of the familiar universalist figure of European Man, with all its structurally necessary 'others' – those humans denied access to humanity and a natural world understood as separate and below it (Yusoff, 2018a). The Anthropocene marks the return of the 'grand narratives' that critical social theory – not least through the work of feminist, queer, postcolonial and Black studies scholars – had done so much to dismantle. In a series of arresting articles Yusoff shows how the new univocal geohistory of Man proffered by the Anthropocene is power-laden, carrying a normative thrust which narrows understanding of human diversity and the possibilities it affords for subjectivity, political freedom and social relations with the Earth (Yusoff, 2013; 2015a; 2016; 2018a). However, Yusoff has also sought to explore the socially differentiated modes of 'geologic subjectivity' that Anthropocene discourses bracket out as an untapped seam of subjective and political potential (Yusoff, 2013; 2015a; 2016).

Yusoff has attempted to think through the possibilities for geologic subjectivity – beyond the homogenising monumentalism of the Anthropos – in relation to the nonhuman and inhuman excesses that are involved in 'becoming human' (Yusoff, 2013; 2015a; 2016). The inhuman forces of the Earth play a material role in the emergence of subjects, Yusoff argues, meaning that 'subjectivity always contains both an anterior and interior nonhuman excess, a surplus of identity that opens to non-normative arrangements' (Yusoff, 2015a: 383). Yusoff contends that opening our thinking to the 'minerological dimension of human composition' (Yusoff, 2013: 780) suggests a 'queer ecology' (Yusoff, 2015a) that undermines human exceptionalism and allows relations to be rethought beyond discrete bounded identities and between human and nonhuman actors, including the inhuman forces of the earth itself (Yusoff, 2013; 2015a). Indeed, in her paper 'Queer Coal' (2015b) Yusoff explores the British Miners' Strikes of 1984–85 as an example of how inhuman intimacies and queer solidarities briefly allowed experimental forms of geologic subjectivity to emerge in the midst of class conflict around fossil fuels.

In 2017 Yusoff and frequent collaborator Nigel Clark edited a special issue of *Theory, Culture & Society* on 'Geosocial Formations and the Anthropocene'. This collection made a major contribution to establishing the emerging field of 'political geology' (see, for example, Bobbette and Donovan, 2018). In their introduction, Yusoff and Clark identified a need for social theory to take the Earth Sciences into consideration in face of the Anthropocene (Clark and Yusoff, 2017). The Anthropocene, they contend, requires 'analysis of relations between geologic forces and social practices' (Yusoff, 2017a), including 'how social and political agency is both constrained and made possible by the forces of the earth' (Clark and Yusoff, 2017: 3). In her own contribution, Yusoff drew on the 'geophilosophy' of **Gilles Deleuze** and Félix Guttari to develop a model of strata that 'conjoins earth and social flows together in planes of interrelated production' highlighting how 'the fossil substratum subtends contemporary forms of social relations' (Yusoff, 2017a: 105). Yusoff employed the concept of 'geosocial strata' to conceptualise how specific arrangements of geologic forces and social processes can be identified, including the forms of social stratification built into particular 'geosocial formations' such as the current planetary configuration of fossil fuel dependency (Yusoff, 2017a). In so doing, Yusoff argues it is possible to consider how alternative geosocial formations might come into being. Thinking in the register of strata has also helped Yusoff to challenge some fundamental spatio-temporal imaginaries. For Yusoff, a stratigraphic imaginary offers an alternative geo-historical analytic that reconfigures how we might understand social relations and the material role of historical processes in the present. For example, approaching socio-environmental relations through the model of stratification rather than through the distribution of territories on the Earth's surface foregrounds the way historic socio-spatial relations materially subtend those of the present. This is important, Yusoff argues, for understanding how the legacies of colonialism fundamentally subtend contemporary environmental injustices.

Yusoff has drawn on the concepts of 'geopower' and 'geoontopower' from Elizabeth Grosz and Elizabeth Povinelli respectively to rethink biopolitics in relation to non-life, the inhuman and the geophysical forces of the earth (Yusoff et al., 2012; Grosz and Yusoff, 2014; Yusoff et al., 2014; Johnson et al., 2019). Building on their work, she has made an analytic shift from *biopower* and *biopolitics* – understood as the governance of life – to *geopower* and *geopolitics* – understood as the governance of non-life (and not in the traditional sense associated with the subfield of political geography). Indeed, she had earlier argued that the new modes of 'geologic life' emerging in the Anthropocene require that social thought consider 'not just our biological (or biopolitical) life, but also our geological (or geopolitical) life and its forms of differentiation' (Yusoff, 2013: 779). In making this conceptual shift Yusoff shows how the governance of non-life is a key operation within biopolitical ordering itself, though often overlooked by the biocentric accounts of biopower developed after **Michel Foucault** (Yusoff, 2013; 2015a; 2018a; 2019; Johnson et al., 2019). This shift from biopolitics to geopolitics, or rather the introduction of *geopolitics into biopolitics*, has important stakes for Yusoff as it reveals how the governance of the Earth cuts into the body politic. Her geopolitical analytic shows how regimes of subjective differentiation place certain lives in proximity to the raw material of the earth, the *geo*, whilst elevating others above it into the realm of the *bio* (Yusoff, 2018a; 2019; 2020).

Key Advances and Controversies

Yusoff's analysis of the relationship between geopower and biopolitical governance has been intensified by a transformative encounter with Black and Indigenous scholarship, particularly Sylvia Wynter's unpublished *Black Metamorphosis* (see McKittrick, 2015). In 2018, Yusoff published *A Billion Black Anthropocenes or None*, a short book which contained a trenchant critique of the racist basis of Anthropocene discourses and the colonial grounds of geology as a discipline. This widely-cited volume has changed the tenor of critical debate around the Anthropocene, in Geography and beyond. The book spoke with an urgency that reflected the highly charged debates around the legacies of colonialism and slavery and their relationship to ongoing settler violence and the policing of Black and brown lives. This anti-racist, geophilosophical tract was forged, Yusoff noted, of the need to critically respond to the 'willful' 'racial blindness' (Yusoff, 2018a: xiii) of Anthropocene discourses and 'too many Anthropocene events where the absence of discussions of race allowed the smooth flow of patriarchal reason to make the earth anew' (Yusoff, 2018a: 109).

In this book, and in two important follow up articles (Yusoff, 2019; 2020), Yusoff elaborates on the analytic category of 'the inhuman', one of her most significant contributions. Yusoff argues that the distinction between life/non-life is used to separate those human lives worthy of subjecthood and being counted as 'life', and those excluded from both subjecthood and 'life' who are placed in the category of the inhuman alongside the earth where they appear as raw materials available for extraction, with race providing the rubric for differentiation (Yusoff, 2018a; 2019; 2020). Hence, in a single conceptual gesture those racialised as non-white are desubjectified whilst being made available for subjugation in regimes of extraction. Whilst this analysis may seem to reflect the relatively common idea that the exploitation of certain groups has historically been justified through processes of dehumanisation, Yusoff's claim goes further in locating the metaphysical and material dimensions of this dual desubjectification/subjugation process within the register of geology. For Yusoff, geology provides a shared conceptual grammar where both earth and Black and brown lives are figured as inhuman, and therefore rendered as material available for extraction, whether mineral or flesh. It is within the logic of geology, where the inhuman is available for extraction, that the racialised regimes of colonial dispossession and enslavement can be located. Here Yusoff reads her earlier engagements with the 'energy-centred social theory' (Yusoff, 2009; 2015a; Clark and Yusoff, 2014; 2017; 2018) of French philosopher George Bataille (Bataille, 1988) back through the geopolitics of colonialism and slavery where Black and brown lives become 'energy for others' – extractable inhuman material used to fuel the development of global capitalism, the legacies of which continue to subtend the colonial present (Yusoff, 2018a; 2020). Yusoff thus embeds the concept of 'racial capitalism' (Robinson, 2000) within the historical development of geology as a scientific field during the expansion of European colonial empires and the growth of the capitalist world system. However, she goes further to suggest that geology provides the 'anti-black ground' of colonialism through its extractive logic (Yusoff, 2018a; 2019; 2020). Yusoff's provocative claim here is that geology is not a 'neutral' science but continues to provide an extractive grammar for ongoing settler colonialism, the exposure of Black and brown lives to the toxic environmental legacies of extraction and climate change, and in the form of the Anthropocene which provides a legitimising framework for futures forged within these extractive logics (Yusoff, 2018a; 2020).

Yusoff has reserved sharp criticism for the emerging interdisciplinary fields of the environmental/geohumanities for the central role they afford to the category of 'the human' and an uncritically assumed 'humanism' (Yusoff, 2020). She argues that the humanities are not innocent of the conceptual baggage of coloniality. For this reason, the historical relationship between humanism and colonialism needs to be interrogated to challenge moves to innocence and efforts to legitimise ongoing patterns of racialised dispossession within the context of the humanities' role in environmental thought and practice. As Yusoff writes:

> In the rush to secure white settler futurity in the context of environmental and climate change, the emergent field of Anthropocene studies and the environmental/geohumanities often assumes rather than problematizing (sic) the human that secures the concept of the humanities at both a philosophical and epistemological level, disregarding the historical colonial geographies that materially delivered humanism, its structures of thought [...] its organisations of value [...] and its scientific and industrial institutions. (Yusoff, 2020: 663–4)

Yusoff's point here is that environmental/geohumanities need to dismantle 'the legitimizing frame of the human' in order to challenge the foundational human/inhuman distinction that underpins the coloniality of dominant environmental discourses (Yusoff, 2020: 672). The environmental and geohumanities partake in the 'rebirth of the humanist subject' (Yusoff, 2020: 672) represented by the Anthropocene. This is important precisely because the 'problem of the future of the world' is being organised around the category of the human and its unspoken inhuman 'others' (Yusoff, 2020: 672). Hence for Yusoff there is 'no environmental or geohumanities without the question of racial justice' (Yusoff, 2020: 672), without a reckoning with the colonial legacies of humanism and its structural dependency on subjugating Black and brown lives as inhuman matter for extraction.

In place of the environmental and geohumanities Yusoff calls for the institution of 'the Inhumanities'. The Inhumanities, Yusoff argues, would register 'a commitment to dismantling the humanist subject and the white supremacy that characterises its geographic project of the differentiation of subjects and earth' (Yusoff, 2020: 672). The inhuman, Yusoff writes, 'is an epistemological site for the undoing' of colonial geographies 'in the present tense across both subjective and earth relations' and hence the Inhumanities couples the projects of environmental and racial justice (Yusoff, 2020: 672).

Yusoff's influence has also been felt beyond the academy, notably within the context of the visual and media arts. She has frequently collaborated with artists on novel forms of knowledge production and creative experiments with presentation formats, picking up aspects of her earlier artistic practice. Recent examples include a lecture-performance with Casper Laing Ebbensgaard, Kerry Holden and Michael Salu entitled *Planetary Portals: Diamond Power* as part of the 2023 Transmediale festival in Berlin (Yusoff et al., 2023), and her leading role in *INFINITY Minus Infinity*, a 2019 film by the award-winning artist collective The Otolith Group (2020). Yusoff received the 2022 Award for Creativity in Geography from the American Association of Geographers in recognition of this work.

Yusoff's analysis of the relationship between geology, race and contemporary forms of capital accumulation has yet to be fully developed in relation to place-based environmental and anti-racist struggles, though there are suggestions of how this might take shape at points in her work (see Braun et al., 2015; Yusoff, 2015b). Yet perhaps the fundamental theoretical challenge posed by Yusoff's most recent work is how to imagine emancipatory geographies, or alternative geosocial formations, in the wake of an environmental politics grounded in the colonial logic of the Anthropocene. By way of an answer Yusoff looks to Sylvia Wynter's evocation of the 'multiple and varying genres of the human' that have always been in excess of the 'monohumanism [...] of overrepresented Western European man' (Yusoff, 2020: 669) but her thoughts have thus far remained understandably provisional. The increasing emphasis Yusoff has placed on Black geopoetics (Yusoff, 2018a; 2020), Black feminist cinema and literature (Yusoff, 2018a; 2019, and the 2018 workshop she co-organised with N.K. Jemisin at the New School, New York) suggest her work may dive further into Black aesthetics as a space of emancipatory experimentation within which to pose these questions, as she continues to develop her geophilosophical explorations.

Yusoff's Key Works

Clark, N. and Yusoff, K. (2017) 'Geosocial formations and the Anthropocene', *Theory, Culture & Society*, 34 (2–3): 3–23.

Yusoff, K. (2009) 'Excess, catastrophe, and climate change', *Environment and Planning D: Society and Space*, 27 (6): 1010–29.

Yusoff, K. (2013) 'Geologic life: Prehistory, climate, futures in the Anthropocene', *Environment and Planning D: Society and Space*, 31 (5): 779–95.

Yusoff, K. (2015a) 'Geologic subjects: Nonhuman origins, geomorphic aesthetics and the art of becoming in human', *cultural geographies*, 22 (3): 383–407.

Yusoff, K. (2015b) 'Queer coal: Genealogies in/of the blood', *PhiloSOPHIA*, 5 (2): 202–29.

Yusoff, K. (2016) 'Anthropogenesis: Origins and endings in the Anthropocene', *Theory, Culture & Society*, 33: 3–28.

Yusoff, K. (2017a) 'Geosocial strata', *Theory, Culture & Society*, 34 (2–3): 105–27.

Yusoff, K. (2018a) *A Billion Black Anthropocenes or None*. Minneapolis, MN: University of Minnesota Press.

Yusoff, K. (2019) 'Geologic realism: On the beach of geologic time', *Social Text*, 37 (1): 1–26.

Yusoff, K. (2020) 'The Inhumanities', *Annals of the American Association of Geographers*, 111 (3): 663–76.

Secondary Sources and References

Bataille, G. (1988) *The Accursed Share: Volume 1*. New York, NY: Zone.

Bobbette, A. and Donovan, A. (eds) (2018) *Political Geology: Active Stratigraphies and the Making of Life*. London: Palgrave Macmillan.

Braun, B., Coleman, M., Thomas., M. and Yusoff, K. (2015) 'Grounding the Anthropocene: Sites, subjects struggles in the Bakken oil fields', *Antipode Online*, November 3. https://antipodeonline.org/2015/11/03/grounding-the-anthropocene/ (accessed November 17, 2023).

Clark, N. and Yusoff, K. (2014) 'Combustion and society: A fire-centred history of energy use', *Theory, Culture & Society*, 31 (5): 203–26.

Clark, N. and Yusoff, K. (2018) 'Queer fire: Ecology, combustion and pyrosexual desire', *Feminist Review*, 118: 7–24.

Elden, S. (2013) 'Secure the volume: Vertical geopolitics and the depth of power', *Political Geography*, 34: 35–5.

Gabrys, J. and Yusoff, K. (2006) 'Time lapses: Robert Smithson's mobile landscapes', *cultural geographies*, 13 (3): 444–50.

Gabrys, J. and Yusoff, K. (2011) 'Climate change and the imagination', *Wiley Interdisciplinary Reviews: Climate Change*, 2 (4): 516–34.

Gabrys, J. and Yusoff, K. (2012) 'Arts, sciences and climate change: Practices and politics at the threshold', *Science as Culture*, 21 (1): 1–24.

Grosz, E. and Yusoff, K. (2014) 'Ontogenesis and the ethics of becoming: An interview with Elizabeth Grosz by Kathryn Yusoff', *Environment and Planning D: Society and Space*, May 22. https://www.societyandspace.org/articles/on-ontogenesis-and-the-ethics-of-becoming (accessed November 17, 2023).

Hird, M. and Yusoff, K. (2016) 'Subtending relations: Bacteria, geology, and the possible', in S. Malik and A. Avanessian (eds), *Genealogies of Speculation*. London: Continuum. pp. 319–41.

Johnson, E.R., Kindervater, G., Todd, Z., Yusoff, K., Woodward, K. and Povinelli, E. (2019) 'Geontographies: On Elizabeth

Povinelli's geontologies: A requiem for late liberalism', *Environment and Planning C: Politics and Space*, 37 (8): 1319–42.

McKittrick, K. (2015) *Sylvia Wynter: On Being Human as Praxis*. Durham, NC: Duke University Press.

Peters, K., Steinberg, P. and Stratford, E. (2018) *Territory Beyond Terra*. Milton Keynes: Rowman and Littlefield.

Robinson, C. (2000) *Black Marxism: The Making of the Black Radical Tradition*. Chapel Hill, NC: University of North Carolina Press.

The Otolith Group (Director) (2020) 'Infinity minus Infinity' [HD Video]. Irish Musuem of Modern Art, Dublin.

Yusoff, K. (2005) 'Visualizing Antarctica as a place in time: From the geological sublime to "real time"', *Space and Culture*, 8 (4): 381–98.

Yusoff, K. (2010) 'Biopolitical economies and the political aesthetics of climate change', *Theory, Culture & Society*, 27 (2–3): 73–99.

Yusoff, K. (2012) 'Aesthetics of loss: Biodiversity, banal violence and biotic subjects', *Transactions of the Institute of British Geographers*, 37 (4): 578–92.

Yusoff, K. (2017b) 'Indeterminate subjects, irreducible worlds: Two economies of indeterminacy', *Body and Society*, 23 (3): 75–101.

Yusoff, K. (2018b) 'Politics of the Anthropocene: Formation of the commons as a geologic process', *Antipode*, 50 (1): 225–76.

Yusoff, K. and Waterton, C. (2017) 'Indeterminate bodies', *Body and Society*, 23 (3): 75–101.

Yusoff, K., Coleman, M. and Povinelli E. (2014) 'Interview with Elizabeth Povinelli with Mat Coleman and Kathryn Yusoff', *Environment and Planning D: Society and Space* Open Site. https://www.societyandspace.org/articles/on-biopolitics-and-the-anthropocene (accessed November 17, 2023).

Yusoff, K., Grosz, E., Clark, N., Saldanha, A. and Nash, C. (2012) 'Geopower: A panel on Elizabeth Grosz's *Chaos, Territory, Art: Deleuze and the Framing of the Earth*', *Environment and Planning D: Society and Space*, 30 (6): 971–88.

Yusoff, K., Holden, K., Ebbensgaard, C.L. and Salu, M. (2023, February 28) 'Planetary Portals: Diamond Power' [Youtube video]. www.youtube.com/watch?v=0ngqisHQqxw [includes transcript] (accessed November 17, 2023).

Index